SUSTAINABLE NANOMATERIALS FOR BIOMEDICAL ENGINEERING

Impacts, Challenges, and Future Prospects

Innovations in Agricultural and Biological Engineering

SUSTAINABLE NANOMATERIALS FOR BIOMEDICAL ENGINEERING

Impacts, Challenges, and Future Prospects

Edited by

Junaid Ahmad Malik, PhD

Megh R. Goyal, PhD, PE

Mohamed Jaffer M. Sadiq, PhD

AAP | APPLE ACADEMIC PRESS

First edition published 2023

Apple Academic Press Inc.
1265 Goldenrod Circle, NE,
Palm Bay, FL 32905 USA

760 Laurentian Drive, Unit 19,
Burlington, ON L7N 0A4, CANADA

CRC Press
6000 Broken Sound Parkway NW,
Suite 300, Boca Raton, FL 33487-2742 USA

4 Park Square, Milton Park,
Abingdon, Oxon, OX14 4RN UK

Library and Archives Canada Cataloguing in Publication

Title: Sustainable nanomaterials for biomedical engineering : impacts, challenges, and future prospects / edited by Junaid Ahmad Malik, PhD, Megh R. Goyal, PhD, PE, Mohamed Jaffer M. Sadiq, PhD.
Names: Malik, Junaid Ahmad, 1987- editor. | Goyal, Megh R., editor. | Sadiq, Mohamed Jaffer M., editor.
Series: Innovations in agricultural and biological engineering.
Description: First edition. | Series statement: Innovations in agricultural and biological engineering | Includes bibliographical references and index.
Identifiers: Canadiana (print) 20220449376 | Canadiana (ebook) 20220449473 | ISBN 9781774911990 (hardcover) | ISBN 9781774912003 (softcover) | ISBN 9781003333456 (ebook)
Subjects: LCSH: Nanomedicine. | LCSH: Nanostructured materials—Therapeutic use. | LCSH: Nanobiotechnology. | LCSH: Biomedical engineering.
Classification: LCC R857.N34 S87 2023 | DDC 610.28—dc23

Library of Congress Cataloging-in-Publication Data

CIP data on file with US Library of Congress

ISBN: 978-1-77491-199-0 (hbk)
ISBN: 978-1-77491-200-3 (pbk)
ISBN: 978-1-00333-345-6 (ebk)

ABOUT THE BOOK SERIES: INNOVATIONS IN AGRICULTURAL AND BIOLOGICAL ENGINEERING

Under this book series, Apple Academic Press Inc. is publishing book volumes over a span of 8–10 years in the specialty areas defined by the American Society of Agricultural and Biological Engineers (<asabe.org>). Apple Academic Press Inc. aims to be a principal source of books in agricultural and biological engineering. We welcome book proposals from readers in areas of their expertise.

The mission of this series is to provide knowledge and techniques for agricultural and biological engineers (ABEs). The book series offers high-quality reference and academic content on agricultural and biological engineering (ABE) that is accessible to academicians, researchers, scientists, university faculty and university-level students, and professionals around the world.

Agricultural and biological engineers ensure that the world has the necessities of life, including safe and plentiful food, clean air and water, renewable fuel and energy, safe working conditions, and a healthy environment by employing knowledge and expertise of the sciences, both pure and applied, and engineering principles. Biological engineering applies engineering practices to problems and opportunities presented by living things and the natural environment in agriculture.

ABE embraces a variety of the following specialty areas (<asabe.org>): aquaculture engineering, biological engineering, energy, farm machinery and power engineering, food, and process engineering, forest engineering, information, and electrical technologies, soil, and water conservation engineering, natural resources engineering, nursery, and greenhouse engineering, safety, and health, and structures and environment.

For this book series, we welcome chapters on the following specialty areas (but not limited to):

- Academia to industry to end-user loop in agricultural engineering;
- Agricultural mechanization;
- Aquaculture engineering;
- Biological engineering in agriculture;

- Biotechnology applications in agricultural engineering;
- Energy source engineering;
- Farm to fork technologies in agriculture;
- Food and bioprocess engineering;
- Forest engineering;
- GPS and remote sensing potential in agricultural engineering;
- Hill land agriculture;
- Human factors in engineering;
- Impact of global warming and climatic change on agriculture economy;
- Information and electrical technologies;
- Irrigation and drainage engineering;
- Nanotechnology applications in agricultural engineering;
- Natural resources engineering;
- Nursery and greenhouse engineering;
- Potential of phytochemicals from agricultural and wild plants for human health;
- Power systems and machinery design;
- Robot engineering and drones in agriculture;
- Rural electrification;
- Sanitary engineering;
- Simulation and computer modeling;
- Smart engineering applications in agriculture;
- Soil and water engineering;
- Micro-irrigation engineering;
- Structures and environment engineering;
- Waste management and recycling;
- Any other focus areas.

For more information on this series, readers may contact:

Megh R. Goyal, PhD, PE
Book Series Senior Editor-in-Chief: Innovations in Agricultural and Biological Engineering
E-mail: goyalmegh@gmail.com

OTHER BOOKS ON AGRICULTURAL AND BIOLOGICAL ENGINEERING BY APPLE ACADEMIC PRESS, INC.

Management of Drip/Trickle or Micro Irrigation
Megh R. Goyal, PhD, PE, Senior Editor-in-Chief
Evapotranspiration: Principles and Applications for Water Management
Megh R. Goyal, PhD, PE and Eric W. Harmsen, PhD Editors

Book Series: Research Advances in Sustainable Micro Irrigation
Senior Editor-in-Chief: Megh R. Goyal, PhD, PE

- **Volume 1:** Sustainable Micro Irrigation: Principles and Practices
- **Volume 2:** Sustainable Practices in Surface and Subsurface Micro Irrigation
- **Volume 3:** Sustainable Micro Irrigation Management for Trees and Vines
- **Volume 4:** Management, Performance, and Applications of Micro Irrigation Systems
- **Volume 5:** Applications of Furrow and Micro Irrigation in Arid and Semi-Arid Regions
- **Volume 6:** Best Management Practices for Drip Irrigated Crops
- **Volume 7:** Closed Circuit Micro Irrigation Design: Theory and Applications
- **Volume 8:** Wastewater Management for Irrigation: Principles and Practices
- **Volume 9:** Water and Fertigation Management in Micro Irrigation
- **Volume 10:** Innovation in Micro Irrigation Technology

Book Series: Innovations and Challenges in Micro Irrigation
Senior Editor-in-Chief: Megh R. Goyal, PhD, PE

- Engineering Interventions in Sustainable Trickle Irrigation: Water Requirements, Uniformity, Fertigation, and Crop Performance
- Management Strategies for Water Use Efficiency and Micro Irrigated Crops: Principles, Practices, and Performance

- Micro-Irrigation Engineering for Horticultural Crops: Policy Options, Scheduling, and Design
- Micro-Irrigation Management: Technological Advances and Their Applications
- Micro-Irrigation Scheduling and Practices
- Performance Evaluation of Micro-Irrigation Management: Principles and Practices
- Potential of Solar Energy and Emerging Technologies in Sustainable Micro-Irrigation
- Principles and Management of Clogging in Micro-Irrigation
- Sustainable Micro-Irrigation Design Systems for Agricultural Crops: Methods and Practices

Book Series: Innovations in Agricultural and Biological Engineering
Senior Editor-in-Chief: Megh R. Goyal, PhD, PE

- Advanced Research Methods in Food Processing Technologies
- Advances in Food Process Engineering: Novel Processing, Preservation and Decontamination of Foods
- Advances in Green and Sustainable Nanomaterials: Applications in Energy, Biomedicine, Agriculture, and Environmental Science
- Advances in Sustainable Food Packaging Technology
- Analytical Methods for Milk and Milk Products, 2-volume set:
    - Volume 1: Sampling Methods, Chemical and Compositional Analysis
    - Volume 2: Physicochemical Analysis of Concentrated, Coagulated and Fermented Products
- Biological and Chemical Hazards in Food and Food Products: Prevention, Practices, and Management
- Bioremediation and Phytoremediation Technologies in Sustainable Soil Management, 4-volume set:
    - Volume 1: Fundamental Aspects and Contaminated Sites
    - Volume 2: Microbial Approaches and Recent Trends
    - Volume 3: Inventive Techniques, Research Methods, and Case Studies
    - Volume 4: Degradation of Pesticides and Polychlorinated Biphenyls
- The Chemistry of Milk and Milk Products: Physicochemical Properties, Therapeutic Characteristics, and Processing Methods
- Dairy Engineering: Advanced Technologies and Their Applications

- Developing Technologies in Food Science: Status, Applications, and Challenges
- Emerging Technologies in Agricultural Engineering
- Engineering Interventions in Agricultural Processing
- Engineering Interventions in Foods and Plants
- Engineering Practices for Agricultural Production and Water Conservation: An Interdisciplinary Approach
- Engineering Practices for Management of Soil Salinity: Agricultural, Physiological, and Adaptive Approaches
- Engineering Practices for Milk Products: Dairyceuticals, Novel Technologies, and Quality
- Enzyme Inactivation in Food Processing: Technologies, Materials, and Applications
- Field Practices for Wastewater Use in Agriculture: Future Trends and Use of Biological Systems
- Flood Assessment: Modeling and Parameterization
- Food Engineering: Emerging Issues, Modeling, and Applications
- Food Process Engineering: Emerging Trends in Research and Their Applications
- Food Processing and Preservation Technology: Advances, Methods, and Applications
- Food Technology: Applied Research and Production Techniques
- Functional Dairy Ingredients and Nutraceuticals: Physicochemical, Technological, and Therapeutic Aspects
- Handbook of Research on Food Processing and Preservation Technologies, 5-volume set:
    - Volume 1: Nonthermal and Innovative Food Processing Methods
    - Volume 2: Nonthermal Food Preservation and Novel Processing Strategies
    - Volume 3: Computer-Aided Food Processing and Quality Evaluation Techniques
    - Volume 4: Design and Development of Specific Foods, Packaging Systems, and Food Safety
    - Volume 5: Emerging Techniques for Food Processing, Quality, and Safety Assurance
- Modeling Methods and Practices in Soil and Water Engineering
- Nanotechnology and Nanomaterial Applications in Food, Health, and Biomedical Sciences

- Nanotechnology Applications in Agricultural and Bioprocess Engineering: Farm to Table
- Nanotechnology Applications in Dairy Science: Packaging, Processing, and Preservation
- Nanotechnology Horizons in Food Process Engineering, 3-volume set:
    - Volume 1: Food Preservation, Food Packaging and Sustainable Agriculture
    - Volume 2: Scope, Biomaterials, and Human Health
    - Volume 3: Trends, Nanomaterials, and Food Delivery
- Novel and Alternative Methods in Food Processing: Biotechnological, Physicochemical, and Mathematical Approaches
- Novel Dairy Processing Technologies: Techniques, Management, and Energy Conservation
- Novel Processing Methods for Plant-Based Health Foods: Extraction, Encapsulation and Health Benefits of Bioactive Compounds
- Novel Strategies to Improve Shelf-Life and Quality of Foods: Quality, Safety, and Health Aspects
- Phytochemicals and Medicinal Plants in Food Design: Strategies and Technologies for Improved Healthcare
- Processing of Fruits and Vegetables: From Farm to Fork
- Processing Technologies for Milk and Milk Products: Methods, Applications, and Energy Usage
- Quality Control in Fruit and Vegetable Processing: Methods and Strategies
- Scientific and Technical Terms in Bioengineering and Biological Engineering
- Soil and Water Engineering: Principles and Applications of Modeling
- Soil Salinity Management in Agriculture: Technological Advances and Applications
- State-of-the-Art Technologies in Food Science: Human Health, Emerging Issues and Specialty Topics
- Sustainable and Functional Foods from Plants
- Sustainable Biological Systems for Agriculture: Emerging Issues in Nanotechnology, Biofertilizers, Wastewater, and Farm Machines
- Sustainable Nanomaterials for Biomedical Engineering: Impacts, Challenges, and Future Prospects
- Sustainable Nanomaterials for Biosystems Engineering: Trends in Renewable Energy, Environment, and Agriculture

- Technological Interventions in Dairy Science: Innovative Approaches in Processing, Preservation, and Analysis of Milk Products
- Technological Interventions in Management of Irrigated Agriculture
- Technological Interventions in the Processing of Fruits and Vegetables
- Technological Processes for Marine Foods, From Water to Fork: Bioactive Compounds, Industrial Applications, and Genomics

ABOUT THE EDITORS

Junaid Ahmad Malik, PhD
Lecturer, Department of Zoology, Government Degree College, Bijbehara, Kashmir (J&K), India

Junaid Ahmad Malik, PhD, is a Lecturer with the Department of Zoology at Government Degree College, Bijbehara, Kashmir (J&K), India, and is actively involved with teaching and research activities. He has more than eight years of research experience. His areas of interest are ecology, soil macrofauna, wildlife biology, conservation biology, etc. Dr. Malik has published 21 research articles and technical papers in international peer-reviewed journals and has authored and edited books, book chapters, and more than 10 popular editorial articles. He also serves as an editor and reviewer of several journals. He has participated in several state, national, and international conferences, seminars, workshops, and symposia and has more than 20 conference papers to his credit. He is a life member of the Society for Bioinformatics and Biological Sciences. Dr. Malik received a BSc (2008) in Science from the University of Kashmir, Srinagar, Jammu and Kashmir; MSc (2010) in Zoology from Barkatullah University, Bhopal, Madhya Pradesh; and a PhD (2015) in Zoology from the same university. He completed his BEd program in 2017 at the University of Kashmir, Srinagar, Jammu and Kashmir, India..

Readers may contact him at: malik.junaidahmad@gmail.com.

ABOUT SENIOR-EDITOR-IN-CHIEF

Megh R. Goyal, PhD, PE

Retired Professor in Agricultural and Biomedical Engineering, University of Puerto Rico, Mayaguez Campus; Senior Acquisitions Editor, Biomedical Engineering and Agricultural Science, Apple Academic Press, Inc.

Megh R. Goyal, PhD, P.E., is, at present, a Retired Professor in Agricultural and Biomedical Engineering from the General Engineering Department in the College of Engineering at the University of Puerto Rico–Mayaguez Campus; and Senior Acquisitions Editor and Senior Technical Editor-in-Chief in Agricultural and Biomedical Engineering for Apple Academic Press Inc. He received his BSc degree in Engineering from Punjab Agricultural University, Ludhiana, India; his MSc and PhD degrees from the Ohio State University, Columbus; his Master of Divinity degree from Puerto Rico Evangelical Seminary, Hato Rey, Puerto Rico, USA.

Since 1971, he has worked as Soil Conservation Inspector (1971); Research Assistant at Haryana Agricultural University (1972–1975) and the Ohio State University (1975–1979); Research Agricultural Engineer/Professor at the Department of Agricultural Engineering of UPRM (1979–1997); and Professor in Agricultural and Biomedical Engineering at General Engineering Department of UPRM (1997–2012). He spent one year of sabbatical leave in 2002–2003 at Biomedical Engineering Department, Florida International University, Miami, USA.

He was the first agricultural engineer to receive a professional license in Agricultural Engineering in 1986 from the College of Engineers and Surveyors of Puerto Rico. On September 16, 2005, he was proclaimed as "Father of Irrigation Engineering in Puerto Rico for the 20th century" by the ASABE, Puerto Rico Section, for his pioneer work on micro-irrigation, evapotranspiration, agroclimatology, and soil and water engineering.

During his professional career of 53 years, he has received awards such as Scientist of the Year, Blue Ribbon Extension Award, Research Paper Award, Nolan Mitchell Young Extension Worker Award, Agricultural Engineer of the Year, Citations by Mayors of Juana Diaz and Ponce, Membership Grand Prize for ASAE Campaign, Felix Castro Rodriguez Academic Excellence, RashtryaRatan Award and Bharat Excellence Award and Gold Medal, Domingo Marrero Navarro Prize, Adopted son of Moca, Irrigation Protagonist of UPRM, Man of Drip Irrigation by Mayor of Municipalities of Mayaguez/Caguas/Ponce and Senate/Secretary of Agriculture of ELA, Puerto Rico.

The Water Technology Center of Tamil Nadu Agricultural University in Coimbatore, India, recognized Dr. Goyal as one of the experts "who rendered meritorious service for the development of micro-irrigation sector in India" by bestowing "*Award of Outstanding Contribution in Micro-Irrigation.*" This award was presented to Dr. Goyal during the inaugural session of the National Congress on "New Challenges and Advances in Sustainable Micro-Irrigation on March 1, 2017, held at Tamil Nadu Agricultural University.

At Annual Meeting on August 01, 2018, in Detroit – MI, the American Society of Agricultural and Biological Engineers (ASABE) bestowed on him Netafim Micro-irrigation Award for his unselfish contribution. VDGOOD Professional Association of India awarded Lifetime Achievement Award at the 12th Annual Meeting on Engineering, Science, and Medicine that was held on 20–21 November 2020 in Visakhapatnam, India.

He has authored more than 200 journal articles and more than 100 books, including: "Elements of Agroclimatology (Spanish) by UNISARC, Colombia"; two "Bibliographies on Drip Irrigation." Apple Academic Press Inc. (AAP) has published his books, namely: "Management of Drip/Trickle or Micro-Irrigation" and "Evapotranspiration: Principles and Applications for Water Management," a ten-volume set on "*Research Advances in Sustainable Micro-Irrigation.*" During 2016–2025, AAP will be publishing book volumes on emerging technologies/issues/challenges under the book series, "*Innovations and Challenges in Micro-Irrigation*" and "*Innovations in Agricultural and Biological Engineering.*"

Readers may contact him at: goyalmegh@gmail.com

Mohamed Jaffer M. Sadiq, PhD

Postdoctoral Researcher, School of Chemical Science and Technology, Yunnan University, Kunming, P.R. China

Mohamed Jaffer M. Sadiq, PhD, is a Postdoctoral Researcher at the School of Chemical Science and Technology at Yunnan University in Kunming, P.R. China. Dr. Sadiq has published 20 research articles and technical papers in international peer-reviewed journals. He is also serving as an editor and reviewer for several journals. He has participated in several state, national, and international conferences, seminars, workshops, and symposia. Dr. Sadiq has more than 10 years of industrial and research experience. His areas of interest are photocatalysis, heterogeneous catalysis, wastewater treatment, biomaterials, bio-nanotechnology, etc. Dr. Mohamed Jaffer Sadiq M received a BSc (2006) in Chemistry from Bharathiyar University, Coimbatore, Tamil Nadu, India; MSc (2008) in Applied Chemistry from National Institute of Technology (NIT), Tiruchirappalli, Tamil Nadu; MTech (2014) in Nanotechnology from Karunya University, Coimbatore, Tamil Nadu; and PhD (2017) in Chemistry from National Institute of Technology Karnataka (NITK), Surathkal, Mangalore, Karnataka, India.

Readers may contact him at: sadiqmsc@gmail.com.

CONTENTS

CONTRIBUTORS

Marghoob Ahmed
Lecturer, Department of Physics, Faculty of Sciences, N.J.C., Chiniot – 35460, Punjab, Pakistan, Mobile: +92-3346641649, E-mail: peermarghoob@gmail.com

Sadia Akter
BSc Candidate, Department of Glass and Ceramic Engineering, Rajshahi University of Engineering and Technology (RUET), Rajshahi – 6204, Bangladesh, Mobile: +88-01312723394, E-mail: sadia1706055@gmail.com

Mintu Ali
Lecturer, Department of Glass and Ceramic Engineering, Rajshahi University of Engineering and Technology (RUET), Rajshahi – 6204, Bangladesh, Mobile: +88-01737605557, E-mail: min2.gce11ruet@gmail.com

Tabasum Ali
PhD Research Scholar, University of Kashmir, Srinagar – 190006, Jammu and Kashmir, India, Mobile: +91-7006147969, E-mail: tabu125101@gmail.com

Muhammad Usman Alvi
Department of Chemistry, Faculty of Sciences, University of Okara, Okara – 56300, Pakistan, Mobile: +92-3347902086, E-mail: usm_alvi@hotmail.com

Anchal Anchal
Research Scholar, University of Delhi, South Campus, New Delhi – 110067, India, Mobile: +91-7014189184, E-mail: deshwal.anchal@gmail.com

Mohd. Anees
Research Scholar, Center for Biomedical Engineering, Indian Institute of Technology Delhi, Hauz Khas – 110016, New Delhi, India, Mobile: +91-9811567740, E-mail: mdanees1992@gmail.com

Manju Arora
Principal Technical Officer, CSIR–National Physical Laboratory, Dr. K.S. Krishnan Marg, New Delhi – 110012, India, Mobile: +91-9868979289, E-mail: manjuarorain@gmail.com

Munir Ashraf
Associate Professor, Chairman of Department of Textile Engineering, School of Textile Engineering, National Textile University, Faisalabad – 37610, Pakistan, +92-3336663743, E-mail: munir.ashraf01@gmail.com

Sabeen Aslam
MPhil Research Scholar, Department of Chemistry, Faculty of Sciences, University of Okara, Okara – 56300, Pakistan, Mobile: +92-3014691539, E-mail: sabeenaslam3040@gmail.com

Ahmed Sidrat Rahman Ayon
Undergraduate Student, Department of Glass and Ceramic Engineering, Rajshahi University of Engineering and Technology (RUET), Rajshahi – 6204, Bangladesh, Mobile: +88-01722202364, E-mail: sidratruet096@gmail.com

Manisha Bal
Assistant Professor, Department of Chemical Engineering, MVJ College of Engineering, Whitefield, Bangalore – 560067, Karnataka, India, Mobile: +91-8617469078, E-mails: chemmanisha1@gmail.com; manishabal@mvjce.edu.in

Afsar Bano
PhD Research Scholar, Department of Physics, Syed Babar Ali School of Science and Engineering, Lahore University of Management Sciences, LUMS – 54792, Lahore, Pakistan, Mobile: +92-3447626043, E-mail: afsar.bano@lums.edu.pk

Rabiah Bashir
PhD Research Scholar, Department of Pharmaceutical Sciences, University of Kashmir, Hazratbal – 190006, Srinagar, Jammu and Kashmir, India, Mobile: +91-7006694821, E-mail: rabiahbashir100@gmail.com

Nitin Bayal
Research Scholar, National Center for Cell Science, Pune University Road, Ganeshkhind – 411007, Pune, Maharashtra, India, Mobile: +91-7857914077, E-mail: nitin.mhg.bhu12@gmail.com

Sajeena Beevi
Associate Professor, Department of Chemical Engineering, Government Engineering College, Thrissur – 680009, Kerala, India, Mobile: +91-9446317516, E-mail: sajeenanazer@gmail.com

Manish Biyani
Professor, Department of Bioscience and Biotechnology, Japan Advanced Institute of Science and Technology, Ishikawa – 9231292, Japan, Mobile: +81-8040720407, E-mail: biyani@jaist.ac.jp

Vijaya Geetha Bose
Assistant Professor (SS), Department of Biotechnology, Rajalakshmi Engineering College (Autonomous), Affiliated to Anna University, Thandalam – 602105, Tamil Nadu, India, Mobile: +91-9884088574, E-mail: vijayageetha.b@rajalakshmi.edu.in

Murthy S. S. S. Chavali
Chief Scientist and Executive Director, NTRC-MCETRC, and Aarshanano Composites Technologies Pvt. Ltd., Guntur – 522201, Andhra Pradesh, India, Mobile: +91-8309337736, E-mail: chavalim@gmail.com

Hagar Fathy Forsan
Researcher, Animal Production Research Institute, Agriculture Research Center, Giza – 12511, Egypt, Mobile: 020-1027133075, E-mail: hagarfathy@pg.cu.edu.eg

Vihaa Sriee Mambullikaavil Ganesan
MTech Student, Department of Biotechnology, Rajalakshmi Engineering College (Autonomous), Affiliated to Anna University, Thandalam – 602105, Tamil Nadu, India, Mobile: +91-7358328726, E-mail: 200411017@rajalakshmi.edu.in

Megh R. Goyal
Senior Editor-in-Chief (Agriculture and Biomedical Engineering) for AAP, Retired Professor in Agricultural and Biomedical Engineering, University of Puerto Rico – Mayaguez, Mayaguez – Puerto Rico, USA, Mobile: 001-787-536-0039, E-mail: goyalmegh@gmail.com

Md. Jahidul Haque
MSc Candidate, Department of Glass and Ceramic Engineering, Rajshahi University of Engineering and Technology (RUET), Rajshahi – 6204, Bangladesh, Mobile: +88-01983125506, E-mail: mjh.ruet26@gmail.com

Peter Hewitson
Senior Lecturer, Department of Chemical Engineering, Brunel University London, UB8 3PH, UK, Mobile: +44(0)7961-172638, E-mail: peter.hewitson@brunel.ac.uk

Tasnuva Humaira
Undergraduate Student, Department of Glass and Ceramic Engineering, Rajshahi University of Engineering and Technology (RUET), Rajshahi – 6204, Bangladesh, Mobile: +88-01759995704, E-mail: tasnuvahumairaruet02@gmail.com

Riaz Hussain
Assistant Professor, Department of Chemistry, Faculty of Sciences, University of Okara, Okara – 56300, Pakistan, Mobile: +92-3336757473, E-mail: riazhussainansari@gmail.com

Svetlana Ignatova
Professor and Director, Advanced Bioprocessing Center, Department of Chemical Engineering, Brunel University London, UB8 3PH, UK, Mobile: +44(0)1895266911, E-mail: svetlana.ignatova@brunel.ac.uk

Rashminiza Abdul Jaleel
MTech Student, Department of Biotechnology, Rajalakshmi Engineering College (Autonomous), Affiliated to Anna University, Thandalam – 602105, Tamil Nadu, India, Mobile: +91-9790938417, E-mail: 200411016@rajalakshmi.edu.in

Vikas B. Kabburi
B.E. Student, Department of Chemical Engineering, MVJ College of Engineering, Whitefield, Bangalore – 560067, Karnataka, India, Mobile: +91-9535810748, E-mail: vikaskabburi1999@gmail.com

Humayan Kabir
Lecturer, Department of Glass and Ceramic Engineering, Rajshahi University of Engineering and Technology (RUET), Rajshahi – 6204, Bangladesh, Mobile: +88-01722831744, E-mail: mhk@gce.ruet.ac.bd

Abdul Kaiyum
Lecturer, Department of Glass and Ceramic Engineering, Rajshahi University of Engineering and Technology (RUET), Rajshahi – 6204, Bangladesh, Mobile: +88-01765988884, E-mail: kaiyum@gce.ruet.ac.bd

Shabnam Kawoosa
PhD Research Scholar, Department of Pharmaceutical Sciences, University of Kashmir, Hazratbal – 190006, Srinagar, Jammu and Kashmir, India, Mobile: +91-7051856940, E-mail: shabnam.scholar@kashmiruniversity.net

Nisar Ahmad Khan
Sr. Associate Professor, Department of Pharmaceutical Sciences, University of Kashmir, Hazratbal, Srinagar – 190006, Jammu and Kashmir, India, Mobile: +91-9419017186; +91-7006886348, E-mail: nakhan2008@gmail.com

Ankita Kumari
Executive, Centyle Biotech Private Limited, Rudrapur – 263153, Uttarakhand, India, Mobile: +91-8460739792, E-mail: ankita.centyle@gmail.com

Tarun Kumar Kumawat
Assistant Professor, Department of Biotechnology, Biyani Girls College, Jaipur – 302039, Rajasthan, India, Mobile: +91-9509185127, E-mail: tarun.kumar1419@gmail.com

Varsha Kumawat
Director, Naturilk Organic and Dairy Foods Pvt. Ltd., Jaipur – 302012, Rajasthan, India, Mobile: +91-9672961627, E-mail: varsha29jun1993@gmail.com

Junaid Ahmad Malik
Lecturer, Department of Zoology, Government Degree College, Bijbehara – 192124, Kashmir, Jammu and Kashmir, India, Mobile: +91-7006317311, E-mail: malik.junaidahmad@gmail.com

Shubhjeet Mandal
Senior Executive, R&D, Centyle Biotech Private Limited, Rudrapur, U.S. Nagar – 263153, Uttarakhand, India, Mobile: +91-8618040998, Email: shubhjeetm@gmail.com

Muhammad Mubin
Associate Professor, CABB, University of Agriculture, Faisalabad – 38000, Punjab, Pakistan, Mobile: +92-3006615706, E-mail: mmubin@uaf.edu.pk

Ahsan Habib Munna
Undergraduate Student, Department of Glass and Ceramic Engineering, Rajshahi University of Engineering and Technology (RUET), Rajshahi – 6204, Bangladesh, Mobile: +88-01521333306, E-mail: ahsanhabibmunna@gmail.com

Kasibatla S. R. Murthy
Associate Professor, Department of Applied Chemistry, University of Petroleum and Energy Studies, Energy Acres, Bidholi Campus, Premnagar, Dehradun – 248007, Uttarakhand, India, Mobile: +91-96547852153, E-mail: Murthy.kasibatlasr@gmail.com

Ghulam Mustafa
Associate Professor, Department of Chemistry, Faculty of Sciences, University of Okara, Okara – 56300, Pakistan, Mobile: +92-3125990695, E-mail: Ghulam.mustafa@uo.edu.pk

Rabeea Muzaffar
Biochemist, Department of Biochemistry, Faculty of Sciences, University of Agriculture, Faisalabad – 38000, Pakistan, Mobile: +92-3377639931, E-mail: rabeeamuzaffar1@gmail.com

Praseetha P. Nair
Associate Professor, Department of Chemical Engineering, Government Engineering College, Thrissur – 680009, Kerala, India, Mobile: +91-9656221371, E-mail: praseetha@gectcr.ac.in

Sharmiladevi Palani
PhD Research Scholar, Faculty of Allied Health Sciences, Chettinad Academy of Research and Education, Kelambakkam – 603103, Tamil Nadu, India, Mobile: +91-9962306290, E-mail: Sharmiladevi.p7@gmail.com

Anjali Pandit
Researcher, Department of Biotechnology, Biyani Girls College, Jaipur – 302039, Rajasthan, India, Mobile: +91-9588957422, E-mail: panditanjali04@gmail.com

Kanagasabai Muruganandam Ponvel
Associate Professor, PG Research Department of Chemistry, V.O. Chidambaram College, Manonmaniam Sundaranar University, Thoothukudi – 628008, Tamil Nadu, India, Mobile: +91-8543514287, E-mail: kmp@gmail.com

Sanzana Tabassum Proma
Undergraduate Student, Department of Glass and Ceramic Engineering, Rajshahi University of Engineering and Technology (RUET), Rajshahi – 6204, Bangladesh, Mobile: +88-01779594456, E-mail: sproma775@gmail.com

Shamimur Rahman
Professor, Head, Department of Glass and Ceramic Engineering, Rajshahi University of Engineering and Technology (RUET), Rajshahi – 6204, Bangladesh, Mobile: +88-01713228532,
E-mail: rahman@gce.ruet.ac.bd

Zubaid-Ul-Khazir Rather
PhD Research Scholar, Department of Chemistry, National Institute of Technology, Hazratbal – 190006, Srinagar, Jammu and Kashmir, India, Mobile: +91-9682189204, E-mail: rather.zubaid007@gmail.com

Rafia Rehman
Assistant Professor (IPFP), Department of Chemistry, Faculty of Sciences, University of Okara, Okara – 56300, Pakistan, Mobile: +92-3326729716,
E-mails: dr_rafia@uo.edu.pk; rehman_rafia@yahoo.co.uk

Adeel Riaz
Shaukat Khanum Cancer Hospital, Department of Radiation Oncology, 7A" Khayaban-e-Firdousi, Block R3 MA Johar Town, Lahore, Punjab, Pakistan, Mobile: +92-3096171712,
E-mail: adeel447@gmail.com

Shagufta Riaz
Assistant professor, Department of Textile Engineering, School of Textile Engineering, National textile University, Faisalabad – 37610, Punjab, Pakistan, Mobile: +92-3063325338,
E-mail: shaguftariaz84@gmail.com

Haleema Sadia
MPhil Research Scholar, Department of Chemistry, Faculty of Sciences, University of Okara, Okara – 56300, Pakistan, Mobile: +92-3317480709, E-mail: Haleemasadiauoe@gmail.com

Mohamed Jaffer M. Sadiq
Postdoctoral Research Fellow, School of Chemical Science and Technology, Yunnan University, 2 North Cuihu Road, Kunming – 650091, P. R. China, Mobile: +91-7904275606,
E-mail: sadiqmsc@gmail.com

G. V. S. Subbaroy Sarma
Associate Professor, Department of Basic Sciences and Humanities, Vignan's Lara Institute of Technology and Science, Guntur – 522213, Andhra Pradesh, India, Mobile: +91-8142354587,
E-mail: subbaroysharma@gmail.com

Sivakala Sarojam
Assistant Professor, Post Graduate and Research Department of Chemistry, Sree Narayana College, Chempazhanthy, Trivandrum – 695587, Kerala, India, Mobile: +91-9400022839,
E-mail: sivakalasnc@gmail.com

Muhammad Shahid
Associate Professor, Department of Biochemistry, Faculty of Sciences, University of Agriculture, Faisalabad – 38000, Pakistan, Mobile: +92-3336629271, E-mail: mshahiduaf@uaf.edu.pk

Zarin Rafa Shaitee
Undergraduate Student, Department of Glass and Ceramic Engineering, Rajshahi University of Engineering and Technology (RUET), Rajshahi – 6204, Bangladesh, Mobile: +88-01799966644,
E-mail: zarinrafashaitee54@gmail.com

Vishnu Sharma
Assistant Professor, Department of Biotechnology, Biyani Girls College, University of Rajasthan, Jaipur – 302039, Rajasthan, India, Mobile: +91-7877034513, E-mail: vishnusharma666@yahoo.com

Ajay Singh
Associate Professor, Department of Physics, GGM Science College (Constituent College of Cluster University of Jammu), Canal Road, Jammu – 180002, Jammu and Kashmir, India,
Mobile: +91-9419154646, E-mail: ajay.dadwal1234@gmail.com

Harpal Singh
Professor, Center for Biomedical Engineering, Indian Institute of Technology Delhi,
Hauz Khas – 110016, New Delhi, India, Mobile: +91-9810030490, E-mail: harpal2000@yahoo.com

Shreenidhi Krishnamurthy Subramaniyan
Assistant Professor, Department of Biotechnology, Rajalakshmi Engineering College (Autonomous), Affiliated to Anna University, Thandalam – 602105, Tamil Nadu, India, Mobile: +91-9444379793,
E-mail: shreenidhi.ks@rajalakshmi.edu.in

Piyush Kumar Tiwari
Manager, Centyle Biotech Private Limited, Rudrapur – 263153, Uttarakhand, India,
Mobile: +91-7291063121, E-mail: piyushkumartiwari@ymail.com

Sana Ullah
Assistant Professor, Department of Zoology, Division of Science and Technology,
University of Education, Lahore – 54000, Pakistan, Mobile: +92-3468981437,
E-mail: sanaullah@ue.edu.pk

Aziz Unnisa
Assistant Professor, Department of Pharmaceutical Chemistry, College of Pharmacy, University of Hail, Hail – 55475, Saudi Arabia, Mobile: +966-537860207, E-mail: khushiazeez@yahoo.co.in

Kusum Upadhyay
PhD Scholar, Mansarovar Global University, Ratnakhedi – 466111, Madhya Pradesh, India,
Mobile: +91-9716044656, E-mail: kusumupadhyay04@gmail.com

Ayesha Younus
Assistant Professor, Faculty of Sciences, Department of Physics, University of Agriculture,
Faisalabad – 38000, Punjab, Pakistan, Mobile: +92-3366084987, E-mail: smilephysics@yahoo.com

ABBREVIATIONS

0D	zero-dimensional
1D	one-dimensional nanomaterial
2D	two-dimensional nanomaterial
3D	three-dimensional
5FU	5-fluorouracil
AAO	anodic aluminum oxide
ADME	absorption, delivery, metabolism, and elimination
ADP	adenosine triphosphate
AD-PEG_8-GRGDS	adamantane-PEG_8-glycine-arginine-glycine-aspartic-serine
AFM	atomic force microscopy
Ag NPs	silver nanoparticles
Ag	silver
$AgNO_3$	silver nitrate
AgNPs	silver nanoparticles
AIE	aggregation-induced emission
AirSEM	air scanning electron microscopy
Al_2O_3	aluminum oxide
APCs	antigen-presenting cells
API	active pharmaceutical ingredient
APTES	amino-propyl-tri-ethoxy-silane
ATF	amino-terminal fragment
Au	gold
AuNPs	gold nanoparticles
BBB	blood-brain barrier
Bi_2S_3	bismuth sulfide
BNMs	bionanomaterials
BNPs	bio nanoparticles
BP	Black phosphorus
BRB	blood-retinal barrier
BRCA1	breast cancer gene1
BRCA2	breast cancer gene2
Ca	calcium
CaO	calcium oxide

CaP_2	calcium phosphate
CaPs NMS	calcium phosphate nanomaterials
CD	cyclodextrin
CDC	Centers for Disease Control and Prevention
CDDSs	cell-based drug delivery systems
CDNSs	CD-based nanosponges
CDs	carbon dots
CdTe	cadmium telluride
Cet	cetuximab
Ch	chitosan
ChT	chitotriosidase
CMC	carboxy methyl cellulose/ critical micellar concentration
CMs	carbon materials
CNH	carbon nanohorns
CNMs	carbon nanomaterials
CNP	carbon nanoparticles
CNT	carbon nanotubes
Con A	concanavalin A
COX-2	cyclooxygenase
CPT	camptothecin
CRCS	calciobiotic root canal sealer
Cs	chitosan
CS	chondroitin sulfate
CS-NCs	CS-nanocapsules
CT Scan	computed tomography
Cu	copper
Cu_3P NWs/CF	phosphide nanowires on porous Cu foam
CuI	copper iodide
CuNPs	copper nanoparticle
CuO	copper oxide
CuS	copper sulfide
$CuSO_4$	copper sulfate
CVD	chemical vapor deposition
CXB	celecoxib
D.R.	diffusion reflection
DD systems	decisive data system
DDS	drug delivery system
DH_2O	distilled H_2O

DLS	dynamic light scattering
DM	Dzyaloshinkii-Moriya
DMF	dimethyl formamide
DMSO	dimethyl sulfoxide
DNA	deoxyribonucleic acid
DOX	doxorubicin
DTX	docetaxel
E. coli	*Escherichia coli*
ECM	extracellular matrix
EDTA	ethylenediaminetetraacetic acid
EDX	X-ray electron microscopy
EGFfr	EGF fragment
EGFR	epidermal growth factor receptor
EMA	European Medicines Association
EMEA	European Medicine Agency
ENMs	engineered nanomaterials
ENP	engineered nanoparticles
EpCAM	epithelial cell adhesion molecule
EPR	electron paramagnetic resonance
EPR	enhanced permeability and retention
ESBL	extended-spectrum beta-lactamases
ESD	emulsification solvent diffusion
FA	folic acid
FCC	face-centered cubic
FDA	Food and Drug Administration
Fe	iron
FE-SEM	field emission scanning electron microscopy
FET	field effect transistor
FI	fluorescent imaging
FMLM	Fe_3O_4@mSiO_2@lipid-PEG-methotrexate nanoparticle
FNPs	ferrite nanoparticles
FRET	Forster resonance energy transfer
FTIR	Fourier transform infrared spectroscopy
GAG	glycosaminoglycans
Gan	gallium nitride
GBM	glioblastoma multiforme
GCS	glucosylceramide synthase
G-CSFR	granulocyte colony-stimulating factor

Gd	gadolinium
GD	Gaucher's disease
Gd-DTPA	gadopentetate dimeglumine
GFLG	gemcitabine, tetrapeptide
Gnrs	gold nanorods
GO	graphene oxide
GQDs	graphene quantum dots
$Gram^-$	gram-negative
$Gram^+$	gram-positive
GSH	glutathione
H_2O_2	hydrogen peroxide
HA	hyaluronic acid
HA	hydroxyapatite
HAase	hyaluronidase
HCC	hepatocellular carcinoma
HER2	anti-human epidermal growth factor receptor 2
HMME	hematoporphyrin monomethyl ether
HOPG	highly oriented pyrolytic graphite
HPAO	3-(4-hydroxyphenyl) propionic acid-OSU
HPLC	high-pressure liquid chromatography
HPMC	hydroxypropyl methylcellulose
HPR	high-pressure homogenization
HRTEM	high-resolution transmission electron microscopy
HSV-1	herpes simplex virus type-1
i.p.i.	intraperitoneal
i.v.	intravenous
ICG	indocyanine green
ICI	immune checkpoint inhibitor
IL	interleukin
IO	iron oxide
IO-NPs	iron oxides nanoparticles
IOP	intra-ocular pressure
IPA	isopropanol
IPM	IR 780 loaded polymeric micelles
iPSCs	induced murine pluripotent stem cells
ISO	International Organization of Standardization
KV	kilo volt
LED	light emitting diode
LSPR	localized surface plasmon resonance

LYS	lysozyme
MAbs	monoclonal antibodies
MDR	multidrug resistance
Mg	magnesium
MgO	magnesium oxide
MHB	muller-Hinton broth
MIC	minimum inhibitory concentration
MIPs	molecularly imprinted polymers
miRNA	micro ribonucleic acid
MMPs	matrix metalloproteinases
MNMR	micro nuclear magnetic resonance
MnO_2	manganese dioxide
MNPs	magnetic nanoparticles
MNPs	metal nanoparticles
MOCVD	metal-organic chemical vapor deposition
MOS	metal oxide semiconducting
MOSFET	metal oxide semiconducting field effect transistors
MPI	magnetic particle imaging
MPs	magnetic particles
MPS	mononuclear phagocyte system
MR	magnetic resonance
MRI	magnetic resonance imaging
MRSA	methicillin-resistant Staphylococcus aureus
MRSE	methicillin-resistant Staphylococcus epidermidis
MSA	methylsulfonic acid
MSCs	mesenchymal stem cells
MSNs	mesoporous silica nanoparticles
MSOT	multispectral optoacoustic tomography
MSSA	methicillin-resistant Staphylococcus aureus
MTMMP	membrane-type MMP
MTX	methotrexate
MWCNTs	multi-walled carbon nanotubes
MWCO	molecular weight cut-offs
MWNT	multi walled nanotubes
NACP	nanoparticles of amorphous calcium phosphate
NBs	nano-bulk materials
NCAM	neural cell adhesion
NCC	ncarboxymethyl chitosan
NEMS	nanoelectromechanical systems

NFs	nanofibers
NIR	Near Infrared
NLC	nanostructured lipid carriers
NM	nanomaterial
NM-Microbes	nanomaterial-microbes
NMP	nanomagnetic particles/ N-methyl-2-pyrrolidone
NMs	nanomaterials
NO	nitric oxide
NPs	nanoparticles
NSAID	non-steroidal anti-inflammatory drugs
NT	nanotechnology
nZnO	nanostructured ZnO
P	phosphate
PA	photoacoustic
PAI	photoacoustic imaging
PAMAM	polyamidoamine
PBASE	1-pyrenbutyric acid N-hydroxysuccinimide ester
PCA	poly cyanoacrylate
PCM	phase change material
Pd	palladium
PdNPs	palladium nanoparticles
PDT	photodynamic therapy
PECA	poly(ethyl cyanoacrylate)
PEEK	polyetheretherketone
PE-FET	piezoelectric-FET
PEG	polyethylene glycol
PEI	polyethyleneimine
PEO-PCL	poly epsilon-caprolactone
PEO-PPO	polyethylene oxide-propylene oxide block copolymer
PET	Positron emission tomography
PFH	perfluoro hexane
P-gp	P-glycoprotein plasma
PGP	polymeric P-glycoprotein
Pgps	P-glycoproteins
PL	photoluminescence
PL	poly-*L*-lysine
PLA	poly lactic acid
PLG	poly(*D,L* glycolide)

PLGA	poly lactic co glycolic acid
PMMA	poly(methylmethacrylate)
PNA	peptide nucleic acid
PNP	purine nucleoside phosphorylase
PO_4	phosphate
PSG	proteasome storage granules
PSMA	prostate-specific membrane antigen
PspA	pneumococcal surface protein A
PSs	photosensitizers
Pt	platinum
PTT	photothermal therapy
PTX	paclitaxel
PVA	polyvinyl alcohol
PVDF	polyvinylidene fluoride
PVP	polyvinylpyrrolidone
QDs	quantum dots
R&D	Research and Development
RCNMV	red clover necrotic mosaic virus
RES	reticuloendothelial system
RF	radio frequency
RGD	arginylglycylaspartic acid
RGO	reduced graphene oxide
RhBMP-2	recombinant human bone morphogenetic proteins-2
RNA	ribonucleic acid
ROS	reactive oxygen species
RPE	retinal monolayer epithelium
RSW	red sandalwood
RTT	radio thermal therapy
RUET	Rajshahi University of Engineering and Technology
SARS-CoV-2	severe acute respiratory syndrome coronavirus
SCLC	small-cell lung cancer
SCs	supercapacitors
SCS	suprachoroidal space
SEM	scanning electron microscopy
SERS	surface-enhanced Raman spectroscopy
SiO_2	silicon dioxide
siRNA	small interfering ribonucleic acid
SLCCNV	squash leaf curl coronavirus
SLN	solid lipid nanoparticles

SP	superparamagnetic
SPECT	single-photon emission computed tomography
SPION	superparamagnetic iron oxide nanoparticle
SPR	surface plasmon resonance
SPRE	surface plasmon resonance energy
SQUIDs	superconducting quantum interference devices
SrO	strontium oxide
SSM	sterically-stabilized micelles
STDs	sexually transmitted diseases
SWCNT	single-wall carbon nanotube
TAA	tumor-associated antigen
TCP	tricalcium phosphate
TDDS	therapeutic drug delivery system
TEM	transmission electron microscopy
TEOS	tetraethyl orthosilicate
TGA	thermogravimetric analysis
THz	terahertz
Ti	titanium
TiO_2	titanium dioxide
TLR	toll-like receptor
TME	tumor microenvironment
TMI	tetra-modal imaging
TNBC	triple-negative breast cancer
TP53	tumor protein53
Tween 80	polysorbate 80
UCNPs	upconversion nanoparticles
uPAR	urokinase-type plasminogen activator receptor
US	ultrasound
UTIs	urinary tract infections
UV	ultraviolet
VB	valance band
VCAM	vascular cell adhesion molecule
VEGF	vascular epithelial growth factor
VNPs	viral nanoparticles
VSM	vibrating sample magnetometry
WHO	World Health Organization
XRD	X-ray diffraction
XRF	X-ray fluorescence spectrometer
Zn	zinc

ZnNPs	Zn nanoparticles
ZnO NPs	Zn oxide nanoparticles
ZnO	zinc oxide
ZOI	zone of inhibition

PREFACE

In recent years, nanomaterials have become one of the most dynamic exploration fields in the areas of engineering, technology, and science. Bionanomaterials are the kind of nanomaterials that are introduced into the body as clinical devices for medical purposes. These materials have various applications in the biomedical fields, such as cancer treatment, in orthopedic surgery for joint replacements, in diagnosis, for bone plates, for wound healing, for nerve regeneration, for breast implants, and so on. In addition, these materials also have a few uses in nonbiomedical fields, for example, growing cells in a culture medium, for blood protein tests in laboratories, and so forth. However, researchers are always looking for further developments and superior features, which can be used for more exciting applications in the biomedical field.

The present book, *Sustainable Nanomaterials for Biomedical Engineering: Impacts, Challenges, and Future Prospects,* has been categorized under three subsections; Nanomaterials for Medical Applications, Nanomaterials for Drug Delivery and Therapy, and Nanomaterials for Antimicrobial and Antibacterial Applications in Medicine. These three sections comprise 17 chapters, including the role of nanomaterials for biological and medical applications, targeted drug delivery systems, dentistry, imaging, cancer theranostics, aiagnostics and cancer treatment, orthopedics applications, ocular diseases, and microbial, antimicrobial, and anti-bacterial applications. Each chapter provides a detailed description of the mechanisms, benefits, and drawbacks of the technologies used. The book emphasizes the use of various nanomaterials that are used in medical engineering applications. This book brings together various topics, representing a new resource for researchers to solve problems in biomedical engineering.

This book is designed for academicians, environmentalists, practitioners, NGOs, and industrialists who are working in the field of biomedical engineering. There is ample information available for all researchers to continue their research on nanomaterials for applications and recent advancements in biomedical engineering based on current and future research.

—Editors

PART I

NANOMATERIALS FOR MEDICAL APPLICATIONS

CHAPTER 1

APPLICATIONS OF NANOMATERIALS IN MEDICINE

HAGAR FATHY FORSAN

ABSTRACT

There are several opportunities for nanotechnology to improve medical research, thereby changing health care practices around the world. It is predicted that several new nanoparticles will be used with an immensely positive effect on human health. The potential uses of nanomaterials in medicine have been extensively studied. Nowadays, with the rapid growth of nanotechnology for medication and gene delivery, biological instruments, nanoelectronic biosensors, or molecular nanotechnology, nanomaterials themselves can be used as imaging agents or therapeutic drugs. Therefore, this chapter describes nanomaterial types, dimensions, chemical composition, and industrial nanomaterials. It also discusses some distinct nanomaterials like silver, gold, phosphorus, and hybrid chitosan for their biological and medical applications and future directions of nanomedicine.

1.1 INTRODUCTION

Nanotechnology is one of the advances which outspreads widely [9, 39, 92]. Nanoscale control allows for characterizing, synthesizing, and manipulating chemical and physical macroscopic characteristics of devices and materials. Over the past few decades, chemistry, engineering, and material science have developed greatly. Recently nanotechnology has been used in medicine, biology, engineering, and chemistry [14, 93].

Sustainable Nanomaterials for Biomedical Engineering: Impacts, Challenges, and Future Prospects.
Junaid Ahmad Malik, Megh R. Goyal & Mohamed Jaffer M. Sadiq (Eds.)

In recent decades, nanotechnology and biotechnology have developed to a great extent. Now it is possible to merge these two exciting fields to get different characteristics. Different nanomaterials of biological applications such as biological studies, diagnosis, and biomedical treatment are used in a wide range and have perceived a great interest [53, 87].

This chapter explores nanomaterial types, scopes, chemical composition, and industrial applications. It also focuses on biomedical applications of certain nanomaterials.

1.1.1 DEFINITIONS AND CONCEPTS REGARDING NANOMATERIALS

Nanomaterials (NMs) are substances that have a very small size less 100 nm. It improved to get new features compared to similar material without nanoscale characteristics, as strength increases, reductive, and conductive chemical. NMs used in biological applications must have certain properties such as being biocompatible and capable of directing biomaterial using a suitable method for drug distribution in complex settings such as animals and humans [50].

Micro-sized carriers of immobilization aptamers or nanoscale enhance building a lot of functional ingredients with preferred properties. It is common to use nanomaterials over covalent and non-covalent bonds [13, 23]. This modification improves identification, reduces analysis time, and improves specificity binding sensitivity and selectivity to target subjects. The average particle size of bulk material is greater than 100 nm. Its physical properties were unaffected by scale, while NMs were affected by form and size.

1.1.2 MAIN CATEGORIZATION OF NANOMATERIALS

Nanomaterials can be categorized into different classes based on different principles and can be divided into two groups: Nanoparticles (NPs) and Nano-Bulk materials (NBs). NPs are particles less than 100 nm and can be seen using an advanced microscope. NMs have better recital and that depends on surface to volume proportion and have great characteristics according to atoms highest ratio and surface molecules. Its surface forces are essential [64].

The most important characteristic of metal NMs is scattering properties. Semiconductor nanoparticles show limited energy states in the electronic group structure. Size and shapes can change chemical and physical properties. The size of the particles tuned NMs features such as changing color for fluorescence when the particle size changed. NMs offer great properties and a diversity of functions of products. NMs marked by high and fast absorption and adsorption of molecules such as gas and liquid phases [64].

On the other side, Bulk materials can be seen using a simple microscope or by the naked eye. Improved performance can occur using a low surface-to-volume percentage. Bulk material properties are owing to the low ratio of atoms and molecules on the surface. Surface forces are not essential for bulk materials. Bulk materials have normal scattering features. Semiconductor nanoparticles did not show any energy states in the electronic group structure. Physical and chemical properties cannot change. Bulk materials have low and slow adsorption and absorption of molecules [7, 8, 68, 77].

Commonly, NMs are characterized related to their morphological properties, dimension, and chemical composition. It is possible to further categorize NMs into four groups: (i) nil dimension (0D); (ii) one dimension (1D); (iii) two dimensions (2D); (iv) three dimensions (3D). NMs are categorized on the basis of their chemical properties, into suspensions forms, colloid forms, dispersed forms, or agglomerated states [40, 59, 68].

NMs can also be categorized based on their chemical composition into different groups; single constituent and nanocomposites (carbonaceous, fullerenes, and graphene). NMs made from metals such as copper, silver, iron, silica, and zinc are called metallic nanomaterials. Branched dendrimers form NMs of various types, which are associated with nanoscale features such as branching [21, 25, 40].

1.1.3 CHARACTERISTICS OF NANOPARTICLES

NMs have significant characteristics as they have a great capacity of loading, biological compatibility, great stability of thermal, constant porosity, and they are characterized by internal and external functionalization [34].

Magnetic NPs have magnetism and selectivity features for imaging substances' utilization, e.g., metals including gold and silver have surface plasmon resonance SPR characteristics used to treat hyperthermia [98], drug transfer and any analysis for magnetic fluid [86].

Silica has a high density, separated easily, and provided internal doping, and biocompatible. Hydrogels can hold large volumes of water and recognition and separate molecular or biological liquids, and it used as transmitting signals [99].

Carbon materials such as single-wall carbon nanotube (SWCNT) have excellent material properties, high surface area and aspect ratio, and transistors may have a major influence on the electrical industry, conductivity of heat, and mechanical strength [15]; carbon materials used in fluorophore [48] and optical protein Near-infrared (NIR) quenchers examine [46].

1.1.4 PROPERTIES OF NANOPARTICLES AND NANO-BULK MATERIALS

Nanoparticles have different physiochemical properties, which are discussed in subsections.

1.1.4.1 PARTICLE FORM AND PERCENTAGE ASPECT

Understanding the interplay between particle size and shape has advanced significantly, allowing more potent nanomaterials to provide a tailored distribution mechanism (Table 1.1) [99].

TABLE 1.1 Nanomaterials' Shapes and Toxic Properties

NMs Shapes	Toxic Properties
Spherical NMs	They are faster and easier compared with rod-shaped. [15, 99].
	Both homo or heterogeneous, they are less toxic compared to other NMs [48], and optical protein near-infrared (NIR) quenchers examine.
Non-spherical NMs	Non-spherical NMs have great properties as they can move through capillaries affecting added biological significances [46].
Rod-shaped NMs	Rod-shaped SWCNT can be more effective than spherical carbon fullerenes 2 to 3 times blocking K^+ ion channels [46].
Gold nanorods	It is slower than spherical nano-spheres [36].
TiO_2 fibers NMs	More cytotoxic than spherical entities [37].

1.1.4.2 SURFACE CHARGE OF NANOPARTICLES

It is influenced by the toxicity of NPs, which primarily explains their interactions with biological systems. NMs have a variety of properties, such as selective adsorption [36]. The surface charge primarily regulates colloidal behavior, binding of plasma protein [67], permeability of transmembrane, and integrity of blood-brain barrier (BBB) [30].

1.1.4.3 COMPOSITION AND STRUCTURE OF CRYSTALLINE

Silver and copper NPs are soluble formulations that can induce toxicity in examined organisms, however, TiO_2 of similar characteristics cannot produce any toxicity effects [31].

1.1.4.4 AGGREGATION AND CONCENTRATIONS OF NANOPARTICLES

It depends on NPs size, surface charge, and composition. It was found that the concentration of toxicity decreased with growth in the concentration of NPs [90].

1.1.4.5 SURFACE COATING AND ROUGHNESS

They may change physicochemical characteristics, for example functions of magnetic, chemical, electrical, and optical responsiveness [32, 53]. Biomaterials have long been used in medical utilization for example, examination and treatment of multiple illnesses [39].

Natural bionanomaterials focus on natural materials. Biocompatible polymers have a different developed type. This is related to polymer functions and biodegradability [3] and is used in drug delivery [10, 70].

We can use inorganic materials like ceramics and glasses and some minerals for biological applications. Inorganic materials have specific properties that make them better than organic materials such as magnetic, optical, thermal, and mechanical properties, in addition to their connected absorbent structures [60]. It has become necessary to use hybrid nanomaterials with organic and inorganic nanomaterials.

Theranostics means the combination of therapeutic and diagnostic functions that allow monitoring and treating disease. It can monitor medicine accretion in targeted tissues and also helps to know therapeutic responses by enhancing medicinal strategies. Due to their novel bio-imaging functions, including (magnetic particle imaging: MPI; computed tomography: CT and optical characteristics) and disease-sensitive targeting ability obtained by inorganic NP and organic polymer compounds [48]. Hybrid nanoparticles can provide very accurate disease diagnosis and medical treatments, such as small-molecule chemotherapeutics like doxorubicin and paclitaxel for cancer treatment. It is also used in biotherapeutics, including DNA, antibodies, and miRNA [17, 84].

1.1.4.6 SIZE OF PARTICLES AND NANOPARTICLES SURFACE AREA

Materials can interact with the biological system. The size of the substance appears to be decreasing. As a result, the surface area grows exponentially faster than the volume [69].

1.1.5 CANCER IMMUNOTHERAPY OF NANOPARTICLES

Nanoparticles play a vital role in cancer immunotherapy and are used as nano-vaccines to inhibit cancer progression for patients. These are also used to administer immunostimulants and chemicals, with the purpose of eliciting an adaptive immune response [49, 78].

Antigens and adjuvant-nanoparticle binding or encapsulation has substantially improved the response of anti-tumor response, including the higher response of T- and B-cell than non-nano particles [76]. We can organize more effectively and improve the immunotherapy effects of hybrids in association with an immune checkpoint inhibitor (ICI), which reduces the inhibition of immune cascade in addition increases the motivation of T cells.

1.1.6 PHOTODYNAMIC THERAPY OF NANOPARTICLES

PDT is an endowed noninvasive policy for care of different conditions such as tumor photosensitizers (PSs) provided by cytotoxic singlet oxygen ($1O_2$) by light irradiation [33, 65]. This approach was limited because, due to their fair tumor selectivity, phototoxicity is unable to select tumors [88]. Several experiments have been carried out to address these limits, such as polymeric fabrication [66], and PS loading and distribution advances by Nano-carrier (silica, polymeric, and liposome NPs) [88]. While these are essential activities, therapeutic efficacy has not yet been achieved. The low tissue diffusion of visible light is one of the important reasons for restricted use [18].

The shock reaction and killing of cancer cells are aided by hot temperatures of roughly 43°C. Cancer cells have a greater heat response than ordinary cells and have begun to gain attention as a treatment for cancer, alternatively, typical hyperthermal procedures with external thermal enhancement. Due to the heating of non-selective tissues, radio frequency (RF) or NIR irradiation may cause significant harm. The maximum temperature that can dissipate during a passage through body tissue is created by external heat. The photothermal therapy (PTT) hybrid Nano platform has evolved cancer-targeting functions to enhance the effects of Indocyanine green (ICG) photothermal and photostability. The PTT hybrid Nano platform has developed cancer-targeting mechanisms to maximize the photostability and photothermal effects of ICG. This system played a vital role in the photothermal cure relative to the 4T1 breast cancer cells combined with NIR laser irradiation [82].

1.1.7 BIOMEDICAL APPLICATIONS OF SILVER NANOMATERIALS

Ag-NPs are widely known against microbes such as fungi, bacteria, and viruses for their antimicrobial characteristics [2]. The benefit of Ag-NPs is that colloidal (coating, enamel, and paint), liquid or solid formulations can be used in various ways, such as Blending Ag-NPs as a polymer with a solid material.

1.1.7.1 ROLE OF AG-NANOPARTICLES IN MEDICINE

The Antibacterial Effects of Ag-Nanoparticles: Silver nanoparticles (AgNPs) play a crucial role in antibacterial action against some species of bacteria, such as pathogenic bacterial (gram "–" and "+" bacteria) [55]. Sondi and Salopeck-Sondi [81] have researched the action of antibacterial for Ag-NPs against *E. coli* on agar surfaces at Luria-Bertani. They used this strain as a representative genus for gram negative bacteria [81]. From results, the concentration effect of Ag-NPs' antibacterial function that was obtained against *E. coli*. It has also been shown the Ag-NPs bind to the gram-negative bacteria cell walls (*E. coli*) which destroy bacterial cells [81].

As well as the silver ions being emitted from the silver nanoparticle, bacterial cells that modify their regular actions, such as breathing and permeability, can bind to the wall. In another study, researchers observed that gram-negative (*E. coli*) bacteria inhibited low levels of Ag-NPs compared with gram-positive (*S. aureus*) bacteria. By controlling Ag-NPs through the pattern of penetration and silver adhesion to bacterial cell wall, the AgNPs antiviral activity mechanism affected by size and dosage, resulting in abnormal action [80]. Pal et al. [62] have studied the structure dependent on morphological antibacterial function of the Ag-NPs.

1.1.8 GOLD NANOMATERIALS BIOMEDICAL APPLICATION

Au NM features, e.g., large surface ratio, compact size, and Localized surface plasmon resonance (LSPR), have unique electrical and optical properties for Au NMs. It is used for bioimaging, biosensing, detection of tumors and gene/drug delivery care [85, 88].

1.1.8.1 PHOTOTHERAPY AND IMAGING

For optical and photographic applications, the plasmon's resonance emanates radiation. It can be used to turn light into heat. The potency of Au NMs in fluorescence yield as optical probes is around a million times greater than dye molecules. In contrast, gold nanomaterials do not seek treatment with photobleaching, exhibiting higher excitation and illumination energies, stable imaging, and longer times of probing. Furthermore, The LSPR absorption spectrum of gold nanomaterials are in the NIR region and can be modified by structure modification, scale, form, and environment making them suitable for thermal therapy and *in vivo* imaging for cancer, since tissues in the NIR region are optically transparent [42, 61].

These capabilities will also provide additional tools of photothermal therapy in real-time visualization of Au NRs. Other studies have shown that the Au NRs in thermal therapy help in tumor cell death mediated by integrity in membrane impairment. The phototherapy-mediated cell damage pathway was also investigated [85].

1.1.8.2 GENE/DRUG DELIVERY

In interacting with biological systems, organic, and inorganic nanomaterials have recently played a crucial role. Any of these delivery pathways, such as silica nanoparticles, micelles, dendrimers, chemical conjugates, liposomes, and microbubbles from ultrasound, have been thoroughly tested for transmission of drugs/genes [4, 35]. Owing to easy synthesis, small size, high biocompatibility and controllable modification, Au NMs have been an enticing choice for different payloads to be carried to their destinations. For the transmission of drugs and genes, Au NMs are also used. Their properties promote their delivery applications, for example elevated surface area, low cytotoxicity, and tunable stability. In addition, efficiency of delivery can be affected by surface charge, scale, and shape [71].

1.2 FOCUS ON NANOMEDICINE

1.2.1 BIO-INORGANIC HYBRID NANOMATERIALS

The interaction between their organic and inorganic components is based on the binding force; organic and inorganic hybrid materials can be broken into groups [53]. The first class of hybrid materials interacts weakly through van der Waals force, creation of the second class of hybrid materials with solid and stable materials. In recent years, good bindings have advanced [74]. For synthesis, a second class of hybrid compounds with covalent bonds may be used [53].

1.2.2 HYBRID NANOMATERIAL THERANOSTICS

For several uses, inorganic and organic-based hybrid NMs produce promising results. Inorganic nanoparticles' superior physical characteristics and the flexibility of surface alteration enable hybrid materials to incorporate several methods for diagnosis or therapies into one device and their bench-to-bedside conversion can be impeded by the complexities of the manufacturing process and biosafety evaluation [89].

1.2.3 CANCER IMMUNOTHERAPY OF HYBRID NANOMATERIALS

The attachment or encapsulation of antigens and adjuvants inside nanoparticles resulted in significantly improved anticancer immune responses (i.e., T-, and B-cell responses) compared to non-nanoparticle-dependent antigens and adjuvants [76].

Nano-vaccines work directly on APCs and also induce tumor-associated antigen (TAA)-specific T-cell activation directly. Unlike nanoparticles which continue to return chemotherapeutics to tumor cells while avoiding detection by antigen-presenting cells (APCs) and other phagocytes, this method works [76].

1.2.4 HYBRID NANOMATERIALS AND THEIR ROLE IN GENE EDITING

It is an effective and ideal technology which causes, in exact target sequences, double-strand breaks and then modifications, exchanges, or additions to the desired gene. A non-viral delivery scheme used in clinical applications is safe and effective. Polymeric and hybrid microcarriers, prepared as polypeptides and polysaccharides by degradable polymers and using all CRISPR-Cas9 components' silica shell delivery system [35, 60].

1.2.5 TRANSDUCERS FOR PHYSICAL STIMULI

Using flexible transducers from various types of inorganic NPs, physical stimuli may transform into new forms. Magnetic NPs, for instance, can exchange magnetic fields for heat or mechanical force. Light can be transformed to heat by Au NPs. It can be operated remotely in transducer nanoparticles because biological tissue can be penetrated without surgical interference; for example, magnetic fields or light [79].

1.3 PHOSPHORUS METAL BIOACTIVE NANOMATERIALS

1.3.1 BIOACTIVE METAL PHOSPHIDE NANOMATERIALS

Compared to nanomaterials of metals, sulfide, and metal oxides, it is too difficult to synthesize metal phosphide nanomaterials, typically due to the great reactivity of the source of phosphorus [11]. However, other metal phosphides are synthesized as new synthesis techniques are developed or by their specific physical and chemical characteristics, metal phosphide NMs typically have the most responsiveness [44, 95].

1.3.2 CLASSIFICATION OF PHOSPHORUS NANOMATERIALS

1.3.2.1 BLACK PHOSPHORUS NANOMATERIALS

BP nanomaterials, for the most part, include Nanodots from BP and nanosheets. Among them, the thin film of two-dimensional BP nanosheets, nowadays phosphorene, has gained significant interest. After years of improvement, different procedures were developed for synthesizing BP nanosheets, as well as the top-down approach as methods of mechanical severance and liquid-phase exfoliate and the method of decomposition of the bottom-up vapors and wet chemistry processes, as the contact force between two layers is little [5].

By mechanical cleavage, BP nanosheets can be made and the output efficacy, however, is relatively low, and morphology is hard to track. Wonderfully, method of exfoliating for the liquid-phase can enhance development and monitor moral morphology. N-methyl-2-pyrrolidone (NMP) containing chemical solvents, pyrrolidone NMP [83], isopropanol (IPA) [52], N, N-dimethyl formamide (DMF), ethanol [91], and dimethyl sulfoxide (DMSO), in the liquid phase exfoliation technique may be used for BP bulk propagation. To even further improve the dispensability of the aqueous deoxygenated surfactant, BP nanosheets were also presented. This also improved the constancy of the content.

1.3.2.2 NANOMATERIALS FOR METAL PHOSPHATE (P)

Phosphate ions have excellent biocompatibility in the human body. Abundant metal phosphate products form a combination of metal elements and phosphate ions. A variety of metal P NMs, typically calcium phosphate nanomaterials – Caps NMs, has been developed by the evolution of nanotechnology (Caps NT) [73, 96].

In contrast using additional varieties of nanomaterials, metal P NMs are commonly developed by hydrothermal technology. An easy two-step procedure containing the system of hydrothermal that was used for producing stable organic and inorganic hybrid CaP_2 with gadopentetate dimeglumine (Gd-DTPA) fused within [58].

Oleamine and oleic acid hydrophobic reagents, for example, used extensively for controlling the uniformity and scale of the synthesis method metal phosphate nanomaterials. Furthermore, MnP for drug delivery and tumor imaging was developed using the technique [94].

1.3.3 PROPERTIES OF PHOSPHORUS NANOMATERIALS

Phosphorus nanomaterials have outstanding characteristics, such as optical, chemical, physical, and biological characteristics, which makes them mainly useful in biomedical uses, as summarized below;

1.3.3.1 PHYSICAL PROPERTIES OF PHOSPHORUS NANOMATERIALS

Black phosphorus (BP) has special characteristics, for example, a phosphorus high surface-to-volume percentage that contains dendrimers. Nano sheets can provide high-capacity loading for efficient molecules that is vital in nanocarriers. Chemotherapy drugs/molecules reacting to phosphorus polymers and gather with final nanoplatforms. It improves the drug loading effect [91].

In order to bind the $[PEEP\text{-}b\text{-}PBYP\text{-}Se]_2$ chain for creating ingenious NMs, used derivative for the Doxorubicin (DOX) enclosing group of azides to create clever nanomaterials, which not only improves the effective loading of the compound, but also reduces and makes it sensitive to the pH drug release [54]. These are used in multi-filled BP nanosheets to recognize distinct utilities, including fluorescent/NIR imaging and treatment [83]. In the meantime, the phosphorus-containing dendrimers will provide timely responses to external stimuli, benefiting from the various structures such as Ultraviolet (UV) light. Since it has a special two-dimensional structure, the BP nanosheets can react to external stimuli in a similar way as NIR light and low pH [83].

1.3.3.2 CHEMICAL PROPERTIES OF PHOSPHORUS NANOMATERIALS

The chemical characteristics of nanomaterials are dependent on phosphorus and have some properties like sensitiveness, degradability, and receptiveness. For biomedical applications of biological systems, it has been important. For example, the Microenvironment with tumors also has low pH, high levels of reactive oxygen species, high concentrations of GSH to perform many functions, many phosphorus-based nanomaterials may react to these structures [94].

1.3.3.3 OPTICAL PROPERTIES OF PHOSPHORUS NANOMATERIALS

Light is a low-cost energy field with which we can assure strength and application in a wide range of study fields. For biomedical applications, the optical properties of BP NMs are very important, such as conversion capabilities and light absorption. This optical asset will really influence the scale, form, and thickness of few-layer BP nanosheets and surface alterations, helping in various applications of medicine. Since metal phosphide and BP nanomaterials can transform a large range of spectrums into heat that targets from UV-visible to NIR light photoacoustic picture (PAI) and PTT. In addition, BP nanomaterials may generate reactive oxygen species (ROS) under light irradiation [83].

1.3.3.4 BIOLOGICAL PROPERTIES OF PHOSPHORUS NANOMATERIALS

In the changed circumstances of cancer and with more intracellular oxidative stress, normal cells exhibit the opposite malignant features [26]. In water conditions and higher oxidative stress, BP nanosheets can degrade to phosphate anions [75]. By interacting with the amyloidogenic procedure, it treats neurodegenerative illness, e.g., Alzheimer's disease [19].

Hydroxyapatite and NMs of calcium phosphate are practically similar to the inorganic bone matrix elements, thereby providing dramatically strengthened osteoconductive and osteoinductive functions relative to typical bone replacements [41]. Metal phosphate nanomaterials were used roughly to rebuild bone tissue with these illustrious biological functions.

1.3.4 FUNCTIONAL PHOSPHORUS NANOMATERIALS FOR BIOMEDICAL APPLICATIONS

In addition to providing access for additional forms of nanomaterial, phosphorus-based nanomaterials have been used to distribute drugs and genomes, biological knowledge tracking, bone structure management, tumor imaging, and therapies [98].

1.3.4.1 NANOCARRIERS PHOSPHORUS NANOMATERIALS

Nanocarriers have played a critical part in cancer growth, developing, and enhancing numerous chemotherapy drugs to cure a variety of cancer forms. In hospitals, chemotherapy plays a key role; on the other hand, the adverse effects of existing chemotherapy medications are possible. Comparing nanocarriers with conventional chemotherapy drugs, nanocarriers containing phosphorus-based nanocarriers have great loading performance, can target tumors, long circulation time, and a vital managed chemotherapy drug release can be achieved and possible toxicities can be minimized by minor side effects. BP quantum dots combined (BPQDs@Lipo) into liposome bilayers. Wonderfully, by inserting biocompatible quantum dots (QDs) for BP in the liposome bilayer, responding to those motives, it is lawful to correctly release chemotherapy drugs. Through the thin lipid film hydration technique, the latest types of nanomaterials were developed [83, 91].

1.3.4.2 THERANOSTICS PHOSPHORUS NANOMATERIALS

In the biomedical field, perfect diagnosis and successful cure of tumors is very necessary. Metal sections, including PAI, Fluorescent Imaging (FI), tumor therapy, and MRI, are related to exposed metal phosphide/phosphate nanomaterials for tumor imaging. Nanomaterials that include BP and phosphorus dendrimers have been used in the management of tumor screening, neurodegenerative disorders, and tumor therapy [21].

1.3.4.3 BIOSENSOR PHOSPHORUS NANOMATERIALS

In recent years, numerous collection methods for biological knowledge have been rapidly established. There are several nanomaterials, such as phosphorus-based nanomaterials, proposed in this area. H_2O_2 is an essential molecule in the body and it is important in the metabolism of cells, proliferation, survival, and signal transduction. It is a great indication to recognize H_2O_2 aggregation in cells sensitively and selectively. Usually, nanoparticle materials that are not notified in the human body are nanomaterials capable of detecting nanomolar H_2O_2 today [23].

Wonderfully, due to their outstanding electrocatalytic reduction efficiency, nanomaterials of the transition metal phosphide are used in the hydrogen production reaction. The topo tactical conversion technique for the production of Cu(I) phosphide nanowires on porous Cu foam (Cu3P NWs/CF) was recognized by Li et al. [51], which has extreme susceptibility to the Nanomolar stage of H_2O_2 concentration. It provides an outstanding view of the compilation of cancer cell outcomes [47, 51].

1.3.4.4 BONE FORMATION REGULATION PHOSPHORUS NANOMATERIALS

Phosphorus-based nanomaterials are used to treat bone-related illnesses, especially metal phosphate nanomaterials such as hydroxyapatite and calcium phosphate are the leading bone components. In addition, the production of phosphate ions after BP degradation also helps to control the structure of bones [38].

1.4 CHITOSAN NANOMATERIALS AS BIOACTIVE SUBSTANCES

Chitosan is classified as a biodegradable non-toxic polymer, with various useful uses such as medicinal applications and photothermal cancer therapy and foodborne pathogens.

1.4.1 NANO-CHITOSAN BIOMEDICAL APPLICATION

There are approximately 48 million people with foodborne infections. In the United States, 128,000 hospitalizations and 3,000 deaths occur annually, as stated in CDC estimates [12]. It causes nearly 600 million patients and 420,000 deaths worldwide each year, according to the WHO. Chitosan-based nanoparticles are also commonly used to treat foodborne pathogens.

1.4.2 ANTIMICROBIAL ACTIVITY OF CHITOSAN NANOPARTICLES

High and low weight of molecules developed by using tripolyphosphate or sodium sulfate cross-linkers with or without sound at altered energy levels, were considered optional compared to *Escherichia coli* O15:H7 [28].

As an edible coating on grapes, chitosan NPs formed by ionic gelation have played a key role in antimicrobial action [57]. Under simulated conditions, washing vegetables added CSNPs with 1% citric acid resulted in a large decrease in the load of bacteria relative to industrial lettuce preparations [63].

1.4.3 ROLE OF CHITOSAN IN CANCER PHOTOTHERMAL THERAPY USAGES

World's biggest health issue in recent decades leading to thousands of deaths every year related to cancer, and it is responsible for every sixth death in the world. For example, chemotherapy, and radiation therapy that have risky effects on neighboring healthy tissues are the latest therapies. In the rapidly dividing effects of noncancerous cells, chemotherapeutic agents play a significant role. Additional concerns with chemotherapeutics also include systemic aggregation, lower successful target site concentrations, and the growth of drug tolerance, leading to adverse cancer disease outcomes [1].

Higher doses of radiation used in radiotherapy have been identified in order to boost cancer cells' invasive properties; some tumor forms have improved radiation resistance. Because of its advantages, like cost efficiency, decreased adverse effects and non-invasive architecture, selective thermal therapies such as photothermal therapy have now reached scientists' treatment [56].

Using NIR light at wavelengths of (700–980 nm) or (1,000–1,400 nm) biological window, nanoparticles with photothermal factors inserted into tumor sites may be triggered. In the NIR window, biological compounds and living tissues absorb light marginally, which results in mild phototoxicity and penetrates deep tissue relative to visible light or UV, rendering it the source of greeting light for cancer therapy [19]. For example, inorganic nanomaterials (copper, silver, and gold nanoparticles (AuNPs)), carbon-based compounds (carbon quantum dots and graphene oxide), and NIR-sensitive dyes are some types of photothermal agents. Restricted features such as high toxicity, low stability, poor internalization, and lower biocompatibility have been added to this photothermal agent.

Chitosan and its water-soluble derivatives must be actively used as it possesses swelling characteristics, cationic attributes, and biocompatibility traits. On many cancer cells, receptors are strongly expressed [41].

1.4.4 CHITOSAN USED WITH METALLIC SUBSTANCES

In contrast with bare MoS_2 particles, tantalum oxide with chitosan-molybdenum disulfide nanosheets has developed higher cytotoxic and photostability effects against cells, where chitosan serves as the surface matrix that makes it possible for formation of end product [29].

In nanocomposite synthesis, chitosan was used in Ag_2Se and Up-Conversion Nanoparticles (UCNPs) with the combination of PTT and Tetra-Modal Imaging (TMI). Chitosan has been integrated into water dispensability, providing UCNPs with stability and biocompatibility, thus providing a forum for the production of Ag_2Se [22]. As precursors via the green, a biocompatible therapeutic agent hydrothermal synthesis, iron cross-linked chitosan-based complexes were also used [97].

1.4.5 NANOCHITOSAN TOXICITY

The toxicity of nanochitosan depends, according to the literature, on the concentration found with certain plants. Potential phytotoxicity of both bulk chitosan and nano-size was reported [6]. In pepper plants (*Capsicum annuum* L.), the lower doses have a growth-promoting effect but a higher effect at 5 to 20 mg/L and 100 mg concentrations [6]. Furthermore, nanochitosan has toxic effects of around 20 nm on the growth of large beans in seedlings at 0.05% and concentrations of 0.1% after seed priming and germination [1].

It is also evident that when nano-chitosan-based material is used, species-dependent toxicity can occur and can be reduced by modifying the dose and scale of the nanoparticles. The risk of breaching biological environments is raised by NPs because of the small size and penetration through biological membranes. Therefore, to avoid any harmful effects, the size, and concentration-dependent properties of nanochitosan and nanochitosan derivatives should be properly studied before application [1].

1.5 PROGRESS AND FUTURE DIRECTIONS FOR NANOMEDICINE

1.5.1 RECENT DEVELOPMENT AND FUTURE DIRECTIONS: THE DELIVERY MECHANISM OF NANO-DRUGS FOR VITILIGO TREATMENT

Vitiligo is a condition of public depigmentation that affects 1% of the population of the world. The white spotting of vitiligo was confused in some places with leprosy and devastating effects on psychological well-being and illness in social incidences. In addition to dermatologists, the treatment and diagnosis of vitiligo also seems to be intractable obstacles for all scientists. Further analysis shows that the origins and pathogenesis of vitiligo are dependably linked to the allergic reaction along with the immune reaction of the skin and melanocytes [43].

Nevertheless, new concepts and insights for the treatment of vitiligo have been provided by the exponential improvement of nano-drug delivery systems. These innovative approaches simplify the ability to modify medications that regulate or exhibit continuous release actions, eliminate side effects, and improve therapeutic effectiveness. There are guaranteed limits, regardless of the significant promise of the present distribution mechanisms of nano-drugs. Most study on new drug delivery mechanisms for the cure of vitiligo gets the lowest regulation of symptoms instead of the best result to have the condition fully cured. According to this result, scientists are required to do and complete research to increase the performance of the distribution mechanism of nano-drugs [43]. Various receptors on keratinocytes surround, e.g., granulocyte colony-stimulating factor (G-CSFR) or melanocyte and melanocortin receptor 1~5 (MC1R~MC5R) surfaces, are improved to increase the feasibility of targeted agents. Their analogs or endogenous ligands show high promise for the lowest side effects for the cure of vitiligo. Cysteine-containing ultrashort peptides that are spontaneously self-gathered into hydrogels were suggested by one group. *In vivo* experiments have demonstrated that there is abundant biocompatibility and the lowest allergenic potential in one of the formulations [43].

Nano-structure design, self-assembly action, phase transition and hydrogenation of these subsequent studies, peptide nanostructures of two tripeptides that were scanned using electron microscopy and investigated by crystalline X-ray diffraction [16].

The key feature of self-gathering is the strong hydrophobic protection of acetylated Leu-Ile-Val-Ala-Gly, and the C-terminal residue has a direct effect on the degree and intensity at which the peptide fibril cooperates with SMD [43]. In view of the enormous progress that has been made and the constant advancement in techniques to combat this disease, it is hoped that vitiligo will be strengthened, managed, governed, and even handled.

1.5.2 FUTURE DIRECTIONS FOR ANTIFUNGAL STUDIES

Fungi have an important cause for hospitals fungal infections [24]. Some reports have stated that AgNPs have been used as an important antimycotic agent to show genius and outstanding antifungal properties against various fungal species. AgNPs antifungal activities using various fungal strains, such as *Candida albanicans* and *Trichophyton mentagrophytes* fungi, and demonstrated strong antifungal activity in Ag-NPs and indicated antifungal mechanism of action of AgNPs [24, 45, 46, 72]. It can be summarized that Ag-NPs are used against many strains (species) of fungi as an antifungal agent and it can be useful in resolving fungi infections.

1.5.3 FUTURE DIRECTIONS FOR ANTIVIRAL STUDIES

These infections will produce mayhem in no time because of the rapid spread (in certain nations, glimpses of damage affected by these viral infections have appeared), potentially causing significant harm to human health and wealth [20].

Ag-NPs play a key function in the regulation of virus-induced infectious diseases and pathogens. Ag-NPs can bind to viral cells' outer proteins, thereby eventually inhibiting the proper function of the viral cells. While valid pathways are yet to be recognized, Ag-NPs play a critical part in the potential management of virus-induced infectious diseases [27].

1.6 SUMMARY

Nanotechnology has developed into a genuinely interdisciplinary field within a short period, undergoing massive growth, and exciting new advances in every conventional scientific discipline. The self-ordering forces and at the nanometer size, material characteristics tend to be distinct from those at the macroscale. In biomedical fields, the application of nanotechnology is one of the key areas of focus that is currently gaining traction, since the concepts of nanotechnology are expressed in all biological systems. The already well-understood nanoscience tools and those that will be built in the future are expected to have a major effect on genetics, biotechnology, and medicine. In biological and medical uses, NMs corresponding size scale and biological elements, including proteins and antibodies, enables their use. The biomedical field has found that the physical features of NMs are distinctive; including their exceptionally high volume to surface area ratio, tunable optical absorption, and peculiar electrical and magnetic activity can be leveraged by a wide variety of biomedical utilities, from selling medications to biosensors. Recognizing enormous potential of nanomaterials in different fields of biology and medicine for applications, we review recent developments in this area and explore future perspectives. Hence, the chapter summarized the biological and medical applications of some nanomaterials, for example, silver nanomaterials, gold nanomaterials, hybrid nanomaterials, phosphorus nanomaterials, and chitosan nanomaterials.

KEYWORDS

- **biomedical applications**
- **chitosan nanomaterials**
- **gold nanomaterials**
- **nanomedicine**
- **nanomaterials**
- **phosphorus nanomaterials**
- **silver nanomaterials**

REFERENCES

1. Abdel-Aziz, H., (2019). Effect of priming with chitosan nanoparticles on germination, seedling growth and antioxidant enzymes of broad beans. *Catrina, 18*(1), 81–86. https://doi.org/10.21608/cat.2019.28609.
2. Ahamed, M., AlSalhi, M. S., & Siddiqui, M. K. J., (2010). Silver nanoparticle applications and human health. *Clinica Chimica Acta; International Journal of Clinical Chemistry, 411*(23–24), 1841–1848. https://doi.org/10.1016/j.cca.2010.08.016.
3. Allen, T. M., & Cullis, P. R., (2004). Drug delivery systems: Entering the mainstream. *Science, 303*(5665), 18–22.
4. Ambrosi, A., Sofer, Z., & Pumera, M., (2017). Electrochemical exfoliation of layered black phosphorus into phosphorene. *Angewandte Chemie, 56*(35), 10443–10445. https://doi.org/10.1002/anie.201705071.
5. Appalakondaiah, S., Vaitheeswaran, G., Lebègue, S., Christensen, N. E., & Svane, A., (2012). Effect of Van Der Waals interactions on the structural and elastic properties of black phosphorus. *Physical Review. Part B, 86*(3). https://doi.org/10.1103/PhysRevB.86.035105, PubMed: 035105.
6. Asgari-Targhi, G., Iranbakhsh, A., & Ardebili, Z. O., (2018). Potential benefits and phytotoxicity of bulk and nano-chitosan on the growth, morphogenesis, physiology, and micropropagation of *Capsicum annuum. Plant Physiology and Biochemistry, 127*, 393–402. https://doi.org/10.1016/j.plaphy.2018.04.013.
7. Asghari, F., Jahanshiri, Z., Imani, M., Shams-Ghahfarokhi, M., & Razzaghi-Abyaneh, M., (2016). *Antifungal Nanomaterials: Synthesis, Properties, and Applications* (3rd edn., pp. 343–383). ON: William Andrew Publishing.

8. Aversa, R., Modarres, M. H., Cozzini, S., Ciancio, R., & Chiusole, A., (2018). The first annotated set of scanning electron microscopy images for nanoscience. *Scientific Data, 5*, 180172. https://doi.org/10.1038/sdata.2018.172.
9. Bae, K. H., Chung, H. J., & Park, T. G., (2011). Nanomaterials for cancer therapy and imaging. *Molecules and Cells, 31*(4), 295–302. https://doi.org/10.1007/s10059-011-0051-5.
10. Bobo, D., Robinson, K. J., Islam, J., Thurecht, K. J., & Corrie, S. R., (2016). Nanoparticle-based medicines: A review of FDA-approved materials and clinical trials to date. *Pharmaceutical Research, 33*(10), 2373–2387. https://doi.org/10.1007/s11095-016-1958-5.
11. Carenco, S., Resa, I., Le Goff, X., Le Floch, P., & Me´ Zailles, N., (2008). White phosphorus as single source of "P" in the synthesis of nickel phosphide. *Chemical Communications, 22*, 2568–2570.
12. Centers for Disease Control and Prevention, (2020). *Burden of Foodborne Illness: Findings*. https://www.cdc.gov/foodborneburden/2011-foodborne-estimates.html (accessed on 14 July 2022).
13. Cha, J. J., Lee, H., Kim, M., Kang, J., Song, H., Kim, M. G., & Lee, J. H., (2020). Immobilized DNA aptamers used as potent attractors for vascular endothelial cell: *In vitro* study of female rat. *Bioscience Reports, 40*(1). https://doi.org/10.1042/BSR20182444, PubMed: 31930392.
14. Chakraborty, S., Dlie, Z. Y., Chakraborty, S., Roy, S., Mukherjee, B., Besra, S. E., Dewanjee, S., et al., (2020). Aptamer-functionalized drug nanocarrier improves hepatocellular carcinoma toward normal by targeting neoplastic hepatocytes. molecular therapy. *Nucleic Acids, 20*, 34–49. https://doi.org/10.1016/j.omtn.2020.01.034.
15. Champion, J. A., & Mitragotri, S., (2006). Role of target geometry in phagocytosis. *Proceedings of the National Academy of Sciences of the United States of America, 103*(13), 4930–4934. https://doi.org/10.1073/pnas.0600997103.
16. Chan, K. H., Xue, B., Robinson, R. C., & Hauser, C. A. E., (2017). Systematic moiety variations of ultrashort peptides produce profound effects on self-assembly, nanostructure formation, hydrogenation, and phase transition. *Scientific Reports, 7*(1), 12897. https://doi.org/10.1038/s41598-017-12694-9.
17. Chen, G., Qiu, H., Prasad, P. N., & Chen, X., (2014). Upconversion nanoparticles: design, nanochemistry, and applications in theranostics. *Chemical Reviews, 114*(10), 5161–5214. https://doi.org/10.1021/cr400425h.
18. Chen, H., Wang, G. D., Chuang, Y. J., Zhen, Z., Chen, X., Biddinger, P., Hao, Z., et al.,, (2015). Nanoscintillator-mediated x-ray inducible photodynamic therapy for *in vivo* cancer treatment. *Nano Letters, 15*(4), 2249–2256. https://doi.org/10.1021/nl504044p.
19. Chen, W., Ouyang, J., Yi, X., Xu, Y., Niu, C., Zhang, W., Wang, L., et al., (2018). Black phosphorus nanosheets as a neuroprotective nanomedicine for neurodegenerative disorder therapy. *Advanced Materials, 30*(3). https://doi.org/10.1002/adma.201703458, PubMed: 29194780.
20. Coker, R. J., Hunter, B. M., Rudge, J. W., Liverani, M., & Hanvoravongchai, P., (2011). Emerging infectious diseases in Southeast Asia: Regional challenges to control. *Lancet, 377*(9765), 599–609. -https://doi.org/10.1016/S0140-6736(10)62004-1.
21. De La Cruz, M. K., Alarcón-Angeles, G., & Merkoçi, A., (2019). Nanomaterial-based sensors for the study of DNA interaction with drugs. *Electroanalysis, 31*(10), 1845–1867. https://doi.org/10.1002/elan.201900286.

22. Du, K., Lei, P., Dong, L., Zhang, M., Gao, X., Yao, S., Feng, J., & Zhang, H., (2020). *In situ* decorating of ultrasmall Ag_2Se on upconversion nanoparticles as novel nanotheranostic agent for multimodal imaging-guided cancer photothermal therapy. *Applied Materials Today, 18*. https://doi.org/10.1016/j.apmt.2019.100497, PubMed: 100497.
23. Elmizadeh, H., Faridbod, F., Soleimani, M., Ganjali, M. R., & Bardajee, G. R., (2020). Fluorescent apta-nanobiosensors for fast and sensitive detection of digoxin in biological fluids using rGQDS: Comparison of two approaches for immobilization of aptamer. *Sensors and Actuators. Part B, 302*. https://doi.org/10.1016/j.snb.2019.127133, PubMed: 127133.
24. Enoch, D. A., Ludlam, H. A., & Brown, N. M., (2006). Invasive fungal infections: A review of epidemiology and management options. *Journal of Medical Microbiology, 55*(7), 809–818. https://doi.org/10.1099/jmm.0.46548–0.
25. Farcal, L., Torres, A. F., Di Cristo, L., Rotoli, B. M., Bussolati, O., Bergamaschi, E., Mech, A., et al., (2015). Comprehensive *in vitro* toxicity testing of a panel of representative oxide nanomaterials: First steps towards an intelligent testing strategy. *PLOS One, 10*(5), e0127174. https://doi.org/10.1371/journal.pone.0127174.
26. Finkel, T., (2003). Oxidant signals and oxidative stress. *Current Opinion in Cell Biology, 15*(2), 247–254. https://doi.org/10.1016/s0955-0674(03)00002-4.
27. Galdiero, S., Falanga, A., Vitiello, M., Cantisani, M., Marra, V., & Galdiero, M., (2011). Silver nanoparticles as potential antiviral agents. *Molecules, 16*(10), 8894–8918. https://doi.org/10.3390/molecules16108894.
28. Garrido-Maestu, A., Ma, Z., Paik, S. Y. R., Chen, N., Ko, S., Tong, Z., & Jeong, K. C., (2018). Engineering of chitosan-derived nanoparticles to enhance antimicrobial activity against foodborne pathogen *Escherichia coli* O157:H7. *Carbohydrate Polymers, 197*, 623–630. https://doi.org/10.1016/j.carbpol.2018.06.046.
29. Geng, S., Wu, L., Cui, H., Tan, W., Chen, T., Chu, P. K., & Yu, X. F., (2018). Synthesis of lipid-black phosphorus quantum dot bilayer vesicles for near-infrared-controlled drug release. *Chemical Communications, 54*(47), 6060–6063. https://doi.org/10.1039/c8cc03423k.
30. Georgieva, J. V., Kalicharan, D., Couraud, P. O., Romero, I. A., Weksler, B., Hoekstra, D., & Zuhorn, I. S., (2011). Surface characteristics of nanoparticles determine their intracellular fate in and processing by human blood-brain barrier endothelial cells *in vitro*. *Molecular Therapy, 19*(2), 318–325. https://doi.org/10.1038/mt.2010.236.
31. Griffitt, R. J., Luo, J., Gao, J., Bonzongo, J. C., & Barber, D. S., (2008). Effects of particle composition and species on toxicity of metallic nanomaterials in aquatic organisms. *Environmental Toxicology and Chemistry, 27*(9), 1972–1978. https://doi.org/10.1897/08-002.1.
32. Gupta, A. K., & Gupta, M., (2005). Cytotoxicity suppression and cellular uptake enhancement of surface modified magnetic nanoparticles. *Biomaterials, 26*(13), 1565–1573. https://doi.org/10.1016/j.biomaterials.2004.05.022.
33. Han, J., Park, W., Park, S. J., & Na, K., (2016). Photosensitizer-conjugated hyaluronic acid-shielded polydopamine nanoparticles for targeted photomediated tumor therapy. *ACS Applied Materials and Interfaces, 8*(12), 7739–7747. https://doi.org/10.1021/acsami.6b01664.
34. He, X., Zhao, Y., He, D., Wang, K., Xu, F., & Tang, J., (2012). ATP-responsive controlled release system using aptamer-functionalized mesoporous silica nanoparticles. *Langmuir, 28*(35), 12909–12915. https://doi.org/10.1021/la302767b.

35. Hilliard, L. R., Zhao, X., & Tan, W., (2002). Immobilization of oligonucleotides onto silica nanoparticles for DNA hybridization studies. *Analytica Chimica Acta, 470*(1), 51–56. https://doi.org/10.1016/S0003-2670(02)00538-X.
36. Hoshino, A., Fujioka, K., Oku, T., Suga, M., Sasaki, Y. F., Ohta, T., Yasuhara, M., et al., (2004). Physicochemical properties and cellular toxicity of nanocrystal quantum dots depend on their surface modification. *Nano Letters, 4*(11), 2163–2169. https://doi.org/10.1021/nl048715d.
37. Hsiao, I., & Huang, Y., (2011). Effects of various physicochemical characteristics on the toxicities of ZnO and TiO_2 nanoparticles toward human lung epithelial cells. *Science of the Total Environment, 409*(7), 1219–1228. https://doi.org/10.1016/j.scitotenv.2010.12.033.
38. Huang, K., Wu, J., & Gu, Z., (2019). Black phosphorus hydrogel scaffolds enhance bone regeneration via a sustained supply of calcium-free phosphorus. *ACS Applied Materials and Interfaces, 11*(3), 2908–2916. https://doi.org/10.1021/acsami.8b21179.
39. Huebsch, N., & Mooney, D. J., (2009). Inspiration and application in the evolution of biomaterials. *Nature, 462*(7272), 426–432. https://doi.org/10.1038/nature08601.
40. Jeevanandam, J., Barhoum, A., Chan, Y. S., Dufresne, A., & Danquah, M. K., (2018). Review on nanoparticles and nanostructured materials: History, sources, toxicity and regulations. *Beilstein Journal of Nanotechnology, 9*(1), 1050–1074. https://doi.org/10.3762/bjnano.9.98.
41. Jensen, T., Baas, J., Dolathshahi-Pirouz, A., Jacobsen, T., Singh, G., Nygaard, J. V., Foss, M., et al., (2011). Osteopontin functionalization of hydroxyapatite nanoparticles in a PDLLA matrix promotes bone formation. *Journal of Biomedical Materials Research. Part A, 99*(1), 94–101. https://doi.org/10.1002/jbm.a.33166.
42. Kelly, K. L., Coronado, E., Zhao, L. L., & Schatz, G. C., (2003). The optical properties of metal nanoparticles: The influence of size, shape, and dielectric environment. *Journal of Physical Chemistry B, 107*(3), 668–677. https://doi.org/10.1021/jp026731y.
43. Azevedo, R. S. S., De Sousa, J. R., Araujo, M. T. F., Martins, F. A. J., De Alcantara, B. N., Araujo, F. M. C., Queiroz, M. G. L., et al., (2018) *In situ* immune response and mechanisms of cell damage in central nervous system of fatal cases microcephaly by Zika virus. *Scientific Reports, 8*(1), 1. https://doi.org/10.1038/s41598-017-17765-5.
44. Kibsgaard, J., Tsai, C., Chan, K., Benck, J. D., Nørskov, J. K., Abild-Pedersen, F., & Jaramillo, T. F., (2015). Designing an improved transition metal phosphide catalyst for hydrogen evolution using experimental and theoretical trends. *Energy and Environmental Science, 8*(10), 3022–3029. https://doi.org/10.1039/C5EE02179K.
45. Kim, K. J., Sung, W. S., Moon, S. K., Choi, J. S., Kim, J. G., & Lee, D. G., (2008). Antifungal effect of silver nanoparticles on dermatophytes. *Journal of Microbiology and Biotechnology, 18*(8), 1482–1484.
46. Kim, S. T., Chompoosor, A., Yeh, Y. C., Agasti, S. S., Solfiell, D. J., & Rotello, V. M., (2012). Dendronized gold nanoparticles for siRNA delivery. *Small, 8*(21), 3253–3256. https://doi.org/10.1002/smll.201201141.
47. Kumar, V., Brent, J. R., Shorie, M., Kaur, H., Chadha, G., Thomas, A. G., Lewis, E. A., et al., (2016). Nanostructured aptamer-functionalized black phosphorus sensing platform for label-free detection of myoglobin, a cardiovascular disease biomarker. *ACS Applied Materials and Interfaces, 8*(35), 22860–22868. https://doi.org/10.1021/acsami.6b06488.

48. Lee, M. K., Lim, S. J., & Kim, C. K., (2007). Preparation, characterization and *in vitro* cytotoxicity of paclitaxel-loaded sterically stabilized solid lipid nanoparticles. *Biomaterials, 28*(12), 2137–2146. https://doi.org/10.1016/j.biomaterials.2007.01.014.
49. Li, S., Feng, X., Wang, J., He, L., Wang, C., Ding, J., & Chen, X., (2018). Polymer nanoparticles as adjuvants in cancer immunotherapy. *Nano Research, 11*(11), 5769–5786. https://doi.org/10.1007/s12274-018-2124-7.
50. Li, L., Xing, H., Zhang, J., & Lu, Y., (2019). Functional DNA molecules enable selective and stimuli-responsive nanoparticles for biomedical applications. *Accounts of Chemical Research, 52*(9), 2415–2426. https://doi.org/10.1021/acs.accounts.9b00167.
51. Li, Z., Xin, Y., Wu, W., Fu, B., & Zhang, Z., (2016). Topotactic conversion of copper(I) phosphide nanowires for sensitive electrochemical detection of H_2O_2 release from living cells. *Analytical Chemistry, 88*(15), 7724–7729. https://doi.org/10.1021/acs.analchem.6b01637.
52. Lin, S., Liu, S., Yang, Z., Li, Y., Ng, T. W., Xu, Z., Bao, Q., et al., (2016). Solution-processable ultrathin black phosphorus as an effective electron transport layer in organic photovoltaics. *Advanced Functional Materials, 26*(6), 864–871. https://doi.org/10.1002/adfm.201503273.
53. Liu, Y., Ji, X., & He, Z., (2019). Organic-inorganic nanoflowers: From design strategy to biomedical applications. *Nanoscale, 11*(37), 17179–17194. https://doi.org/10.1039/c9nr05446d.
54. Ma, G., Liu, J., He, J., Zhang, M., & Ni, P., (2018). Dual-responsive polyphosphoester-doxorubicin prodrug containing a diselenide bond: Synthesis, characterization, and drug delivery. *ACS Biomaterials Science and Engineering, 4*(7), 2443–2452. https://doi.org/10.1021/acsbiomaterials.8b00429.
55. Marambio-Jones, C., & Hoek, E. M. V., (2010). A review of the antibacterial effects of silver nanomaterials and potential implications for human health and the environment. *Journal of Nanoparticle Research, 12*(5), 1531–1551. https://doi.org/10.1007/s11051-010-9900-y.
56. Martin, C., Schulz, R., Rose, J., & Heusch, G., (1998). Inorganic phosphate content and free energy change of ATP hydrolysis in regional short-term hibernating myocardium. *Cardiovascular Research, 39*(2), 318–326. https://doi.org/10.1016/s0008–6363(98)00086–8.
57. Castelo, B. M. N. F., De MendonçaSoares, B. L., Marques, D. K., Ferreira, L., C., Canto, D., Flores, M. A. P., et al., (2018). Effects of fungal chitosan nanoparticles as eco-friendly edible coatings on the quality of postharvest table grapes. *Postharvest Biology and Technology, 139*, 56–66. https://doi.org/10.1016/j.postharvbio.2018.01.014.
58. Mi, P., Kokuryo, D., Cabral, H., Kumagai, M., Nomoto, T., Aoki, I., Terada, Y., et al., (2014). Hydrothermally synthesized pegylated calcium phosphate nanoparticles incorporating Gd-DTPA for contrast-enhanced MRI diagnosis of solid tumors. *Journal of Controlled Release, 174*, 63–71. https://doi.org/10.1016/j.jconrel.2013.10.038.
59. Mornet, S., Vekris, A., Bonnet, J., Duguet, E., Grasset, J., Choy, H., & Portier, J., (2000). DNA-magnetite nanocomposite materials. *Materials Letters, 42*, 183–188.
60. Nicole, L., Rozes, L., & Sanchez, C., (2010). Integrative approaches to hybrid multifunctional materials: From multidisciplinary research to applied technologies. *Advanced Materials, 22*(29), 3208–3214. https://doi.org/10.1002/adma.201000231.

61. Oldenburg, S. J., Averitt, R. D., Westcott, S. L., & Halas, N. J., (1998). Nanoengineering of optical resonances. *Chemical Physics Letters, 288*(2–4), 243–247. https://doi.org/10.1016/S0009-2614(98)00277-2.
62. Pal, S., Tak, Y. K., & Song, J. M., (2007). Does the antibacterial activity of silver nanoparticles depend on the shape of the nanoparticle? A study of the gram-negative bacterium *Escherichia coli. Applied and Environmental Microbiology, 73*(6), 1712–1720. https://doi.org/10.1128/AEM.02218-06.
63. Paomephan, P., Assavanig, A., Chaturongakul, S., Cady, N. C., Bergkvist, M., & Niamsiri, N., (2018). Insight into the antibacterial property of chitosan nanoparticles against *Escherichia coli* and *Salmonella typhimurium* and their application as vegetable wash disinfectant. *Food Control, 86*, 294–301. https://doi.org/10.1016/j.foodcont.2017.09.021.
64. Park, K. H., Chhowalla, M., Iqbal, Z., & Sesti, F., (2003). Single-walled carbon nanotubes are a new class of ion channel blockers. *Journal of Biological Chemistry, 278*(50), 50212–50216. https://doi.org/10.1074/jbc.M310216200.
65. Park, W., Cho, S., Han, J., Shin, H., Na, K., Lee, B., & Kim, D. H., (2017). Advanced smart-photosensitizers for more effective cancer treatment. *Biomaterials Science, 6*(1), 79–90. https://doi.org/10.1039/c7bm00872d.
66. Park, W., Park, S. J., & Na, K., (2011). The controlled photoactivity of nanoparticles derived from ionic interactions between a water-soluble polymeric photosensitizer and polysaccharide quencher. *Biomaterials, 32*(32), 8261–8270. https://doi.org/10.1016/j.biomaterials.2011.07.023.
67. Pietroiusti, A., Massimiani, M., Fenoglio, I., Colonna, M., Valentini, F., Palleschi, G., Camaioni, A., et al., (2011). Low doses of pristine and oxidized single-wall carbon nanotubes affect mammalian embryonic development. *ACS Nano, 5*(6), 4624–4633. https://doi.org/10.1021/nn200372g.
68. Pokropivny, V. V., & Skorokhod, V. V., (2007). Classification of nanostructures by dimensionality and concept of surface forms engineering in nanomaterial science. *Materials Science and Engineering, 27*(5–8), 990–993. https://doi.org/10.1016/j.msec.2006.09.023.
69. Powers, K. W., Palazuelos, M., Moudgil, B. M., & Roberts, S. M., (2007). Characterization of the size, shape, and state of dispersion of nanoparticles for toxicological studies. *Nanotoxicology, 1*(1), 42–51. https://doi.org/10.1080/17435390701314902.
70. Qiu, L. Y., & Bae, Y. H., (2006). Polymer architecture and drug delivery. *Pharmaceutical Research, 23*(1), 1–30. https://doi.org/10.1007/s11095-005-9046-2.
71. Qiu, M., Ren, W. X., Jeong, T., Won, M., Park, G. Y., Sang, D. K., Liu, L. P., et al., (2018). Omnipotent phosphorene: A next-generation, two-dimensional nanoplatform for multidisciplinary biomedical applications. *Chemical Society Reviews, 47*(15), 5588–5601. https://doi.org/10.1039/c8cs00342d.
72. Tran, Q. H., Nguyen, V. Q., & Le, A., (2013). Silver nanoparticles: Synthesis, properties, toxicology, applications and perspectives. *Advances in Natural Sciences: Nanoscience and Nanotechnology, 4*(3). https://doi.org/10.1088/2043–6262/4/3/033001, PubMed: 033001.
73. Rim, H. P., Min, K. H., Lee, H. J., Jeong, S. Y., & Lee, S. C., (2011). Ph-tunable calcium phosphate covered mesoporous silica nanocontainers for intracellular controlled release of guest drugs. *Angewandte Chemie, 50*(38), 8853–8857. https://doi.org/10.1002/anie.201101536.

74. Sanchez, C., Ribot, F., & Lebeau, B., (1999). Molecular design of hybrid organic-inorganic nanocomposites synthesized via sol-gel chemistry. *Journal of Materials Chemistry, 9*(1), 35–44. https://doi.org/10.1039/a805538f.
75. Shao, J., Xie, H., Huang, H., Li, Z., Sun, Z., Xu, Y., Xiao, Q., et al., (2016). Biodegradable black phosphorus-based nanospheres for *in vivo* photothermal cancer therapy. *Nature Communications, 7*, 12967. https://doi.org/10.1038/ncomms12967.
76. Shao, K., Singha, S., Clemente-Casares, X., Tsai, S., Yang, Y., & Santamaria, P., (2015). Nanoparticle-based immunotherapy for cancer. *ACS Nano, 9*(1), 16–30. https://doi.org/10.1021/nn5062029.
77. Shiau, B. W., Lin, C. H., Liao, Y. Y., Lee, Y. R., Liu, S. H., Ding, W. C., & Lee, J. R., (2018). The characteristics and mechanisms of Au nanoparticles processed by functional centrifugal procedures. *Journal of Physics and Chemistry of Solids, 116*, 161–167. https://doi.org/10.1016/j.jpcs.2018.01.033.
78. Shin, H., & Na, K., (2019). *In situ* vaccination with biocompatibility controllable immuno-sensitizer inducing antitumor immunity. *Biomaterials, 197*, 32–40. https://doi.org/10.1016/j.biomaterials.2019.01.015.
79. Shin, T. H., & Cheon, J., (2017). Synergism of nanomaterials with physical stimuli for biology and medicine. *Accounts of Chemical Research, 50*(3), 567–572. https://doi.org/10.1021/acs.accounts.6b00559.
80. Shrivastava, S., Bera, T., Roy, A., Singh, G., Ramachandrarao, P., & Dash, D., (2007). Characterization of enhanced antibacterial effects of novel silver nanoparticles. *Nanotechnology, 18*(22). https://doi.org/10.1088/0957-4484/18/22/225103, PubMed: 225103.
81. Sondi, I., & Salopek-Sondi, B., (2004). Silver nanoparticles as antimicrobial agent: A case study on *E. coli* as a model for gram-negative bacteria. *Journal of Colloid and Interface Science, 275*(1), 177–182. https://doi.org/10.1016/j.jcis.2004.02.012.
82. Song, S., Shen, H., Yang, T., Wang, L., Fu, H., Chen, H., & Zhang, Z., (2017). Indocyanine green loaded magnetic carbon nanoparticles for near-infrared fluorescence/magnetic resonance dual-modal imaging and photothermal therapy of tumor. *ACS Applied Materials and Interfaces, 9*(11), 9484–9495. https://doi.org/10.1021/acsami.7b00490.
83. Tao, W., Zhu, X., Yu, X., Zeng, X., Xiao, Q., Zhang, X., Ji, X., et al., (2017). Black phosphorus nanosheets as a robust delivery platform for cancer theranostics. *Advanced Materials, 29*(1). https://doi.org/10.1002/adma.201603276, PubMed: 27797119.
84. Taylor-Pashow, K. M., Della, R. J., Huxford, R. C., & Lin, W., (2010). Hybrid nanomaterials for biomedical applications. *Chemical Communications, 46*(32), 5832–5849. https://doi.org/10.1039/c002073g.
85. Tong, L., Zhao, Y., Huff, T. B., Hansen, M. N., Wei, A., & Cheng, J. X., (2007). Gold nanorods mediate tumor cell death by compromising membrane integrity. *Advanced Materials, 19*, 3136–3141. https://doi.org/10.1002/adma.200701974.
86. Veiseh, O., Gunn, J. W., & Zhang, M., (2010). Design and fabrication of magnetic nanoparticles for targeted drug delivery and imaging. *Advanced Drug Delivery Reviews, 62*(3), 284–304. https://doi.org/10.1016/j.addr.2009.11.002.
87. Wang, H., Yang, R., Yang, L., & Tan, W., (2009). Nucleic acid conjugated nanomaterials for enhanced molecular recognition. *ACS Nano, 3*(9), 2451–2460. https://doi.org/10.1021/nn9006303.
88. Wei, X., Chen, H., Tham, H. P., Zhang, N., Xing, P., Zhang, G., & Zhao, Y., (2018). Combined photodynamic and photothermal therapy using cross-linked polyphosphazene

nanospheres decorated with gold nanoparticles. *ACS Applied Nano Materials, 1*(7), 3663–3672. https://doi.org/10.1021/acsanm.8b00776.
89. Yan, L., Zhao, F., Wang, J., Zu, Y., Gu, Z., & Zhao, Y., (2019). A safe-by-design strategy towards safer nanomaterials in nanomedicines. *Advanced Materials, 31*(45). https://doi.org/10.1002/adma.201805391, PubMed: 1805391.
90. Yang, S. T., Wang, X., Jia, G., Gu, Y., Wang, T., Nie, H., Ge, C., et al., (2008). Long-term accumulation and low toxicity of single-walled carbon nanotubes in intravenously exposed mice. *Toxicology Letters, 181*(3), 182–189. https://doi.org/10.1016/j.toxlet.2008.07.020.
91. Yang, Y., Gao, J., Zhang, Z., Xiao, S., Xie, H. H., Sun, Z. B., & Yu, X. F., (2016). Black phosphorus based photocathodes in wideband bifacial dye-sensitized solar cells. *Advanced Materials, 28*(40), 8937–8944.
92. Ye, L., Yong, K. T., Liu, L., Roy, I., Hu, R., Zhu, J., Cai, H., et al., (2012). A pilot study in non-human primates shows no adverse response to intravenous injection of quantum dots. *Nature Nanotechnology, 7*(7), 453–458. https://doi.org/10.1038/nnano.2012.74.
93. Yin, H., Too, H. P., & Chow, G. M., (2005). The effects of particle size and surface coating on the cytotoxicity of nickel ferrite. *Biomaterials, 26*(29), 5818–5826. https://doi.org/10.1016/j.biomaterials.2005.02.036.
94. Yu, J., Hao, R., Sheng, F., Xu, L., Li, G., & Hou, Y., (2012). Hollow manganese phosphate nanoparticles as smart multifunctional probes for cancer cell-targeted magnetic resonance imaging and drug delivery. *Nano Research, 5*(10), 679–694. https://doi.org/10.1007/s12274-012–0252-z.
95. Zhang, G., Wang, G., Liu, Y., Liu, H., Qu, J., & Li, J., (2016). Highly active and stable catalysts of phytic acid-derivative transition metal phosphides for full water splitting. *Journal of the American Chemical Society, 138*(44), 14686–14693. https://doi.org/10.1021/jacs.6b08491.
96. Zhang, M., & Kataoka, K., (2009). Nano-structured composites based on calcium phosphate for cellular delivery of therapeutic and diagnostic agents. *Nano Today, 4*(6), 508–517. https://doi.org/10.1016/j.nantod.2009.10.009.
97. Zhang, M., Wang, W., Zhou, N., Yuan, P., Su, Y., Shao, M., Chi, C., & Pan, F., (2017). Near-infrared light-triggered photo-therapy, in combination with chemotherapy using magnetofluorescent carbon quantum dots for effective cancer treatment. *Carbon, 118*, 752–764. https://doi.org/10.1016/j.carbon.2017.03.085.
98. Zhao, X., Tapec-Dytioco, R., Wang, K., & Tan, W., (2003). Collection of trace amounts of DNA/MRNA molecules using geomagnetic nanocapturers. *Analytical Chemistry, 75*(14), 3476–3483. https://doi.org/10.1021/ac034330o.
99. Zhu, Z., Wu, C., Liu, H., Zou, Y., Zhang, X., Kang, H., Yang, C. J., & Tan, W., (2010). An aptamer cross-linked hydrogel as a colorimetric platform for visual detection. *Angewandte Chemie, 49*(6), 1052–1056. https://doi.org/10.1002/anie.200905570.

CHAPTER 2

SUSTAINABLE ZNO NANOMATERIALS IN MEDICINE: SYNTHESIS, APPLICATIONS, IMPACTS, AND CHALLENGES

G. V. S. SUBBAROY SARMA, KANAGASABAI MURUGANANDAM PONVEL, KASIBATLA S. R. MURTHY, and MURTHY S. S. S. CHAVALI

ABSTRACT

Nanotechnology is an innovation of design and the applications of nanoscale materials with their new properties and functions. Nanomaterials-based devices provide very fast response, high-efficiency, long-lifetime, easy to use, and low-cost. With the latest developments in nanoscience and nanotechnology, zinc oxide nanomaterials were synthesized by various means by chemical, physical, and bio-based methods. Biological methods of synthesizing metal oxide nanoparticles (especially ZnO) using microorganisms, enzymes, and plants or plant extracts are possible eco-friendly alternatives to chemical and physical methods. Green methodologies and ZnO as a sustainable material is discussed here in this chapter. ZnO for being biosafe and biocompatible has attracted much interest because of its usefulness for intracellular measurements of biochemical species by using its semiconducting, electrochemical, and catalytic properties. Recently, ZnO nanomaterials have received more concerns because of their prominent biological characteristics and biomedical applications. Zinc oxide nanoparticles as one of the most important metal oxide nanoparticles are popularly active in various fields due to their peculiar physical and chemical properties. In this chapter, various interesting and multifunctional properties of ZnO in

Sustainable Nanomaterials for Biomedical Engineering: Impacts, Challenges, and Future Prospects.
Junaid Ahmad Malik, Megh R. Goyal & Mohamed Jaffer M. Sadiq (Eds.)

areas like sensing, tissue engineering, wound healing, their antioxidant, anti-inflammatory, antimicrobial, antibacterial, antidiabetic, and anticancer properties are also discussed along with their biocompatibility and safety evaluation.

2.1 INTRODUCTION

Each development and utilization by nanocomposites having functional properties between molecules as well as its composites becomes nanotechnology. In general, nanotechnology, by its new reactivity's, would be a technique for both the development and execution using nanocomposites. The characteristics of composites vary greatly with that of molecules along with those of larger particles because the measurements of its materials were in micro-scale. Nanotechnology is a rapidly evolving scientific field concerned with both the preparation of nanoparticles, including nanostructured materials across multiple areas to its uses. Another very accessible aspect of nanotechnology is nanoparticles, almost always referred to as designed nanomaterials [1]. Due to vast uses through multiple fields of scientific research, oxide nanoparticles, especially nano-scale ZnO, have also increased throughout the world in current years [2, 3]. ZnO is such an important substance; many techniques are being established to achieve Zinc oxide micro and nanoparticles of several compositions. Furthermore, zinc oxide is an environmentally safe material, as are several organic and inorganic processes that have been used to obtain diverse morphological ZnO micro nanoparticles and other nanoparticles [4]. For the synthesis of ZnO nanomaterials [5–7], different methods like chemical vapor deposition, electrochemical oxidation, hydrothermal synthesis, or even sol-gel manufacturing were established.

Zinc oxide had been categorized throughout materials science as a semiconductor through group II-VI. In the hexagonal wurtzite structure (Figure 2.1), where each ion has been enclosed with 4 metal ions just at the edge of a tetrahedron, zinc oxide crystallizes at standard temperature and pressure. Two interlinking sub-lattices with Zn^{2+} as well as O^{2-} ions at that every zinc ion has been enveloped with a tetrahedral of 4 oxygen ions describe a certain ZnO space group.

Interestingly, both zinc, as well as oxygen ions were positioned along its c-axis which show high positively and negatively polar planes through Zn^{2+} and O^{2-}. Each polar configuration across the hexagonal axis [8, 9] has been the basis of such Zn^{2+} and O^{2-} tetrahedral arrangement. Its inhomogeneous development in the 1-dimensional ZnO crystalline phase is liable to that c-axis asymmetry. The *wurtzite* arrangement having four surface terminations exhibits its main typical and stable ZnO crystal; its polar Zn terminated (0001), and O terminated ($000\bar{1}$) surfaces, including its non-polar ($10\bar{1}0$) facets possessing identical numbers from atoms of Zn and O [8–11].

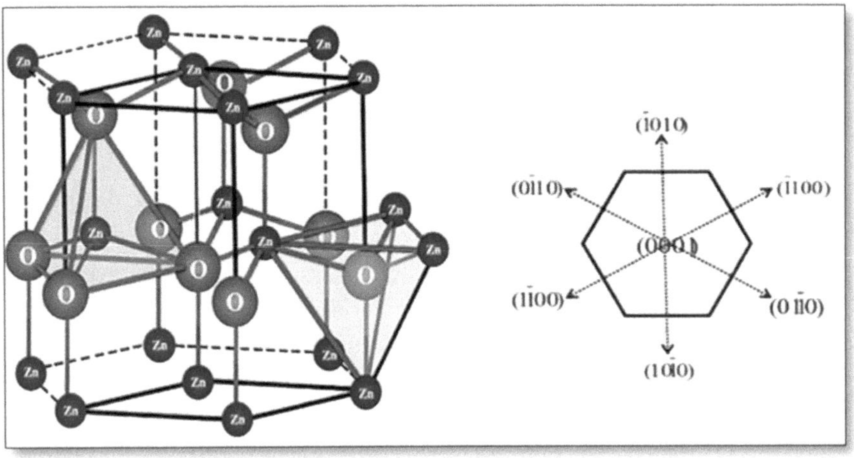

FIGURE 2.1 Schematic illustration with certain dimensions throughout the zinc oxide hexagonal *wurtzite* arrangement as well as for their polar but also non-polar as its atomic design.

ZnO nanoparticles became desirable substitutes to most devices due to their significant efficiency, including advanced electronics, optics, and photonics, including Ultra Violet Lasers, LED's, Nanogenerators, battery storage, sensing applications, photodetectors, including photocatalysts. Within certain technologies, ZnO nanoparticles were widely used as photocatalysts that can remove bacteria and viruses and under sufficient light beam for the destruction by climate change contaminants such as pigments, chemicals, and hazardous chemical substances. Both piezo and pyro electrical properties of ZnO indicate how this has been utilized through power generation for a sensor, converter, energy generator, and photocatalyst [12, 13]. This would also be a significant substance throughout the ceramics industrial sector owing to high toughness, stiffness, and piezoelectric stability, even its low toxicity, excellent biocompatibility allow itself a substance with significance, including biomedicine, and pro-ecological systems [14–16].

This chapter describes green methodologies towards ZnO as a sustainable material. ZnO has attracted much interest because of its prominent biological characteristics, biomedical applications, practicality for intracellular measurements of biochemical species by using its semiconducting, electrochemical, and catalytic properties. Besides, several interesting and multifunctional properties of ZnO in areas like sensing, tissue engineering, wound healing, their antioxidant, anti-inflammatory, antimicrobial, antibacterial, antidiabetic, and anticancer properties are also discussed along with their biocompatibility and safety evaluation.

2.2 CHARACTERISTICS OF ZNO NANOPARTICLES

2.2.1 PHYSICAL PROPERTIES

ZnO nanomaterials have amazing physical characteristics. With a decrease in its dimensions among semiconductor materials, some of its physical properties exhibit variations based on quantum structural parameters. Its acceptance with specific physical characteristics becomes important for the consistent development of device applications. To improve its capability even as basic materials to prospective nanoscale devices, analysis into certain characteristics of single ZnO nanostructures was important. Updated investigation upon its physical properties of ZnO nanostructures, namely mechanical, electromagnetic, and photoluminescence features is discussed [17].

2.2.2 MECHANICAL PROPERTIES

Accurate estimation of its mechanical behavior of selected nanomaterials seems to be very difficult because its conventional composite materials evaluation system will not operate. Researchers specified its folding compressive strength to ZnO nanoparticles utilizing TEM based on battery-powered modulated stimulation. A specific TEM test specimen has been designed for the application of an alternating electric field in between ZnO nanobelt as well as a static electrode for this process. The movement of the nanobelt regulates the electric field, but as a consequence of adjusting each guiding frequency, the resonant oscillation was acquired. Mostly as a nanoresonator and nanocantilever, ZnO nanobelt seems to always be a good option. The exploitation of ZnO nanobelt to either the required width and location was stated that demonstrates the possibility of any use as a cantilever with extremely active atomic force microscopy (AFM) [18].

2.2.3 *ELECTRICAL PROPERTIES*

The Zn^{2+} and O^{2-} ions packed out throughout polar planes with opposite charges give ZnO excellent electric features that could be also accurately adjusted. ZnO semiconductor material has its fairly high exciton binding energy becomes specific as well as deep bandgap substrate that is suitable for several electronics and optoelectronic applications. That is because substances including big band gaps might have higher breaking voltages, lower noise production, and the ability to maintain huge electric fields, yet could work under maximum voltage over high temperatures. With a certain reasonably maximum and minimum electric field, every movement of electrons within ZnO varies.

ZnO does have sufficient potential that sustains high temperatures including greater electric fields, greater break-down voltages but also high power activity here because of the result of a specific and broad bandgap semiconductor. Each existence of zinc interstitials including oxygen vacant positions gives way to N-type conductivity to ZnO [19]. Because of the self-compensation induced by the point flaws, its manufacture with p-type ZnO stays a challenge [20]. Certain electrical properties of ZnO have highly affected through its ambient air adsorbent with oxygen species that in turn influences their density of the carrier. That total resistance of ZnO materials is modified by both physisorption and chemisorption. It is also recognized that ZnO provides high resistivity in its clearest state related to lower charge density, which has a great importance of memristor applications including several gas sensing applications. Reduced ZnO resistivity, on the other hand, could be attained through enhancing oxygen vacancy even with doping, which often affects their mobility of electrons [21]. Along its specific ZnO nanorods and nanowires, electric transfer calculations were taken out [22, 23]. Regarding various techniques, the solitary ZnO nanowire was designed on field-effect transistors. For characterization, vary of interface electrodes, photolithography can be used and the degenerately doped Si surface acted here as a back gate electrode. ZnO nanowires were recorded to disclose n-type semiconductor activity due to habitant flaws such as oxygen vacant positions with zinc substitution. The electrical characteristics are also based on doping ions. Resistive switching behavior was demonstrated by pure

ZnO, showing that its defects within ZnO are quite a crucial role in causing charge storage behavior.

2.2.4 OPTICAL PROPERTIES

Each semiconductor's optical characteristics were attributed, including it's intrinsically and extrinsically influences. Intrinsic optical characteristics identify its relationship between particles in the absorption group as well as excitation group holes. Extrinsic characteristics were connected by semiconductor doping or flaws that create distinct electronic structure among the Conduction band and VB. Different testing approaches like light illumination, propagation, reflection, photoluminescence, cathodoluminescence, etc., are being tested for optical transitions throughout ZnO. To research its electrical characteristics within ZnO nanomaterials, all photoluminescence methodology has been commonly adopted by such techniques. ZnO's environment temperatures PL variability usually displays Ultraviolet emission but each as well as more apparent defect-induced discharges, like vacancies, interstitials, antisites, including complicated defects [24, 25]. ZnO displays a high and instant band-gap of 3.37 eV near lower temperatures around slightly higher exciton energy around 60 meV. This amount for 60 meV is significantly greater with those by GaN (25 meV) and nearby low temperatures radiant efficiency (26 meV), that almost below lower excitation energy, guarantees an effective excitation-emission at room temperature. Thus in the blue region, ZnO is among the strong photonic materials [26]. Photoluminescence spectroscopy also being commonly utilized with ZnO nanorods that examine optical properties, providing details like band distance, defects, even performance of its crystal [26, 27]. Low-temperature PL analysis for ZnO nanorods reports revealed that near band edge Ultraviolet emissions including wider band deep-level emissions were identified by ZnO nanorods, but just one ultraviolet emission has become identified within lower impurity compositions for a better feature of ZnO nanorods. For an indicator of the optical efficiency of the ZnO nanostructures, their comparative strength towards both the NBE emissions and the DLE emissions are utilized. Thus their ratio, including its amplitude with that relatively close interface emissions for their sharp emissions frequency can be tested for the optical efficiency from their ZnO nanostructures. The high ratio indicates lower deep-level emission concentrations [28].

2.2.5 PHOTOLUMINESCENCE PROPERTIES

Photoluminescence characteristics for ZnO and carbon nanoparticles were examined by Rauwel et al. [79] based on the route of synthesis, scale, shape, deep scale, including substrate faults, ZnO exhibits its emissions of photoluminescence throughout the Ultraviolet-visible region. Once ZnO nanoparticles are combined between carbon nanomaterials, large pores are transformed. Whereas ZnO enables the control of such characteristics within photoluminescence can generate, for example, visible lighting. In addition, its effective transfer of energy through ZnO to carbon nanomaterials never merely allows their ideal targets for energy storage uses, but also for biomaterials, photo sensors and lower temperatures thermal imaging apps. Furthermore, its implementation by ZnO nanoparticles in a metal oxide system was observed may generate variations throughout the PL effects due to its oxidation of substrate defects [29].

2.2.6 PIEZOELECTRIC PROPERTIES

Through their control of exterior forces, a physical property named strain was defined as the deformation of solids and stress by their intrinsic mechanical force; this removes elongation that aims towards returning each crystal into the initial condition [30]. Piezoelectric characteristics across its atomic-scale were related to polarization. The mechanical strain may influence their transfer of charged particles throughout substances. Three mechanisms can modify their transport properties of a uniform substance, such as the structural shift, the piezoresistive effect, and also the piezotronic effect.

Basis of piezoelectric materials was defined for wurtzite ZnO nanostructures as: imagine each atom has its negative charge which is tetrahedrally enclosed through positive charges where only oxygen atoms and zinc atoms were tetrahedrally connected. Owing to intrinsic defects, if an additional force is applied upon these crystals in the path of the tetrahedron, each center of the positive charge as well as the negative charge could be moved. Such distortion will enable their positive charge center and negative charge may change from one another, leading to localized dipole moments would be caused. Those structures would provide macroscopic dipole moments when that entire structure has this identical orientation when their external pressure and external stress is encountered [31, 32]. The functional characteristics of nanoparticles of wurtzite ZnO are listed in (Table 2.1).

ZnO nanomaterial consisted of alternate surfaces of atoms Zn^{2+} and O^{2-} stacked throughout each c-axis with a lacking of similarity center in ZnO, contributing towards ZnO having a good piezoelectric behavior that provides a sufficiently wide electromechanical binding. It provides considerable possibilities with innovations including nano-electromechanical systems, sensor growth, including sensors and transducers that are electromechanically coupled [33, 34].

TABLE 2.1 The Functional Characteristics of Nanoparticles of Wurtzite ZnO

Physical Parameters	Quantity
Density	5.606 g/cm^3
Melting point	2,248 K
Refractive index	2.008, 2.029
Energy gap	3.37 eV
Exciton binding energy	60 meV
Electron effective mass	0.24
Hole effective mass	0.59
Relative dielectric constant	8.66

2.3 SYNTHESIS

2.3.1 *TRADITIONAL TECHNIQUES*

2.3.1.1 *METALLURGICAL PROCESS*

Metallurgical methodologies were concentrated upon its roasting of zinc ore throughout order to produce zinc oxide. ZnO is either known as class A, extracted through a specific method, or type B, produced through some intermediate method. The specific method includes reducing zinc ore by coal-fired warming, accompanied by zinc vapor corrosion within the identical furnace throughout a simple manufacturing process. Samuel Wetherill built an above-mentioned approach that participated within a furnace, where a coal sheet illuminated by the heat resulting from the prior charges comprises the initial layer. Any secondary layer throughout the type of zinc ore combined by coal was located beyond this sheet. Providing each of such surfaces than with additional heating that carries carbon monoxide besides zinc removal, blow wind has been pumped along via beneath. This resultant zinc oxide includes impurities by zinc ore in the type of compositions with several materials.

Each ZnO nanoparticles that arise seem to be primarily needle-shaped, and often spheroidal. Its oxides of lead, iron, and cadmium which are available will be modified somehow to sulfates to acquire material with quite a stable white color. Improving its color combinations continuity has been associated with increasing its material with water-soluble substances, and as well as boosting the manufacturer's acidity. In each situation of rubber technology development, acidity is attractive because it prolongs the prevulcanization time by allowing its secure production of the mixtures [35]. Metallic zinc gets heated in a heater during the intermediate method then vaporized with a rate of approx. 910°C. ZnO is formed by the instant response, including its zinc vapor through oxygen in the atmosphere. The zinc oxide was transferred through a conditioning pipe and is stored in a bag filtering facility. Each substance is composed of aggregates varying in average crystallite around 0.1 through just micrometers [36]. ZnO nanoparticles class B seems to have a greater level of purity than class A.

2.3.1.2 MECHANO-CHEMICAL PROCESS

The cheapest and easy way to acquire nanomaterials across its high level was its mechano-chemical procedure. That alone requires large hot grinders at low temperatures that start reactivity inside a grinding machine via ball-powder effects. In the case of solid, which serves as a reaction medium that divides the nanomaterials were produced, a "thinner" was applied to those methods. It is the continuous crushing of the substance and their limitation of particles into the appropriate size that is a major problem in this process that reduces by increasing machining process times and energy. It's higher grinding duration results in a higher amount of contaminants. Lower cost, tiny particles dimensions as well as a negligible potential with particles can accumulate, but also the higher uniformity of such crystals size and shape, are often the benefits of that kind of approach.

Anhydrous $ZnCl_2$ and Na_2CO_3 were essential as beginning substances utilized during this mechanochemical process and then NaCl is introduced into this process; it functions here as a mechanism of response that extracts the nanomaterials. With its temperatures around 800°C, the ZnO nanoparticles substrate developed, $ZnCO_3$ is calcinated.

For an entire, this mechanism includes the succeeding reactions:

$$ZnCl_2 + Na_2CO_3 \longrightarrow ZnCO_3 + 2NaCl$$

$$ZnCO_3 \xrightarrow{\text{Temperature}} ZnO + CO_2$$

The mechanochemical process; they synthesized ZnO at an estimated particle size around 21 nm [37]. They noticed that even a grinding period around 4 h is necessary with a process to occur among its substances, providing its substrate $ZnCO_3$ that generated synthesized ZnO crystalline phase at an estimated size of about 26 nm once calcinated around 400°C.

2.3.1.3 CONTROLLED PRECIPITATION

Controlled precipitation seems to be a commonly adopted process of gaining zinc oxide since it enables the material for replicable characteristics to be obtained. This process includes its rapid as well as random diffusion, through employing a reduction agent, for a sufficient mixture of zinc salt that restricts the formation of specimens of defined sizes, accompanied by the accumulation for its solution through a substrate of ZnO. The above substrate needs to undergo heat treatment during its following step, preceded by grinding to remove contaminants. In all aggregates, the shape is quite hard to break apart, so the calcinated powders provide some significant levels with particles size. Characteristics like pH, temperature, and time of precipitation regulate its phase of precipitation. Zinc oxide was even precipitated by zinc chloride and zinc acetate aqueous treatments [38]. Its concentrations with the solvents, maximum amount with inclusion for substances and the reaction temperature were the regulated variables throughout this method. By a single-modal scale of particles and a large substrate region, zinc oxide was created.

2.3.1.4 SOL-GEL METHOD

Despite the flexibility, low-cost, durability, reproducibility, and comparatively mild circumstances of fabrication that allow for the surface modification of zinc oxide using specified chemical materials, the acquisition of ZnO nanopowders by the sol-gel process seems to be with significant concern. This sol-gel process was being utilized by Ristic et al. [82] to generate nanocrystalline zinc oxide [39, 40]. Elevated, homogeneous, organized ZnO nanomaterials were effectively produced onto the ultra-thin AAO (anodic aluminum oxide) membranes by the sol-gel process. Incorporating ultra-thin AAO membranes through the sol-gel process will help to generate and expand the production of high 1-dimensional nanoparticles as a model towards each development of nanotubes.

2.3.1.5 HYDROTHERMAL METHOD

In this process, the surface solution is steadily warmed towards high temperatures of 100–300°C then kept over many days. Crystal particles were produced with their consequence of warming proceeded through cooling, and then expand.

Hydrothermal synthesis of zinc oxide powders provides several benefits:

- Its process will be taken place within reasonable circumstances;
- That was possible to achieve nanometer-sized powders through this methodology;
- Even by changing its reaction conditions, powders containing different morphologies were formed; and
- Its powders being produced have different characteristics towards that produced during higher temperatures.

Zinc oxide through the resulting reactions utilized the hydrothermal process [41]:

$$Zn\,(CH_3\,COO)_2 + 2NaOH \longrightarrow Zn\,(OH)_2 + 2CH_3COONa$$

$$Zn\,(OH)_2 \xrightarrow{\text{Temperature}} ZnO + H_2O$$

Both time, as well as the temperature of the hydrothermal process, often influences its form of the particles. Its particle size varies for each rise of time, temperature as well as penetration enhancer concentrations. Based upon the circumstances during synthesis, hydrothermal processing of the substrate accompanied through drying generated as spherical ZnO particles for dimensions mostly in 55–110 nm region.

2.3.1.6 MICROEMULSION ENVIRONMENT

Microemulsions comprise an aqueous layer, an oil surface and a surfactant which were steady, translucent, isotropic fluids. In a microemulsion, its drop scale was slightly lower than within an emulsion and has been over that

0.0015–0.15 μm range [42–44]. Microemulsions shape randomly throughout optimal circumstances, whereas opposed to emulsions. During an interfacial synthesis of zinc oxide containing sodium hydroxide, emulsified methods where zinc oxide has occurred. This method overall contained the reaction [45, 46].

$$Zn(C_{17}H_{33}COO)_3 + 2NaOH \rightarrow ZnO + H_2O + 2C_{17}H_{33}COONa$$

2.3.2 GREEN SYNTHESIS

In any synthesis of nanomaterials, the green synthesis method makes usage of relatively pollutant-free materials. This involves the need for eco-friendly and sustainable solvents, like plant sources and water. Thus, biological strategies towards its synthesis of metal nanoparticles utilizing microorganisms including plant as well as plant extracts are being proposed as healthy replacements to conventional processes. So many biological processes, like bacteria, fungus, and yeast, will be utilized successfully throughout the biogenic processing of nanoparticles [47]. Distinct leaf extracts are utilized during its controlled manufacturing of zinc oxide nanomaterials in the biosynthesis system of ZnO nanoparticles, employing the aqueous extract of several plant leaves like a reducing agent to establish a more eco-acceptable procedure by biocompatible strategies instead of normal techniques using alternative pathways towards ZnO nanoparticle synthesis.

- **Benefits with Green Synthesis:** Throughout its synthesis of nanoparticles, the "green" approach has become a major subject of importance (Figure 2.2) since traditional chemical methods remain costly that involve extensive utilization of chemical compounds for hazardous reduction agents [48].

At the origin stage, green chemistry decreases the risk of contamination and is improved to avoid contamination instead of handling or cleaning it up contamination, once it has been created. These theories focus mostly on environmentally choice among materials. While physical and chemical approaches were quicker but simpler towards its synthesis of nanoparticles, their biogenic methodology becomes safer with more environmentally friendly [49, 50].

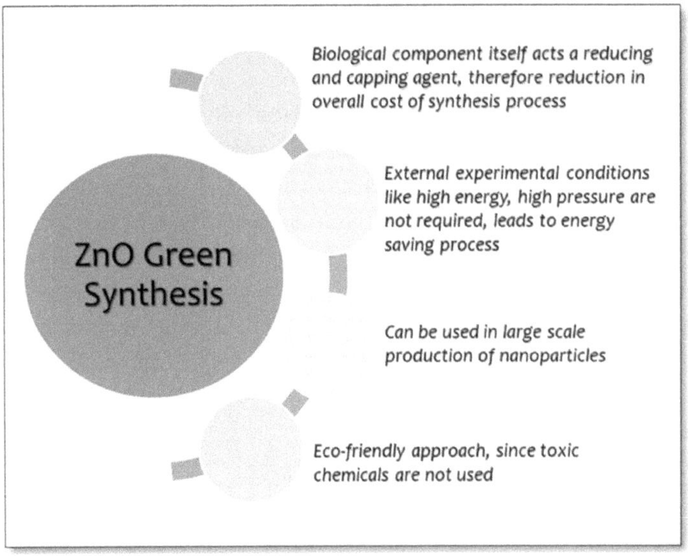

FIGURE 2.2 Green synthesis of ZnO nanomaterials.

2.3.2.1 SAFFRON LEAF EXTRACT

Saffron is a triploid sterile plant that corresponds with the sub-family Crocoideae. This originated through the nearby Behshar area within its region of Mazandaran, Iran. Saffron has utilized throughout the food processing industry mostly flavoring as well as a coloring agent. Analyzes studies suggest saffron contains anti-tumor, anti-inflammatory, anti-depressant, antibiotic including lowering characteristics of insulin sensitivity. In addition to regular mostly, this has many uses there in neurobiology and cosmetic products. Even if saffron leaves are not considered to be a reasonable source of bioactive substances, their existence of phenolic molecules within them encourages their consumption towards a relatively viable choice.

Polyphenols, primarily glycosides like kaempferol, luteolin including quercetin, that was stated might possess certain antioxidant, metal-chelating, and anticancer properties, become its key components within the leaves [51, 52]. Its phenolic compounds present throughout any plant may help to establish NPs [53]. Due to its powerful capacity can oxidize with low toxicity, metal oxide NPs like ZnO NPs were termed a photocatalyst. ZnO NPs has multiple activated locations and a strong response rate, enabling them to produce H_2O_2 effectively, making them an ideal photocatalyst. Synthetic dyes including certain compounds including strong illumination stabilization were highly toxic. These were still non-degradable but also oxidizing substances.

The leaves were first cleaned using deionized liquid then processed over two days inside a furnace at about 50°C to make that saffron leaves extract (Figure 2.3). To acquire the powder form, these leaves are thus smashed. Approximately 5 g of the powder were added into 100 mL with deionized water, mixed with 60 min around 70°C then centrifuged to 20 min at 6,000 rpm. Thus, the solution was filtered by filter paper extracted the precipitate. Ultimately, these extracted were collected and stored for later study through an airtight container inside a refrigerator.

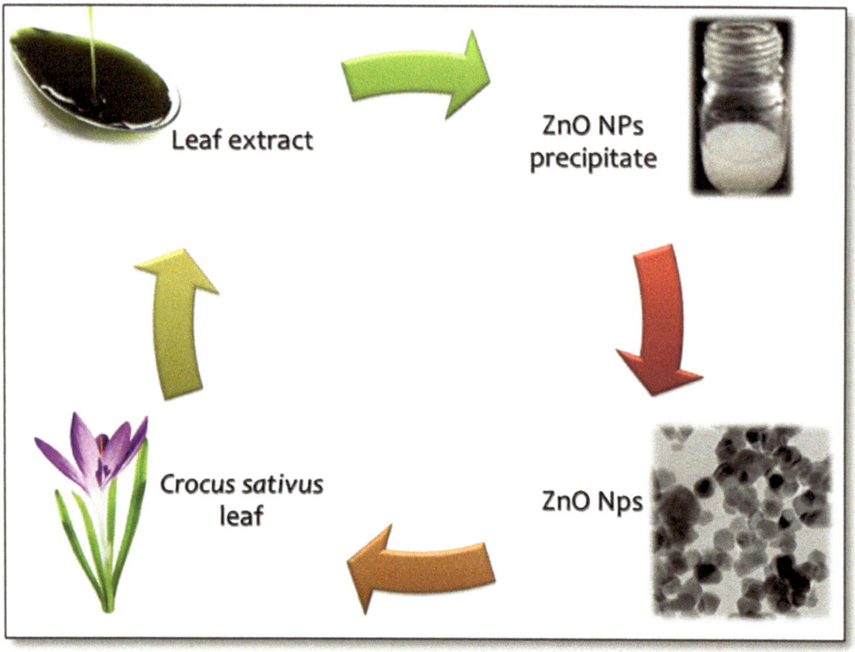

FIGURE 2.3 Fabrication of zinc oxide nanoparticles (ZnO NPs) using saffron leaf extract.

2.3.2.2 *MIKANIAMICRANTHA PLANT*

Mikaniamicrantha, typical across Central and South America, was a sprawling perennial plant of the Asteraceae family. If these weeds have been developed, they drench, penetrate crowns that destroy neighboring plants.

Its growing percentage over 24 hours is extremely high, around 8–9 centimeters. Those plants are regarded as compatible with tea, rubber, bamboo, teak, and eucalyptus. This even impacts rotational crop varieties, plant vegetation, including upsets forest area. Literature reveals that at this plant of Mikaniamicrantha, phytochemicals it is possible to differentiate sterols, polyphenols, coumarins, flavonoids, and diterpenes. Such phytochemicals display behaviors which are showing antimicrobial, antibacterial, antifungal, antiviral, allelopathic, analgesic, phytotoxic, and genotoxic behaviors [54, 55].

Leaves were collected from Mikaniamicrantha trees, washed with distilled water then drying in a heated furnace around 60°C. About 10 g of the substance are applied into 100 ml of purified water yet then heated over 10 mins. That leaves are crushed and filtered. This was instead purified and centrifuged about 10 min around 4,000 rpm using filter paper. For better analysis, this supernatant was placed within that freezer [56–58]. Utilizing filtered water, about 0.5 M mixture with zinc acetate were made after stirring at 4,000 rpm for 10 min. About 5 mL of plant extracts are applied to such a mixture gradually, by vigorous agitation. With a 2 M NaOH solution, its pH was held around 12. This resulting white reaction mixture has been held over 2 hrs. with a magnetic stirring. Around 8,000 rpm, this pale white reaction mixture produced was then collected by centrifugation, cleaned repeatedly using Deionized water, accompanied by ethanol. That reaction mixture is then stored for 12 hours only inside the heated furnace at 60°C being crushed into tiny powder [59]. Nanoparticles synthesis of zinc oxide utilizing Mikaniamicrantha aqueous extract that serves as a capsulating agent as well as a reduction agent.

2.3.2.3 MENTHA EXTRACTS

Mentha becomes a common herbaceous plant located across Europe, North America, and North Africa in the Mediterranean region but still includes a broad history of traditional medicinal usage. Its flowers and leaves of such a plant had also often used to treat diseases over the decades. *M. spicata* is often employed for food and drinks, including with both fragrance and flavoring agent. Owing to its antioxidant, antimicrobial, and sensory properties, it is widely used throughout the food industry [60].

Mentha is a herbaceous plant that was cultivated in Turkey on a commercial scale. It includes flavonoids, quinines, organic acids, also unpredictable components [61]. The Mentha leaf is collected and cleaned 7–8 times by running water even with distilled H_2O (dH_2O). Clean leaves are placed within 100 mL of dH_2O that has been heated over 100C for 60 min. Then extractions were cool around 25°C during that treatment. With filter paper, its resulting mixture was diluted. Its color throughout this extraction appeared light yellow, so it was preserved until subsequent experiments in a refrigerator at 3°C. The preparation procedure for ZnO-NPs is illustrated in Figure 2.4.

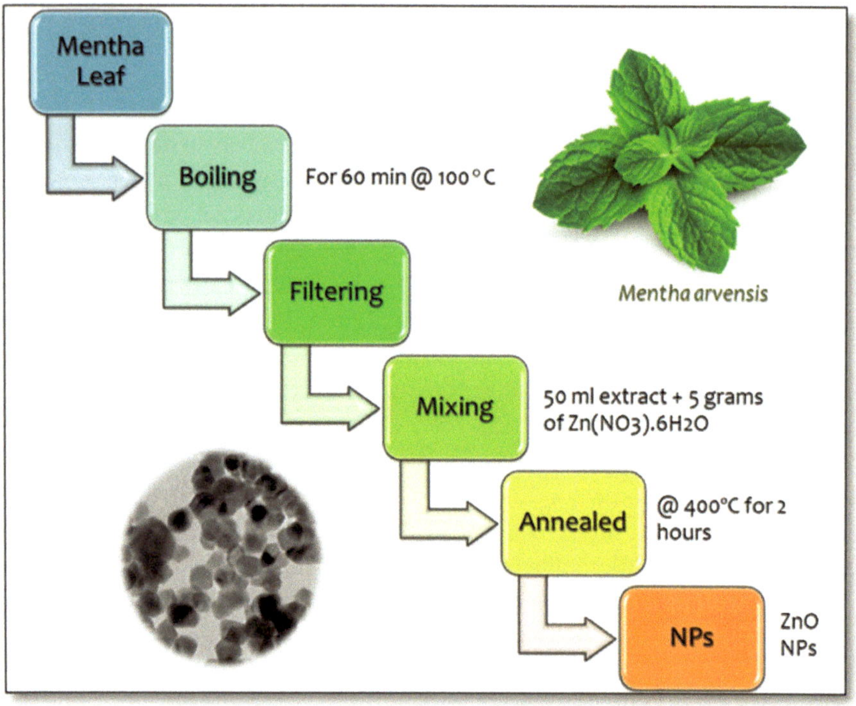

FIGURE 2.4 The preparation process of ZnO-NPs using Mentha leaf.

2.3.2.4 LEAF EXTRACT OF *CORIANDRUM SATIVUM*

Coriander (*Coriandrum sativum L.*) was often a plant belonging to Umbelliferae. Throughout this Mediterranean region, this seems to be a common plant and is extensively cultivated across Asia, Central Europe, and South America. Having green lanceolate-shaped leaves, white or pink umbellate flowers with a prominent seedpod develops around 20–70 cm tall. Having several linear curved layers, the seeds were spheroidal dried with the bottom hole. There seems to be triglyceride oil, petroselinic acid and monounsaturated fatty acid within the sativum crop. This plant is itself a potential cause for lipids. In addition, coriander leaves comprise anthocyanins which could be biosynthesized or improved via their usage of salicylic acid and microelements, with particular zinc [62]. Coriander is being utilized by cooking, medicine including flavors during the old period. These leaves have also observed that demonstrate a broad variety of biological effects even amongst medical products, like anticancer, antioxidant features. Antifungal, Antiprotozoal, post-coital, and decreasing of cholesterol are additional medicinal properties. In the food industry, coriander seeds and leaves were widely utilized for seasoning to add flavor to a large variety of consumer products, including meats, pickles, liqueurs, and teas.

ZnO NPs can be synthesized by utilizing extraction of Coriandrum sativum plant leaves. About 50 mL of purified water was collected in this process as well as 0.02 M of aqueous dihydrate of zinc acetate is introduced over its during continuous mixing. Its aqueous leaf extract of Coriandrum are being added throughout its following mixture in various sets during 10 minutes of mixing. To obtain pH 12, with a white colored organic compound like that, 2.0 M NaOH was often applied. Upon mixing, a pale white precipitate is extracted and thoroughly washed using purified water via ethanol which keeps this clear of contaminants. Further, this was placed on a magnetic stirrer for about two hours. While dried night-time at 60°C in a tubular furnace, a light powdery substance of ZnO nanoparticles is produced.

2.3.2.5 ALBIZIA LEBBECK STEM BARK

Albizia lebbeck seems to be a tropical species belonging to the group Albizia, including some that expand about its medium heights of 24 m, the trunk size about 50 cm, and seed pods containing 6–12 seeds. In Australia, Asia, Africa, and South America, these plants have been extensively spreading [63]. The plant seeds show antitumor, antifungal, and antibacterial activity towards the hepatoma cells of HepG2, the fungi *Rhizoctonia solani* and bacterium *E. coli* [64]. Crop leaves comprise phytochemicals with therapeutic value, such as flavonoids, tannins, triterpenoid saponins, and cardiac glycosides [65]. Freshly stems barks of *A. lebbeck* has been accurately cleaned by deionized water, drying in the light, which finally pulverized via grinding and strainer through fine particles.

Along with 90 mL filtered water, containing different molarities of $Zn(NO_3)_2.6H_2O$ solution; afterwards, 10 mL of prepared *A. lebbeck* extracts being mixed dropwise towards their zinc nitrate concentrations (Figure 2.5) by continuous agitation, about 60°C about 5 hrs, while NaOH has been introduced with the liquid in most of the mixing phase that changes the pH. Both the extracts and the solution later are calcinated in a muffle oven at 350°C ± 10°C over 2 hours to acquire ZnO NPs. Its antibacterial behavior is calculated using the specified quantity of ZnO NPs, extracts, including dimethyl sulfoxide (DMSO).

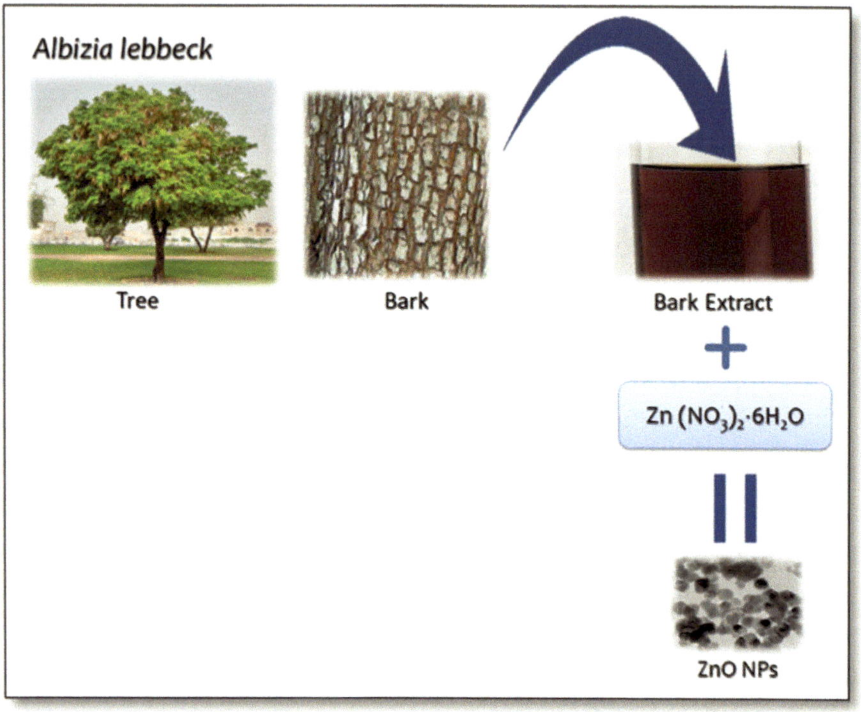

FIGURE 2.5 *Albizia lebbeck* stem bark extract Zn $(NO_3)_2{\cdot}6H_2O$ ZnO NPs.

2.3.2.6 LEAVES EXTRACTS OF GUAVA, LEMON, AND OLIVE

Through decades, zinc oxide seems to have proven a critical product to manufacturers and is currently a focus with a major modern development. Over past decades, nanomaterials biosynthesis, therefore, acquired considerable interest which is now one among its widely favored synthesis methods.

The advancement with an understanding of organic manufacturing and many natural approaches has subsequently related with the development of quick, cost-effective as well as environmentally friendly zinc oxide nanomaterial biosynthesis utilizing its plant leaf extraction method to synthesis of nanomaterials.

Extracts by plant leaves are obtained, including new olive, guava, as well as lemon plants leaf (Figure 2.6). For concentrations of 0.2 M, 70 mL of $Zn\ (NO_3)_2.6H_2O$ is decreased to 30 mL in leaf extraction. Such leaves are gathered, cleaned of dust removal using water, including distilled water, which finally drying around 70°C throughout the night inside a furnace and then grinded leaves to a powder form. After depositing 12.5 g of drying leaf inside a 250 mL beaker filled in 200 mL of water, its extraction, which could be needed in minimizing zinc ions (Zn^{2+}) to (ZnO) nanomaterials were made. This liquid is therefore heated at 60–80°C over 2 hrs through a stirring furnace until the aqueous solution color modified watery towards light yellow, indicating that complete conversion of zinc hydroxide to zinc oxide and calcinated nearby 2 hours above 500°C within a glass furnace. Upon calcination, its substance was ground for their processing of powder form. Utilizing filter paper, this extraction was cooled until room temperatures and then purified. Kept in a freezer, the collected mixture has been utilized like a reduction agent [66].

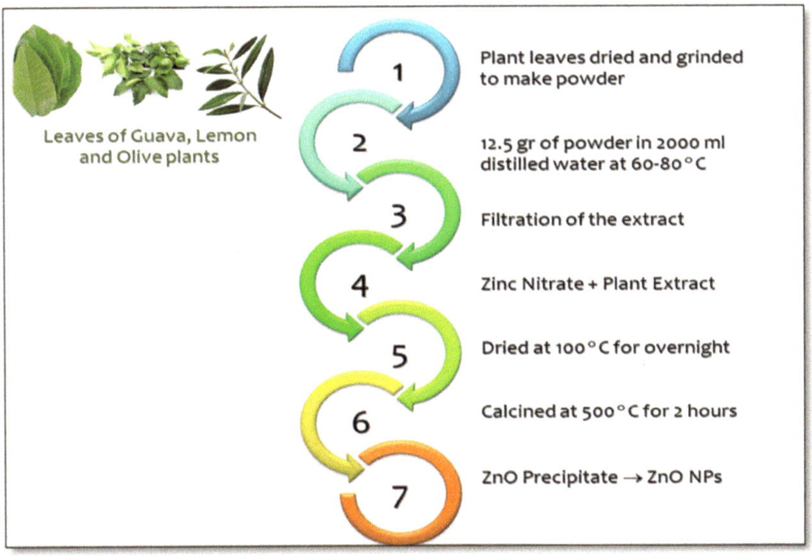

FIGURE 2.6 Biosynthesis of ZnO nanoparticles.

Some of the Green synthesized ZnO nanoparticles and their biological applications are listed in Table 2.2 along with selective ZnO extraction procedures in Table 2.3. According to its major antibacterial activities, photocatalysis, including metal ion adsorbent functions, several organic synthesized ZnO nanoparticles were exploited across different areas like biomedical application. Besides that, according to its functional groups onto their interfaces which arrive via phytochemicals, synthesized nanoparticles by the green method show improved antibacterial efficiency. Another big factor with its antibacterial effect of ZnO nanoparticles was cell membrane structure destruction. Through its lack of phospholipid bilayer stability as well as leaking of intracellular parts of the cell, ingestion through ZnO nanoparticles leads to cell destruction. Although Gram-positive bacterial provide a thicker surface of peptidoglycan, teichoic acid, including lipoteichoic acid within its cell membranes, multiple coating of peptidoglycan is present among Gram-positive bacterial. Specific method nanomaterials enter through the cell membranes results in distinct arrangement throughout the cell membranes of certain two groups of bacterial. We based during each portion on the biomedical operation of ZnO nanoparticles, and the plant extracts, zinc proteins, medicinal activities and associated biomolecules have been summarized in Table 2.2.

TABLE 2.2 Green Synthesized ZnO Nanoparticles and Their Biological Applications

Plant	Precursor	Size (nm)	Treated with	Biological Field
Eclipta alba	Acetate	6	*E. coli*	Antibacterial
Vitex negundo	Nitrate	38	Human serum albumin	Protein binding
Pongamia pinnata	Acetate	21	*C. maculatus*	Anti-pesticide
Albizia saman	Nitrate	15–80	*D. indica*	Genotoxicity
Tradescantia pallida	Acetate	25	HeLa cervical cancer cell	Anticancer
Sechium edule	Acetate	36	*B. subtilis* and *K. pneumoniae*	Antibacterial
Stevia	Acetate	10–90	*E. coli* and *S. aureus*	Antibacterial Antiparasitic

Zinc oxide was listed throughout materials science with a semiconductor among group II-VI, the co-valence of which is mostly on the border among covalent and ionic semiconductors. A broad energy range, high bond energy and a high degree of physical and chemical strength at low temperature make it more desirable to possible device applications, optoelectronic devices and laser technology. Because of its piezo including pyroelectric characteristics of ZnO's indicate which they would be employed in hydrogen manufacturing such as a detector, transformer, power generation and photocatalyst. This suits a particularly essential product in the ceramics industries due to their toughness, stiffness, and piezoelectric stability, although their lower toxicity, biocompatibility as well as biodegradability making them a source of importance towards biomedicine and pro-ecological processes. Over a relatively broad number of structures, zinc oxide exists and provides an enormous variety of uses. Different techniques for the manufacture of ZnO, like vapor deposition, hydrothermal method, the mechanism of sol-gel, and properties containing particle varies in form, width, and the following methods explain these approaches in detail (Table 2.3).

TABLE 2.3 Selective ZnO Extraction Procedures

Method	Precursors	Synthesis Conditions	Properties	References
Mechano-chemical	$ZnCl_2$, Na_2CO_3, NaCl	400–800°C	Hexagonal structure; particles diameter: 18–35 nm	[67]
Sol-gel	$Zn(CH_3COO)_2$, diethanol-amine, ethanol	Room temperature; annealed of sol: 2 h, 500°C	Hexagonal wurtzite structure; particles: nanotubes of 70 nm	[68]
Solvothermal hydrothermal	$Zn(CH_3COO)_2$, $Zn(NO_3)_2$, LiOH, KOH, NH_4OH	10–48 h, 120–250°C	Hexagonal (wurtzite) structure, size of micro crystallites: 100 nm–20 μm	[69]

TABLE 2.3 *(Continued)*

Method	Precursors	Synthesis Conditions	Properties	References
Microemulsion	$Zn(NO_3)_2$, oxalic acid, isooctane, benzene, ethanol, diethyl ether, chloroform, acetone, methanol, aerosol OT	1 h; calcination: 3 h, 300°C	Equivalent spherical diameter: 11.7–12.9 nm, size: 11–13 μm	[70]
Other methods	$Zn(CH_3COO)_2$	Thermal decomposition: 350–800°C	Uniform size of particles 20–30 nm	[71]
	Diethylzinc (DEZ), oxygen	Helium as carrier gas	Wurtzite structure; average particle size: 9 nm	[72]

2.4 APPLICATIONS OF GREEN ZNO NANOMATERIALS

2.4.1 BIOSENSING

Biosensors were operational testing instruments allowing various analytes to be selectively identified. There is some available for a small surface in the biosensing which always interacts between its targeted compound as well as its transducer, converting each biological activity as a visible signal. Each transducer category describes their biosensing measuring technique, evaluation requirements, including drawbacks. A system with optical biosensors has drawn interest. Good accuracy, including symbol principles was its significant benefits among optical investigative techniques. The future century of detection systems with daily usage was optical biosensing systems focused on the absorption spectrum. Through the transmission of its successful signals from biological interaction towards a mechanical signal, its biosensing system needs unique materials containing specialized design, electrical, and optical properties. In specific, ZnO's optical properties access up a good potential with biosensing implementations to be utilized. Its main aspects were identified in the immobilization method, which is essential towards efficiency via their bio-sensing.

1. **The Method Regarding Sensory Including Optical Biosensing Forms:** Each biological sensor module was linked with any optical

transmitter device throughout electrochemical imaging and the signaling is focused by absorption, luminosity or specular reflection. Optical biosensors will typically be separated into two major categories: specific optical and conditional optical detection. Specific detector mechanisms are presented by biosensors focused on ellipsometry, spectrometry, mass spectroscopy, and luminescence. Fluorescence, photoluminescence quantum dots (QDs), etc., were focused for conditional biosensing. Variation among both categories was that the one studies its transducer's characteristics towards each context of current optical inspection; within the other category, their connected marks may be utilized to identify its aim sample. This biosensitive film is created through immobilization across its surfaces to that transducer with their specific biologically feature. In order to capture its targeted sample matrix, such a bio-recognition surface provides a source. Each major objective for biosensors is to achieve accurate as well as quick recognition through strong tolerance and specificity through its desired biomolecules.

2. **Functionalization of the Substrate of ZnO and Biosensitive Material Shaping:** By its production of biosensors with better efficiency, their creation on any reliable bioselective surface being quite interesting activity. This should be noted as the structure including mechanical characteristics of ZnO oxide plays very significant roles throughout its immobilization [76]. Its morphology of the nanostructure of ZnO, therefore, determines the transport properties, the number of positions with adsorption, including their level of proteins adsorbent. Two main techniques with bioselective surface formation mostly on the substrate of ZnO is available:
   i. Specific immobilization of organic compounds; and
   ii. Covalent bonding between the surface of zinc oxide and organic compounds.

   Owing to several advantages, particularly non-toxicity, biosafety, bio-compatibility, high electron-transfer speeds, and even interaction on immobilized enzymes, ZnO nanostructures have currently gained focus within biosensor technologies [73]. Likewise, ZnO has a significant level of ionic binding, which melts relatively easily during basic physiological pH levels. An amperometric glucose sensor dependent upon arranged films of ZnO nanorods developed specifically mostly over ITO glass substrate [74]. The manufacture with ZnO-nanorod-based biosensing of good reproducibility as well as specificity has been shown by Ibupoto et al. [34] can quickly track penicillin only through immobilization of the penicillinase enzymes via a normal biological adsorption mechanism [75]. A ZnO nanocrystalline field-effect-transistor dependent biosensor was identified by Choi et al. [16] towards their identification at lowest grade biological substances.

2.4.2 ANTI-INFLAMMATORY

The infection becomes one of the organ tissue's complicated reactions towards harmful stimuli, such as bacteria, tissues that are harmed, even allergens. Its anti-inflammatory actions of ZnO Nanoparticles have also drawn considerable interest since the emergence of nanomaterials, including the consideration for certain biological properties of zinc ions. Atopic dermatitis becomes a severe infectious skin condition that has been associated with complicated interactions among genomic and atmospheric factors, characterized by impaired skin membrane function [77, 78].

Atopic dermatitis model, different-sized ZnO NPs could reach damaged tissue or even wounded sensitive tissue [79]. Whose tests specifically showed why just nanostructured ZnO (nZnO) can enter deeper surfaces of sensitive tissue, however, bulk-sized ZnO persisted in the surface layer of allergies or even injured tissue. nZnO exhibited greater anti-inflammatory potential as an alternative against nZnO after significantly reducing proinflammatory cytokines in the Atopic dermatitis mouse study. These studies showed how simple ZnO NPs were good effects on their reduction of skin inflammation for samples of atopic dermatitis. ZnO NPs anti-inflammatory role is not limited to its care with atopic dermatitis; however, this had even been found to be quite effective in many various inflammatory infections.

2.4.3 ANTIMICROBIAL AGENTS

ZnO nanoparticles had numerous usages which involve the function of wound healing. Microorganism infection is indeed a dangerous concern to tissue transplants and implants, i.e., where the nanoparticles of antimicrobials and antibacterial arrive within the illustration. Which were known as antimicrobial agents that were natural as well as artificial? Its reactivity against infection is slightly interrupted by microbial antimicrobial agents, whereas natural antimicrobial agents respond rather quickly

[80]. Natural antimicrobials, like instability over extreme temps, do have disadvantages.

This directed to certain bacteria may grow drug resistance towards natural antimicrobials but also there was hardly any establishment including its basic idea of that too. From its opposite side, synthetic microbes such as metal oxides were willing to create suitable disinfection against every form of by-product explosion, while slower during the effect. This showed even as contrasts to the latter, synthetic antimicrobial nanoparticles are much quite comfortable. ZnO is among those generally recognized for secure identification of products by the US Food and Drug Administration (FDA). This could be used throughout its fortification of cereal-based products to a food additive but as a good feature with zinc. ZnO nanoparticle antibacterial activity may be attributed to their concentration with NPs on the external surface within microbial tissues that induce excessive ROS which releasing Zn^{2+} [81].

2.4.4 ANTIBACTERIAL

Due to its excellent characteristics and large specific surfaces area concentration, ZnO NPs can be chosen as an antimicrobial substance to prevent a broad variety of infectious organisms. That antibacterial operation can include the deposition of ZnO NPs inside the external layer of bacterial cells then their releasing of Zn^{2+} will allow disintegration of its microbial cellular membrane, harm to the cellular proteins causing genetic instability, resulting in microbial cellular destruction [82–84]. To evaluate its antibacterial Property of Zinc Oxide nanoparticles, gram-negative Escherichia coli even gram-positive Staphylococcus aureus were currently used primarily with design microbes [85, 86].

ZnO NPs were found to be quite successful in inhibiting the growth of *V. cholera's* El Tor biotype that is strongly correlated with the development of ROS. Such findings will harm its cell membranes, raise binding but also alter its structure substantially. Moreover, to avoid infection by adhering to, transmitting, or breeding in biomedical equipment, this would be covered onto various surfaces.

2.4.5 DIABETES TREATMENT

Around 1934, scientists started to believe how Zn, insulin, and diabetes could be closely related since this is noticed whether the insulin crystals included zinc [87]. It is now understood that Zn associates within certain components during certain sensing mechanism of insulin which thus influences the synthesis of insulin. In contrast, its existence to these ions often affects many forms of diabetic disorder, like β-cell activity, glucose tolerance, insulin sensitivity, and diabetic pathogenesis [88]. Its various reports on the occurrence with ZnO Quantum Dot's in the control of diabetes had expanded dramatically because its deterioration of ZnO resulted during the exposure of Zn^{2+} ions throughout the news. Their decreases of non-esterified fatty acids, triglycerides including glucose within the blood but indeed the rise in serum insulin influence [89]. The combination effect between ZnO and drugs was also investigated, but the anti-diabetic component was covalently linked onto its substrate through their QDs within that instance. The potent natural extracted anti-diabetic product, Red Sandalwood (RSW), is selected as the active ingredient. These reports indicated whether ZnO-RSW QDs were most successful with basic murine pancreatic glycosidase than either with each two recently controlled components [90].

2.4.6 BIOMEDICAL

ZnO clearly shows excellent activity within this biomedical sector across this case of nanoscale metal oxides because of their specific electronic, mechanical, and pharmaceutical characteristics. ZnO nanoparticles demonstrate antibacterial activity among such a wide range of pathogenic bacteria, and similar nanomaterials follow a different process which includes generation of superoxide, disturbance of the stability of its cellular membranes, development of biofilms, including activation of enzymes [91]. ROS, such as superoxide ions, hydroxide ions, singlet oxygen species including peroxide molecules, were generated during Ultraviolet light. That peroxide ions created would penetrate deeply via the cell surface, which leads to the deaths of cells. Figure 2.7 shows the possible mechanisms for producing ROS as well as their impact mostly against the bacterial membrane. With their loss of phospholipid bilayer stability including a release with intracellular components, insertion with ZnO nanoparticles causes cell death. Although gram-positive bacteria provide some thin coating of peptidoglycan, teichoic acid, and lipoteichoic acid throughout their membrane surface, a triple film of peptidoglycan is present among Gram-positive bacteria. This particular method of nanoparticle entry through the cellular membrane resulted in the distinct cell memory system of such two species of bacteria.

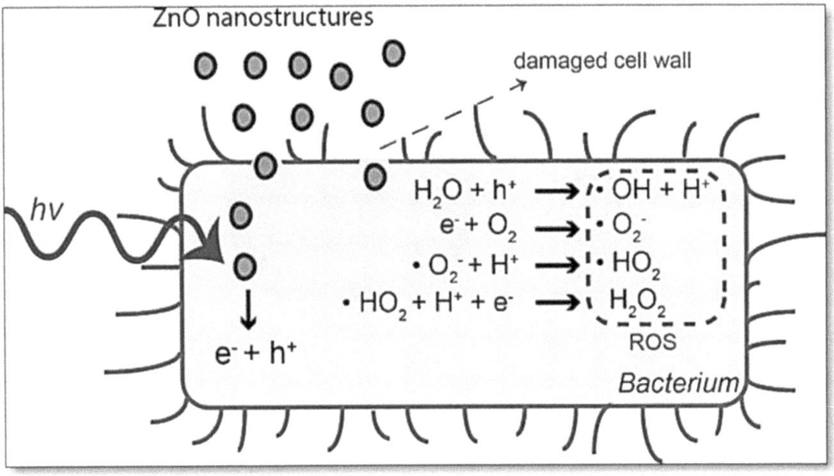

FIGURE 2.7 ROS mechanism of ZnO nanoparticles.

2.4.7 PHOTOCATALYTIC

According to its optical and electronic properties, ZnO nanoparticles have been used in photocatalytic uses. Valence band electrons were stimulated towards their conduction band that causes holes away whenever its ZnO nanoparticles get exposed by UV radiation. Then by oxidizing H_2O and OH^-, that produced holes produce hydroxyl radicals, as well as those excitons, were collected mostly in the atmosphere by oxygen (Figure 2.8). The resultant anionic radical becomes extremely unstable that transform CO_2 and H_2O via natural chemicals.

Various quantities of *Camellia sinensis* impact extraction mostly on synthesized ZnO nanoparticles by photocatalysts oxidation of methylene blue dye, their synthesized nanomaterials are examined at which nanoparticles provided MB degradation to around 84% for 120 min [92]. In this same some other study, Parkiaroxburghii leaf extract are used in the growth of Zinc Oxide nanomaterials and has also found degeneration besides neither Methylene Blue but also Rhodamine B dyes at almost 98% effectiveness in 8 min [93]. Through synthesizing nanoparticles, its aqueous leaf extract from *Coriandrum sativum* was utilized including its corresponding material were using in anthracene photocatalytic dissolution with 96% performance within 240 min [94].

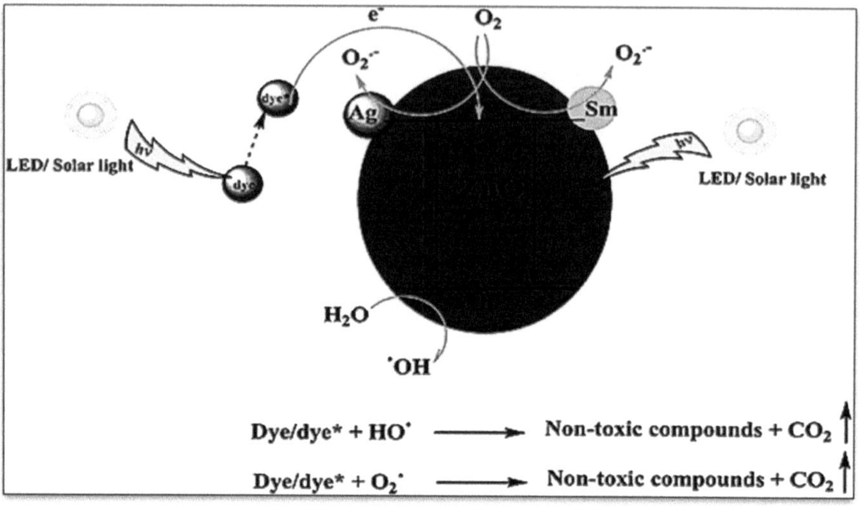

FIGURE 2.8 Schematic view for zinc oxide nanostructures dye deterioration.

2.4.8 FIELD-EFFECT TRANSISTOR

Standard field-effect transistor nanowires comprised by its semiconductor nanowire such as ZnO coupled across two ends by two electrodes then mounted on a flat surface which functions like a gate electrode. Neither applicable gate voltage nor synthetic materials coated onto the layer of the nanowires regulate its current passage from the sink towards the origin through the nanomaterial. Piezoelectric-FET (PE-FET) also currently suggested either with the pairing of ZnO's semiconductor or piezoelectric materials identified through its piezotronic effect [95–98]. PE-working FET's theory depends upon the piezoelectric capacity of stressed nanowire, which acts as a voltage base by regulating the flowing of electricity through their drain towards its source [99].

2.4.9 SUPERCAPACITORS

Electrochemical supercapacitors had its ability that substitutes current electricity equipment but is commonly accepted as just an eco-sustainable energy storage unit. For connector hybrid vehicles, including emergency power grids, supercapacitors (SCs) are used. Transition metal oxides but also phosphates have been commonly employed in supercapacitors and many various energy-related devices in recent years with possible interface materials [92, 93]. ZnO is commonly accepted as a rechargeable batteries component for a favorable energy density of 650 Ahg^{-1} amongst those transition metal oxides. In particular, ZnO provides strong electrical conductivity that is greater than certain metal oxide substances. Composite materials dependent on ZnO have many advantages, such as cost-effectiveness, eco-friendliness, reasonable electrochemical reproducibility, excellent specific capacitance, higher capacity, power output, proper cycling reliability and simpler manufacturing, making them desirable electrochemical supercapacitors competitor.

2.4.10 WOUND HEALING

Wound healing was indeed a strongly established procedure for repairing damaged tissue comprising four biological phases which are simultaneous but merging: hypertrophy, swelling, replication, including renovating. The nanoscale and large surface-area-to-volume proportion usually allows, interacting effectively with its injury site but also penetrating the layer of skin at its injured area effectively. Therefore, nanoparticles will not only act as healing factors for wound healing, yet could often provide continuous and regulated delivery of therapies towards the site of injury. An excellent reality there within the clinic is the shortage of zinc production that prevents wound healing. That impact with clinically proven zinc over legs tissue repair and even its influence on certain wound healing processes had been observed during 1990, employing standardized animal models [100]. Afterwards, many clinical and research experiments involving natural Zn and zinc oxide are carried out [101]. Results revealed as topical ZnO-based therapies have provided many advantages, amongst many, in re-epithelialization, wound healing, infections, and ulceration [102]. Researchers were now mainly concentrating their attention on developing modern wound treatment products which combine both antimicrobials as well as medicinal benefits of ZnO QDs.

Throughout fact, a hydrogel formed with chitosan has been integrated through ZnO QDs. Analysis indicates whether an antimicrobial property with good impact with improving wound healing and high porosity was seen in such nanoparticles bandages. Chitosan was also widely used during biomedicine high thermal stability, namely bioactivity, hemostatic, bactericidal, and healing ability [103]. To incorporate antimicrobial and wound healing properties materials dependent upon chitosan and including ZnO QDs. The wound healing ability throughout its system was tested by filled excision in rat specimens. Upon 6 days, their reports demonstrated a noticeable reduction in the wounded relative against its test wound [104].

2.4.11 TISSUE-ENGINEERING

Simple polymeric substances are among its main promising strategies towards tissue technology throughout its development with synthetic substrates. Coatings were designed with artificial flexible surfaces which facilitate their development of fresh tissues but to sustain cellular development from its start until their completion during each cycle of reconstruction. Those should be allowed to colonize its entire scaffold structure while seeding the cell. It is perhaps important to encourage cellular uptake also on the external layer but also its interpretation inside this system of scaffolds. When they have gained adhesion, its subsequent propagation could always become sought. Ultimately, for instance, the transfer through resources, including production agents for their inflowing cells, an adequate arrangement inside each scaffolding system should be established.

The goal behind tissue engineering may be towards developing novel substances that replace defective tissues, thereby preventing any variety involving transfusions otherwise extensive but also costly interventions [105]. Within metallic oxides, due to their antimicrobial activities and their function through stimulating cell formation, proliferation, and segregation, ZnO becomes one of those main examined with tissue engineering applications [106]. Various biopolymers have been included throughout tissue engineering applications along association using ZnO; of these, poly (ε-caprolactone) provides numerous advantages, such as bioactivity and biodegradability [107], but also was licensed by the FDA and are employed throughout medical applications [108].

2.4.12 ANTICANCER

Zinc oxide is a semiconductor ionic material that was commonly formed through water by alkali metal crystals that seem to be practically insoluble.

ZnO is very much an affordable substance being utilized in cosmetics items such as moisturizers where this consumes 350 to 380 nanometers of UV radiation. In biochemical and preclinical studies, ZnO particles, particularly nanometer particles, had equally being utilized, even in biosensing applications and drug development. For example, Particulates smaller than 100 nm with maximum absorbance have been recorded that being the most effectively distributed *in vivo*. Particularly, ZnO nanoparticles were often found to *in vitro* impact several tumor cells, possibly as Zn^{2+} causes the formation of reactive oxygen species. ZnO particles through valence band to conduction band will also be promoted by UV radiation to generate photocatalysts superoxide. Researchers claim, though, that ZnO shields neutrophils from cytotoxicity with its cancer therapy by accumulating intravenous Zinc oxide over multiple cells, especially lung tissues. Jointly, such results indicate how ZnO will kill tiny lung cancer cells through a mechanism that is different from existing chemotherapy.

2.5 SAFETY EVALUATION

Nanotechnology seems to be a creation that can be used within its smaller level feasible, in substances. So numerous technological and manufacturing industries, namely pharmacy, food safety, and several more others have been vastly enhanced but also revolutionized by nanomaterials. In many uses, including the photography and drug transmission, this association of nanoparticles towards tissues including certain macromolecular materials is important, but those similar characteristics seem to be a problem for their security. Security analysis with the nanomaterials becomes important when its application from ZnO-NPs rises but becomes integrated across different consumer goods. Research has confirmed that ZnO-NPs have been harmful to several organisms during their past century, such as bacteria, algae, plants, and animal cells. Throughout comparison, ecosystem rates with ZnO-NPs were projected for improving continuously which was projected for varying between 76–760 μg/L to water and 3.1–31 mg/kg through the soil depends on consumer penetration [109]. It is therefore important that consider accounts that toxicity testing of biosynthesized nanomaterials with a security evaluation. Dependent upon its findings with the genotoxicity analysis, further application of ZnO-NPs with public health is being practiced with precaution.

2.6 SUMMARY

The increasing widespread acceptance of modern detectors utilizing nanomaterials has been promoted by the development of semiconductor devices. When biological and technological disciplines converge, their production of biosensors would become extremely prevalent. It had developed a modern area of nano-biosensing that shows why nanometer-scale technologies and biologically converge. For sensor technologies, this advent of nanotransducers allows this conventional distinction of sensors and bioreceptors redundant. Through its advanced level identification of infection indicators with reasonable rates, nano-biosensing has become crucial in growing countries towards medical care. A continuous tracking of diabetes mellitus influencing even its study and industrial production of insulin detectors illustrates everything. Nanotechnology does have the ability to create elevated, interconnected electrochemical detectors by developing highly reactive sensors such as nanowires that can achieve tolerance out to simple molecules. Zinc oxide nanomaterials demonstrate major biosensing ability for effective positions. That happens why ZnO nanomaterials provide morphologies including high substrate area, enabling systems containing complex designs to be manufactured. Zinc oxide materials within various sizes, particularly electrochemical biosensing but also its corresponding signaling efficiency, have been examined.

Zero Dimension ZnO nanostructures are commonly utilized preceding the creation with One Dimension, Two Dimension and Three Dimension nanomaterials related to their flexibility with processing. Despite certain issues that lower latency encountered in Zero Dimensional nanocrystals, nevertheless, their emphasis shifted onto One Dimensional ZnO. Nevertheless, rapid development in this area with nanotechnology today has made it possible to explore ZnO QDs differently. Consequently, their progress with Zero Dimensional ZnO-based electrical and chemical biosensing nanomaterials becomes optimistic. Each longitudinal and lateral One Dimensional ZnO-based FET biosensing nanomaterials have also been shown to improve the biosensor's efficiency. Besides this, two dimensional ZnO nanomaterials have the maximum surface-to-volume proportion, including a particular facet that provides an effective loading mechanism for immobilization.

In addition, to achieve optimum flexibility; the perfect situation was to produce a planar ecosystem of limited scattered locations. Likewise, this had been rendered desirable to amperometric, conductometric, and colorimetric dependent biosensing by the permeability of 3D zinc oxide nanomaterials.

Furthermore, the focus had never been given to the usage of such three-dimensional ZnO nanostructure for an effective stream within FET-based biosensors. Biosensing technology was driving the development of simpler, better, and quicker detectors that could one day lead to the convergence between digital and biological processes. To build extremely responsive, highly precise, multi-analysis, including nanoscale biosensing as well as biotechnology, a collection of quantum mechanics, surface physics, genetics, bioinformatics, and electrical technology might be expected, that could significantly help advanced diagnoses and medical care.

KEYWORDS

- **antibacterial**
- **anticancer**
- **atomic force microscopy**
- **biosensing**
- **nanomaterials**
- **supercapacitors**
- **wound healing**
- **ZnO**

REFERENCES

1. Abel, J. J., (1926). Crystalline insulin. *Proceedings of the National Academy of Sciences of the United States of America, 12*(2), 132–136. https://doi.org/10.1073/pnas.12.2.132.
2. Agren, M. S., (1990). Studies on zinc in wound healing. *Acta Dermato-Venereologica. Supplementum, 154*, 1–36.
3. Akgul, G., Akgul, F. A., Attenkofer, K., & Winterer, M., (2013). Structural properties of zinc oxide and titanium dioxide nanoparticles prepared by chemical vapor synthesis. *Journal of Alloys and Compounds, 554*, 177–181. https://doi.org/10.1016/j.jallcom.2012.11.158.
4. Alagumuthu, G., & Kirubha, R., (2012). Green synthesis of silver nanoparticles using *Cissus quadrangularis* plant extract and their antibacterial activity. *International Journal of Nanomaterials and Biostructures, 2*(3), 30–33.
5. Alenezi, M. R., Alshammari, A. S., Jayawardena, K. D. G. I., Beliatis, M. J., Henley, S. J., & Silva, S. R., (2013). Role of the exposed polar facets in the performance of thermally and UV activated ZnO nanostructured gas sensors. *Journal of Physical Chemistry: C, Nanomaterials and Interfaces, 117*(34), 17850–17858. https://doi.org/10.1021/jp4061895.

6. Alim, K. A., Fonoberov, V. A., Shamsa, M., & Balandin, A. A., (2005). Micro-Raman investigation of optical phonons in ZnO nanocrystals. *Journal of Applied Physics, 97*(12). https://doi.org/10.1063/1.1944222, PubMed: 124313.
7. Anagnostakos, K., Hitzler, P., Pape, D., Kohn, D., & Kelm, J., (2008). Persistence of bacterial growth on antibiotic-loaded beads: Is it actually a problem? *Acta Orthopaedica, 79*(2), 302–307. https://doi.org/10.1080/17453670710015120.
8. Ao, W., Li, J., Yang, H., Zeng, X., & Ma, X., (2006). Mechanochemical synthesis of zinc oxide nanocrystalline. *Powder Technology, 168*(3), 148–151. https://doi.org/10.1016/j.powtec.2006.07.014.
9. Asghari, F., Samiei, M., Adibkia, K., Akbarzadeh, A., & Davaran, S., (2017). Biodegradable and biocompatible polymers for tissue engineering application: A review. *Artificial Cells, Nanomedicine, and Biotechnology, 45*(2), 185–192. https://doi.org/10.3109/21691401.2016.1146731.
10. Bhattacharyya, S., & Gedanken, A., (2008). A template-free, sonochemical route to porous ZnO nano-disks. *Microporous and Mesoporous Materials, 110*(2, 3), 553–559. https://doi.org/10.1016/j.micromeso.2007.06.053.
11. Boguniewicz, M., & Leung, D. Y., (2011). Atopic dermatitis: A disease of altered skin barrier and immune dysregulation. *Immunological Reviews, 242*(1), 233–246. https://doi.org/10.1111/j.1600-065X.2011.01027.x.
12. Bordbar, M., Negahdar, N., & Nasrollahzadeh, M., (2018). *Melissa officinalis* L. Leaf extract assisted green synthesis of CuO/ZnO nanocomposite for the reduction of 4-nitrophenol and rhodamine b. *Separation and Purification Technology, 191*, 295–300. https://doi.org/10.1016/j.seppur.2017.09.044.
13. Boxall, A. B., Tiede, K., & Chaudhry, Q., (2007). Engineered nanomaterials in soils and water: How do they behave and could they pose a risk to human health? *Nanomedicine, 2*(6), 919–927. https://doi.org/10.2217/17435889.2.6.919.
14. Cahú, T. B., Silva, R. A., Silva, R. P. F., Silva, M. M., Arruda, I. R. S., Silva, J. F., Costa, R. M. P. B., et al., (2017). Evaluation of chitosan-based films containing gelatin, chondroitin 4-sulfate and ZnO for wound healing. *Applied Biochemistry and Biotechnology, 183*(3), 765–777. https://doi.org/10.1007/s12010-017-2462-z.
15. Chaari, M., & Matoussi, A., (2012). Electrical conduction and dielectric studies of ZnO pellets. *Physica. Part B, 407*(17), 3441–3447. https://doi.org/10.1016/j.physb.2012.04.056.
16. Choi, A., Kim, K., Jung, H. I., & Lee, S. Y., (2010). ZnO nanowire biosensors for detection of biomolecular interactions in enhancement mode. *Sensors and Actuators. Part B, 148*(2), 577–582. https://doi.org/10.1016/j.snb.2010.04.049.
17. Coleman, V. A., & Jagadish, C., (2011). Chapter 1; Basic properties and applications of ZnO. In: Jagadish, C., & Pearton, S. J., (eds.), *Zinc Oxide Bulk, Thin Films and Nanostructures: Processing, Properties, and Applications* (pp. 1–20). Elsevier.
18. Cowan, M. M., (1999). Plant products as antimicrobial agents. *Clinical Microbiology Reviews, 12*(4), 564–582. https://doi.org/10.1128/CMR.12.4.564.
19. Croisier, F., & Jérôme, C., (2013). Chitosan-based biomaterials for tissue engineering. *European Polymer Journal, 49*(4), 780–792. https://doi.org/10.1016/j.eurpolymj.2012.12.009.
20. Cui, J., (2012). Zinc oxide nanowires. *Materials Characterization, 64*, 43–52. https://doi.org/10.1016/j.matchar.2011.11.017.

21. Dejene, F. B., Ali, A. G., Swart, H. C., Botha, R. J., Roro, K., Coetsee, L., & Biggs, M., (2011). Optical properties of ZnO nanoparticles synthesized by varying the sodium hydroxide to zinc acetate molar ratios using a SOL-GEL process. *Open Physics, 9*(5), 1321–1326. https://doi.org/10.2478/s11534-011-0050-3.
22. Dem'yanets, L. N., Li, L. E., & Uvarova, T. G., (2006). Zinc oxide: Hydrothermal growth of nano-and bulk crystals and their luminescent properties. *Journal of Materials Science, 41*(5), 1439–1444. https://doi.org/10.1007/s10853-006-7457-z.
23. Djurisić, A. B., & Leung, Y. H., (2006). Optical properties of ZnO nanostructures. *Small, 2*(8, 9), 944–961. https://doi.org/10.1002/smll.200600134.
24. Dutta, R. K., Nenavathu, B. P., Gangishetty, M. K., & Reddy, A. V. R., (2013). Antibacterial effect of chronic exposure of low concentration ZnO nanoparticles on *E. coli*. *Journal of Environmental Science and Health, Part A, 48*(8), 871–878. https://doi.org/10.1080/10934529.2013.761489.
25. Fan, Z., & Lu, J. G., (2005). Zinc oxide nanostructures: Synthesis and properties. *Journal of Nanoscience and Nanotechnology, 5*(10), 1561–1573. https://doi.org/10.1166/jnn.2005.182.
26. Fei, P., Yeh, P. H., Zhou, J., Xu, S., Gao, Y., Song, J., Gu, Y., et al., (2009). Piezoelectric potential gated field-effect transistor based on a free-standing ZnO wire. *Nano Letters, 9*(10), 3435–3439. https://doi.org/10.1021/nl901606b.
27. Fulati, A., Ali, S. M. U., Asif, M. H., Alvi, N. H., Willander, M., Brännmark, C., Strålfors, P., Börjesson, S. I., Elinder, F., & Danielsson, B., (2010). An intracellular glucose biosensor based on nanoflake ZnO. *Sensors and Actuators: Part B, 150*(2), 673–680. https://doi.org/10.1016/j.snb.2010.08.021.
28. Gao, Z., Zhou, J., Gu, Y., Fei, P., Hao, Y., Bao, G., & Wang, Z. L., (2009). Effects of piezoelectric potential on the transport characteristics of metal-ZnO nanowire-metal field effect transistor. *Journal of Applied Physics, 105*(11), (1–6). https://doi.org/10.1063/1.3125449, PubMed: 113707.
29. Gayen, R. N., & Paul, R., (2018). Phosphorous doping in vertically aligned ZnO nanorods grown by wet-chemical method. *Nano-Structures and Nano-Objects, 13*, 163–169. https://doi.org/10.1016/j.nanoso.2016.03.007.
30. Ghanem, M. A., Arunachalam, P., Amer, M. S., & Al-Mayouf, A. M., (2018). Mesoporous titanium dioxide photoanodes decorated with gold nanoparticles for boosting the photoelectrochemical alkali water oxidation. *Materials Chemistry and Physics, 213*, 56–66. https://doi.org/10.1016/j.matchemphys.2018.04.037.
31. Hassan, S. S. M., El Azab, W. I. M. E., Ali, H. R., & Mansour, M. S. M., (2015). Green synthesis and characterization of ZnO nanoparticles for photocatalytic degradation of anthracene. *Advances in Natural Sciences: Nanoscience and Nanotechnology, 6*(4). https://doi.org/10.1088/2043–6262/6/4/045012, PubMed: 045012.
32. Hsu, J. C., & Chiang, Y. S., (2013). Influence of oxygen on zinc oxide films fabricated by ion-beam sputter deposition. *ISRN Materials Science, 2013*, 1–7. https://doi.org/10.1155/2013/710798, PubMed: 710798.
33. Hughes, W. L., & Wang, Z. L., (2003). Nanobelts as nanocantilevers. *Applied Physics Letters, 82*(17), 2886–2888. https://doi.org/10.1063/1.1570497.
34. Ibupoto, Z. H., Ali, S. M., Khun, K., Chey, C. O., Nur, O., & Willander, M., (2011). ZnO nanorods based enzymatic biosensor for selective determination of penicillin. *Biosensors, 1*(4), 153–163. https://doi.org/10.3390/bios1040153.

35. Ilves, M., Palomäki, J., Vippola, M., Lehto, M., Savolainen, K., Savinko, T., & Alenius, H., (2014). Topically applied ZnO nanoparticles suppress allergen induced skin inflammation but induce vigorous IgE production in the atopic dermatitis mouse model. *Particle and Fiber Toxicology, 11*, 38. https://doi.org/10.1186/s12989-014-0038-4.
36. Iravani, S., (2011). Green synthesis of metal nanoparticles using plants. *Green Chemistry, 13*(10), 2638–2650. https://doi.org/10.1039/c1gc15386b.
37. Ismail, A. A., El-Midany, A., Abdel-Aal, E. A., & El-Shall, H., (2005). Application of statistical design to optimize the preparation of ZnO nanoparticles via hydrothermal technique. *Materials Letters, 59*(14–15), 1924–1928. https://doi.org/10.1016/j.matlet.2005.02.027.
38. Jacob, S. J. P., Finub, J. S., & Narayanan, A., (2012). Synthesis of silver nanoparticles using *Piper longum* leaf extracts and its cytotoxic activity against Hep-2 cell line. *Colloids and Surfaces. B, Biointerfaces, 91*, 212–214. https://doi.org/10.1016/j.colsurfb.2011.11.001.
39. Jagadish, C., & Pearton, S. J., (eds.). (2011). *Zinc Oxide Bulk, thin Films and Nanostructures: Processing, Properties, and Applications* (1st edn., p. 600). Elsevier.
40. Janotti, A., & Van De, W. C. G., (2009). Fundamentals of zinc oxide as a semiconductor. *Reports on Progress in Physics, 72*(12). https://doi.org/10.1088/0034-4885/72/12/126501, PubMed: 126501.
41. Jiang, J., Li, Y., Liu, J., Huang, X., Yuan, C., & Lou, X. W., (2012). Recent advances in metal oxide-based electrode architecture design for electrochemical energy storage. *Advanced Materials, 24*(38), 5166–5180. https://doi.org/10.1002/adma.201202146.
42. Jiang, Y., Zhang, L., Wen, D., & Ding, Y., (2016). Role of physical and chemical interactions in the antibacterial behavior of ZnO nanoparticles against *E. coli. Materials Science and Engineering. C, Materials for Biological Applications, 69*, 1361–1366. https://doi.org/10.1016/j.msec.2016.08.044.
43. Jiang, Y. C., Jiang, L., Huang, A., Wang, X. F., Li, Q., & Turng, L. S., (2017). Electrospun polycaprolactone/gelatin composites with enhanced cell-matrix interactions as blood vessel endothelial layer scaffolds. *Materials Science and Engineering. C, Materials for Biological Applications, 71*, 901–908. https://doi.org/10.1016/j.msec.2016.10.083.
44. Jurakić, T. R., & Marinović, B., (2016). The role of impaired epidermal barrier function in atopic dermatitis. *Acta Dermatovenerologica Croatica, ADC, 24*(2), 95–109.
45. Khan, Z. U. H., Sadiq, H. M., Shah, N. S., Khan, A. U., Muhammad, N., Hassan, S. U., Tahir, K., et al., (2019). Greener synthesis of zinc oxide nanoparticles using *Trianthema portulacastrum* extract and evaluation of its photocatalytic and biological applications. *Journal of Photochemistry and Photobiology. B, Biology, 192*, 147–157. https://doi.org/10.1016/j.jphotobiol.2019.01.013.
46. Kitture, R., Chordiya, K., Gaware, S., Ghosh, S., More, P. A., Kulkarni, P., Chopade, B. A., & Kale, S. N., (2015). ZnO nanoparticles-red sandalwood conjugate: A promising anti-diabetic agent. *Journal of Nanoscience and Nanotechnology, 15*(6), 4046–4051. https://doi.org/10.1166/jnn.2015.10323.
47. Klingshirn, C., (2007). ZnO: From basics towards applications. *Physica Status Solidi (B), 244*(9), 3027–3073. https://doi.org/10.1002/pssb.200743072.
48. Kogan, S., Sood, A., & Garnick, M. S., (2017). Zinc and wound healing: A review of zinc physiology and clinical applications. *Wounds: A Compendium of Clinical Research and Practice, 29*(4), 102–106.

49. Kołodziejczak-Radzimska, A., Jesionowski, T., & Krysztafkiewicz, A., (2010). Obtaining zinc oxide from aqueous solutions of KOH and $Zn(CH_3COO)_2$. *Physicochemical Problems of Mineral Processing, 44*, 93–102.
50. Kukreja, L. M., Barik, S., & Misra, P., (2004). Variable band gap ZnO nanostructures grown by pulsed laser deposition. *Journal of Crystal Growth, 268*(3, 4), 531–535. https://doi.org/10.1016/j.jcrysgro.2004.04.086.
51. Kumar, R., Kumar, M. S., & Mohadevan, N., (2012). Multiple emulsions: A review. *International Journal of Recent Advances in Pharmaceutical Research, 2*, 9–19.
52. Kwon, S. S., Hong, W. K., Jo, G., Maeng, J., Kim, T., Song, S., & Lee, T., (2008). Piezoelectric effect on the electronic transport characteristics of ZnO nanowire field-effect transistors on bent flexible substrates. *Advanced Materials, 20*(23), 4557–4562. https://doi.org/10.1002/adma.200800691.
53. Lam, S. K., & Ng, T. B., (2011). First report of an anti-tumor, anti-fungal, anti-yeast and antibacterial hemolysin from *Albizia lebbeck* seeds. *Phytomedicine, 18*(7), 601–608. https://doi.org/10.1016/j.phymed.2010.08.009.
54. Lansdown, A. B. G., Mirastschijski, U., Stubbs, N., Scanlon, E., & Ågren, M. S., (2007). Zinc in wound healing: Theoretical, experimental, and clinical aspects. *Wound Repair and Regeneration, 15*(1), 2–16. https://doi.org/10.1111/j.1524-475X.2006.00179.x.
55. Laribi, B., Kouki, K., M'Hamdi, M., & Bettaieb, T., (2015). Coriander (*Coriandrum sativum L.*) and its bioactive constituents. *Fitoterapia, 103*, 9–26. https://doi.org/10.1016/j.fitote.2015.03.012.
56. Li, Y., Li, J., Li, Y., Wang, X. X., & Cao, A. C., (2013). Antimicrobial constituents of the leaves of mikania micrantha H.B.K. *PLOS One, 8*(10), e76725. https://doi.org/10.1371/journal.pone.0076725.
57. Lissan, K. J., (1974). *Emulsion and Emulsion Technology, Part I:Surfactant Science Series* (Vol. 6, p. 440). Marecel Dekker, Inc.
58. Liu, X., Hu, Q., Wu, Q., Zhang, W., Fang, Z., & Xie, Q., (2009). Aligned ZnO nanorods: A useful film to fabricate amperometric glucose biosensor. *Colloids and Surfaces. B, Biointerfaces, 74*(1), 154–158. https://doi.org/10.1016/j.colsurfb.2009.07.011.
59. Lowry, J. B., (1992). Cattle in tropical woodlands obtain late dry season feed supplementation from the native tree *Planchonia careya. Tropical Grasslands, 26*, 243–247.
60. Ludi, B., & Niederberger, M., (2013). Zinc oxide nanoparticles: Chemical mechanisms and classical and non-classical crystallization. *Dalton Transactions, 42*(35), 12554–12568. https://doi.org/10.1039/c3dt50610j.
61. Mahmud, S., Johar, A. M., Putrus, G. A., Chong, J., & Karim, M. A., (2006). Nanostructure of ZnO fabricated via French process and its correlation to electrical properties of semiconducting varistors. *Synthesis and Reactivity in Inorganic, Metal-Organic, and Nano-Metal Chemistry, 36*(2), 155–159. https://doi.org/10.1080/15533170500524462.
62. Malikmammadov, E., Tanir, T. E., Kiziltay, A., Hasirci, V., & Hasirci, N., (2018). PCL and PCL-based materials in biomedical applications. *Journal of Biomaterials Science. Polymer Edition, 29*(7–9), 863–893. https://doi.org/10.1080/09205063.2017.1394711.
63. Mason, C., Vivekanandhan, S., Misra, M., & Mohanty, A. K., (2012). Switchgrass (<i>Panicum virgatum</i>) extract mediated green synthesis of silver nanoparticles. *World Journal of Nano Science and Engineering, 2*(2), 47–52. https://doi.org/10.4236/wjnse.2012.22008.

64. Huang, M. H., Mao, S., Feick, H., Yan, H., Wu, Y., Kind, H., Weber, E., et al., (2001). Room-temperature ultraviolet nanowire nanolasers. *Science, 292*(5523), 1897–1899. https://doi.org/10.1126/science.1060367.
65. Morkoc, H., & Ozgur, M., (eds.). (2009). *Zinc Oxide: Fundamentals, Materials and Device Technology* (1st edn., p. 477). Wiley-VCH Verlag GmbH & Co. KGaA.
66. Nava, O. J., Luque, P. A., Gómez-Gutiérrez, C. M., Vilchis-Nestor, A. R., Castro-Beltrán, A., Mota-González, M. L., & Olivas, A., (2017). Influence of *Camellia sinensis* extract on zinc oxide nanoparticle green synthesis. *Journal of Molecular Structure, 1134*, 121–125. https://doi.org/10.1016/j.molstruc.2016.12.069.
67. Noté, O. P., Jihu, D., Antheaume, C., Zeniou, M., Pegnyemb, D. E., Guillaume, D., Chneiwess, H., Kilhoffer, M. C., & Lobstein, A., (2015). Triterpenoid saponins from *Albizia lebbeck* (L.) benth and their inhibitory effect on the survival of high grade human brain tumor cells. *Carbohydrate Research, 404*, 26–33. https://doi.org/10.1016/j.carres.2014.12.004.
68. Ozgür, U., Alivov, Y. I., Liu, C., Teke, A., Reshchikov, M. A., Dogan, S., Avrutin, V., Cho, S. J., & Morkoç, H., (2005). A comprehensive review of ZnO materials and devices. *Journal of Applied Physics, 103*. PubMed: 041301.
69. Özgür, Ü., Hofstetter, D., & Morkoç, H., (2010). ZnO devices and applications: A review of current status and future prospects. *Proceedings of the IEEE, 98*(7), 1255–1268. https://doi.org/10.1109/JPROC.2010.2044550.
70. Padmanabhan, J., & Kyriakides, T. R., (2015). Nanomaterials, inflammation, and tissue engineering. *Wiley Interdisciplinary Reviews: Nanomedicine and Nanobiotechnology, 7*(3), 355–370. https://doi.org/10.1002/wnan.1320.
71. Park, W. I., Kim, J. S., Yi, G. C., Bae, M. H., & Lee, H. J., (2004). Fabrication and electrical characteristics of high-performance ZnO nanorod field-effect transistors. *Applied Physics Letters, 85*(21), 5052–5054. https://doi.org/10.1063/1.1821648.
72. Patolsky, F., Zheng, G., & Lieber, C. M., (2006). Nanowire sensors for medicine and the life sciences. *Nanomedicine, 1*(1), 51–65. https://doi.org/10.2217/17435889.1.1.51.
73. Paul, B., Vadivel, S., Dhar, S. S., Debbarma, S., & Kumaravel, M., (2017). One-pot green synthesis of zinc oxide nano rice and its application as sonocatalyst for degradation of organic dye and synthesis of 2-benzimidazole derivatives. *Journal of Physics and Chemistry of Solids, 104*, 152–159. https://doi.org/10.1016/j.jpcs.2017.01.007.
74. Paul, B. K., & Moulik, S. P., (2001). Uses and applications of microemulsions. *Current Science, 80*(8), 990–1001.
75. Pyskło, L., Niciński, K., Piaskiewicz, M., Bereza, M., & Łojkowski, W., (2007). Synthesis of zinc oxide with nanometric particle size, its characteristics and influence on the properties of rubber compounds. *Elastomer, 11*, 10–19.
76. Raj, L. F. A., & E, J., (2015). Biosynthesis and characterization of zinc oxide nanoparticles using root extract of *Zingiber officinale*. *Oriental Journal of Chemistry, 31*(1), 51–56. https://doi.org/10.13005/ojc/310105.
77. Rajabi, H. R., Naghiha, R., Kheirizadeh, M., Sadatfaraji, H., Mirzaei, A., & Alvand, Z. M., (2017). Microwave-assisted extraction as an efficient approach for biosynthesis of zinc oxide nanoparticles: Synthesis, characterization, and biological properties. *Materials Science and Engineering, 78*, 1109–1118. https://doi.org/10.1016/j.msec.2017.03.090.
78. Ranasinghe, P., Pigera, S., Galappatthy, P., Katulanda, P., & Constantine, G. R., (2015). Zinc and diabetes mellitus: Understanding molecular mechanisms and clinical implications. *Daru, 23*, 44. https://doi.org/10.1186/s40199-015-0127-4.

79. Rauwel, P., Salumaa, M., Aasna, A., Galeckas, A., & Rauwel, E., (2016). A review of the synthesis and photoluminescence properties of hybrid ZnO and carbon nanomaterials. *Journal of Nanomaterials, 2016*, 1–12. https://doi.org/10.1155/2016/5320625.
80. Reddy, K. M., Feris, K., Bell, J., Wingett, D. G., Hanley, C., & Punnoose, A., (2007). Selective toxicity of zinc oxide nanoparticles to prokaryotic and eukaryotic systems. *Applied Physics Letters, 90*(213902), 2139021–2139023. https://doi.org/10.1063/1.2742324.
81. Reed, S. M., & Hutchison, J. E., (2000). Green chemistry in the organic teaching laboratory: An environmentally benign synthesis of adipic acid. *Journal of Chemical Education, 77*(12), 1627–1628. https://doi.org/10.1021/ed077p1627.
82. Ristić, M., Musić, S., Ivanda, M., & Popović, S., (2005). Sol-gel synthesis and characterization of nanocrystalline ZnO powders. *Journal of Alloys and Compounds, 397*(1, 2), L1–L4. https://doi.org/10.1016/j.jallcom.2005.01.045.
83. Rufatto, L. C., Gower, A., Schwambach, J., & Moura, S., (2012). Genus mikania: Chemical composition and phytotherapeutical activity. *Revista Brasileira De Farmacognosia, 22*(6), 1384–1403. https://doi.org/10.1590/S0102–695X2012005000099.
84. Sánchez-Vioque, R., Rodríguez-Conde, M. F., Reina-Ureña, J. V., Escolano-Tercero, M. A., Herraiz-Peñalver, D., & Santana-Méridas, O., (2012). *In vitro* antioxidant and metal chelating properties of corm, tepal and leaf from saffron (*Crocus sativus L.*). *Industrial Crops and Products, 39*, 149–153. https://doi.org/10.1016/j.indcrop.2012.02.028.
85. Sánchez-Vioque, R., Santana-Méridas, O., Polissiou, M., Vioque, J., Astraka, K., Alaiz, M., Herraiz-Peñalver, D., et al., (2016). Polyphenol composition and *in vitro* antiproliferative effect of corm, tepal and leaf from *Crocus sativus* L. on human colon adenocarcinoma cells (Caco-2). *Journal of Functional Foods, 24*, 18–25. https://doi.org/10.1016/j.jff.2016.03.032.
86. Shanthi, P., & Kothai, S., (2015). Synthesis and characterization of chitosan with incorporated herb—A novel bionano composite. *International Journal of ChemTech Research, 8*(8), 208–214.
87. Shi, L. E., Li, Z. H., Zheng, W., Zhao, Y. F., Jin, Y. F., & Tang, Z. X., (2014). Synthesis, antibacterial activity, antibacterial mechanism and food applications of ZnO nanoparticles: A review. *Food Additives and Contaminants. Part A, Chemistry, Analysis, Control, Exposure and Risk Assessment, 31*(2), 173–186. https://doi.org/10.1080/19440049.2013.865147.
88. Singh, R., Shushni, M. A. M., & Belkheir, A., (2015). Antibacterial and antioxidant activities of *Mentha piperita* L. *Arabian Journal of Chemistry, 8*(3), 322–328. https://doi.org/10.1016/j.arabjc.2011.01.019.
89. Singh, S., Thiyagarajan, P., Mohan, K. K., Anita, D., Thirupathiah, S., Rama, N., Tiwari, B., et al., (2007). Structure, microstructure and physical properties of ZnO based materials in various forms: Bulk, thin film and nano. *Journal of Physics. Part D, 40*(20), 6312–6327. https://doi.org/10.1088/0022-3727/40/20/S15.
90. Singhai, M., Chhabra, V., Kang, P., & Shah, D. O., (1997). Synthesis of ZnO nanoparticles for varistor application using Zn-substituted aerosol OT microemulsion. *Materials Research Bulletin, 32*(2), 239–247. https://doi.org/10.1016/S0025-5408(96)00175-4.
91. Singhal, P., Rani, M., Puri, S. S., Dhot, P. S., & Sehgal, R., (2018). The association of haemoglobin and ferritin concentration in newborn and cord blood with maternal hemoglobin and ferritin concentration in three trimesters. *Journal of Evolution of Medical and Dental Sciences, 7*(19), 2375–2379. https://doi.org/10.14260/jemds/2018/535.

92. Sone, B. T., Manikandan, E., Gurib-Fakim, A., & Maaza, M., (2015). Sm_2O_3 nanoparticles green synthesis via *Callistemon viminalis'* extract. *Journal of Alloys and Compounds, 650*, 357–362. https://doi.org/10.1016/j.jallcom.2015.07.272.
93. Stanković, A., Veselinović, L. J., Škapin, S. D., Marković, S., & Uskoković, D., (2011). Controlled mechanochemically assisted synthesis of ZnO nanopowders in the presence of oxalic acid. *Journal of Materials Science, 46*(11), 3716–3724. https://doi.org/10.1007/s10853-011-5273-6.
94. Theerthagiri, J., Thiagarajan, K., Senthilkumar, B., Khan, Z., Senthil, R. A., Arunachalam, P., Madhavan, J., & Ashokkumar, M., (2017). Synthesis of hierarchical cobalt phosphate nanoflakes and their enhanced electrochemical performances for supercapacitor applications. *Chemistry Select, 2*(1), 201–210. https://doi.org/10.1002/slct.201601628.
95. Tong, Y., Liu, Y., Dong, L., Zhao, D., Zhang, J., Lu, Y., Shen, D., & Fan, X., (2006). Growth of ZnO nanostructures with different morphologies by using hydrothermal technique. *Journal of Physical Chemistry. B, 110*(41), 20263–20267. https://doi.org/10.1021/jp063312i.
96. Tsai, M. L., Wu, C. T., Lin, T. F., Lin, W. C., Huang, Y. C., & Yang, C. H., (2013). Chemical composition and biological properties of essential oils of two mint species. *Tropical Journal of Pharmaceutical Research, 12*(4), 577–582. https://doi.org/10.4314/tjpr.v12i4.20.
97. Tundo, P., & Anastas, P., (eds.). (2004). *Green chemistry: Challenging perspectives* (1st edn., p. 282). Oxford University Press.
98. Umrani, R. D., & Paknikar, K. M., (2014). Zinc oxide nanoparticles show antidiabetic activity in streptozotocin-induced type 1 and 2 diabetic rats. *Nanomedicine, 9*(1), 89–104. https://doi.org/10.2217/nnm.12.205.
99. Vijayakumar, S., Vaseeharan, B., Malaikozhundan, B., Shobiya, M., & Leaf, L., (2016). Extract mediated green synthesis of ZnO nanoparticles. Characterization and biomedical applications. *Biomedicine & Pharmacotherapy Journal, 84*, 1213–1222.
100. Vorobyova, S. A., Lesnikovich, A. I., & Mushinskii, V. V., (2004). Interphase synthesis and characterization of zinc oxide. *Materials Letters, 58*(6), 863–866. https://doi.org/10.1016/j.matlet.2003.08.008.
101. Wang, X., Zhou, J., Song, J., Liu, J., Xu, N., & Wang, Z. L., (2006). Piezoelectric field effect transistor and nanoforce sensor based on a single ZnO nanowire. *Nano Letters, 6*(12), 2768–2772. https://doi.org/10.1021/nl061802g.
102. Wang, Z. L., (2004). Zinc oxide nanostructures: Growth, properties and applications. *Journal of Physics: Condensed Matter, 16*(25), R829–R858. https://doi.org/10.1088/0953-8984/16/25/R01.
103. Wang, Z. L., (2008). Splendid one-dimensional nanostructures of zinc oxide: A new nanomaterial family for nanotechnology. *ACS Nano, 2*(10), 1987–1992. https://doi.org/10.1021/nn800631r.
104. Wen, B., Huang, Y., & Boland, J. J., (2008). Controllable growth of ZnO nanostructures by a simple solvothermal process. *Journal of Physical Chemistry C, 112*(1), 106–111. https://doi.org/10.1021/jp076789i.
105. Willander, M., Nur, O., Zhao, Q. X., Yang, L. L., Lorenz, M., Cao, B. Q., Zúñiga, P. J., et al., (2009). Zinc oxide nanorod based photonic devices: Recent progress in growth, light emitting diodes and lasers. *Nanotechnology, 20*(33), (332001). https://doi.org/10.1088/0957-4484/20/33/332001.

106. Yue, S., Yan, Z., Shi, Y., & Ran, G., (2013). Synthesis of zinc oxide nanotubes within ultrathin anodic aluminum oxide membrane by sol-gel method. *Materials Letters, 98*, 246–249. https://doi.org/10.1016/j.matlet.2013.02.037.
107. Yue, S. B., Trujillo, R. D., Tang, Y., O'Gorman, W. E., & Chen, C. Z., (2011). Loop nucleotides control primary and mature miRNA function in target recognition and repression. *RNA Biology, 8*(6), 1115–1123. https://doi.org/10.4161/rna.8.6.17626.
108. Zhang, Z. Y., & Xiong, H. M., (2015). Photoluminescent ZnO nanoparticles and their biological applications. *Materials, 8*(6), 3101–3127. https://doi.org/10.3390/ma8063101.
109. Zhao, X., Zheng, B., Li, C., & Gu, H., (1998). Acetate-derived ZnO ultrafine particles synthesized by spray pyrolysis. *Powder Technology, 100*(1), 20–23. https://doi.org/10.1016/S0032-5910(98)00047-3.

CHAPTER 3

GRAPHENE-BASED NANOMATERIAL CONJUGATES: IMPORTANCE, CLASSIFICATION, AND APPLICATIONS

VIKAS B. KABBURI and MANISHA BAL

ABSTRACT

Biomaterial has emerged as an exciting area that brings attention to more scientific research and its enormous prospect in many applications stimulating the broader interest in biomaterial technology. Bionanomaterials can be integrated into graphene 2-D structures to get exceptional properties in composite form. In the present scenario, the advancement of nanomaterials has excellent potential in different applications such as biosensors, electrochemical devices, rechargeable batteries, etc. Due to its large surface area, high mechanical strength, excellent electron transfer rate, outstanding carrier mobility, and capacitance. In this chapter, graphene, along with its structure and properties, is discussed in detail. Also, this chapter presents an overview of the advanced synthesis process of Graphene-Nanoparticles composites and the issues faced so far. The importance of Graphene-Nanomaterial composite in the medical field and other applications are also highlighted here. Moreover, significant challenges and future prospects are discussed.

3.1 INTRODUCTION

In the last 40–50 years, rapid industrialization in society has led to the development of many new technologies and processes that make our lives significantly more accessible and advanced today. Among these advancements of

Sustainable Nanomaterials for Biomedical Engineering: Impacts, Challenges, and Future Prospects.
Junaid Ahmad Malik, Megh R. Goyal & Mohamed Jaffer M. Sadiq (Eds.)

technologies, nanotech is the major one. Nanotechnology first appeared in the 1980s when the invention of the scanning tunneling microscope (1981), the discovery of fullerenes (1985), and other great works. However, implementation of nanotechnology to the commercial applications leisurely started later in the early '20s [9]. These nanomaterials can be engineered with other materials to form a composite that improves and enhances nanomaterial properties; these can implement composites in various applications such as biosensing, DNA origami [12], and many more. Now, the materials attached to the nanomaterials can be of many types depending upon the application. Among all the materials, graphene turns out to be the best one due to its exceptional and admirable characteristics such as high electrical conductivity, high heat conduction, high surface area, antibacterial effect, etc.

More than 1 lakh papers have been published regarding graphene and its applications in various fields in the last three decades. These papers show how graphene can be an excellent alternate for the other components in various streams [33]. Even implementation of graphene in various devices started deliberately in the early '20s [37]. Graphene possesses admirable properties, which makes it an extraordinary component. To take advantage of this component, a gold rush has been started among scientists [19]. Theses Graphene's remarkable properties have led to a flip-flop in the material science stream [5]. Recent studies showed that graphene-based composites have a very high demand and potential in advanced technologies and processes. Graphene-nanomaterials composite has limitless applications, which makes it a wonder material. The recent technologies are mainly looking for graphene composites and additive manufacturing. Due to its unconventional properties, many automotive companies are trying hard to bring graphene-based composites into their production line.

In this chapter, various available synthesis techniques of both Graphene and graphene composites and their respective properties and applications have been discussed. This chapter also elaborates on various applications of graphene composites such as photocatalysis, enhanced catalysis, water purification, and other energy storage practices. Challenges and future prospective of graphene material are also highlighted here.

3.2 BIONANOMATERIALS

Bionanomaterials are the ones that are built of biological components such as RNA, DNA, and proteins, and have dimensions in nanoscale, i.e., anyone dimensions of material stand in the range of 1 to 100 nanometer. These bionanomaterials can be combined or engineered with other particles to obtain extraordinary properties in composite form. For example, bionanomaterials are combined with gold particles to obtain an 'interface element' implemented in sensors to determine the interaction between the test piece and bionanomaterial. Bionanomaterials have a wide range of applications, and these can be used as Adhesives in 'Protein Engineering' to stick materials together, even some of them can be used as Novel fiber, i.e., source of dietary fiber which majorly contains carbohydrates in it [12]. As nanomaterials have good mechanical properties and high porosity, there is a high demand for them in various fields; they are used to improve food production methods, in nanomedicine, in water purification, etc.

3.2.1 CLASSIFICATION OF BIONANOMATERIALS

Bionanomaterials are broadly categorized into three different groups, namely – nanoparticles, nano-clays, and nano-emulsions [21].

3.2.1.1 NANOPARTICLES

As discussed above, nanoparticles are excellent particles whose properties differ from those of macro-sized particles. These are further divided into two classes based on their origin. The first one is Inorganic nanoparticles which are either engineered in the laboratory or obtained incidentally; these include QDs, gold nanoparticles (AuNPs), super-para-magnet nanoparticles. These are majorly utilized for sensing applications and drug delivery. The second one is Organic nanoparticles, which occur naturally in the environment; these include antibodies, liposomes, viruses, etc. This type of nanoparticle finds applications in food industries, improving sensory quality, modifying physical properties, etc. [12].

3.2.1.2 NANOCLAYS

Nano-clays are the nanoparticles of layered mineral silicates with a high aspect ratio, i.e., high width to the height ratio. These are the minerals that are known for their cation exchange capacity. As these are not flammable, they can be used as additives in plastics to provide fire protection.

The other applications are used as modifiers in paints, inks, and greases to enhance performance. Also used in concrete to improve self-compacting and mechanical properties [21].

3.2.1.3 NANOEMULSIONS

Nano-emulsions, also called mini-emulsion, is generally defined as oil in water emulsion with mean droplet diameter within the range of 50 to a few 100 nanometers. This type of nanomaterials finds application in the field of cosmetics, cell culture technology, drug delivery, non-toxic disinfectants, etc. [21].

3.2.2 PROPERTIES

Nanomaterials are the foremost in many advanced technologies; before implementing these materials in the system, it is crucial to understand their properties and lifecycle. For example, some nanomaterials are harmless, some are not, some are stable at high temperatures, and some start decomposing. Therefore, these should be studied as individual cases and put into practice accordingly [12]. Some of the significant properties are:

- Aerodynamic size;
- Solubility;
- Agglomeration;
- Hydrophilicity;
- Exhibit different colors depending on size;
- Lower melting point;
- High reactivity.

3.2.3 APPLICATIONS

Nanotechnology has a high demand in the field of biochemistry, material science, engineering, etc. Also, as there is a pressing medical need, this technique is majorly implemented in the medical field for various pertinence [12] such as:

1. **Biosensor:** Here, the nanomaterials are used to detect an analyte's presence, where nanomaterials themselves are used as sensing

elements that come in contact with the analyte. This interaction is measured, and output is generated by a peripheral component called interface [12].

2. **DNA Origami:** This is the technique of folding DNA to create a three-dimensional structure at the nanoscale. These can be implemented in 'Detection of target molecules' where these scaffolds are used to immobilize the enzymes on the substrate. Also, these are used in the studies of fluorescence [7, 12].
3. **Nanofabrication:** This technique does nothing but coat or positions the particular functional group in a predetermined space. Fabrication is very helpful in protein engineering, where nano-sized virus structures are used to determine the concentration of antigens that are helpful in vaccine development, etc. [12].

3.3 GRAPHENE

Graphene, a wonder material that is considered an essential breakthrough in materials since the plastics revolution more than a century ago, is a 2-dimensional carbon allotrope that is 10^7 times thinner than a hair but 200 times stronger than stainless steel. It is one of the allotropes of carbon found in the earth's layer in a large quantity [3]. Even though it has a mono-atomic layer (i.e., the thickness of 0.345 nanometers), it is the most robust material known. Surprisingly, graphene has the same carbon structure as graphite, which we use in our pencils to write or draw every day. If we consider 0.03 inches of graphite, there are about 3 million graphene layers in it [8].

Novoselov and Geim first discovered the wonder material while examining the efficiency of graphite as a transistor in 2004 at the University of Manchester. For this work, they were awarded the Noble prize in 2010 [8].

Graphene, also known as an exotic material, has various exceptional properties, due to which it has received attention worldwide. It is astonishingly stretchy, i.e., it can stretch about 25% of its original length [8]. Graphene can transfer electricity more precisely and more efficiently than any other material available. These are because graphene has a higher density than copper and better intrinsic mobility than silicon, as its electrons have no resistance when they move throughout the layer. This property makes graphene batteries 10 times more efficient than ordinary ones. One of the exciting features of graphene is that it starts expanding when cooled and starts shrinking when heated, which is precisely the opposite compared to the other materials [8].

Some of the ideas which seem to be impossible may finally get off the ground with the discovery of graphene, such as:

- A space elevator;
- Purification of ocean water;
- Irresistible body armor;
- Recharging gadgets;
- Flexible displays, etc.

3.3.1 STRUCTURE OF GRAPHENE

A defect-free, single-atom layer of graphite layer is known as graphene. It is a carbon allotrope that is made up of a single layer of graphite. Its structure is like an extended network of benzene rings where carbon atoms are positioned in a hexagonal design due to which it looks like a honeycomb (Figure 3.1) [8].

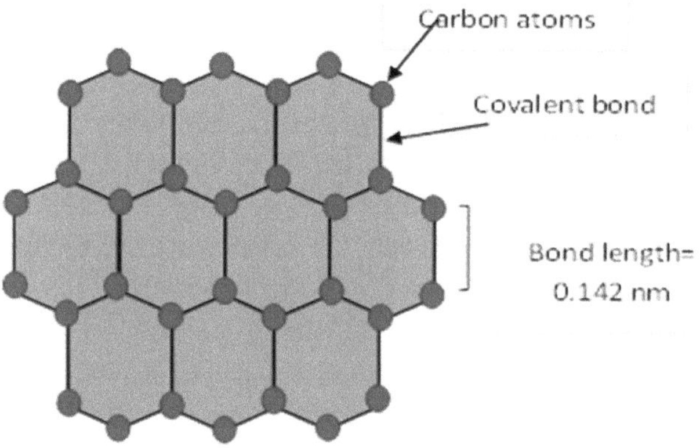

FIGURE 3.1 Structure of graphene.

Graphene is a 2-dimensional nanomaterial that has a thickness of about 0.345 nanometer. As discussed above, graphene comprises only carbon atoms, which are SP^2 hybridized and connected by strong covalent bonds. The bond length is approximately 0.142 nanometer, and the bond-angle is about 120° [8].

Graphene is incredibly stretchy and flat, due to which it can access from both sides. The single carbon atom in the graphene sheet bonds with the other three carbon atoms even though it can bond with four atoms [8]. This ability makes it very suitable to use in composite materials. In addition, the graphene layer has good relation with electricity, is irresistible, flexible, and even transparent. Due to its multi-functionality mentioned above, it can be employed in a wide range of applications.

3.3.2 PROPERTIES OF GRAPHENE

Graphene has impressive versatile properties, which makes it the world's thinnest, lightest, and strongest material. It is an ideal conductor of heat and electricity, due to which it finds application in various fields like physics, material science, electronic information, computers, aerospace, etc. The Discovery of graphene is a great revolution, although numerous materials and minerals have been discovered in human history [8]. These are because of the extreme variety of properties of graphene, which is not seen in any other known material. Some of the significant properties of graphene [8] are:

1. **Simple Structure:** Although graphene is nonmetal as it is obtained from carbon, it acts similarly to metal. Its' straightforward, organized, and tight inter atomic connection system is the main reason behind these. It makes graphene a unique semi-metal.
2. **Strength and Stiffness:** In graphene, the carbon atoms are connected with a powerful bond which is even more potent than the bond between carbon atoms in diamond. These are what make graphene indeed the most robust material known. It has been found that graphene is 200 times stronger than conventional steel.
3. **Elasticity and Flexibility:** Another exciting property of graphene is that it can be stretched to almost 20% to 25% of its original length without any damage. The 2-dimensional and flat structure makes graphene highly flexible as well.
4. **Thin and Light:** Graphene is very light in weight and only one atom thick.
5. **High Thermal Conductivity:** Apart from being incredibly durable and lightweight at the same time, graphene also revealed excellent heat conductance properties. Studies showed that the high thermal conductivity of graphene made this material a superior heat conductor to silver and copper.
6. **High Electrical Conductivity:** Graphene has a free single carbon electron, due to which it acts as a highly high conductor of electrical energy. It allows the free flow of electricity because of its flat structure, which shows minimum resistance to electrons.

7. **Inert Material:** Graphene does not readily react with other atoms due to its inert nature. But it can absorb different atoms and molecules, which in turn accordingly changes the electronic conductance. Due to this property, graphene is very useful in making sensors and biodevices.
8. **High Impermeability:** Graphene is the most impermeable material. If we try, we won't compress even a single atom of helium through it for its hardness. With the help of this property, we can develop effective filters which can even filter out salt from ocean water.

In addition, graphene also comprises properties such as – corrosion resistance, transparent, antibacterial effect, low electricity consumption, and also ability to generate electricity by exposure to sunlight [8].

3.3.3 *ETHICAL ISSUES*

Prediction revealed graphene would almost instantly facilitate the products and technologies that we're used to seeing in sci-fi movies, but that still hasn't happened. These are because of some hold-ups and drawbacks [4] as mentioned below:

- Commercially manufacturing graphene in a large quantity is difficult and expensive;
- graphene loses the property of conductance of electricity when it reacts with oxygen at high temperatures;
- graphene has a suitable conductance property, but it isn't easy to switch it off due to its zero bandgap;
- Large graphene sheets comprise impurities and other alien particles;
- graphene is not stable if its size is less than 20 nanometers.

3.4 WHY GRAPHENE CALLED SMART MATTER?

Smart materials are the ones that have shape memory and the ability to respond to multiple external conditions. Some of the currently used smart materials are metal alloys, piezoelectric ceramics, and polymers [38] which have various advantages but are limited in practical applications due to their low energy efficiency, complex motion control, and other secondary deficiencies. To overcome these limitations, graphene can be used as an alternative.

Graphene has multifunctional properties, which proves the advantages of graphene. The best materials for transport and aerospace applications are preferred to be very strong and light, and graphene has both these attributes. Graphene comprises admirable properties such as high conductivity, high impermeability, and high surface area, due to which it has gained attention throughout the world. In addition to these inherent properties, hybridization, and operationalization can be done to enhance the performance of this material. Also, it holds excellent sensitivity towards optical, electrical, and thermal excitation. This material finds potential applications in various streams such as strain or pathogen sensing, controlled drug delivery, photo-thermal therapy, etc. Therefore, due to this high-performance stimuli-response, it is called a Smart Material [38].

Apart from technical applications, it can also implement this material for making sports kits such as Formula 1 or Skis. It is even being used in Tennis rackets that perform better than conventional rackets. Also, we can foresee graphene being used in motor vehicles, flexible electronics, aerospace, and many applications in medical and biomedical devices. These admirable and extraordinary properties of graphene, which were examined in the laboratory, makes it a promising material for our future [26].

3.5 SYNTHESIS OF GRAPHENE

As discussed in the above section, graphene has a wide range of applications in various fields. Thus, mass production of graphene is necessary [1]. Further, graphene can be processed into various novel materials with customed morphological characteristics [6].

One of the significant holdups of graphene is that it is costly to manufacture; it cost a whopping $1,100 to synthesis enough graphene to cover the head of a pin. However, in the year 2015, the price of 0.35oz graphene was reduced to $1,000. This is possible because of using the optimized production methods, and specialists expect the price of this substance to reduce even more. Surprisingly, it is observed that cost becomes significantly less than its raw material graphite cost. To date, we don't have any manufacturing technique that would yield just single-layer structures reproducibly. However, some processes are used to produce different graphene structures. Some of the significant techniques are discussed in subsections.

3.5.1 MECHANICAL EXFOLIATION

The graphene was first produced with the help of a mechanical exfoliation technique by Novoselov and Geim in the year 2004 while investigating how efficient the graphite is as a transistor. They stuck a sticky tape to a piece of graphite and pulled free a single layer of the material. These left them with the monolayer of the so-called graphite graphene [30].

Exfoliation is a simple method where the layers of graphite are peeled off to form graphene. Here the Highly Oriented Pyrolytic Graphite (HOPG) is taken as raw material and etched into small buttes. These trim pieces are substituted onto a photoresistor and using the scotch tape, the layers of graphite are peeled off. This procedure is repeated until the thin film of graphite is obtained on the resistor. The left-out thin film is nothing but graphite or graphene monolayer, which is washed with acetone and transferred to the silicon substrate or wafer. Using this technique single layer or a small number of layers of graphite can be extracted, which are suitable for nanofabrication of electronic devices. But this technique fails to scale up the production. The major problem faced in this technique is the formation of irreversible clusters or agglomerates, i.e., the samples appear randomly distributed with uneven shapes and sizes. In some cases, the layers restack to form graphite. As most of the salient features of graphene are observed in monolayers, it is crucial to prevent the formation of clusters [30].

An alternate to this technique can be the chemical oxidation of graphite with the subsequent peeling process. Using this method, bulk-quantity of graphene can be produced with ease. However, the disadvantage of this method is the high cost of chemicals, and structural defects are observed due to the invasive use of chemicals. Therefore, physical exfoliation is much desirable to form a defect-free monolayer of graphite [30]. But this is not an industrial way to make graphene; hence chemistry came along to help with the fabrication process.

3.5.2 CHEMICAL VAPOR DEPOSITION (CVD)

Chemical vapor deposition (CVD) is a promising method that helps the mass production defect-free, high-quality graphene structures [30]. CVD is nothing but the pyrolysis of hydrocarbon used to make HOPG, but here the goal is to obtain layers as thin as possible. CVD is of many types, namely:

- Plasma enhanced CVD;

- Low-pressure CVD;
- Atmospheric CVD.

Although discovered the CVD process earlier, the graphene layers were first synthesized in 2006 by Somani and Co. They diffused the carbon atoms onto the nickel wafer using camphor as the precursor, and then the graphene layer was detached from the substrate with the help of chemical etching. This method is most desirable than any other method available. Issues observed in other techniques such as the unmanageable thickness of the sheet, number of layers, and folding of graphene can be controlled using this method [30].

The CVD method is a procedure that achieves the growth of thin films of materials. The typical CVD process for the production of graphene starts with a copper wafer. It is first placed in a CVD chamber [30] immersed in the dilute hydrocarbon gas (usually methane) and heated to high temperatures. If we observe at the molecular level, when the hydrocarbon molecules strike or hit the copper surface, the carbon atoms get trapped while the remaining hydrogen atoms keep moving around. These effectively create a single layer or few layers of carbon atoms that can be several centimeters long. This layer of graphene can be further transported to the other substrates using an organic polymer. Here the graphene attached to copper is firstly covered with organic polymer using the spin-coating technique. Then the wafer is suspended on acid that dissolves the copper and does nothing to the graphene layer until all the copper has been removed. Finally, the graphene, substituted on the polymer, is washed with water to remove any leftovers or residues from the acid. When the graphene is ready to be used in a device or an experiment, the polymer is removed using acetone and other solvents. Therefore, we obtain a two-dimensional, defect-free graphene material [30].

3.6 GRAPHENE-NANOMATERIALS COMPOSITE

The multi-functional properties of graphene composite open up an entirely new path that we've not yet thought of. Composite materials are nothing but a mixture of 2 materials. One is called reinforcement, and another is called matrix, where the reinforcement usually is the more complex or the more substantial part that holds the mixture together. The most crucial aspect of graphene for composite materials is its functional properties such as sensing, thermal, superconductivity, etc. The recent technologies are mainly looking for graphene composites and additive manufacturing. Due to its admirable properties, many automotive companies are trying hard to bring graphene-based composites into their production line. Therefore, graphene-based composites and additive manufacturing assisted by quantum technology, artificial intelligence, and machine learning would play the most crucial role in various fields and industries in the future [14].

Graphene has a higher efficiency than any other carbon allotrope such as fullerene and nanotube, but as discussed above, the production of individual graphene sheets is complicated; overcoming this is an issue Graphene Oxide (GO) is used as an alternative. These are prepared from graphite films by Hummer's method or modified by Hummer's method. Colloidal GO is not efficient as a 2-dimensional graphene sheet but comprises the unique properties of graphene. In some cases, the GO is transformed into Reduced Graphene Oxide (RGO) as required using reducing agents. GO and RGO can be widely used in the preparation and processing of graphene-nanomaterial composite. Here graphene gives excellent support for metal nanomaterials because of its unique structural and physicochemical properties [11]. With the help of composites, can implement properties of both nanomaterials and graphene, and the deficiencies observed in individuals can be eliminated. The performance of individual material can also be improved [14, 25].

In addition to the inherent properties of graphene, functionalization or operationalization can be done to impart new extraordinary properties [38]. The nanomaterials can incorporate graphene effortlessly just with the help of linker molecules; in some cases, even molecular linkers are not required. Amino-propyl-tri-ethoxy-silane (APTES) and 1-pyrene butyric acid N-hydroxysuccinimide ester (PBASE) [28] are some of the essential linker molecules used in various devices and experiments. In many recent technologies and processes, various bionanomaterials are decorated on 2-dimensional graphene sheets to develop new frontiers in their respective fields, especially in the medical field. Due to graphene's flat structure and remarkable properties, the nanomaterials can be bounded to it from both sides, making it a brilliant drug carrier. In recent years, graphene has gained more importance in neuroscience and other biomedical applications [22].

Some of the commonly used nanomaterials [14, 22] are:

- Semiconductor nanoparticles that have a small bandgap can be attached to graphene and used as photocatalysts.
- Biomolecules attached to the graphene layer can easily absorb or emit light in visible regions, hence can be used in sensing and electrochemical applications.

- Also, the bionanomaterials such as antibodies, antigens, and other proteins linked to graphene can be used to detect and sense pathogens, chemicals, and other strains.

3.7 SYNTHESIS OF GRAPHENE-NANOPARTICLES COMPOSITE

Nowadays, scientific research on macro-sized materials has reached a saturation stage; as a result, researchers are now looking to develop nanomaterials [21]. Several approaches have been developed in the last few years to effectively synthesize the graphene-nanomaterials composites (Figure 3.2). Each method have their advantages, along with some disadvantages, such as uneven deposition of materials on the graphene layer, complex mechanism, etc.

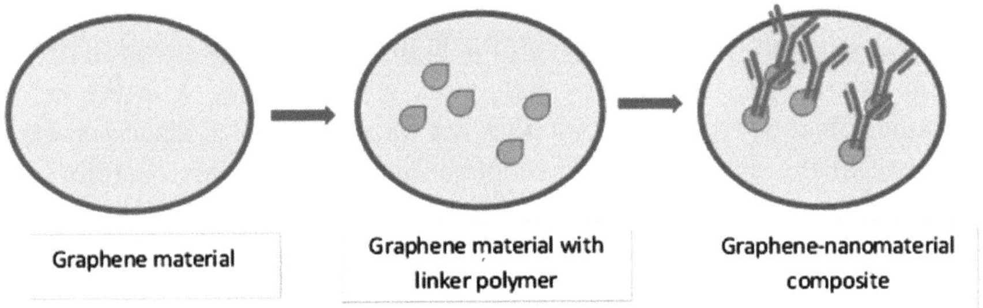

FIGURE 3.2 Synthesis of graphene nano composite.

These approaches mainly involve three methodologies: Pre-graphenization, Post-graphenization, and Syn-graphitization [30]. 'N' number of composites were developed using these strategies where the nanoparticles were functionalized onto the graphene, GO, and even on RGO. These strategies are discussed in detail in subsections.

3.7.1 PRE-GRAPHENIZATION

Pre-graphenization is nothing but a simple mixing process. As graphene is highly hydrophobic, it is insoluble in organic solvents; instead, it forms clusters that restack to form graphite structures [14]. Hence GO is considered as an alternative and transformed to RGO. In pre-graphenization, RGO is mixed with nanomaterials to form composites; in addition, RGO can also be mixed with multiple nanomaterials to form complex composites [30].

Before addressing this methodology, one must consider its crucial parameters like the ability of nanomaterial to incorporate onto the surface, RGO's solubility in various solvents, etc. [30].

3.7.2 POST-GRAPHENIZATION

Post-graphenization involves thorough mixing of pre-synthesized GO or RGO with various nanoparticles or salts followed by reduction. Hydrazine monohydrate and Sodium borohydride are some of the commonly used reducing agents in this technique. This method is performed to integrate various metal nanoparticles, metal-oxide nanoparticles, semiconducting nanoparticles, and even metal salts to the GO suspension [30].

The typical post-graphitization method involves functionalizing desired nanoparticles on GO or RGO with the help of physical absorption or electronic interaction, followed by reduction, which is done with the help of organic solvents. Here the RGO does not form solvents as discussed earlier because of the attachment of nanomaterials to RGO film. Therefore, this ability makes methodology more desirable than any other methods available [30].

3.7.3 SYN-GRAPHENIZATION

Syn-graphenization is the third type of methodology which is also called One-pot approach. This method involves thoroughly mixing two or more nanomaterials depending on the application with the graphene sheets. Here the extra nanoparticle added to the graphene acts as a stabilizer which helps in enhancing the performance of composite properties [30]. In some cases, particles are added only to give support and to neutralize the composite structure.

3.8 ISSUES IN GRAPHENE-NANOMATERIAL COMPOSITE SYNTHESIS

Although the graphene-nanomaterials composites are well established, some limitations have to be eliminated [30]. Before addressing the mass production of graphene-nanomaterial composites, it is essential to understand and overcome the holdups. Some of the significant drawbacks [14, 30] are:

- It is tough and expensive to produce individual sheets of graphene. To date, we don't have any manufacturing technique that would yield just single-layer structures reproducibly.
- The functional groups randomly get distributed with uneven shapes and sizes on the graphene structure.
- The complex mechanism of functionalization.
- Development of disregarded property within the composite due to the unintentional interaction between graphene and nanomaterials.
- Due to the invasive use of chemicals, there are chances of formation of structural defects, which may reduce the accuracy of the composite.
- As graphene is highly hydrophobic, it is insoluble in organic solvents instead of forms clusters that stack back to form graphite via Van der Waals force.
- The use of complex linker polymers in between graphene and nanoparticles would affect the accuracy of the composites.

3.9 APPLICATIONS OF GRAPHENE-NANOMATERIAL COMPOSITES

In recent technologies and processes, demand for graphene-nanomaterial composites has been very high. Due to its multifunctionality, it can implement in all the significant areas of science. Graphene or GO or RGO are adhered to composites to extract exceptional properties such as controlled aggregation, high surface area, superior tolerance property, high performance, etc. These are suitable for various applications. Among these applications, some of the emerging areas have been highlighted in subsections [14].

3.9.1 BIOSENSORS

Graphene has exceptional and unique electrochemical and physical characteristics, making it a sensitive material that can implement for sensing various stains, chemicals, and antigens, etc. [18]. Bio-sensing is nothing but to diagnose or detect particular molecules, pathogens, chemicals, or strains qualitatively and quantitatively in the given sample. Due to graphene's excellent electronic conductance and high surface area can be used as an efficient electrode material [22] in biosensors.

Recently this technology was implemented by scientists from the Republic of Korea to diagnose the infectious SARS-CoV-2 (Severe Acute Respiratory Syndrome Corona Virus 2) in the sample. The group of scientists used the electronic conductance property of graphene and Field-Effect Transistor (FET) to develop a biosensor. Nanofabrication techniques were used to make a graphene-based transistor. To deposit graphene onto a silicon dioxide substrate, generally used the wet-transfer method. Orderly patterns of graphene were formed on silicon dioxide substrate using UV lithography. Then film deposition method substitutes the metal electrodes.

Further, the graphene was soaked with a P-BASE solution, which acts as a linker. Then added the spike protein antibody of SARS-CoV-2, which reacted with the linker to form a bond. After the graphene was derivatized with antibodies, they could test their device's sensitivity. Their design used aqueous solution gated FET, a buffered water droplet with an electrode as the transistor's gate. Charge flow through the graphene can be measured as a function of the gate voltage if the voltage is applied across the source electrode and drain electrode. Then a group of scientists tested whether the device would respond to the spike antigen of SARS-CoV-2 (Figure 3.3). As spike antigens have a high affinity towards their antibodies, they are bound to them, due to which the charge distribution throughout the graphene changes, thus changing its conductivity. The amount of current that can flow between the source and drain electrodes can be altered and measured due to a change in conductivity. They took a swab sample from a COVID-19 infected patient for the real-time test and demonstrated that their device could diagnose the infection. The biosensor shows the signal for the presence of the concentration of virus in the swab sample. Higher the signals produced more the concentration of the sample [28].

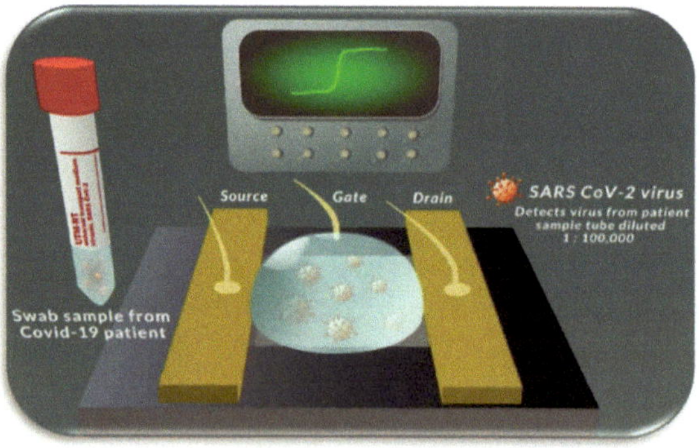

FIGURE 3.3 Biosensor to detect COVID-19 (swab test).

3.9.2 BIOMEDICAL APPLICATIONS

Other than energy conservation and bio-sensing applications, it can also implement in tissue engineering and neurology, which gives rise to the new stream of science known as neuroscience [13].

3.9.2.1 TISSUE ENGINEERING

In a laboratory, tissue engineering refers to strive to make functional human tissues from cells. The graphene or GO can be implemented and its composites as supporting structures due to its strength, endurance, and antibacterial effects. In some cases, the composites can be used as stem cells. These composites can also be used as scaffolds that dissolve over time or remain to provide support. Some examples of tissues that have been successfully tissue-engineered using composites are – mesenchymal stem cells (MSCs) developed on fluorinated graphene or GO substrates [10, 22] and so on.

3.9.2.2 NEUROSCIENCES

Aerodynamic size, chemical stability, and high electricity conductance properties make graphene- and carbon-based nanomaterials useful in neurology. Graphene composites can be implemented to diagnose the damages on brain tissues and nerve cells, and in some cases, also acts as stimulating technology to repair the tissues. As nanomaterials can be bounded on both sides of graphene, it acts as an excellent drug carrier which will be a helpful tool in neurosurgery [22].

3.9.3 EFFICIENT PHOTOCATALYST

Maintaining a clean and pleasant environment is a common aspiration for all of us. Hence, the development of photocatalysis is crucial. In this application, graphene functionalized with semiconductor nanoparticles can decompose organic compounds and hazardous pollutants dispersed in the atmosphere. Graphene-TiO_2 composite has been widely used as a photocatalyst in many environmental applications due to its high stability, low cost, efficiency, and safety measures. According to studies, then bare TiO_2, graphene-TiO_2 composites have higher photoactivity and efficiency [14]. When light energy from the sun is irradiated onto the TiO_2 composite material, real-time applications will develop a solid potentiality to decompose across the surface. This ability kills or inactivates microorganisms, viruses, fungus, and even offensive odors by turning them into water or other harmless substances.

The only limitation associated with this composite is that it activates only under UV light. These are because of its large bandgap. To overcome this issue, the other composites of graphene and nanomaterials such as Ag_3PO_4 and CdS have been developed, which can activate and absorb under visible light [14].

3.9.4 *ENHANCED CATALYST*

As we know, metal, and metal oxide nanoparticles can be widely used as catalysts in many organic reactions such as addition, pericyclic, redox, etc. On the other hand, even graphene with negligible or zero bandgap is also used to enhance different organic conversion reactions. Therefore, graphene-nanoparticles composite can be used as an efficient catalyst which helps in increasing the conversion yield. GO coated or integrated with iron oxide and silver nanoparticles (AgNPs) by post-graphitization method can be used as a catalyst for high yielding click reactions, oxidation reaction, and A^3 coupling reaction, i.e., aldehyde-alkyne-amine reaction. Similarly, composites of graphene and metal salts such as copper salts can enhance the performance of the arylation reaction [14].

1. **Arylation Reaction:**

   $$ROH + R'X \xrightarrow{\text{G-Cu}} = R - O - R'$$

2. **Oxidation:**

   $$R'C_6H_4CH_2OH \xrightarrow{\text{G-y–Fe}_2\text{O}_3} PhCHO + PhCOOH$$

3. **Click Reaction:**

   $$R^2C_6H_4CH_2Br + PhCCH + NaN_3 \xrightarrow{\text{G-y–Fe}_2\text{O}_3} \text{1,4 substituted 1,2,3 triazole}$$

4. **A^3 Coupling:**

   $$RCHO + HNR_2 + PhCCH \xrightarrow{\text{G.Ag}} RCH(NR_2)CCPh$$

In addition, graphene provides a stable reacting medium on which reactants react to form a product effectively. Graphene acts as a support material for the reactants, and due to its high surface area and superior tolerance property, higher catalytic performance can be achieved [14].

3.9.5 ADSORBENT FOR WATER PURIFICATION

Graphene-based composites may be the solution to all the water crises which many nations are facing. If a membrane from graphene material is made, it will let water through while filtering out impurities and other salts simultaneously. This would be a revolution in water purification technology. To date, there are many methods developed to purify the wastewater, such as filtration, sedimentation, distillation, ion exchange, and other biological and chemical processes such as slow sand filtration, chlorination, etc. Also, biologically active carbon and its composites can be used as an effective adsorbent of impurities from water [14].

In this field, graphene composites can be used as adsorbents due to their highly hydrophobic nature and high surface area. Graphene or GO or RGO coated with iron oxide nanoparticles can be used to remove various pollutants from the wastewater. Graphene in the composite exhibits high surface area while iron oxide exhibits magnetic property, which offers high separation efficiency. These composites can extract organic substrates such as 1-naphthol, dibutyl phthalate, bisphenol A, atrazine the graphene surface via weak Van der Waals forces from the contaminated water. Also, hormonally active agents, i.e., Endocrine disruptors, can be removed using graphene-iron oxide composite from wastewater. According to the studies, these composites shows better performance than conventional activated carbon [14].

3.9.6 SOME OF THE OTHER APPLICATIONS

- Apart from the above applications, graphene-based bionanomaterials can also be used in various mass transfer operations, implemented in promising areas such as food processing, disease diagnostics, water purification techniques, etc. [32, 34].
- The graphene's unique chemical and physical properties can be utilized for efficiently storing and accessing the data produced by information technologies. This can also be used to improve the existing storage technologies [2].
- According to the studies, graphene-based bionanomaterials even facilitate the advanced storage system for energy [16]. This is because of its superior electric conductivity, thermal conductivity, and chemical stability [42].
- The 2D structure of graphene helps in designing reconfigurable terahertz (THz) devices [20].

- Most of the papers published based on graphene in the last decade show interest for electrical double-layer capacitor applications [27].

3.10 CHALLENGES AND FUTURE PERSPECTIVES

As discussed above, graphene has excellent characteristics, due to which it is utilized in almost all fields. By integrating this material with different nanomaterials, excellent properties can be extracted, which can be used for various applications such as pathogen sensing, bioimaging, neurology, catalysis, water purification, and even in sports goods. Graphene composites have multifunctional properties, which makes them intelligent and wonder material. Due to these exceptional qualities, graphene-based composites may soon take over the world, hence called Future material [14].

Although many advancements and improvisation are done in this field, some setbacks still have to be fixed and cleared up. Among all the issues discussed in the above section, the mass production of graphene is the major one. Most of the methodology discovered and implemented to produce graphene is suitable for only laboratory scale and is ineffective. Producing individual sheets of graphene reproducibly using these methods is almost impossible. In addition, most of these synthesis methods cannot solve the agglomeration issue, resulting in structural defects. Hence, it should develop simple and effective processes to produce 2-dimensional graphene sheets in upcoming years, which helps scale up the production. It should reduce the invasive use of chemical treatments and linker polymers in graphene-based composites to obtain clean and irresistible composites. New methods that can control the loading of nanomaterials on graphene surfaces should be discovered to avoid unintentional interaction between graphene and nanomaterials. New composites should be formed by functionalizing various nanomaterials to graphene and should study their combined properties. Therefore, further studies and experiments should be conducted to overcome the current issues and open up new opportunities that we had never thought of [14].

In the future, these multifunctional composites play a significant role in the development of technologies. Its unconventional properties could be a backbone for innovations in various fields. We can probably design structures, components, sports cars, flexible electronics, and many more using these technologies. It can be used in medical and biomedical fields to develop nanomedicines, biosensors, nanofabrication, and even protein engineering, which is very useful in vaccine development. Storing energy for the future is crucial nowadays; hence graphene-based batteries can be developed, which will charge quickly and last longer than conventional batteries. We can foresee graphene composites being used in aerospace, automobiles, energy storage, photocatalysis [14].

3.11 SUMMARY

This chapter gives an overview of recently developed technologies and processes developed based on graphene-nanomaterial composites. In this chapter, we have given a brief outline of bionanomaterials and graphene and their respective properties and applications. The various synthesis methods of graphene developed across the world are explained step by step. Further, the graphene-based composites are highlighted with various synthesis methodologies. The graphene-nanomaterial composite's applications such as biosensing, tissue engineering, enhanced catalysis, photocatalysis, and water purification methods are also explained along with their specific merits and demerits. Finally, the expectations of the future from these composite materials are discussed.

ACKNOWLEDGMENT

Our study was supported by Department of Chemical Engineering, MVJ College of Engineering, Whitefield, Bangalore – 560067, Karnataka, India.

KEYWORDS

- **bionanomaterials**
- **biosensing**
- **catalysis**
- **chemical vapor deposition**
- **exfoliation**
- **future material**
- **graphene**
- **graphene-based composites**
- **graphitization**
- **synthesis**

REFERENCES

1. Aïssa, B., Memon, N. K., Ali, A., & Khraisheh, M. K., (2015). Recent progress in the growth and applications of graphene as a smart material: A review. *Frontiers in Materials Journal, 2* (58), 1–19.
2. Bertolazzi, S., Bondavalli, P., Roche, S., San, T., Choi, S. Y., Colombo, L., &Samorì, P., (2019). Nonvolatile memories based on graphene and related 2D materials. *Advanced Materials Journal, 31*(10), 1–81.
3. Bharech, S., & Kumar, R., (2015). A review on the properties and applications of graphene. *Journal of Material Science and Mechanical Engineering, 2*(10), 70–73.
4. Biswas, (2021). *Top 5 Drawbacks of Graphene | How to Overcome it?* (p. 1). https://zedore.com/top-5-drawbacks-of-graphene-how-to-overcome-it/ (accessed on 14 July 2022).
5. Catania, F., Marras, E., Giorcelli, M., Jagdale, P., Lavagna, L., Tagliaferro, A., & Bartoli, M., (2021). A review on recent advancements of graphene and graphene-related materials in biological applications. *Applied Sciences Journal, 11*(1), 1–21.
6. Compton, O. C., & Nguyen, S. T., (2010). Graphene oxide, highly reduced graphene oxide, and graphene: versatile building blocks for carbon-based materials. *Nano-Micro Small Journal, 6*(6), 711–723.
7. Contributors, W., (2021). *DNA Origami* (p. 41). https://en.wikipedia.org/w/index.php?title=DNA_origami&oldid=1012332865 (accessed on 14 July 2022).
8. Contributors, W., (2021). *Graphene*. (p. 45). https://en.wikipedia.org/w/index.php?title=Graphene&oldid=1035721841 (accessed on 14 July 2022).
9. Contributors, W., (2021). *History of Nanotechnology* (p. 48) https://en.wikipedia.org/w/index.php?title=History_of_nanotechnology&oldid=1017438105 (accessed on 14 July 2022).
10. Chen, G.Y., Pang, D. W. P., Hwang, S. M., Tuan, H. Y., &Hu, Y. C., (2012). A graphene-based platform for induced pluripotent stem cells culture and differentiation. *Biomaterials Journal, 33*(2), 418–427.
11. He, K., Zeng, Z., Chen, A., Zeng, G., Xiao, R., Xu, P., & Chen, G., (2018). Advancement of Ag–graphene-based nanocomposites: An overview of synthesis and its applications. *Nano-Micro Small Journal, 14*(32), 1–56.
12. Honek, J. F., (2013). Bionanotechnology and bionanomaterials: John Honek explains the good things that can come in very small packages. *BMC Biochemistry Journal, 14*(29), 1471–2091.
13. Hoseini-Ghahfarokhi, M., Mirkiani, S., Mozaffari, N., Sadatlu, M. A., Ghasemi, A., Abbaspour, S., & Karimi, M., (2020). Applications of graphene and graphene oxide in smart drug/gene delivery: Is the world still flat? *International Journal of Nanomedicine, 15*(1), 9469–9496.
14. Jana, N. R., & Mondal, A., (2014). Graphene-nanoparticle composites and their applications in energy, environmental and biomedical science. *Reviews in Nanoscience and Nanotechnology Journal, 3*(1), 177–192.
15. Jang, H., Park, Y. J., Chen, X., Das, T., Kim, M. S., & Ahn, J. H., (2016). Graphene-based flexible and stretchable electronics. *Advanced Materials Journal, 28* (22), 4184–4202.
16. Ji, L., Meduri, P., Agubra, V., Xiao, X., & Alcoutlabi, M., (2016). Graphene-based nanocomposites for energy storage. *Advanced Energy Materials Journal, 6* (16), 1–73.

17. Kim, J., Cote, L. J., & Huang, J., (2012). Two-dimensional soft material: New faces of graphene oxide. *Accounts of Chemical Research Journal, 45*(8), 1356–1364.
18. Kumar, P. S., Bukkitgar, D. S., Singh, S., Pratibha, Singh, V., Reddy, D. K., & Shetti, P. N., (2019). Electrochemical sensors and biosensors based on graphene functionalized with metal oxide nanostructures for healthcare applications. *Chemistry Europe Journal, 4*(18), 5322–5337.
19. Luo, B., Liu, S., & Zhi, L., (2012). Chemical approaches toward graphene-based nanomaterials and their applications in energy-related areas. *Nano-Micro Small Journal, 8*(5), 630–646.
20. Luo, Y., Zeng, Q., Yan, X., Jiang, T., Yang, R., Wang, J., & Zhang, X., (2019). A graphene-based tunable negative refractive index metamaterial and its application in dynamic beam-tilting terahertz antenna. *Microwave and Optical Technology Letters Journal, 61*(12), 2766–2772.
21. Mageswari, A., Srinivasan, R., Subramanian, P., Ramesh, N., & Gothandam, K. M., (2016). Nanomaterials: Classification, biological synthesis and characterization. *Nanoscience in Food and Agriculture Journal, 3*(1), 31–71.
22. Monaco, A. M., & Giugliano, M., (2016). Graphene-based smart nanomaterials: Novel opportunities for biology and neuroengineering. *Carbon Nanostructures Journal, 7*(1), 191–218.
23. Ovid'ko, I. A., (2013). Mechanical properties of graphene. *Revolution of Advanced Material Science Journal, 34*(1), 1–11.
24. Thabitha, P. D. S., Danielle, M., Asok, K. D., & Paul, B. T., (2018). A Review on graphene-based nanomaterials in biomedical applications and risks in environment and health. *Nano-Micro Letters, 1*(1), 10–53.
25. Palermo, V., Kinloch, I. A., Ligi, S., & Pugno, N. M., (2016). Nanoscale mechanics of graphene and graphene oxide in composites: A scientific and technological perspective. *Advanced Materials Journal, 28*(29), 6232–6238.
26. Rudrapati, R., (2020). Graphene: Fabrication methods, properties, and applications in modern industries. *IntechOpen Journal, 16*(1), 1–14.
27. Salunkhe, R. R., Lee, Y. H., Chang, K. H., Li, J. M., Simon, P., Tang, J., & Yamauchi, P. D., (2014). Nanoarchitectured graphene-based supercapacitors for next-generation energy-storage applications. *Chemistry Europe Journal, 20*(43), 13838–13852.
28. Seo, G., Lee, G., Kim, M. J., Baek, S. H., Cho, M., Ku, K. B., & Kim, S. I., (2020). Rapid detection of COVID-19 causative virus (SARS-CoV-2) in human nasopharyngeal swab specimens using field-effect transistor based biosensor. *ACS Nano Journal, 14*(1), 5135–5142.
29. Silva, A. D., Paschoalino, D. W., Damasceno, D. J., & Kubota, P. T., (2020). Structure, properties, and electrochemical sensing applications of graphene-based materials. *Chemistry Europe Journal, 7*(22), 4508–4525.
30. Singh, V., Joung, D., Zhai, L., Das, S., Khondake, S. I., & Seal, S., (2011). Graphene based materials: Past, present and future. *Progress in Materials Science Journal, 56*(1), 1178–1271.
31. Song, H., Zhang, X., Liu, Y., & Su, Z., (2019). Developing graphene-based nanohybrids for electrochemical sensing. *The Chemical Record, 19*(2, 3), 534–549.
32. Sun, P., Wang, K., & Zhu, H., (2016). Recent developments in graphene-based membranes: Structure, mass-transport mechanism and potential applications. *Advanced Materials Journal, 28* (12), 2287–2310.

33. Tiwari, S. K., Mishra, R. K., Ha, S. K., & Huczko, A., (2018). Evolution of graphene oxide and graphene from imagination to industrialization. *Chemnanomat Journal, 4*(7), 598–620.
34. Tung, T. T., Nine, M. J., Krebsz, M., Pasinszki, T., Coghlan, C. J., Tran, D. N., & Losic, D., (2017). Recent advances in sensing applications of graphene assemblies and their composites. *Advanced Functional Materials Journal, 27*(46), 1–156.
35. Wan, S., Peng, J., Jiang, L., & Cheng, Q., (2016). Bioinspired graphene-based nanocomposites and their application in flexible energy devices. *Advanced Materials Journal, 28*(36), 7862–7898.
36. Yu, W., Wong, C. L., Kiran, K. M., Priscilla, K. A., Jiong, L., Yan, P. L., ChweeTeck, L., & Kian, P. L., (2012). Fluorinated graphene for promoting neuro-induction of stem cells. *Advanced Materials Journal, 24*(31), 4285–4290.
37. You, R., Liu, Y. Q., Hao, Y. L., Han, D. D., Zhang, Y. L., & You, Z., (2020). Laser fabrication of graphene-based flexible electronics. *Advanced Materials Journal, 32*(15), 577–588.
38. Yu, X., Cheng, H., Zhang, M., Zhao, Y., Qu, L., & Shi, G., (2017). Graphene-based smart materials. *Nature Reviews Materials Journal, 2*(1), 1–13.
39. Yuan, Z., Xiao, X., Li, J., Zhao, Z., Yu, D., & Li, Q., (2018). Self-assembled graphene-based architectures and their applications. *Advanced Science Journal, 5*(2), 1–18.
40. Zhang, C., Zhang, Z., Yang, Q., & Chen, W., (2018). Graphene-based electrochemical glucose sensors: Fabrication and sensing properties. *Electroanalysis Journal, 30*(11), 2504–2524.
41. Zhang, Y., Xia, X., Liu, B., Deng, S., Xie, D., Liu, Q., & Tu, J., (2019). Multiscale graphene-based materials for applications in sodium-ion batteries. *Advanced Energy Materials Journal, 9*(8), 1–18.
42. Zhu, J., Yang, D., Yin, Z., Yan, Q., & Zhang, H., (2014). Graphene and graphene-based materials for energy storage applications. *Nanomaterials for Energy and Water Management Journal, 10*(17), 3480–3498.

CHAPTER 4

USAGE OF NANOMATERIALS FOR ORTHOPEDICS, TISSUE, AND 3D CELL CULTURES

VIJAYA GEETHA BOSE, SHREENIDHI KRISHNAMURTHY SUBRAMANIYAN, VIHAA SRIEE MAMBULLIKAAVIL GANESAN, and RASHMINIZA ABDUL JALEEL

ABSTRACT

The healthcare sector is one of the most demanding sectors due to its sensitivity and need. Every second, there is someone who is falling sick or someone who requires immediate medical attention. Though there is the treatment being found for diseases, time, and efficiency always need to be prioritized. The usage of bionanomaterials can overcome these shortcomings. Bionanomaterials consist of molecular substances in the Nanoscale dimension. The molecular substances include biological substances such as antibodies, proteins/enzymes, DNA, RNA, lipids, oligosaccharides, viruses, and cells. The rate of joint replacement surgeries has been tremendously increasing in recent years, leading to a rise in demands for orthopedic implant materials. Since the implantation surgery has the risk of failure and is prone to several other clinical problems, a chief alternative is required. This requirement can be met by bio nanomaterial synthesized based on the study of structure and organization of bones. The same condition applied to the tissues also. Engineering scaffolds achieve the solution to this problem with specific properties. With the application of 3D cell cultures in the clinical field and *in-vitro* studies, the use of bionanomaterials leads to better prediction of biological mechanisms and therapeutic effects. Thus as part of

Sustainable Nanomaterials for Biomedical Engineering: Impacts, Challenges, and Future Prospects.
Junaid Ahmad Malik, Megh R. Goyal & Mohamed Jaffer M. Sadiq (Eds.)

novel challenges in clinical research, the use of bionanomaterials can show an effective strategy in tissue engineering.

4.1 INTRODUCTION

Bio-nanotechnology is a technique used to replicate the structure and function of biological systems. In the same way, tissue-engineered materials have their importance in the clinical field. The application of nanotechnology in tissue engineering has paved the way to make implants more biocompatible. These prevent the complications that occur during orthopedic implantation and the cause of implant failure [1]. The application of nanoscience in tissue engineering helps achieve better compatibility, adhesion, and prevents other infectious factors. The unique properties of nanomaterials make them fit to invent new things with improved regenerative properties. It helps to fabricate the mimics of tissue scaffolds with desirable properties [33]. Stem cells combined with biomaterials are essential for tissue engineering approaches. Implanting cells are either seeded or encapsulated in biomaterials. Orthopedics is a field of medicine that is used to diagnose and treat musculoskeletal disease and trauma. The restoration of the physiological function of diseased tissues is restored in orthopedics using nanomaterials [40]. Nano-textured substrates help in obtaining improved regulation of cell adhesion and vascularization [38].

This chapter briefly explains the use of nanomaterials in the construction of scaffolds and implants in orthopedic tissues and 3D cell cultures – its properties and applications.

4.2 CONCEPT OF BONE REPAIR PROCESS

The bone repair process is the action of healing fractured bones in several stages [10], as shown in Figure 4.1.

- **Stage 1: The Hematoma Formation:** It occurs straight away at the site of injury of fractured bone because of disruption of blood vessels [3]. When the blood vessel is disrupted, hematoma formation occurs, and the healing process is initiated [34].
- **Stage 2: The Soft Callus Formation: This** subsequently involves angiogenesis, the new blood vessel formation from the already existing internal and external callus [3].

- **Stage 3: In the Primary Bone Formation Stage:** Complex callus formation will take place [13]. The internal soft callus gets mineralized by calcium hydroxyapatite, which results in the hard callus formation [3].
- **Stage 4: During Bone Remodeling:** The fractured callus is restored by secondary lamellar bone formation followed by reverting vascular supply to the normal state. These occur at the end, which is called bone remodeling [3].

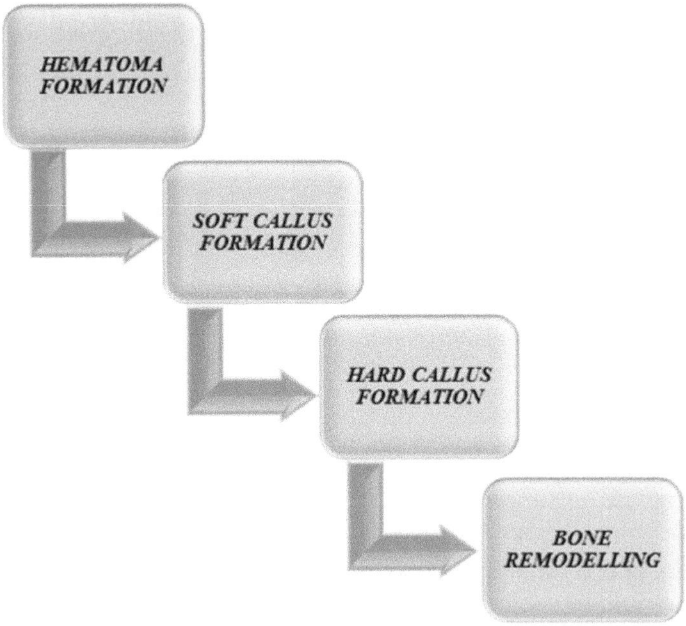

FIGURE 4.1 Schematic diagram of the bone repair process.

4.3 PROPERTIES TO BE CONSIDERED FOR THE BONE HEALING PROCESS

1. **Osteogenity:** It is nothing but the occasion of the generation of bone-forming cells [9].
2. **Osteoinduction:** It is the process of progenitor cell formation, which gives rise to new bone formation [9].
3. **Osteoconduction:** It is the formation of matrix-like support from the bone-forming cells for further bone formation [9].

4.4 THE ORTHOPEDICS IMPLANTS AND THEIR DRAWBACKS

Apart from the natural healing process, some severe trauma necessitates orthopedic implant materials to better heal fractured sites, like femur fracture repair. The materials that are implanted at the site fracture may be either Permanent or Temporary Orthopedic Implants. The implant system may include [1].

Clinical replacements that come under permanent implants, including shoulder, elbow, wrist, hip, knee, and ankle, whereas fixing broken or fractured bones during the healing process are termed temporary implants [16].

The materials that are used for the construction of orthopedic implants are represented in Figure 4.2.

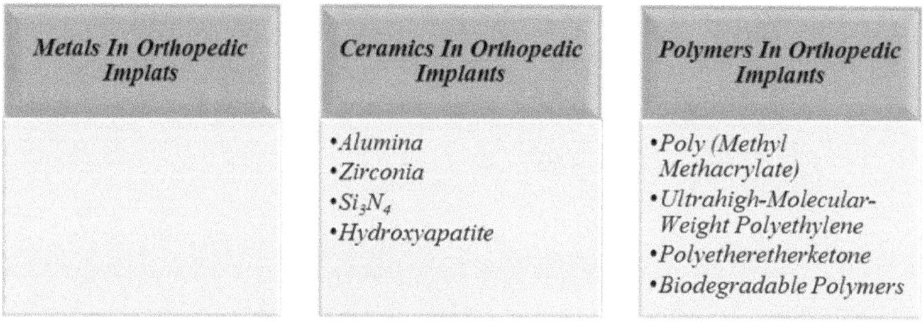

FIGURE 4.2 Orthopedic implant materials.

4.5 BENEFITS OF NANOTECHNOLOGY IN TISSUE ENGINEERING

Tissue engineering is an interdisciplinary field that helps create scaffolds or biomaterials for repairing, creating, and replacing tissue or organs using a combination of cells with biomaterials. Tissue engineering employs nanotechnology to provide the best microenvironment to the cells in combination with nanofabrication techniques to achieve the requirements. The benefits of nanotechnology are listed in Figure 4.3.

4.6 DEMERITS OF CLINICALLY AVAILABLE IMPLANTS

The clinically available implants are highly prone to implant-associated infections due to biofilms formed over the implants [44]. This biofilm formation provides the basic survival for the bacteria to exist and cause the disease [7]. The implant's biocompatibility is determined based on the tissue response to the biomaterial [28] and foreign body reaction against wear particles [36].

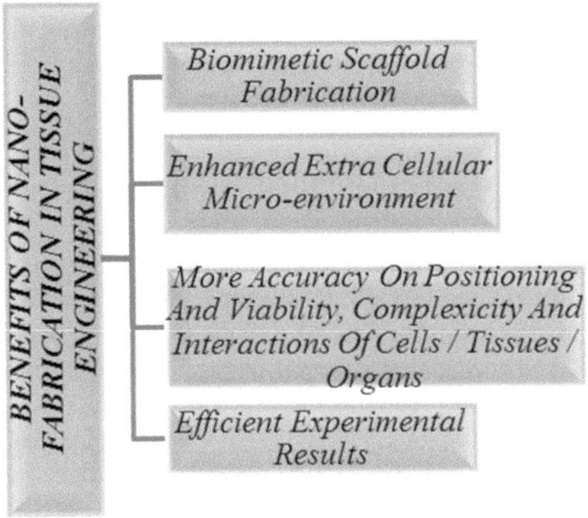

FIGURE 4.3 Benefits of nano-fabrication in tissue engineering.

4.7 NANOTECHNOLOGY ENHANCED ORTHOPEDIC BIOMATERIALS

1. **Metals:** Stainless steel, cobalt-chrome alloys, titanium, and its alloys were used as biomaterials in orthopedics [48].
2. **Ceramics:** Most common ceramics used in orthopedic applications are metallic oxides such as alumina, zirconia, titania, calcium phosphates – hydroxyapatite (HA), tricalcium phosphate (TCP), calcium tetraphosphate ($Ca_4P_2O_9$), glass ceramics – bioglass, and ceravital [18].
3. **Polymers:** These can be easily converted into desired shape and structure with desired physical properties of proteins present in soft and hard tissues. They can also be modified and functionalized via chemical and biochemical reactions [50].

4.7.1 PROPERTIES OF NANOMATERIALS

1. **Physical Properties:** Melting point, elastic constants, diffusivity, surface/magnetic, electric/optical.
2. **Mechanical Properties:** Hardness, yield strength, ductility, toughness, creep.
3. **Chemical Properties:** Based on the compound we choose [33].

4.8 TECHNOLOGY INVOLVED

4.8.1 CELL CULTURE

In vivo, drug delivery responses to too many destinations and routes can be predicted using 2D cell models. It is greatly involved in drug delivery. 2D cell cultures suffer from the loss of tissue-specific structures, biochemical cues, mechanical cues, cell-cell interactions, and cell-matrix interactions. It is not a good choice for predicting drug response to a particular disease, such as cancer. Even before clinical trials, 3D cell cultures cultured in stem cells and primary batteries may be the best choice for predicting drug response. 3D cell culture and co-culture models are advantageous because they pay more attention to evaluating drug safety and efficacy. It also allows for direct drug testing, eliminating certain differences that often interfere with the interpretation of preclinical results [11].

4.8.2 3D CELL CULTURE

A wide range of 3D cell culture technologies has been developed through cell biology, microtechnological techniques, and tissue engineering advances. 3D cell culture technologies include multicellular spheroids, organelles, scaffolds, hydrogels, organic chips, and 3D bioprinting. Although it has advantages and disadvantages, it is used to restore the morphological, functional, and microenvironmental characteristics of human organs and tissues [11].

4.8.3 SPHEROIDS

Spheroids have been tested on cell types such as stem cells, hepatocytes, and nerve cells. As is known, spheroids have a well-defined shape and have optimal physiological cell-cell and cell-extracellular matrix interactions. Spheroids created a heterogeneous cell population with gradients such as oxygen, nutrients, metabolites, and soluble signals. Growing and storing uniformly sized spheroids in specific proportions of different cell types with a minimum number of seeds is complex. They make up for many of the defects observed in monolayer cultures [11].

Spheroids were cultivated in four different approaches:

1. Cell self-aggregation was converted to spheroids using a geometrically shaped low adhesion plate and a very low adhesion surface coating to minimize cell adhesion. It can be formed, cultured, and probed in the same plate, enabling high-throughput or high-content screening [47].
2. The hanging drop plate method is used to develop multicellular spheroids. Multiple cell types are added first or in sequence at the top of the HDP well. This method is similar to a board with weak adhesive force. However, the spheroid could not be probed on the same plate and must be transferred to a second plate for assay.
3. Using a bioreactor for large-scale production is possible. Under dynamic conditions, cells self-aggregated themselves to form spheroids. The spheroids thus formed were of non-uniformity size and resulted in fluidic flow-induced shear stress [57].
4. The surface of micro/nanostructures is used as a scaffold to control cell adhesion and migration. It can print various nanoscale scaffolds on a flat substrate to select the appropriate pattern and adhesion properties and can damage the surface of the microstructure by removing air bubbles that are rapidly formed during culture [53].

4.8.4 ORGANOID

Organoids are shell-based 3D developmental tissues. Organ-specific cell types that arise from stem cells or organ precursors are organized by cell selection to form organoids. Organoids cannot perform all the functions of an organ. It mimics some of the structure and function of organs. Organoids lack blood vessels to carry nutrients and waste products inside and outside the body. Some key cell types are missing. In some cases, only the early stages of organ development are replicated by organoids [27].

Organoids are divided into tissue organoids and stem cell organoids. Tissue organoids are primarily associated with epithelial cells because they can organize into tissue-like structures. Stem cell organoids grow from embryonic stem cells, induced pluripotent stem cells, or primary stem cells. They can also be produced from adult tissue stem cells or neonatal stem cells. Various techniques have been used to build organoids that resemble tissues, thyroid, pancreas, liver, stomach, intestines, angiogenic heart, cerebral cortex, thymus, kidneys, and lungs [11].

Initially, cells were cultured on a bed of feeder cells forming a monolayer or an ECM-coated surface. With the help of mechanically supported cultures, the primary tissue was able to differentiate. Subsequent hanging drop cultures or lower adhesive plates create embryoid bodies. Finally, serum-free suspension culture of embryoid body aggregates that rapidly reaggregate on plates containing poorly adherent cultured organoids [11].

4.8.5 SCAFFOLDS AND HYDROGELS

Synthetic 3D structures that mimic specific tissues are known as scaffolding. They can be biological or polymer. Scaffolds were manufactured using materials with a variety of porosity, permeability, surface chemistry, and properties. Biological scaffolds use naturally occurring ECMs to create a suitable environment for molecules to interact with cells. Other natural polymers used in 3D culture are fibrin, hyaluronic acid, chitosan, alginate, or silk fibril. In tissue engineering and regenerative medicine, the scaffold must be biodegradable. Scaffold properties can regulate cell adhesion, proliferation, activation, and differentiation. Mesenchymal stem cells were neurogenic in softer matrices, myogenic in harder matrices, and osteogenic in harder matrices [11]. When treated with soluble factors, MSCs differentiate on specific lineage scaffolds constructed using a variety of techniques. 3D printing, particle leaching, gas foaming, etc., are part of the adopted approach when building scaffolding. Leaching or solvent casting processes manufacture porous scaffolds, and electrospinning is used to build fibrous scaffolds. The 3D printing process is used to create a defined shape and shape framework [35].

4.8.6 ORGANS ON-CHIPS

The organs on the chip represent a small artificial person whose purpose is to restore living human organs' structural and functional complexity. Most organ son chip designs capture only the important characteristics of the organ type or disease model. A chip consists of several well-defined structures, patterns, or frameworks. Therefore, researchers can reproduce the phenotype and pharmacological response of clinically relevant diseases [23]. Numerous tissue-on-chips were available for the application: skin, lungs, heart, vascular system, muscles, liver, and intestines.

4.8.7 THREE-DIMENSIONAL BIOPRINTING

It refers to printing cells onto complex 3D living tissue with the desired cell/organoid structure, topology, and function. The advantages of 3D bioprinting are bespoke microarchitecture, high throughput capacity, and co-culture capability. Compared to 3D cell culture, there are additional challenges [32]. In 3D bioprinting, biological materials, biochemicals, and living cells are arranged in layers.

The three approaches used for bioprinting are biomimetics, autonomous self-organization, and the creation and assembly of mini-organization building blocks. For transplants, use 3D bioprinting to create multi-layered skin, bone, vascular grafts, tracheal splints, and more. They are used not only to create scaffolds for 3D cell culture but also to build 3D bioprinting tissue models for drug screening and profiling [24].

4.8.8 3D CELL CULTURES IN DRUG DISCOVERY

3D cell culture begins with disease modeling research, target identification and validation, screening, lead selection, efficacy, safety assessment, and drug discovery.

4.8.8.1 DISEASE MODELING

If there is no drug available for the disease or clinical condition, drug discovery begins. Various disease mechanisms have been identified with the help of 3D cell culture. While 2D models cannot detect illness, dormancy, recurrence, etc., 3D models have played an important role in discovering tumor biology [43].

Spheroid cultures help model tissue structure, signal transduction, microenvironment, cancer infiltration, and immune behavior, and study and expand cancer stem cells. Cancer cell line spheroids have been used to study various aspects of the cancer infiltration process. Organoid cultures have developed models for cancer, developmental disorders, infectious diseases, and neurodegeneration. You can model many hereditary diseases that cannot be modeled in animals. Organon chips are used for cancer modeling [11].

4.8.8.2 TARGET IDENTIFICATION AND VALIDATION

The rate-determining step of preclinical discovery was to identify the target and analyze its validation. Compared to 2D models, 3D models may discover targets and their validations [11].

4.8.8.3 SCREENING FOR HIT IDENTIFICATION

The target-based HTS-compatible cell assay was simple and highly efficient. It's cheap. Three factors determine target identification: an innovative strategy for discovering active substances from research and development, the generation of top-notch active substance phenotypes, and subsequent active substance, the molecular mechanism of its action is known [42].

4.8.8.4 EFFICACY PROFILING FOR LEAD IDENTIFICATION

The hit was identified and further assessed synthetic controllability, freedom of action, drug similarity, and possible toxicity. Medicinal chemistry is optimized to produce suitable lead compounds with improved efficacy, low cost, and desirable physicochemical and metabolic properties. 3D cell culture models play an important role in identifying key substances and eliminating animal studies for large preclinical studies [11].

4.8.8.5 TOXICITY PROFILING FOR LEAD SELECTION

Adverse drug reactions often occur when off-target interactions or excessive binding of drug molecules to toxic cells occur. The 3D cell culture model is used as a powerful tool for assessing drug-induced toxicity. Drug-induced toxicity in critical organs can lead to drug abrasions [51].

4.8.8.6 PHARMACOKINETICS AND PHARMACODYNAMICS PROFILING FOR LEAD SELECTION

Poor pharmacokinetics and pharmacodynamics are important factors in drug failure. Liver spheroids, organoids, etc., constructed in 3D cell culture, help determine the pharmacokinetic profile of drug molecules. Liver spheroids and organoids are used to study the metabolism of drug molecules [39].

4.9 APPLICATIONS OF NANOMATERIALS

4.9.1 CLINICAL APPLICATIONS

4.9.1.1 REGENERATIVE MEDICINE AND IMMUNE RESPONSE IN BIOMATERIALS

Biomaterials have been used in cheap implants to treat advanced diseases, but regenerative medicine has prevented tissue loss. It also used regenerative medicine in the early stages of the disease. Tissue engineering techniques use stem cells with biomaterials through germ cell encapsulation or transplantation to achieve the desired results [40].

When exogenous cells are transplanted into the host tissue, most cells die in a short time. These can interfere with the regeneration of tissue lost during the degeneration process of trauma. However, the immune response occurs due to inflammation of the implantation site [21].

4.9.1.1.1 BIOMATERIALS FOR REGENERATION

Biomaterials are intended to restore the lost function and structure of damaged tissue. Biomaterials should decompose over time as the host tissue regenerates. The first M1 reaction causes a foreign body reaction. The next transition to the M2 phase is essential to facilitate organizational remodeling. Changes in biomaterials need to take into account the unwanted effects of other cell types that contribute to tissue regeneration [21].

4.9.1.1.2 BIOMATERIALS FOR REPLACEMENT

Biologically inert implants were used to minimize cell implant interactions. Naive proteins are usually adsorbed on implants, paving the way for the formation of temporary matrices. It acts as a buffer between the biomaterial and the host. Precise surgical procedures have been performed to minimize the relative movement between the implant and the host tissue. Recent advances in bioengineering have left no visible scars between recycled biomaterials and alternatives. Many coatings on replacement implants are functionally similar to those used in regenerative medicine. Under certain conditions, can achieve improved immune tolerance and implant integration. Titanium material is used in joint replacements and exhibits higher Osseointegration by reshaping the surface to induce osteoblast migration and attachment [6].

4.9.1.1.3 STEM CELLS

Stem cells are involved in the regeneration, maintenance, and proliferation of human tissues. They are capable of cell regeneration and can differentiate into different cell types of the human body. With differentiation, the program functions as a molecular tool in biomedical applications and regenerative medicine [6].

Stem cells come from three major sources: embryonic origin, mesenchymal origin, and so-called induced pluripotent stem cells. Cells of embryonic origin are obtained from the inner cell mass of the blastocyst. They can differentiate into any adult cell type and are used in regenerative medicine and cell therapy. Mesenchymal cells are derived from adults. Pluripotent mesenchymal stem cells are used to repair cartilage. Osteoarthritis of the carpometacarpal joint is common in postmenopausal women and can be treated with mesenchymal cells and avoid surgery [18].

The combination of nanoparticles and stem cells improved the proliferation and differentiation of cells used in the treatment of various diseases such as ischemic stroke, spinal cord, multiple sclerosis, Parkinson's disease, and Alzheimer's disease [57].

4.9.1.1.4 BIO-NANO COMPOSITES

Administering a drug in a nanostructured system has many advantages, thereby preventing drug degradation and improving pharmacokinetics and specificity at the nanoscale level. Organic and inorganic are mutual classifications of nanomaterials. Organic materials are derived from polymers such as polysaccharides, collagen, and chitosan. They have been used in biomedical applications and stem cell differentiation [31]. It can combine electrospinning technology with electrostatic attraction to produce porous nanoscale 3D fibers from multiple sources. A voltage source that passes through the solution produces droplets, and surface tensions are collected at different distances, producing different forms of fibers [30].

4.9.1.2 SPINAL IMPLANT

Nanotechnology facilitates spinal fusion to overcome some complications associated with recombinant human bone morphogenetic proteins-2 (RhBMP-2). By adding nanoparticles such as titanium oxide and zirconium oxide, it is possible to modify the surface of titanium spinal implants. It is believed to reduce resorption and increase bone formation compared to traditional implants. Nano block with Titan Spine Technology shows more bone formation, and angiogenic growth factors than traditional titanium PEEK cages. The use of RhBMP-2 is often associated with side effects from hyper physiological doses. One form of nanofiber structure, known as an amphipathic extracellular filament, induces cell regeneration [21].

4.9.1.3 ORTHOPEDIC ONCOLOGY THERAPEUTIC APPLICATIONS – DRUG DELIVERY

Nanotechnology can find solutions for cytotoxicity, reduce chemotherapy, and improve drug resistance and pharmacokinetic problems by using a unique drug delivery carrier molecule. When a drug is loaded into a nano molecule, it binds to a specific ligand and penetrates cancer cells. This allows chemotherapy to act directly on the target of interest while reducing the collateral toxicity of non-cancerous cells.

The lipid nanoparticle carrier has been regarded as the best option for treating osteosarcoma because of its excellent bioavailability and oral delivery capability [25]. Many metals, such as titanium, gold, calcium phosphate, and chitosan, were considered nanocarriers [37]. Custom-made nanocarriers loaded with cisplatin, used for tumor cells, resulted in less renal accumulation and fewer side effects than the free form. Maintains optimal antitumor activity against osteosarcoma cells [14].

4.9.1.4 MUSCULOSKELETAL TRAUMA

Surface modification improves the simulation of the natural bone environment over traditional implants and is a successful procedure for nanostructured implants. Efforts have been made to show that nanofiber scaffolds support cell migration and growth during the healing of bone damage [41].

4.9.2 DIAGNOSTIC APPLICATIONS

Nanotechnology in cancer diagnosis is based on the binding of nanoparticle-ligand complexes to specific gene mutations, enabling detailed imaging at the cellular level. It can visualize tumors with specific mutations by adding a contrast agent to the complex. This technique recognizes the earliest possible metastasis of malignant tumors. It can start Chemotherapy combined with drug delivery by nanotechnology in the early stages of the disease before clinical symptoms lead to reduced morbidity. Fluorescent probes based on nanomaterials help assess the response of cancer after treatment [27].

4.9.2.1 ARTHROPLASTY

The use of nanotechnology in internal prostheses enables the development of longer-lived, portable materials that are safe and effective and also help prevent infection. The nanotextured implant surface has become more important [20]. Using ultra-high molecular weight polyethylene implant materials by increasing mechanical strength with nanotechnology is suitable for implants. Adding carbon nanotubes to ultra-high molecular weight polyethylene improves mechanical strength. This new connection makes it more useful as a lining or tibial component of the acetabular cup [17]. Modifying the surface of the nanostructures increases resistance and improves implant viability.

4.9.2.2 CEMENTS

The addition of antibiotics such as polymethylmethacrylate to bone cement is a commonly used phenomenon. Antibiotics can only last for a shorter period. Polymethylmethacrylate is a common material to which nanotechnology-based antibiotic carriers such as lipid nanoparticles, silicon dioxide, and clay nanotubes are added. It may improve drug delivery and allow the timely release of the drug. PMMA provokes a strong autoimmune response, causes inflammation, and causes implant failure. Ceramic particles such as zirconium oxide and barium sulfate are often added to cement to allow visualization by x-rays. However, this does affect biocompatibility at the bone-implant interface. Nanoscale modifications of these particles, when added to bone cement, increase cell compatibility and reduce mechanical failure.

4.9.2.3 SPORTS MEDICINE

The response to complete cartilage tissue regeneration in adults is poor, resulting in progressive degeneration to osteoarthritis. Nanotechnology is used to support MSC therapy by developing biocompatible scaffolds that enhance native cartilage repair [32].

4.9.2.4 TENDON HEALING

Hydrosol nanoparticles as a drug carrier provide a controlled release of mitomycin C, a chemotherapeutic agent that can reduce postoperative adhesions. This allows naturally healed tendons to maintain mechanical strength and reduce postoperative adhesions. Improved healing and mechanical stability are observed in rat shoulders during supraspinatus repair surgery treated with autologous nanoscaffolding [29].

4.9.2.5 ORTHOPEDIC INFECTIONS

Antibiotic implants with nanoparticles are designed to help overcome bacterial infections. A titanium femoral stem with a new vancomycin drug delivery system showed long-term drug release [42]. Nanophase Silver is more effective in wound healing in infection prevention and healing than traditional wound dressings. Titanium screws coated with nanosilver particles suppress the formation of biofilm. Nanotechnology infection control program prevents acute postoperative infections in trauma, spine implants, and joint replacements [39].

4.10 NEWLY EMERGING SMART NANOMATERIALS

The new intelligent nanomaterials include intelligent scaffolding and stem cell structures for bone tissue engineering applied to maxillofacial, periodontal, and pulp formation, intelligent drug delivery methods using biomaterials, and non-drug resistant bright colors. Biofilm species modulation with resin included custom-made materials, intelligent pH-sensitive materials for colonization of acid-forming bacteria to avoid drug resistance, with infection control at the wound site Biomaterials suitable for bone construction and intelligent biomaterials for dental treatment that can be used in arthritis [56].

4.11 NANO-ENGINEERED MATERIALS FOR SKIN WOUND HEALING

4.11.1 SKIN WOUND HEALING

Our body has a natural mechanism for healing wounds, closing the wound when tissue engineering therapy fails. The regenerated skin is in great need for people with diabetes and burns [11].

4.11.2 DIABETIC FOOT AND TISSUE ENGINEERING

Diabetic foot ulcers due to peripheral arterial disease and sensory neuropathy affect the diabetic foot. Tissue engineering implants replace traditional treatments such as systematic antibacterial therapy and combine with other surgical procedures. The wound healing process is constrained by inflammation as a permanent condition, which is a never-ending problem [25].

4.11.3 NANOMATERIALS FOR SKELETAL MUSCLE INJURIES

Irreversible scarring is inevitable when treating an injury. The application of nanomaterials can fill gaps in the recovery or regeneration of skeletal muscle tissue. Its physical and chemical properties determine the function of nanomaterials. It can adjust the function of nanomaterials by adding growth factors [35].

4.12 SUMMARY

The use of nanotechnology in tissue engineering helps to make transplants more successful, achieve better resistance, adhesion, and prevent other infections. Bone conduction is a process that exhibits bone healing with the formation of scaffolds and the formation of new blood vessels. Some severe trauma requires orthopedic implants for better fracture healing. Clinical alternatives to the hips, shoulders, wrists, knees, elbows, knuckles, and ankles are permanent implants, but plates, pins, screws, wires, and nails hold the fracture or broken bone in place. Clinically available implants are at risk due to the formation of biofilms by microorganisms. To overcome these complications, requirements include orthopedic transplantation using appropriate biocompatible resources, efficient fighting of microorganisms, wear resistance, corrosion resistance, mechanical properties, osseointegration, etc.

Tissue engineering helps to mimic naive tissue through biomaterials. Tissue engineering finds its way in hepatocytes, stem cells, chondrocytes, nerve cells, bone cells, and more. Cell-based assessments have been widespread in drug discovery. The 2D cell culture model has been replaced by 3D cell culture and co-culture that are more effective in predicting drug release. 3D cell culture techniques include spheroids, chips, scaffolds, organoids, hydrogels, and 3D bioprinting. Spheroids have a well-defined shape and can produce heterogeneous cell populations. Organoids are tissue or stem cell-based organs developed by a particular cell type. It does not perform all the functions of an organ, but it mimics some structures. Scaffolds are materials designed to enable cellular interactions in the formation of new functional tissues. The framework with fixed shapes and shapes was created using 3D bioprinting. Organonchip is an unnatural human organ model for recording the phenotypes of clinically relevant diseases and important characteristics of organ types or disease models from pharmacological test tubes. 3D bioprinting consists of three approaches: imitation, independent self-organization, and small tissue components. It helps create scaffolds for 3D cell culture and 3D bioprinting tissue models for profiling and active ingredient screening. 3D cell culture models such as spheroids and organoids are used to study the pharmacokinetics of drug particles. The 3D cell culture model is used as a powerful model in assessing drug-induced toxicity. Liver spheroids and organoids are used to study the metabolism of drug molecules.

Tissue engineering technology uses stem cells in combination with biomaterials through germ cell encapsulation or transplantation. Inexpensive implant biomaterials have been used to treat advanced diseases. Regenerative medicine has been used to treat the early stages of the disease. Regeneration and promotion of the immune system are inversely proportional. The opportunity to start and balance each macrophage is important for tissue healing. Implants are usually adsorbed on the outside with a naive protein that acts as a buffer between the biomaterial and the host. Advances in biotechnology allow scar-free transplantation. Joint replacement with titanium material shows high osseointegration due to surface modification. Nanoparticles and stem cells are effectively combined with increased proliferation, differentiation, and treatment of various diseases.

The role of nanotechnology in cancer diagnosis helps to self-identify the potential for metastasis of malignant tumors at an early stage. Initiation of chemotherapy-associated with drug release with nanotechnology reduces morbidity. Adding antibiotics to bone cement improves drug delivery and allows the drug to be released on time. The application of nanomaterials can fill many gaps in the recovery and regeneration of skeletal muscle tissue.

KEYWORDS

- **bionanomaterial**
- **orthopedic implants**
- **recombinant human bone morphogenetic proteins-2**
- **RNA**
- **tissue engineering**
- **tricalcium phosphate**

REFERENCES

1. Beale, B., (2004). Orthopedic clinical techniques femur fracture repair. *Clinical Techniques in Small Animal Practice, 19,* 134–150.
2. Bettinger, C. J., Orrick, B., Misra, A., Langer, R., & Borenstein, J. T., (2006). Microfabrication of poly (glycerol–sebacate) for contact guidance applications. *Biomaterials, 27,* 2558–2565.
3. Bielby, R., Jones, E., & McGonagle, D., (2007). *The Role of Mesenchymal Stem Cells in Maintenance and Repair of Boneinjury, 38,* S26–S32. Pub Med.
4. Brown, B. N., Ratner, B. D., Goodman, S. B., Amar, S., & Badylak, S. F., (2012). Macrophage polarization an opportunity for improved outcomes in biomaterials and regenerative medicine. *Biomaterials, 33,* 3792–3802.

5. Carano, R. A., & Filvaroff, E. H., (2003). Angiogenesis and bone repair. *Drug Discovery Today, 8,* 980–989.
6. Cheng, H., Chawla, A., Yang, Y., Li, Y., Zhang, J., Jang, H. L., & Khademhosseini, A., (2017). Development of nanomaterials for bone-targeted drug delivery. *Drug Discovery Today, 22,* 1336–1350.
7. Costerton, J. W., (1999). Bacterial biofilms: A common cause of persistent infections, *Science, 284,* 1318–1322.
8. Cruz-soto, M. E., Fernandez, C., & Carrillo, J. L., (2018). Articulate cartilage enriched with autologous bone marrow aspirate as an approach of pain amelioration and functionality restoration in osteoarthritis patients older than 50 years old. *Medical Clinical Research and Reviews, 2,* 1–3.
9. De Long, W. G., Einhorn, T. A., Koval, K., Michael, M., Wade, S., Roy, S., & Tracy, W., (2007). Bone grafts and bone graft substitutes in orthopedic trauma surgery. *A Critical Analysis: Journal of Bone and Joint Surgery, 89,* 649–658.
10. Einhorn, T. A., (1998). The cell and molecular biology of fracture healing. *Clinical Orthopedics and Related Research, 355,* S7–S21.
11. Fang, Y., & Eglen, R. M., (2017). Three-dimensional cell cultures in drug discovery and development. *SLAS Discovery: Advancing Life Sciences R&D, 22,* 456–472.
12. Gatoo, M. A., Naseem, S., Arfat, M. Y., Mahmood, D. A., Qasim, K., & Zubair, S., (2014). Physicochemical properties of nanomaterials: Implication in associated toxic manifestations, *BioMed Research International, 2014,* 1–8.
13. Gerstenfeld, L. C., Cullinane, D. M., Barnes, G. L., Graves, D. T., & Einhorn, T. A., (2003). Fracture healing as a post-natal developmental process: molecular, spatial, and temporal aspects of its regulation. *Journal of Cellular Biochemistry, 88,* 873–884.
14. Gillani, R., Ercan, B., Qiao, A., & Webster, T. J., (2010). Nano functionalized zincoria and barium sulfate particles as bone cement additives. *International Journal of Nanomedicine, 5,* 1–111.
15. Grasman, J. M., Zayas, M. J., Page, R. L., & Pins, G. D., (2015). Biomimetic scaffolds for regeneration of volumetric muscle loss in skeletal muscle injuries. *Acta Biomaterials, 25,* 2–15.
16. Grosse, S., Haugland, H. K., Lilleng, P., Ellison, P., Hallan, G., & Høl, P. J., (2014). Wear particles and ions from cemented and uncemented titanium-based HIP prostheses-a histological and chemical analysis of retrieval material, *Journal of Biomedical Materials Research - Part B: Applied Biomaterials, 103,* 709–717.
17. Gusic, N., Ivkovic, A., Vafaye, J., Vukasovic, A., Ivkovic, J., Hudetz, D., & Jankovic, S., (2014). Nanobiotechnology and bone regeneration: A mini-review. *International Orthopedics, 38,* 1877–1884.
18. Hayashi, H., Uchida, A., Hamada, H., Yoshikawa, H., Shinto, Y., & Ono, K., (1992). Alumina ceramic prostheses for bone tumor surgery. *Archives of Orthopedic and Trauma Surgery, 112,* 1–4.
19. Hazer, D. B., Sakar, M., Dere, Y., Altinkanat, G., Ziyal, M. I., & Hazer, B., (2016). Antimicrobial effect of polymer-based silver nanoparticle coated pedicle screws. *SPINE, 41,* E323–E329.
20. Hennig, S., Van De, L. S., Lummer, M., Simonis, M., Huser, T., & Sauer, M., (2015). Instant live-cell super-resolution imaging of cellular structures by nano-injection of fluorescent probes. *Nano Letters, 15*(2), 1374–1381.

21. Hsu, W. K., Goldstein, C. L., Shamji, M. F., Cho, S. K., Arnold, P. M., Fehlings, M. G., & Mroz, T. E., (2017). Novel osteo-biologics and biomaterials in the treatment of spinal disorders. *Neurosurgery, 80,* S100–S107.
22. Huegel, J., Kim, D. H., Cirone, J. M., Pardes, A. M., Morris, T. R., Nuss, C. A., Mauck, R. L., et al., (2016). Autologous tendon-derived cell-seeded nanofibrous scaffolds improve rotator cuff repair in an age-dependent fashion. *Journal of Orthopedic Research, 35,* 1250–1257.
23. Huh, D., Hamilton, G. A., & Ingber, D. E., (2011). From 3D cell culture to organs-on-chips. *Trends in Cell Biology, 21,* 745–754.
24. Ingher, D. E., Mow, V. C., & Butler, D., (2006). Tissue engineering and developmental biology going biomimetic. *Tissue Engineering, 12,* 3265–3283.
25. Kannan, A., Dodwad, S. N. M., & Hsu, W. K., (2015). Biologics in spine arthrodesis. *Journal of Spinal Disorders & Techniques, 28,* 163–170.
26. Katsikogianni, M., & Missirlis, Y. F., (2004). Concise review of mechanisms of bacterial adhesion to biomaterials and of techniques used in estimating bacteria material interactions. *European Cells and Materials, 8,* 37–57.
27. Lancaster, M. A., & Knoblich, J. A., (2014). Organogenesis in a dish: Modelling development and disease using organoid technologies. *Science, 345,* 1247125. doi: 10.1126/science.1247125.
28. Long, P. H., (2008). Medical devices in orthopedic applications. *Toxicologic Pathology, 36*, 85–91.
29. Mahboudi, H., Kazemi, B., Soleimani, M., Hanaee-Ahvaz, H., Ghanbarian, H., Bandehpour, M., Enderami, S. E., Kehtari, M., & Barati, G., (2018). Enhanced chondrogenesis of human bone marrow mesenchymal stem cell (BMSC) on nanofiber-based polyethersulfone (PES) scaffold. *Gene- Elsevier, 643,* 98–106.
30. Mengistu, L. S., Bossard, F., & Rinaudo, M., (2016). Preparation of pure and stable chitosan nanofibers by electrospinning in the presence of poly(ethylene oxide). *International Journal of Molecular Sciences, 17*, 1790. doi: 10.3390/ijms17111790.
31. Mocan, L., Ilie, & Mocan, T., (2012). Influence of nanomaterials on stem cell differentiation: Designing an appropriate nano bio-interface. *International Journal of Nanomedicine, 7*, 2211–2215.
32. Murphy, S. V., & Atala, A., (2014). 3D bioprinting of tissues and organs. *Nature Biotechnology, 32*, 773–785.
33. Murty, B. S., Shankar, P., Raj, B., Rath, B. B., & Murday, J., (2012). Unique properties of nanomaterials. *Textbook of Nanoscience and Nanotechnology* (pp. 29–65). Berlin, Heidelberg: Springer.
34. Opal, S. M., (2000). Phylogenetic and functional relationships between coagulation and the innate immune response. *Critical Care Medicine, 28*, S77–S80.
35. Papadimitropoulos, A., Piccinini, E., Brachat, S., Braccini, A., Wendt, D., Barbero, A., Jacobi, C., & Martin, I., (2014). Expansion of human mesenchymal stromal cells from fresh bone marrow in a 3D scaffold-based system under direct perfusion. *PLoS One, 9*(7), e102359. doi: 10.1371/journal.pone.0102359.
36. Park, G. E., & Webster, T. J., (2005). A review of nanotechnology for the development of better orthopedic implants. *Journal of Biomedical Nanotechnology, 1,* 18–29.
37. Savvidou, O. D., Bolia, I. K., Chloros, G. D., Goumenos, S. D., Sakellariou, V. I.,Galanis, E. C., & Papagelopoulos, P. J., (2016). Applied nanotechnology and nanoscience in orthopedic oncology. *Orthopedics, 39*, 280–286.

38. Schindler, M., Ahmed, I., Kamal, J., Nur-E-Kamal, A., Grafe, T. H., Young, C. H., & Meiners, S., (2005). A synthetic nanofibrillar matrix promotes *in vivo*-like organization and morphogenesis for cells in culture. *Biomaterials, 26*(28), 5624–5631.
39. Schwartz, M. P., Hou, Z., Propson, N. E., Zhang, J., Engstrom, C. J., Costa, V. S., Jiang, P., et al., (2015). Human pluripotent stem cell-derived neural constructs for predicting neural toxicity. *Proceedings of the National Academy of Sciences, 112*, 12516–12521.
40. Sridharan, R., Cameron, A. R., Kelly, D. J., Kearney, C. J., & O'Brien, F. J., (2015). Biomaterial based modulation of macrophage polarization: A review and suggested design principles. *Materials Today, 18*, 313–325.
41. Susa, M., Milane, L., Amiji, M. M., Hornicek, F. J., & Duan, Z., (2010). Nanoparticles: A promising modality in the treatment of sarcomas. *Pharmaceutical Research, 28*, 260–272.
42. Swinney, D. C., & Anthony, J., (2011). How were new medicines discovered?. *Nature Reviews Drug Discovery, 10*, 507–519.
43. Tanner, K., & Gottesman, M. M., (2015). Beyond 3D culture models of cancer. *Science Translational Medicine, 7*, 283ps9. doi: 10.1126/scitranslmed.3009367.
44. Trampuz, A., Osmon, D. R., Hanssen, A. D., Steckelberg, J. M., & Patel, R., (2003). Molecular and antibiofilm approaches to prosthetic joint infection. *Clinical Orthopedics and Related Research, 414*, 69–88.
45. Uçkay, I., Aragón-Sánchez, J., Lew, D., & Lipsky, B. A., (2015). Diabetic foot infections: What have we learned in the last 30 years?. *International Journal of Infectious Diseases, 40*, 81–91.
46. Vial, S., Reis, R. L., & Oliveira, J. M., (2017). Recent advances using gold nanoparticles as a promising multimodal tool for tissue engineering and regenerative medicine. *Current Opinion in Solid State and Materials Science, 21*, 92–112.
47. Vinci, M., Gowan, S., Boxall, F., Patterson, L., Zimmermann, M., Court, W., Lomas, C., Mendiola, M., Hardisson, D., & Eccles, S. A., (2012). Advances and establishment and analysis of three-dimensional tumor spheroid-based functional assays for target validation and drug evaluation. *BMC Biology, 10*, 29. doi: 10.1186/1741-7007-10-2.
48. Wang, M., (2004). *Bioactive Materials and Processing* (pp. 1–82). Springer.
49. Watnick, P., & Kolter, R., (2000). Biofilm, city of microbes. *Journal of Bacteriology, 182*, 2675–2679.
50. Webster, T., (2000). Enhanced functions of osteoblasts on nanophase ceramics. *Biomaterials, 21*, 1803–1810.
51. Wilke, R. A., Lin, D. W., Roden, D. M., Watkins, P. B., Flockhart, D., Zineh, I., Giacomini, K. M., & Krauss, R. M., (2007). Identifying genetic risk factors for serious adverse drug reactions: Current progress and challenges. *Nature Reviews Drug Discovery, 6*, 904–916.
52. Xin, T., Greco, V., & Myung, P., (2016). Hardwiring stem cell communication through tissue structure. *Cell, 164*, 1212–1225.
53. Yoshii, Y., Waki, A., Yoshida, K., Kakezuka, A., Kobayashi, M., Namiki, H., Kuroda, Y., et al., (2011).The use of nanoimprintedscaffolds as 3d culture models to facilitate spontaneous tumor cell migration and well-regulated spheroid formation. *Biomaterials, 32*, 6052–6058.
54. Youn, B. S., Sen, A., & Behie, I. A., (2006). Scale up of breast cancer stem cell aggregate cultures of suspension bioreactor. *Biotechnology Progress, 22*, 801–810.

55. Yu, B., Chou, P., Sun, Y., Lee, Y., & Young, T., (2006). Topological micropatterned membranes and its effect on the morphology and growth of human mesenchymal stem cells (hMSCS). *Journal of Membrane Science, 273*, 31–37.
56. Zhang, G., Khan, A. A., Wu, H., Chen, L., Gu, Y., & Gu, N., (2018). The application of nanomaterials in stem cell therapy for some neurological diseases. *Current Drug Targets, 19*, 279–298.
57. Zhang, K., Wang, S., Zhou, C., Cheng, L., Gao, X., Xie, X., Sun, J., et al., (2018). Advanced smart biomaterials and constructs for hard tissue engineering and regeneration. *Bone Research, 6*, 31. doi: 10.1038/s41413-018-0032-9.
58. Zhou, H., Wang, G., Lu, Y., & Pan, Z., (2016). Bio-inspired cisplatin nanocarriers for osteosarcoma treatment. *Biomaterials Science, 4*, 1212–1218.

CHAPTER 5

NANOFORMULATIONS FOR THE TREATMENT OF OCULAR DISEASES

RABIAH BASHIR, SHABNAM KAWOOSA, TABASUM ALI, and NISAR AHMAD KHAN

ABSTRACT

The study of tiny structures with sizes ranging from 0.1 to 100 nanometers is known as nanotechnology. Nanomedicine is a rapidly growing discipline that employs nanotechnology in medical applications, including targeted therapeutic operations at the molecular level to treat disease or repair cells. The employment of nanomedicine in the management of ocular illnesses has progressed significantly in recent years. The application of nanomedicine based on nanotechnology to cure ocular diseases has become the hope of millions of patients. Nanomaterial-based drug delivery systems are taken into account by an increasing number of scientists. The employment of nano-based medicines presents a novel opportunity to develop drug systems that can go through the protective barriers and sustain sufficient saturation of tissues. This chapter will consider and highlight the current application of diverse bio nanomaterials like electrospun nanofibers, nanoparticles, nanocrystals, liposomes, niosomes, dendrimers, polymeric micelles, hydrogels, and cyclodextrins as ophthalmic methods for drug delivery for the enhancement of ocular bioavailability.

5.1 INTRODUCTION

When a new drug delivery system is developed, the issues ADME of drugs are considered [12]. The eye acts as a challenging and opportunistic organ

Sustainable Nanomaterials for Biomedical Engineering: Impacts, Challenges, and Future Prospects.
Junaid Ahmad Malik, Megh R. Goyal & Mohamed Jaffer M. Sadiq (Eds.)

for the delivery of pharmaceuticals. The transfer of drugs to the ophthalmic system is the most unique and exciting challenge that formulation scientists and researchers have to deal with [80]. The interior and exterior anatomy of the eye makes it very impervious to foreign substances. The hardships faced by formulation scientists in curing ocular diseases to evade the defensive obstacles without affecting the eye tissues [41]. The formulation of innovative, extra sensitive, problem-solving methods and healing substances renders firmness to advancing highly productive and progressive ophthalmic dosage forms. The drug is delivered to the eyes generally for localized therapy and not for systemic therapy. It is helpful to prevent them from injury due to high drug concentrations not desirable for the eye [63, 66]. Typical dosage forms for the eyes include eye drops, ointment, eye gel, solution, eye injection, irritant solution, suspension, and sol-gel system accounting for 80% of the total number of eye drop dosage forms and the most widely used formulations [16]. It is easy to instill in the dosage form of eye drops. Still, there are inherent disadvantages such as the need for repeated instillation, variable dosage, rapid withdrawal, loss of drainage, swollen eyelids stabbing, and patients have decreased confidence, blurred vision, and irritability [23, 56, 60]. The intraocular bioavailability of topical agents is very low, generally agreed in 5–10% of the whole drug administered [10, 35, 44]. Because ocular eye drops suffer from the drawback of having reduced ophthalmic bioavailability, there arises the need for repeated intervallic administration of eye drops to retain a constant continued dose of the therapeutic moiety. It has negative effects on the eyes and the rest of the body, causing the eyes to get an enormous and unpredictable dose of medicine, due to greater drug concentrations, unfortunately [19]. The prime solutions, suspensions, are not adequate to fight specific current infectious ailments of the eye [80].

The constraints of these traditional ocular treatments are solved by employing innovative ways to drug administration through the eye, which are currently being researched in order to develop controlled-release and sustained-release delivery systems. Successful and efficient management of diseases about eyes is an arduous task for pharmaceutical researchers in this arena, mainly depending on the character of diseases and the existence of the optical hurdles, particularly in posterior ocular segments. Several attempts have been undertaken to change the product composition, such as viscosity enhancers and mucoadhesive polymers, to increase the bioavailability of ophthalmic pharmaceutical formulations. These methodologies help in prolonging corneal contact time and lead to self-effacing enhancement in ocular bioavailability. Therefore, it gives the good impression to consider the novel approach like nanotechnology, utilizing nanocarriers like nanomicelles, nanocrystals, nanoparticles, liposomes, prodrug methods, *in situ* gelling systems, and iontophoresis for active transfer of medicaments and resulting in enhancement of ocular bioavailability and lessen toxic effects. In recent times nanocarriers have been studied for their relevance in the delivery of drugs to ophthalmic tissues. Nanosystems can overcome the ocular barriers that now limit the efficacy of traditional therapies, as well as provide further regulated drug release, lower administration rates, and improved patient compliance [19].

In this chapter, novel nano biomaterial-based formulations for the cure of diseases of the eyes will be discussed in detail, highlighting the drawbacks faced by conventional ophthalmic dosage forms currently available and the need for nanotechnology-based nanomedicines.

5.2 CONVENTIONAL TYPES OF OPHTHALMIC DRUG DELIVERY

Different conventional ocular dosage forms (Figure 5.1) are discussed in subsections.

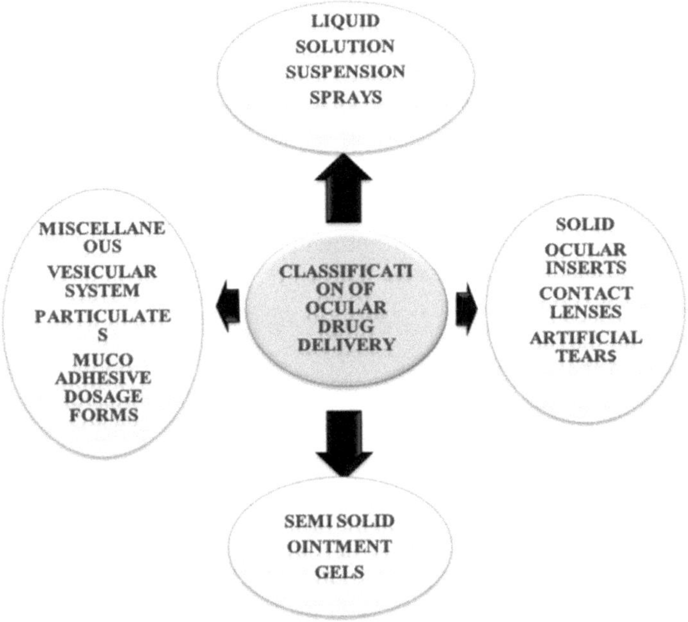

FIGURE 5.1 Classification of ocular drug delivery system [69].

5.2.1 SOLUTIONS

Non-obtrusive type of medication having fluid arrangements comprises a more significant part of easily reachable topical ophthalmic preparations. Following the installation of a topical eye drop, the eye arrangement provides strike medicine saturation, after which its focus soon decreases. Accordingly, various excipients may be added to topical solutions like viscosity enhancers, buffering operators, permeability enhancers, etc. Polymers such as polyvinyl alcohol (PVA), hydroxyl methylcellulose, sodium carboxyl methylcellulose, hydroxypropyl methylcellulose (HPMC), carboxymethyl cellulose and carbomers, as well as some regular polymers (e.g., Hyaluronic acid (HA), guar gum, gellan gum, and so on) have been used to impact the thickness of the visual description and thus enhance drug bioavailability [57].

5.2.2 SUSPENSIONS

Suspensions are another non-obtrusive medication option. They are characterized as scattered, lightly isolated unsolvable medications in a dissolution medium like water comprised of suitable suspending and distribution agents [22]. Non-homogeneity of the ocular measurement structure, settling, cake development, collection of suspended particles, effective conservation, re-dependability, and ease of production are just a few of the challenges that a formulator faces while developing a suspension. Suspensions are actively stable in contrast to thermodynamically unstable frameworks, bearing in mind that if left unattended for an extended period of time, they will accumulate particles, silt, and finally harden [1].

5.2.3 OINTMENTS

Eye ointment is a semi-solid dosage form for topical application and mainly contains solid or semi-solid hydrocarbon groups with dissolution or softening points close to the temperature of the human body. The slope of the base is based on the clinical recommendations of the balm. Different types of treatment substrates are absorption substrates (such as beeswax, lanolin), water-soluble substrates (such as polyethylene glycol 200, 300, 400), hydrocarbon substrates (such as hard paraffin, refined paraffin, microcrystalline wax), oils from vegetables (e.g., coconut oil, olive oil, almond oil, sesame oil,), bases of emulsifying (e.g., emulsifying wax, cetriol amine) [2]. The determination of hydrocarbons depends on biocompatibility. These treatments help increase visual bioavailability and maintain the supply of medicines. Compared to various ophthalmic solutions, it has a longer eye contact time. The ophthalmic ointment breaks down into minute droplets after application and stays in the conjunctiva for a longer length of time, increasing the drug's bioavailability [2].

5.2.4 EMULSION

Ocular emulsion-based dosage forms offer the benefit of advancing the bioavailability and dissolvability of medications. Two types of eye emulsions are financially employed as carriers for dynamic drugs and medicines: water in oil w/o based emulsions and oil in water o/w frameworks. For eye usage, oil in water emulsions is generally favored compared to w/o emulsions because of minor disturbance and better visual pliability of o/w emulsion. Emulsions have been shown to improve precorneal residence duration, drug corneal penetration discharge, and visual bioavailability in a number of trials. Mainly ophthalmic emulsions are prepared by blending or scattering the aqueous phase into an oil stage with suitable emulsifying and blending to shape a homogeneous o/w emulsion. Prior to or while rushing into the mixing vessel, each step is often cleansed. High-shear homogenization might be used to reduce oil bead size to sub-micron size, which could indicate an increase in the physical reliability of the oil micelles, preventing them from condensing [24].

5.2.5 GELS

Gels for ocular use are made up of polymers with mucoadhesive properties, e.g., CMC, polycarbophil, sodium alginate, etc., helping confine drug delivery to the eye. The gels can extend pre-corneal times as they remain in the conjunctiva sac for a prolonged period and help in increasing drug bioavailability. PVA, hyaluronic acid, cellulose derivatives are used as gelling agents [55].

5.2.6 DRAWBACKS OF CONVENTIONAL OCULAR DOSAGE FORM

Conventional ocular dosing formulations have a number of disadvantages as below (Figure 5.2):

1. **Blurred Vision:** The eye's natural vision might be affected after applying conventional types of ocular dosage forms, including drops, gels, and ointments, as they cover the outer surface of the eyes after application [55].
2. **Irritation:** The excipients used in eye drop dosage forms like anti-oxidants, surfactants, buffering agents, etc., may irritate the tissues of the eyes.
3. **Patient Non-Compliance:** Various ocular formulations, such as ointments, gels, and the like, are not tolerated by some patients due to the complex installation.
4. **Drug Loss:** This, from the eye, occurs as a result of the blinking reflexes, provoked lachrymation, and normal tear turnover.
5. **Non-Sustain Action:** Because there are no controlled release polymers in traditional ocular dosage forms, medication release is very quick.

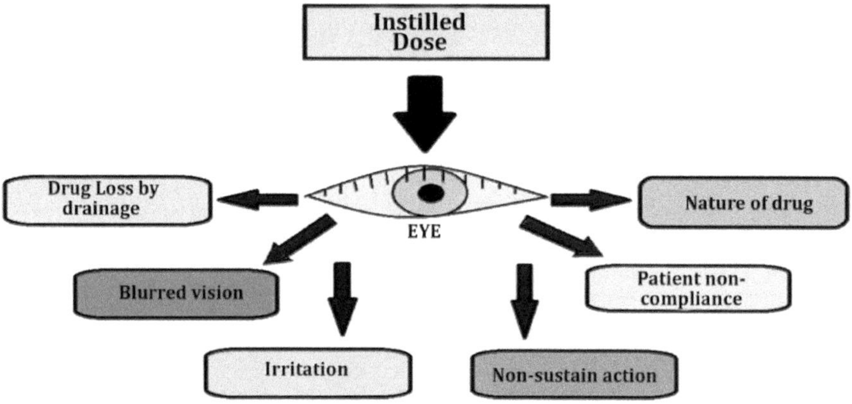

FIGURE 5.2 Different drawbacks of conventional ocular dosage form

Source: Reprinted from Ref. [69]. Open access..

5.3 DIFFERENT ANATOMICAL BARRIERS TO RESTRICT OCULAR DOSAGE FORMS

5.3.1 TEAR AS BARRIER

Because of weakening by tear turnover, increased freedom, and medicine tying with tear proteins, the tear film is the primary pre-corneal hindrance of the eye, reducing the absorbing convergence of the supplied pharmaceuticals. Lipids, mucins, and water are the main ingredients of the tear film. The tear film can moisten the surface of the eye due to a coating of mucin attached to the corneal and conjunctival epithelia. The tear layer is composed of a sloppy mucin gel enclosed by a slim layer of lipids and is very thin, about 5 microns. During irritation or emotional stress, the lacrimal production water content of tears is enormously increased [67].

5.3.2 CORNEA AS BARRIER

The cornea comprises five layers, including epithelium of cornea layer of bowman, stroma, Descemet's film, storm cellar layer, and endothelium. For drug pervasion, each layer has an alternative extremity and a rate-limiting course of action. The corneal epithelium is lipid-soluble by nature, and stretched junctions with cells are structured to keep para-cellular drug pervasion to a minimum, similar to the tear film. Corneal obstruction is an imperative involuntary and synthetic obstacle limiting the entrance of external substances in the visual site and ensuring intraocular tissues. The cornea is a transparent vascular structure having a breadth of 12 mm and thickness of 520 mm [72].

5.3.3 CONJUNCTIVA BARRIER

As a suspicious blockage on the visual external, the conjunctiva plays an important role. It helps to organize and secure the tear film during the production of body fluid glycoproteins, and it has a lot of capillaries and lymphatics. Even when the conjunctiva intersects, a significant portion of the medicine is lost to systemic distribution. The residual medicine, which generally comprises collagen and mucopolysaccharides, might enter the sclera. The medicine must pass through the bulbar conjunctiva, which is permeable to various extremity and size pharmaceuticals, in order to reach the leading eye through the non-corneal route. Administrated medications can be flushed out of the conjunctiva or sclera spaces by blood and lymph. Because the veins of the conjunctiva do not form a tight junction barrier, drug atoms can enter the bloodstream via pinocytosis [62].

5.3.4 BLOOD-RETINAL BARRIER

BRB restricts the drug transport from the blood to the retina. It is surrounded by endothelial cells from the retinal vein and colored epithelial cells from the retina. Multilayer neurons are migrated distally through the sub-chondral space of the retinal monolayer epithelium (RPE), circumscribing the outer surface of the neural retina away from the choroid. In a physiological situation, the retina is permanently connected to the RPE [13]. The RPE plays an important role in maintaining the neural retina's reasonability and capability. It is in charge of removing fluids from the sub-retinal gap with the purpose of maintaining the retinal connection and continuing through to the end retina in a dehydrated state [78].

5.4 CONCEPT OF NANOTECHNOLOGY AND NANOBIOMATERIALS IN OCULAR DELIVERY OF DRUGS

Nanotechnology applies molecular and submicron level substances in engineering, electronics, physics, materials science, medicine, and manufacturing at the molecular and submicron levels. According to Albert Franks, nanotechnology is science and technology whose dimensions span from 0.1 to 100 nanometers. Nanotechnology is the science of controlling matter at the microscopic level [29]. Nanotechnology is a branch of science concerned with the study, processing, and use of a variety of materials, electronics, and functional systems at the nano-scale level. The term "nanomedicine" refers to nanotechnology in the diagnosis, cure, prevention, and monitoring of various diseases. Nanotechnology is a field that studies, designs, synthesis, and manipulates nano-scale materials. According to the National Nanotechnology Initiative, the core of nanotechnology is the ability to work at the molecular level, atom by atom, to develop a massive structure with a fundamentally novel molecular organization. It enables applications for illness diagnosis, treatment, prevention, and monitoring.

According to the World Health Organization, about 1.3 billion individuals worldwide had some visual impairment in 2018, with untreated refractive errors and cataracts being the most common causes. Cataract, trachoma, corneal scarring, glaucoma, retinopathy due to diabetes, macular degeneration related to ages, and congenital disabilities account for about 36 million blind people. It is estimated that 80% of the population is illiterate. The bioavailability of ophthalmic drugs from typical topical formulations is well known, with only 1–5% of the topically applied drug penetrating the cornea, due to different defense mechanisms and several barriers to drug penetration, including rapid nasal drainage due to the high turnover of tear fluid and eyelid blinking and hydrophilic fat in the cornea [21, 40].

In addition to the anatomical and physiological limitations of the eye, any other limiting factor encountered with anti-inflammatory or immunosuppressive agents is poor water solubility [37, 40, 51]. They, therefore, require a complex formulation that follows regulatory specifications due to their attractiveness, low bioavailability, and low solubility in water. Therefore, certain pharmaceutical bases are advertised with anti-infective molecules [27]. Nanotechnology software, mainly based on the comprehensive treatment of ophthalmic diseases, is the wish of thousands of people with eye diseases. The subject of ocular drug delivery has been strongly influenced in the past decade by technological advances, especially nanostructured drug delivery systems [81]. Ocular preparation is one of the nanotechnology-based ophthalmic therapy approaches. To ensure low irritation, sufficient bioavailability, and tissue compatibility, suitable particle sizes are sought for anterior and posterior drug delivery systems. Nano-assistants, such as nanoparticles, nano-administrators, liposomes, nano micelles, and dendrimers, are used to deliver drugs through the eye.

5.4.1 BENEFITS OF OCULAR DELIVERY BASED ON NANOTECHNOLOGY

Nanotechnology-based drug delivery systems have various benefits over other drug delivery methods, especially in the ophthalmic delivery of drugs using nanoparticle-based delivery systems. They act more quickly than controlling the rate of drug release and protect the drug from denaturing agents due to the unique makeup of the eye. Today's most popular nanotechnology-based ophthalmic drug delivery systems are nanoparticles, liposomes, nanosuspensions, microemulsion, and iontophoresis. These methods can penetrate the protective barrier of the eye. A few of their merits are mentioned below:

- Conventional ophthalmic systems have the problem of pulsed dosing, which is overcome by nanotechnology-based ophthalmic drug delivery systems because they have more precise dosing.
- Nanocarriers have an advantage in targeted drug delivery [26].
- The sustained release effect is achieved due to the nano-molecular nature of the drug as evidenced by the increased residence time of the drug in the surrounding environment. Nanoparticles protect drugs from degenerative agents [47].

- A nanoscale has a faster onset of action and changes the investment dose, altering pharmacokinetic parameters, thereby improving the safety and efficacy of the drug [33].
- Nanotechnology may also be employed in medicine delivery and gene therapy by leveraging nanodevices and self-assembled materials [31].
- Nano-transporters improve the interaction with the corneal and conjunctiva epithelium, improving their bioavailability [73].
- Increase the residence time of related drugs on the ophthalmic surface and reduce the degradation of unmetabolized drugs [73].

5.5 BIO-NANOMATERIAL-BASED OPHTHALMIC DELIVERY OF DRUGS

As for the treatment of ophthalmic diseases, the development of science in recent years is conducive to new advances in the treatment of ophthalmic diseases. Current anterior and posterior ocular therapy procedures are based on the use of nanotechnology in conjunction with leading pharmaceutical substances. With fine-tuning, the nano-molecular system can guarantee a reduction in side effects and an increase in bioavailability, and better absorption. In current practice, several examples of nano-support are used: dendrimers, liposomes, nanoparticles, nano-delivery, and nano-bacterial cells are just a few examples that can use to treat problems diseases of the eye (Figure 5.3) [20, 71, 73].

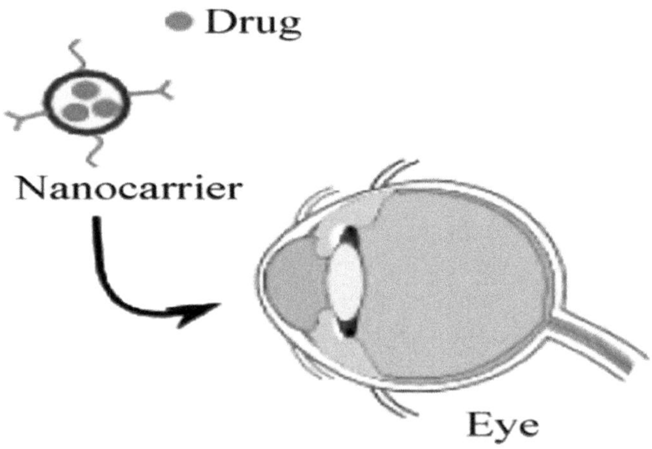

FIGURE 5.3 Eye and the nanocarriers [69].

5.5.1 NANOPARTICLES

These treatments must be prepared specifically for use in ocular tissues in order to be safe and effective. Several researchers have created nanoparticles containing pharmaceuticals for delivering medications to the anterior and posterior eye tissues, bringing attention to the delivery of ophthalmic drugs. Nanoparticles are colloidal supports with nanoscale sizes from 10 nm to 1,000 nm. Polymers such as albumin, sodium alginate, chitosan, polylactidecoglycolide, polylactic acid, and polycaprolactone prepare ophthalmic beads. Nanoparticles containing a drug can be of two types. The drug is enclosed in a polymer shell called a nanocapsule or a drug evenly dispersed in the polymer matrix called a nanoparticle [36]. The factors related to the adherence of the nanoparticles are their small size, which is a very cool feature given their promising properties, reduced corneal tissue irritation, and ability to maintain drug delivery by avoiding additional damage from frequent injections. For this, nanoparticles with mucosal adhesion properties have been prepared with the aim of being used topically, with the result of improved residence time in the pre-reproductive compartment. To improve the survival of the nanoparticles in the compartment above, some compounds are currently used as large molecules, without water-soluble molecules like chitosan, hyaluronic acid, or polyethylene glycol (PEG) water from the visual system [17, 64]. The superficial barrier prevents a systematic and direct approach to drugs for a particular site. The advantageous biological properties of the huge drug-loaded NPS comprise the high capacity of reducing toxic substances; possess high penetrating strength through eye surfaces, ability to transform lipophilic substances to hydrophilic substitutes [3, 34]. Some NP loading methods can benefit a mixture of long-term ophthalmic drugs natamycin encapsulated nanoparticles as chitosan and lecithin have high biometric use in rabbit eyes [4, 28, 30, 45]. Top of form bottom of form another study by Musumeci et al. [53] reported that nanoparticles made from a mixture of PLGA PEG are more sensitive to melatonin and were successful in lowering intraocular pressure in rabbits with nanoparticles designed with PLGA alone and purified with melatonin or even with the other aqueous solutions. One proposal to interpret the results is that the nanoparticles generated with PLGA PEG reach a reduced zeta potential than the PLGA nanoparticles. For this reason, substances may interact with the surface of the eye, producing a longer-lasting hypotensive effect.

5.5.2 LIPOSOMES

Liposomes of their brilliant biocompatibility are considered as perfect drug delivery mechanisms for ophthalmic applications. Both hydrophilic and hydrophobic drugs can be encapsulated in liposomes. Liposomes can deliver drugs to both posterior and anterior segments of the eye. Liposomes are the lipid-containing vesicular framework comprising one or more phospholipid bilayers encasing a watery center. The liposomes may be of unilamellar vesicles with a size of 10–100 nm, and giant measured unilamellar vesicles of size 100–300 nm, and multilamellar estimated vesicles containing more than one bilayer [6]. Liposomes assist improve contact time with corneal and conjunctival surfaces, which is important for poorly swallowed drugs; nevertheless, the negatively charged mucin on the corneal epithelium may interact with liposomes with a positive charge [48]. Mishra et al. [50] prepared and evaluated timolol maleate-loaded chitosan-coated liposomes to enhance ocular bioavailability and permeation, increasing precorneal residence time. Corneal permeation was increased 3.18-fold as compared to commercial eye drops; significant mucin adhesion was also seen by liposome formulation. A recent study by Natarajan et al. [54] of latanoprost administration via encapsulated liposomes. This formulation was applied in the anterior segment of the eyeball, and it confirmed that this liposomal formulation was capable of reducing IOP in rabbit eyes for 50 days. A limited number of liposomal formulations are submitted for preclinical testing, and an even smaller number are marketed. Still, currently, a wide variety of pharmaceutical formulations using liposome technology are being developed [8]. The formulations already in the market are Tears again® and Visudyne®. The first product was formulated to treat dry eyes that outperformed isotonic saline solutions and triglyceride-containing gels and was formulated as a liposome-based spray of a phospholipid nature [9, 18]. The latter is used to treat age-related macular degeneration as a therapy to recirculate choroidal tissue using the liposomal formulation of osteporfin, a light irritant used in phototherapy motion. Other diseases to which it can be applied are myopia or histiocytosis of the eye [5, 61].

5.5.3 NANOSUSPENSION

Nanosuspensions are characterized by drugs with poor aqueous solubility dispersed in a suitable medium and are developed as a potential methodology for transporting poorly soluble medications. Hydrophobic drugs are encapsulated in a pretty appropriate way by forming nanosuspensions. Nanosuspensions offer plenty of favorable circumstances for ophthalmic drug delivery, e.g., disinfection, minor disturbance, simplicity of eye drop detailing, increment in precorneal arrangement time, and altering the bioavailability of ophthalmic medications insoluble in tear liquid [76]. The feasibility of nanosuspensions in improving the ocular bioavailability of the glucocorticoid family of drug has been demonstrated in a few exploratory investigations. Nanosuspensions do not affect the eyes, especially on or conjunctiva, cornea, or iris; therefore, they act as inert carriers for ocular drugs [15, 32]. To control a stimulating environment impacting the front segment of the eye, glucocorticoids such as hydrocortisone, dexamethasone, and prednisolone are commonly used. The ocular bioavailability of prednisolone, hydrocortisone, and dexamethasone from nanosuspensions and microcrystalline suspensions has been investigated. Compared with other conventional eyedrop delivery methods, the application of nano-supporting formulations to the ocular tract gives optimal results such as reduced irritation, better eyedrop configuration, increased half-life in presynaptic tissue, sterilizing the product, and improving the solubility of hydrophobic drugs in tear fluid [15, 70].

5.5.4 NANOFIBER-BASED NANOPATCH

In recent times, nanofiber-based nano patches have been more efficient for the delivery of ocular drugs. It can successfully insert the nanofibrous patch into the conjunctiva sac of the eye [74]. The role of polymer in the formulation of nanofiber patches is most important. The nature of polymers should be biodegradable and biocompatible to degrade the nanofiber patch with time slowly. Due to slow degradation, the release of drug molecules is sustained. The physical characteristics and specific practical uses of the selected polymer depend on the degree of hydrolysis and polymerization [74]. In the field of ophthalmology, electrospun nanofibers are of interest because the soft materials come in the form of eye patches, and the nanofiber patches can quickly adapt to the corneal and sclera surfaces and stay on the eye surfaces for a reasonable amount of time and act as sustainable platforms [7, 14, 25]. Although there are several methods like solvent casting, a glass substrate, and melt extrude methods used to fabricate ocular inserts. Still, the ocular inserts which are prepared by using electrospun nanofibers as a matrix for drug loading have numerous advantages like increased ocular residence, the possibility of releasing the drug at a slow constant rate, accurate dosing and increased shelf life, possess high surface-to-volume ratio, highly porous, more than one drug can be encapsulated directly into the fibers. Electrospinning is the most commonly used technique to formulate nanofibers. Its advantages include continuous and straightforward techniques making it possible to produce nanofibers from a wide range of polymers with industrial scale-up. Therefore, compared with liquid and semi-solid preparations, electrospun nanofiber inserts can better overcome barriers that oppose ocular bioavailability [25].

5.5.5 MICRONEEDLES

Microneedle-based strategy is a negligibly intrusive medication delivery mechanism to back visual tissues and is developing as a new method in the field of ophthalmology. The danger and inconveniences related to intravitreal infusions are lessened by the microneedle-based organization methodology [75]. Microneedles deliver suspensions, nanoparticles, and microparticles into the sclera. Microneedles are specially designed to enter only a few microns into the sclera, avoiding damage to deeper visual tissues. They can also store medicine with a carrier framework in the sclera or in the "suprachoroidal space (SCS)" a small gap between the sclera and the choroid [58].

5.5.6 NANOMICELLES

The most commonly used transporter frameworks to specify the helpful experts as exact watery configurations are nanomicelles. Nano micelles are prepared using both hydrophilic and lipophilic components. These particles might have a polymeric structure in nature or maybe surfactants also. In no time, a significant hobby is being demonstrated to improve nanomicelles detailing-based innovation for ophthalmic drug conveyance [46]. Some efforts are made to take advantage of nanomicelles to deliver medicines to the rear segments of the eyes. It is expected that predators can help improve the improved transmission effects in new vascularized locations. The pre-planning of the United States has shown that it is mainly accumulated at the level of larger pathological neonatal vessels than typical tissues [24, 49].

5.5.7 DENDRIMERS

Dendrimers can be used as an effective medium for the delivery of ophthalmic drugs. Dendrimers are characterized by supermolecular compounds having an inner core and a series of branches around that core. Dendrimers have become striking systems for drug transport mechanisms due to their properties ranging from their nanoscale size, easily prepared and functionalized. They have the ability to show numerous copies of surface groupings. Polymers, such as poly (acrylic acid), are bio-adhesive and are used to advance the delivery of drugs and release by increasing the contact time of the drug with the absorption region to prolong its duration stay and reduce medication frequency. Bio-adhesive polymers have disadvantages such as blurred vision and fog formation in the corneal area. To avoid these problems, dendrimers such as polyamidoamine are used [43, 49].

5.5.8 NANOCRYSTALS

Nanocrystals are made up of a simple system for their production and use. Nanocrystals are nanoparticles that do not contain any matrix material and are 100% drug compounds, ranging in size from 200 to 500 nm. Several methods, such as top-down technology and bottom-up method, are used to reduce the particle size of drugs. Nanocrystals can be applied in a variety of uses, and this is where they stand out. Nanocrystals are used to deliver ophthalmic drugs to increase retention time. Several studies have been carried out on the use of NSAIDs in ophthalmic formulations as nanocrystals [42, 59].

5.5.9 NANOSYSTEMS OF IN SITU GELS

The notation used refers to a solution with macromolecular properties, which can undergo a phase change from liquid phase to gel and then form a viscous gel under the influence of the medium. *In situ* gelling also occurs by changes, e.g., pH change, temperature change, whether ions are present or absent and frequency of UV radiation. The main factor chosen for the development and application of such a system to ocular structures was the temperature at which heat-sensitive gels further grow [39, 68]. The available literature describes a range of polymers sensitive to temperature changes and can be used in the eye. Among these compounds are PEG, polylactide, and polyglycolide and their derivatives, polycaprolactone or even chitosan. The mechanism used by these chemicals depends on the development of micellar agglomerates when the temperature is reaching a definite threshold, i.e., the formation of colloids by encapsulation or agglomeration. These systems are designed for use as drug delivery agents and can be manufactured into gels. After mixing both the solutions with the desired drug, the gel will continue to form in the administration medium as a function of temperature between physiological values [52, 68, 77, 79]. These heat-sensitive preparations are attached to both segments of the eye and have shown promising efficacy in improving drug bioavailability [11, 65].

5.6 SUMMARY

The requirement of novel sustained and controlled release ocular drug delivery systems are met by including polymers in the ocular dosage forms, which help in constant and controlled release of medication at targeted sites in the eyes. Polymers help to increase the bioavailability of ocular dosage forms, which otherwise are rapidly eliminated from the eyes. The bioadhesive polymers help retain the formulation in the eyes for an extended duration by adhering it to the eyes' surface. Polymers can coat the drugs to control their release and deliver it in a better way. Nanotechnology is the study of microscopic structures ranging from 0.1 to 100 nm. Nanomedicine is a growing field that uses nanotechnology in medical applications, with specific medical interventions at the molecular level to heal or repair tissue. The application of nanomedicine to treat eye diseases has made great strides in recent years. Nanotechnology and nanomedicine have many appliances in the field of ophthalmology. The use of these devices and nano-formulations favors the bioavailability of the drug, allows diffusion through the anatomical barrier, and can reduce the adverse reactions attributed to the use of traditional topical ophthalmic medications and can be lowered greatly to a certain extent point, the invasive intervention is in the posterior pole, and the complications due to the use of certain drugs require surgical operations to be implanted. In the end, the benefits of drugs and their harmful effects were reduced, opening a large window within the scope of the so-called personalized drugs. It will likely require unique designs for people with individual characteristics, new research, and ongoing research in different animal and laboratory models. The use of nanocarriers in the treatment of ophthalmic illnesses has sparked new interest in developing fresh and highly developed delivery methods that carry drugs more effectively to inaccessible parts of the eyes. By nanotechnology, scientists have focused on designing new strategies for ophthalmic drug delivery, which help remove the drawbacks and problems associated with topical and intravitreal routes that are not considered safe and effective. This chapter focuses on new drug delivery mechanisms to ocular surfaces with particular emphasis on nanotechnology-based multidisciplinary approaches like nanosuspensions, nanocrystals, dendrimers, and microneedle.

KEYWORDS

- **nanofibers**
- **nanomaterials**
- **nanomedicine**
- **nanoparticles**
- **ocular drug delivery**
- **retinal monolayer epithelium**

REFERENCES

1. Achouri, D., Alhanout, K., Piccerelle, P., & Andrieu, V., (2013). Recent advances in ocular drug delivery. *Drug Development and Industrial Pharmacy, 39*(11), 1599–1617.
2. Al-Amri, A. M., (2014). Long-term follow-up of tacrolimus ointment for treatment of atopic keratoconjunctivitis. *American Journal of Ophthalmology, 157*(2), 280–229.
3. Ammar, H. O., Salama, H. A., Ghorab, M., & Mahmoud, A. A., (2009). Nanoemulsion as a potential ophthalmic delivery system for dorzolamide hydrochloride. *AAPS PharmSciTech, 10*(3), 808–819.
4. Baba, K., Tanaka, Y., Kubota, A., Kasai, H., Yokokura, S., Nakanishi, H., & Nishida, K., (2011). A method for enhancing the ocular penetration of eye drops using nanoparticles of hydrolyzable dye. *Journal of Controlled Release, 153*(3), 278–287.
5. Bakri, S. J., & Kaiser, P. K., (2004). Verteporfin ocular photodynamic therapy. *Expert Opinion on Pharmacotherapy, 5*, 195–203.
6. Bhattacharjee, A., Das, P. J., Adhikari, P., Marbaniang, D., Pal, P., Ray, S., & Mazumder, B., (2019). Novel drug with special reference to liposomal ocular delivery. *European Journal of Ophthalmology, 29*(1), 113–126.
7. Bhattarai, R. S., Das, A., Alzhrani, R. M., Kang, D., Bhaduri, S. B., & Boddu, S. H., (2017). Comparison of electrospun and solvent cast polylactic acid (PLA)/poly(vinyl alcohol) (PVA) inserts as potential ocular drug delivery vehicles. *Materials Science and Engineering: C, 77*, 895–903.

8. Bulbake, U., Doppalapudi, S., Kommineni, N., & Khan, W., (2017). Liposomal formulations in clinical use: An updated review. *Pharmaceutics, 9*, 12.
9. Calvao-Santos, G., Borges, C., Nunes, S., Salgado-Borges, J., & Duarte, L., (2011). Efficacy of 3 different artificial tears for the treatment of dry eye in frequent computer users and/or contact lens users. *European Journal of Ophthalmology, 21,* 538–544.
10. Chien, Y. W., (1992). *Novel Drug Delivery Systems* (2nd edn., pp. 1–42). New York: Marcel Dekker, Inc.
11. Cholkar, K., Patel, S. P., Vadlapudi, A. D., & Mitra, A. K., (2013). Novel strategies for anterior segment ocular drug delivery. *Journal of Ocular Pharmacology and Therapeutics, 29*, 106–123.
12. Chrai, S. S., Makoid, M. C., Erikson, S. P., & Robinson, J. R., (1974). Drop size and initial dosing frequency problems of topically applied ophthalmic drugs. *Journal of Pharmaceutical Sci*ences, *64*, 333–338.
13. Cunha-Vaz, J., Marques, F. B., Fernandes, R., Alves, C., & Velpandian, T., (2016). Drug transport across blood-ocular barriers and pharmacokinetics. In: Velpandian, T., (ed.), *Pharmacology of Ocular Therapeutics* (pp. 37–63). Springer.
14. Da Silva, G. R., Lima, T. H., Fernandes-Cunha, G. M., Oréfice, R. L., Da Silva-Cunha, A., Zhao, M., & Behar-Cohen, F., (2019). Ocular biocompatibility of dexamethasone acetate loaded poly("-caprolactone) nanofibers. *European Journal of Pharmaceutics and Biopharmaceutics, 142*, 20–30.
15. Das, S., & Suresh, P. K., (2011). Nanosuspension: A new vehicle for the improvement of the delivery of drugs to the ocular surface. Application to amphotericin B. *Nanomedicine: Nanotechnology, Biology and Medicine, 7*, 242–247.
16. Davis, J. L., Gilger, B. C., & Robinson, M. R., (2004). Novel approaches to ocular drug delivery. *Current Opinion in Molecular Therapeutics, 6*, 195–205.
17. Diebold, Y., & Calonge, M., (2010). Applications of nanoparticles in ophthalmology. *Progress in Retinal and Eye Research, 29*(6), 596–609.
18. Essa, L., Laughton, D., & Wolffsohn, J. S., (2018). Can the optimum artificial tear treatment for dry eye disease be predicted from presenting signs and symptoms?. *Contact Lens and Anterior Eye, 41,* 60–68.
19. Friedrich, S. W., Saville, B. A., Cheng, Y. L., & Rootman, D. S., (1996). Pharmacokinetic differences between ocular inserts and eye drops. *Journal of Ocular Pharmacology and Therapeutics, 12*(5), 18.
20. Fuente De La, M., Ravina, M., Paolicelli, P., Sanchez, A., Seijo, B., & Alonso, M. J., (2010). Chitosan-based nanostructures: A delivery platform for ocular therapeutics. *Advanced Drug Delivery, 62*, 100–117.
21. Gause, S., Hsu, K. H., Shafor, C., Dixon, P. Q., Powell, K. C., & Chauhan, A., (2016). Mechanistic modeling of ophthalmic drug delivery to the anterior chamber by eye drops and contact lenses. *Advances in Colloid and Interface Sci*ence, *233*, 139–154.
22. Ghate, D., & Edelhauser, H. F., (2006). Ocular drug delivery. *Expert Opinion on Drug Delivery, 3*(2), 275–287.
23. Gibaldi, M., & Perrier, D., (1993). *Pharmacokinetics* (2nd edn., p. 494). New York: Marcel Dekker, Inc.
24. Gote, V., Sikder, S., Sicotte, J., & Pal, D., (2019). Ocular drug delivery: Present innovations and future challenges. *Journal of Pharmacology and Experimental Therapeutics, 370*, 602–624.

25. Göttel, B., De Souza, E. S. J. M., Santos De, O. C., Syrowatka, F., Fiorentzis, M., Viestenz, A., Viestenz, A., & Mäder, K., (2020). Electrospun nanofibers A promising solid *in-situ* gelling alternative for ocular drug delivery. *European Journal of Pharmaceutics and Biopharmaceutics, 146,* 125–132.
26. Hans, M. L., & Lowman, A. M., (2002). Biodegradable nanoparticles for drug delivery and targeting. *Current Opinion in Solid State and Materials Science, 6*, 319–327.
27. Hoffman, R. S., Braga-Mele, R., Donaldson, K., Emerick, G., Henderson, B., Kahook, M., Mamalis, N., et al., (2016). Cataract surgery and non-steroidal anti-inflammatory drugs. *Journal of Cataract and Refractive Surgery, 42,* 1368–1379.
28. Ibrahim, H. K., El-Leithy, I. S., & Makky, A. A., (2010). Mucoadhesive nanoparticles as carrier systems for prolonged ocular delivery of gatifloxacin/prednisolone bitherapy. *Molecular Pharm*aceutics, *7,* 576–585.
29. Jitendra, S. P., Banik, A., & Dixit, S., (2011). A new trend: Ocular drug delivery system. *International Journal of Pharmaceutical Sciences, 2*, 457–461.
30. Jung, H. J., & Chauhan, A., (2012). Temperature sensitive contact lenses for triggered ophthalmic drug delivery. *Biomaterials, 33*(7), 2289–2300.
31. Kabanov, A. V., (2006). Polymer genomics: An insight into pharmacology and toxicology of nanomedicines. *Advanced Drug Delivery Review, 58*(15), 1597–1621.
32. Kassem, M. A., Abdel-Rahman, A. A., Ghorab, M. M., Ahmed, M. B., & Khalil, R. M., (2007). Nanosuspension as an ophthalmic delivery system for certain glucocorticoid drugs. *International Journal of Pharmaceutics, 340*, 126–133.
33. Kayser, O., Lemke, A., & Hernandez-Trejo, N., (2005). The impact of nanobiotechnology on the development of new drug delivery systems. *Current Pharmaceutical Biotechnology, 6*(1), 3–5.
34. Kesavan, K., Balasubramaniam, J., Kant, S., Singh, P. N., & Pandit, J. K., (2011). Newer approaches for optimal bioavailability of ocularly delivered drugs: Review. *Current Drug Delivery, 8*(2), 172–193.
35. Khar, R. K., & Vyas, S. P., (2004). *Targeted and Controlled Drug Delivery: Novel Carrier Systems* (1st edn., p. 594). New Delhi: C.B.S. Publishers and Distributors, Inc.
36. Kim, J., & Chauhan, A., (2008). Dexamethasone transport and ocular delivery from poly hydroxyethyl methacrylate gels. *International Journal of Pharmaceutics, 353*, 205–222.
37. Koç, F. E., & Senel, M., (2013). Solubility enhancement of non-steroidal anti-inflammatory drugs (NSAIDs) using polypolypropylene oxide core PAMAM dendrimers. *International Journal of Pharmaceutics, 451*, 18–22.
38. Kompella, U. B., Kadam, R. S., & Lee, V. H. L., (2010). Recent advances in ophthalmic drug delivery. *Therapeutic Delivery, 1*, 435–456.
39. Kouchak, M., (2014). *In situ* gelling systems for drug delivery. *Jundishapur Journal of Natural Pharmaceutical Products, 9(3*), e20126. doi: 10.17795/jjnpp-20126.
40. Lallemand, F., Perottet, P., Felt-Baeyens, O., Kloeti, W., Philippoz, F., Marfurt, J., Besseghir, K., & Gurny, R., (2005). A water-soluble prodrug of cyclosporine A for ocular application: A stability study. *European Journal of Pharmaceutical Sciences, 26*, 124–129.
41. Lee, V. H., & Robinson, J. F., (1986). Review: Topical ocular drug delivery; recent developments and future challenges. *Journal of Ocular Pharmacology, 2*, 67–108.
42. Li, V. H., & Patton, F., (1986). Ocular drug delivery of progesterone using nanoparticles. *Journal of Microencapsulation, 3*, 213–218.

43. Loftsson, T., Jarho, P., Masson, M., & Jarvinen, T., (2005). Cyclodextrins in drug delivery. *Expert Opinion on Drug Delivery, 2*, 335–351.
44. Mainardes, R. M., Urban, M. C., Cinto, P. O., & Chaud, M. V., (2005). Colloidal carriers for ophthalmic drug delivery. *Current Drug Targets, 6*, 363–371.
45. Maiti, S., Paul, S., Mondol, R., Ray, S., & Sa, B., (2011). Nanovesicular formulation of brimonidine tartrate for the management of glaucoma: *In vitro* and *in vivo* evaluation. *AAPS PharmSciTech, 12*(2), 755–763.
46. Mandal, A., Patel, P., Pal, D., & Mitra, A. K., (2019). Multi-layered nanomicelles as self-assembled nanocarrier systems for ocular peptide delivery. *AAPS PharmSciTech, 20*(2), 66–76.
47. Marcato, P. D., & Duran, N., (2008). New aspects of nanopharmaceuticals delivery systems. *Journal of Nanoscience and Nanotechnology, 8*, 1–14.
48. Meng, T., Kulkarni, V., Simmers, R., Brar, V., & Xu, Q., (2019). Therapeutic implications of nanomedicine for ocular drug delivery. *Drug Discovery Today, 23*, 1075–1091.
49. Milhe, O. M., Myles, C., Yamakawa, J., & Kawase, M., (2000). Polyamidoamine starburst dendrimers as solubility enhancers. *International Journal of Pharmaceutics, 197*, 239–241.
50. Mishra, G. P., Bagui, M., Tamboli, V., & Mitra, A. K., (2011). Recent applications of liposomes in ophthalmic drug delivery. *Journal of Drug Delivery, 2011*, 1–14.
51. Moya-Ortega, M. D., Messner, M., Jansook, P., Nielsen, T. T., Wintgens, V., Larsen, K. L., Amiel, C., et al., (2011). Drug loading in cyclodextrin polymers: Dexamethasone model drug. *Journal of Inclusion Phenomena and Macrocyclic Chemistry, 69*, 377–382.
52. Mundada, A. S., & Avari, J. G., (2009). *In situ* gelling polymers in ocular drug delivery systems: A review. *Critical Reviews in Therapeutic Drug Carrier Systems, 26*, 85–118.
53. Musumeci, T., Bucolo, C., Carbone, C., Pignatello, R., Drago, F., & Puglisi, G., (2013). Polymeric nanoparticles augment the ocular hypotensive effect of melatonin in rabbits. *International Journal of Pharmaceutics, 440*, 135–140.
54. Natarajan, J. V., Ang, M., Darwitan, A., Chattopadhyay, S., Wong, T. T., & Venkatraman, S. S., (2012). Nanomedicine for glaucoma: Liposomes provide sustained release of latanoprost in the eye. *International Journal of Nanomedicine, 7*, 123–131.
55. Nayak, A. K., & Bera, H., (2019). *In situ* polysaccharide-based gels for topical drug delivery applications. *Polysaccharide Carriers for Drug Delivery, 2019*, 615–638.
56. Neefe, C. W., (1974). *Contact Lens for Ocular Drug Delivery* (Vol. 3, pp. 786–812). US Patent.
57. Patel, A., Cholkar, K., Agrahari, V., & Mitra, A. K., (2013). Ocular drug delivery systems: An overview. *World Journal of Pharmacology, 2*(2), 47.
58. Patel, S. R., Berezovsky, D. E., McCarey, B. E., Zarnitsyn, V., Edelhauser, H. F., & Prausnitz, M. R. (2012). Targeted administration into the suprachoroidal space using a microneedle for drug delivery to the posterior segment of the eye. *Investigative Ophthalmology & Visual Science, 53*(8), 4433–4441.
59. Patton, T. F., & Robinson, J. R., (1975). Ocular evaluation of polyvinyl alcohol vehicle in rabbits. *Journal of Pharmaceutical Sciences, 64*, 1312–1316.
60. Robinson, J. C., (1993). Ocular anatomy and physiology relevant to ocular drug delivery. In: Mitra, A. K., (ed.), *Ophthalmic Drug Delivery Systems* (pp. 29–57). New York: Marcel Dekker.
61. Rosenfeld, P. J., Saperstein, D. A., Bressler, N. M., Reaves, T. A., Sickenberg, M., Rosa, R. H. Jr., Sternberg, P. Jr., Aaberg, T. M. Sr., & Aaberg, T. M. Jr., (2004). Photodynamic

therapy with verteporfin in ocular histoplasmosis: Uncontrolled, open-label 2-year study. *Ophthalmology, 111*(9), 1725–1733.
62. Ruponen, M., & Urtti, A., (2015). Undefined role of mucus as a barrier in ocular drug delivery. *European Journal of Pharmaceutics and Biopharmaceutics, 96*, 442–446.
63. Saettone, M. F., & Salminen, L., (1995). Ocular inserts for topical delivery. *Advanced Drug Delivery Reviews, 16*, 95–106.
64. Sanchez-Lopez, E., Espina, M., Doktorovova, S., Souto, E. B., & Garcia, M. L., (2017). Lipid nanoparticles (SLN, NLC): Overcoming the anatomical and physiological barriers of the eye—part II—ocular drug-loaded lipid nanoparticles. *European Journal of Pharmaceutics and Biopharmaceutics, 110*, 58–69.
65. Sánchez-López, E., Egea, M. A., Cano, A., Espina, M., Calpena, A. C., Ettcheto, M., Camins, A., et al., (2016). PEGylated PLGA nanospheres optimized by design of experiments for ocular administration of dexibuprofen- *In vitro, ex vivo* and *in vivo* characterization. *Colloids and Surfaces B: Biointerfaces, 145*, 241–250.
66. Schoenwald, R. D., (2016). Ocular pharmacokinetics/pharmacodynamics. In: Mitra, A. K., (eds.), *Ophthalmic Drug Delivery Systems* (pp. 135–179). New York: Marcel Dekker.
67. Sharma, O. P., Patel, V., & Mehta, T., (2016). Nanocrystal for ocular drug delivery: Hope or hype. *Drug Delivery and Translational Research, 6*(4), 399–413.
68. Sheshala, R., Kok, Y. Y., Ng, J. M., Thakur, R. R., & Dua, K., (2015). *In situ* gelling ophthalmic drug delivery system: An overview and its applications. *Recent Patents on Drug Delivery and Formulations, 9*, 237–248.
69. Singh, A., Dogra, T. S., Mandal, U. K., & Narang, R. K., (2019). Novel approaches for ocular drug delivery: A review. *International Journal of Bio-Pharma Research, 8*(7), 2722–2735.
70. Singh, A., Rath, G., Singh, R., & Goyal, A. K., (2018). Nanofibers: An effective tool for controlled and sustained drug delivery. *Current Drug Delivery, 15*(2), 155–166.
71. Singh, T. R. R., & Jones, D., (2014). Advances in ophthalmic drug delivery. *Journal of Pharmacy and Pharmacology, 66*, 487–489.
72. Skalicky, S. E., Petsoglou, C., Gurbaxani, A., Fraser, C. L., & McCluskey, P., (2015). New agents for treating dry eye syndrome. In: Babizhayev, M. A., Li, D. W. C., Kasus-Jacobi, A., Zoric, L., & Alio, J. L., (eds.), *Studies on the Cornea and Lens* (pp. 115–129). Springer.
73. Souza, J. G., Dias, K., Pereira, T. A., Bernardi, D. S., & Lopez, R. F., (2014). Topical delivery of ocular therapeutics: Carrier systems and physical methods. *Journal of Pharmacy and Pharmacology, 66*, 507–530.
74. Stevenson, C. L., Santini, Jr. J. T., & Langer, R., (2012). Reservoir-based drug delivery systems utilizing microtechnology. *Advanced Drug Delivery Reviews, 64*(14), 1590–1602.
75. Thakur, S. R. R., Tekko, I., Mc Avoy, K., McMillan, H., Jones, D., & Donnelly, R. F., (2017). Minimally invasive microneedles for ocular drug delivery. *Expert Opinion on Drug Delivery, 14*(4), 525–537.
76. Van-Erdenbrugh, B., Froyen, L., & Vanden-Hotter, G., (2008). The importance of surface hydrophobicity on dissolution behavior upon redispersion of dried crystalline nano-suspensions. *European Journal Pharmaceutical Science, 35*, 127–135.

77. Van-Tomme, S. R., Storm, G., & Hennink, W. E., (2008). *In situ* gelling hydrogels for pharmaceutical and biomedical applications. *International Journal of Pharmaceutics, 355*, 1–18.
78. Yellepeddi, V. K., & Palakurthi, S., (2016). Recent advances in topical ocular drug delivery. *Journal of Ocular Pharmacology and Therapeutics, 32*(2), 67–82.
79. Zahir-Jouzdani, F., Wolf, J. D., Atyabi, F., & Bernkop-Schnurch, A., (2018). *In situ* gelling and mucoadhesive polymers: Why do they need each other? *Expert Opinions on Drug Delivery, 15*, 1007–1019.
80. Zaki, I., Fitzgerald, P., Hardy, J. G., & Wilson, C. G., (1986). Comparison of effect of viscosity on the precorneal residence of solution in rabbit and man. *Journal of Pharmacy and Pharmacology, 38,* 463–466.
81. Zimmer, A., & Kreuter, J., (1995). Microspheres and nanoparticles used in ocular drug delivery systems. *Advanced Drug Delivery Reviews, 16*(1), 61–73.

CHAPTER 6

APPLICATIONS OF NANOMATERIALS IN DENTISTRY

MD. JAHIDUL HAQUE, AHSAN HABIB MUNNA, AHMED SIDRAT RAHMAN AYON, ZARIN RAFA SHAITEE, SANZANA TABASSUM PROMA, SADIA AKTER, TASNUVA HUMAIRA, HUMAYAN KABIR, MINTU ALI, ABDUL KAIYUM, and SHAMIMUR RAHMAN

ABSTRACT

The advancement of nanomaterials has shown a new world full of possibilities in dentistry. The vast and effective applications of these advanced materials give us the message that nanotechnology can take us to the future of dentistry. Nanoparticles have an extremely small size and relatively larger surface area compared to larger particles. Nanomaterials used in dentistry also have outstanding biocompatibility. It is for this reason they are so effective in treating dental issues compared to traditional materials. Today Nanomaterials are also being used in complex dental operations and deadly disorders like oral cancer. But even after possessing great benefits, nanomaterials have some drawbacks too. The tiny size of these materials makes them very toxic, corresponding to their bulk counterpart. Sometimes they can also hamper genetic activities creating major problems. Therefore, more and more research has no alternatives to flourish in our dental sector by using nanotechnology. The chapter clarifies nanomaterials' concept and depicts nanomaterials' outstanding advancement in various dental fields along with their shortcomings.

Sustainable Nanomaterials for Biomedical Engineering: Impacts, Challenges, and Future Prospects.
Junaid Ahmad Malik, Megh R. Goyal & Mohamed Jaffer M. Sadiq (Eds.)

6.1 INTRODUCTION

Biomaterial scientists show a vast interest in nanomaterials to achieve dental application benefits because they have distinctive properties. The dentistry sector is all about dental and oral treatment and medicine. Dentistry is a predominant sector in medical science, and applications of nanomaterials have enriched this sector. Compared to past achievements, nanoparticles have changed this sector remarkably. Previously, different kinds of mistreatments were seen in dentistry. The Noble Prize winner named Richard P. Feynman first examined nanosized devices in 1959. In 2000, R. A. Freitas first used the term 'nano dentistry,' which means using nanoparticles in dentistry through modern technology [1]. He had a great interest in science fiction. Some of his ideas were how nanorobots could be useful in orthodontics, dentition regeneration, nanoparticles, and dentirobots (robots in dentifrices). Today, these ideas are being appreciated in practice. Due to the advancement of nanomedicine, dentistry is also starting to expand in the world of nanotechnology. It is not too far away that one-day nanotechnology will be a prominent candiate in the fields associated with medicine, surgery, materials as well as restorative dentistry. The inspiring contemporary areas today such as nanorobotics, nano-drugs, nanodiagnosis, nanosurgery as well as nanomaterials will have a far-reaching impact on clinical dentistry then.

Among the 10 most common diseases, 60% are oral problems [2]. The extremely smallsized nanoparticles of some nanomaterials can trigger the tooth remineralization process. As a result, dental caries can be treated much easier these days. Once, it was unimaginable that the arrangement of teeth can be rearranged, but the orthodontic process made it possible. Nanomaterials are gradually making the orthodontic process much easier. Such as Ni-P film impregnated with IF-WS_2 nanoparticles [3], the time of proceeding the orthodontic process can be reduced, making it easier for people. Presently, for root canal treatment Gutta-percha is used. It is a standard sealer nanomaterial, which makes this treatment more spontaneous [4]. Then Periodontal treatment has found a new approach only because of nanomaterials. Nanoparticles of $CaSO_4$ is being used in the periodontal process as they can conduct the antibacterial activity without doing any damage to the host [4]. Oral cancer can be detected by nanotechnology very precisely and can be cured by nanoparticles. Nowadays, nanotechnology has brought root canal treatment, endodontic, and prosthodontic treatment within our reach with much ease.

This chapter describes the concept of nanomaterials, noteworthy use of nanomaterials in dentistry, their effectiveness in those systems and finally, their limitations.

6.2 CONCEPT AND CLASSIFICATION

The term 'nano' came from the Greek word Nanos meaning 'very short man.' Nanotechnology studies the tiniest things in the world, like atoms and molecules. It deals with the manipulation of matters in nanoscale. Nanoscale ranges from 1–100 nanometers, whereas a nanometer is equal to 10^9 meters. Human hair has a diameter of 80,000 to 100,000 nanometers, whereas the helix of DNA can be seen to have a diameter of 2 nanometers. From these two examples, we can get an idea of how small a nanoscale is. Now let's jump into our topic of nanomaterial.

According to the International Organization of Standardization (ISO), the material has any of its external dimensions in nanoscale or has a nano-scaled internal or surface structure. A particulate form of nanomaterial may include agglomerates or aggregates. But it is mandatory to define their size along with their average particle sizes. The question is, what if any of the external dimensions of the aggregates and agglomerates of nanoparticles is more than 100 nanometers. Will they be treated as nanomaterial? Our answer would be 'yes.' It's because the greater specific surface area allows them to retain nanomaterials' properties despite having a larger particle size.

Nanomaterials can be found naturally or engineered with sets of specialized properties. They can possess properties which may not be present in their bulk counterpart. A massive variation in properties is observed when a material is cracked down to its nanoscale. Variations in electrical, mechanical, thermal, pharmacological, and catalytic properties are evident, which is attributable to the larger surface area to volume ratio. An increase in this ratio results in an increasing trend in surface reactivity. Crystal structures associated with nanomaterials get changed when they interact with dispersion media. Nanomaterials are more energy efficient. Since they can be used in nano-ranged sites, they have a versatile application in dentistry and other major fields.

Nanomaterials can be categorized based on different factors. Based on the dimension of nanomaterial, it can be classified into four categories. They are:

1. **Zero Dimensional:** This type of nanomaterials possesses all three dimensions in nanoscale, such as QDs, nanoparticles.
2. **One Dimensional:** It has one of its dimensions beyond the nanoscale. Such as – nanotubes, nanorods, nanowires.
3. **Two Dimensional:** It has only one dimension in the nanoscale. That means two of its dimensions are greater than the nanoscale range. Examples of such nanomaterials are graphene and nanolayers.
4. **Three Dimensional:** All of the dimensions are greater than the nanoscale. But they are still called nanomaterials because of typically show other signs of engineered nanomaterial, such as bundles of nanowires, nanorods, nanostructured materials, etc.

6.3 APPLICATION OF NANOMATERIAL IN DENTISTRY

6.3.1 IN DENTAL REMINERALIZATION

Dental remineralization is a reformatory process, and it occurs when there is any injury in the teeth. Acid is produced from our everyday food, which assaults our teeth and causes a cavity. But this acid gets neutralized by a neutralization reaction (acid-base reaction) caused by calcium, phosphate, and fluoride secreted from human saliva [5]. So, this is a naturally occurring remineralization process. Whenever acid is formed from food, it neutralizes by occurring buffer reaction, and our tooth becomes cavity-free.

If it is possible to increase the remineralization process in our mouth, it can resist various kinds of acid attacks in our teeth. Besides natural occurrences, this process can be created by using nanotechnology. Nowadays, nanotechnology is the most promising sector for increasing remineralization.

Nanoparticles like NACP (Nanoparticles of amorphous calcium phosphate) are very efficient in this case. Its diameter is 116 nm and adjusted by spray drying (it is a process by which we can get dry powder from the liquid by fast-drying, including hot gas).

6.3.1.1 PROCESS

NACP ionizes in the cavity, releasing calcium (Ca) and phosphate (PO_4) ions. After that, if any acid reaches the hole, it becomes neutralized [6]. It is the basic process for remineralization. The main purpose is to create a neutralization reaction environment in the mouth. NACP can also be used to make tablets or syrup-type medicine; then, it will be easier for people, especially children, to fear dentists. It is necessary to keep in mind that calcium and phosphate should be perfect in making medicine [22]. For this, proper pharmaceutical knowledge has no alternatives. Apart from calcium and phosphate, the other chemicals that react against acid can be taken for remineralization. Here two pictures are added; Figure 6.1 is about the situation before remineralization, and Figure 6.2 is after remineralization. In Figure 6.1, there is a deep cavity on the teeth, so natural remineralization, will not fully cure it. Therefore, it is required to do filling from a dentist. Instead of getting treated with machines, taking medicines is a simpler way for patients.

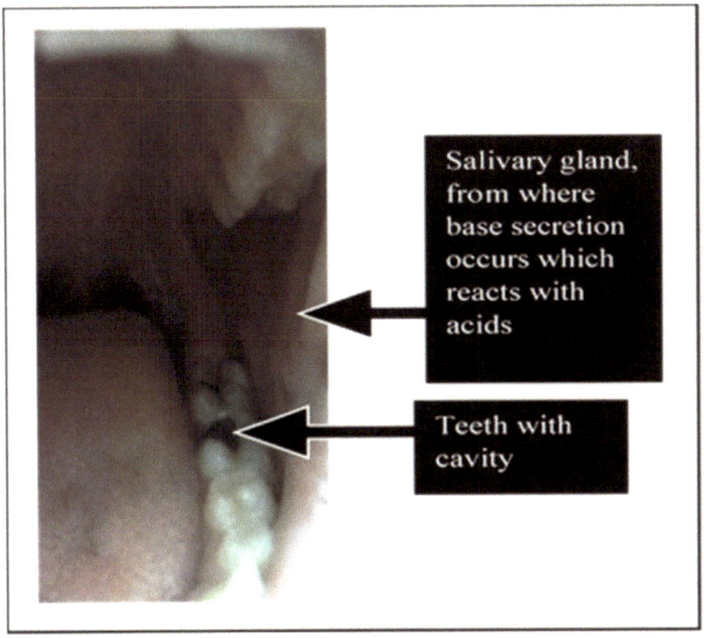

FIGURE 6.1 Teeth with the cavity.

Figure 6.2 indicates the condition after the remineralization process, where teeth have become cavity-free. So, the main target is increasing the neutralization reaction, which can prevent the dental cavity. Most of the time, the foods like cookies cause dangerous cavities in tooth and then sometimes the situation may need surgical approaches in serious cases. So, at this point of our discussion of tooth remineralization, it is needless to say nanocomposites have made our life easier by playing their noble role in fighting against cavities [21].

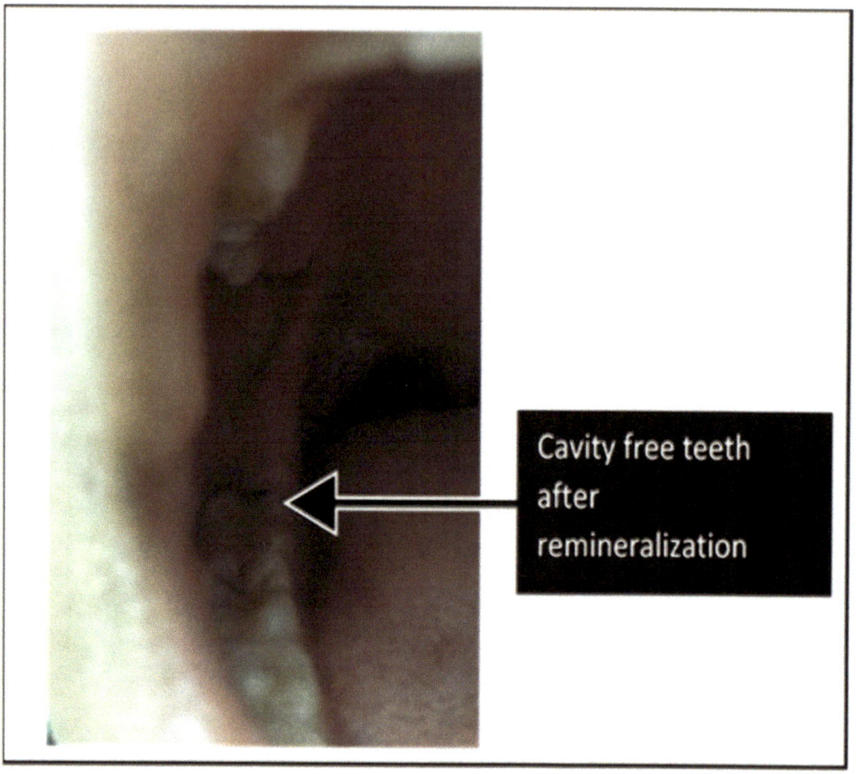

FIGURE 6.2 Cavity free teeth after remineralization.

6.3.2 *IN ORTHODONTICS*

Sometimes the spatial arrangement of our tooth is not correct. Some teeth have meandered in the cheek, and some teeth remain above or below the cheek. Orthodontics is a modern process in dentistry by which this zigzag arrangement of teeth can be repaired. This treatment applies braces over the tooth with surgery's help and utilizes the braces; the teeth are aligned perfectly after a definite period. It is a lengthy process, and this treatment causes a lot of pain for the patients. Sometimes teeth need to be removed if necessary.

It is known that nanotechnology and nanoparticles have made dentistry much easier for people. By using nanoparticles, this orthodontic process will be much facile for us.

6.3.2.1 PROCESS

Table 6.1 shows that Ni-P film impregnated with IF-WS_2 nanoparticles used in the braces. Along with them, there also remain wire (stainless-steel). The stainless-steel wire creates the mechanical pressure, and teeth are then forced to align perfectly and correct malocclusion (contact between upper and lower teeth), and spacing.

TABLE 6.1 Required Nanomaterial for Orthodontic Process

Material Studied	Nanoparticle	Parameters Assessed	Results	References
Orthodontic wire (stainless-steel)	Ni-P film impregnated with IF-WS_2 nanoparticles	• Frictional forces measured on coated and uncoated wires • Friction coefficient	• Reduced to 54% on coated wires • Friction coefficient reduced one-third from 0.25 to 0.08	[3]

When an object touches another object, friction produces at the interphase, and for this resistive force, teeth ultimately move to their ideal position. Here the produced forces are proportional to the frictional forces existing between the surfaces. If frictional forces are decreased between the orthodontic wire and brackets, there will be an increase in teeth' velocity, causing treatment time to decrease.

Here nanoparticles can be called dry lubricants coated over the orthodontic wire (stainless steel). Ni-P (nickel-phosphorus) film with IF-WS_2 impregnates up the wire. Their job is to decrease friction. So, the main target is to reduce friction, then the increase of velocity reduces the treatment time [7].

We can see the orientation not being right for a massive number of people around the globe. There is a high chance of forming cavities between the teeth in this condition, and it will produce lethal damage to the teeth. So this orthodontic process is required if our spatial arrangement of teeth is not correct. As we have already told that this is a long-time treatment, we can reduce this treatment time by using nanoparticles.

6.3.3 ENDODONTIC TREATMENT

Nanodentistry is classified into some applications for better understanding. The endodontic application is one of them. The term 'endo' means internal, and 'dontist' means treatment of tooth. So, the whole meaning is the internal treatment of the tooth. Endodontics is mainly a dentistry section that deals with dental pulp and tissues adjacent to the tooth's root. It repairs the internal tissues and pulp. Inflammation or infection is mainly caused by deep decay, cracks, and the damaged pulp in the tooth. In such a case, endodontic treatment is applied. The phenomenon can be summarized that when the internal tissues are damaged or in some cases, bacteria are created in the pulp or cavities, and the filling is not enough to preserve the tooth. In that case, one must need the endodontic treatment for the betterment of the tooth [24].

6.3.4 ROOT CANAL

One of the major endodontic treatments is the root canal, which is highly related to the tooth's root. The treatment can perform in the tooth's hollow-core portion, dental pulp, and repair the tooth's internal tissues. The treatment can protect the tooth by removing infected or inflamed blood vessels and nerve tissues. The dentist always recommended that treatment as permanent filling or act as a permanent bridge for which bacteria or substances cannot penetrate to the roots. This treatment is applied when tooth decay has reached the pulp of the tooth and ruins the tissues [23].

As the infected teeth carry away cavities, some sealer materials must be needed to fill that voids. Gutta-percha, a standard sealer material for the root canal, is commonly applied in root fillers, and also mineral trioxide is used in this treatment [4]. In endodontic treatment, dentists use sealer materials that are nanomaterial, and without these materials, treatment becomes incomplete, so this branch must be included in nano dentistry from this criterion.

At first, in a root canal process, the infected tooth towards the cavity means anyhow the action can appear to be able to reach the tooth systemically's pulp surface. Then destruct the infected or inflamed tissues and clean up the whole cavity. Then the canal can be filled with sealer materials. The most used sealer material is gutta-percha. Finally, covered the upper side of the tooth with a crown made of ceramic materials. After doing this whole operation, one can feel at ease and comfort with one's oral health. Endodontic treatment has some benefits. It can remove the pain of the tooth. When eating food, a patient feels comfortable as the treatment can remove cavities from the tooth. Endodontic treatment can preserve the nerves and blood vessels of the tooth (Table 6.2) [8].

TABLE 6.2 Some Widely Used Nanomaterial Root Canal Sealers

Substance	About	Benefits	Problems	Examples
Zinc oxide eugenol cement sealer	• Frequently utilized sealer • Forms mixing vehicle • The powder contains ZnO that has been finely sifted to enhance the flow • The setting process is a chemical reaction with the physical embedding of ZnO in a matrix of zinc eugenol	• Absorbed if extruded into periradicular tissue • The antimicrobial effect through ZnO or additives such as rosin and Canada balsam or corticosteroids	• Shrinks on setting • stains the tooth (Roth's sealer was developed to be as a nonstaining ZnOE sealer)	• Pulp canal sealer (SybronEndo) • Preschool (Procosol, Inc., Philadelphia) • Roth (Roth's Pharmacy, Chicago, IL)
Calcium hydroxide sealers	• It was developed for two main reasons: o Antimicrobial effect o Osteogenic Cementogenic potential	• Great antibacterial property • Debases the toxic proteins found in the root canal • Starts the calcium-dependent ADP reaction; responsible for tissue hardening [9]	• Needs to be soluble to leach out the hydroxyl group responsible for its activity. • Poor cohesive strength • Poor dentin adhesion	• Calciobiotic Root Canal Sealer (CRCS) • Sealapex (Sybron Endo) • Apexit (Ivoclar Vivadent)
Noneugenol sealers	• Developed from periodontal dressings	• No irritating effect of eugenol	–	• Nogenol (G.C. America, Alsip, IL)

TABLE 6.2 *(Continued)*

Substance	About	Benefits	Problems	Examples
Glass ionomer sealers	• Developed to take advantage of their bond to dentin, fluoride release, antimicrobial activity, and biocompatibility	• Dentin bonding: • Minimal tissue • Irritation • Low toxicity	Solubility concern: • Inadequate • bonding with gutta-percha • Minimal antibacterial effect	• Ketac-Endo (3M ESPE, Minneapolis, MN) • Activ G.P. (Brassler USA, Savannak, GA)
Resin	• Introduced because it provides good adhesion • AH-26 and A.H. plus are epoxy resins. • Epiphany is a dual curable dental resin	• Good adhesion ability • Does not contain eugenol	• AH-26 contains formaldehyde, which is toxic when freshly mixed	• AH-26 • AH-Plus • Epiphany • EndoREZ
Silicone-based sealers	• Two materials are available: i. RoekoSeal (Langenau, Germany): which is a polyvinylsiloxane or polydimethylsiloxane root canal sealer. ii. GuttaFlow (Cuyahoga Falls, OH): a. Cold, flowable, self-curing obturation material for root canals combines gutta-percha and polydimethylsiloxane sealer (RoekoSeal) into one injectable system. Fills canal irregularities. b. Working time is 15 min and cures in 25–30 min.			
Bioceramics (calcium silicate) sealers	• Composed of zirconium oxide and calcium silicates	• Hydraulic sets by utilizing moisture • Antimicrobial and biocompatible	Retreatment might difficult	• iRoot S.P. or Endosequence BC Sealer • MTA Fillapex

6.3.5 *PROSTHODONTIC TREATMENT*

The term 'Prostho' means a replacement, and 'dontist' means treatment of tooth. That means the full meaning of prosthodontic treatment is the replacement of the tooth. In this treatment, the main discussion concerns the artificial replacement of the lost or highly infected tooth. That means the infected tooth is picked off and inlaid with an artificial tooth. The implantation of the false tooth is permanent, and that's why for prosthodontic treatment, various nanomaterials are used to make the artificial tooth strong enough.

Acrylic is the most used material for making artificial teeth. It is based on polymethyl methacrylate (PMMA) resin and metals like Ti (biocompatible) or cobalt alloy. Also, ZrO_2 or Al_2O_3 are used to make false teeth [9]. The advantages of using acrylic are creating face esthetics, and bonding with the denture base as the teeth cannot detach from the denture base. Metal is used to make a strong skeletal structure of the tooth and make the crown, and bridge ceramic materials such as ZrO_2 or Al_2O_3 must be needed. Acrylic, metal, and ceramics are all needed to make a complete strong artificial tooth.

Dental implantation is another name of prosthodontic treatment because the treatment operates implantation of the tooth. The first material used in this process is Titanium (Ti), which is also regarded as the best metal for prosthodontic treatment. It is generally adjusted in the jawbone. Since Ti can create a solid base, it becomes easy to replace the crown of the tooth. A junction is created where the titanium screw, the implant and the tooth crown are adjoined. When the root implantation is settled down, the crown placement is set up, and finally, it looks like a natural tooth.

In prosthodontic treatment, varieties of materials are used, which have nanomaterial characteristics. Nano-range materials such as metals (Ti), nano-ceramic materials, and resin are used in this treatment. It is because nano-ranged materials have some characteristics like modulus elasticity, surface hardness, polymerization shrinkage and filler loading [10]. These properties are very important for prosthodontic treatment. Ordinary bulk materials lack these properties. As the branch works with nano ranged particles, it is included in the nano dentistry.

Prosthodontics gives back our face esthetics and a wonderful smile. After doing this treatment, one can receive improved jaw structure and recover the jaw's function. With this treatment, the pain from the denture can be gone forever.

6.3.6 PERIODONTICS

Periodontics is a dentistry branch that deals with the dental specialty that involves the inflammatory diseases affecting the gums and other nearby tissues holding up the teeth. It also involves the precautionary measures, detection, and treatment of these disorders. Minimizing severity, improving oral cavity hygiene and regenerating the destroyed tissues when needed are periodontal treatment aims. Periodontology is very a challenging dental subject requiring both knowledge and experience expertise.

However, the standard primary treatment of these diseases is a bit casual. It usually involves the mechanical crushing and elimination of infective biofilm, concretion from the teeth and other nearby unwanted substances like teeth-plaque with root planning and ultrasonic scaling. But this method is not always convenient due to various limitations. For example, the diagnosis and treatment of some regions like root depression or furcation turn out impractical. To improve clinical outcomes in certain periodontal conditions, various chemotherapeutic agents have proved their worth. Nevertheless, it has some side effects too and also tends to develop other pathogens [11].

At present, nanotechnology has brought more suitable formulas and techniques to periodontal treatment. Nano delivery systems are developed using NPs, colloidal carriers or liposomes, which can be served as an active anti-infective part of periodontal treatment. These methods can be able to reach the inflamed tissues by breaking the hydrophobic barrier pertaining to the oral biofilm. The biggest interest in recent years is NPs as they have an extremely small size that helps to penetrate through the junctional epithelium, bioavailability, and stability. NPs–$CaSO_4$ has the potential to be used as an antibacterial cement for the treatment of periodontics. They also can run their operation without causing issues to the host tissue cells. Tetracycline hydrochloride-loaded particles in polycaprolactone nanofibers can be very efficient in healing affected periodontal pockets [4].

Triclosan-loaded nanoparticles are also used for the remedy of periodontic diseases as a delivery system. In fact, it's a new delivery system. *In vivo* studies in dogs have indicated that triclosan-loaded nanoparticles possess inflammation-reducing capability. Drugs can be organized into nanospheres, which are made by a biodegradable polymer. Biodegradable polymer Nanosphere drug carriers can be used because they can properly supply drugs in required areas. For instance, Restininin has a microstructure of tetracycline for delivering the drug to the periodontal pocket [12].

QDs (Quantum dots) are examples of another astonishing and promising nanostructures for diagnostic purposes. These are low toxic and stable semiconductors. These materials also show fluorescence attributes. They might be used as fluorescent levels for biomolecules because of having strong light absorbance properties [1]. QDs can be sticking with dental resins to tune the resin's radiation color. In periodontal therapy, cadmium-free, and lead-free QDs are applied to improve the remedy of affected periodontal tissues [10]. Many research types showed that drugs could be boosted by co-operating the active moiety with a carrier system [13]. Table 6.3 depicts a few noteworthy carrier systems along with the influence on treating such problems.

Many kinds of complex dental disorders are seen in patients. Oral cancer is by far one of the most threatening problem. Oral cancer or oral cavity cancer, or mouth cancer, is a type of head and neck cancer which is caused by hostile cells present in the lips or mouth. About one-quarter of patients who are younger than age 55 have oral cavity cancer. But children are rarely diagnosed with this disorder. However, oral cancer is in men is twice as common as in women. As people age, however, they may be more able to oral cavity cancer. The main factors for oral cancer are using tobacco, having alcohol, sunlight or artificial light over a long period. However, we can prevent oral cancer by maintaining these risk factors. But for now we will be discussing about the treatments for oral cancer.

The summarization of some articles in the PubMed database of different nanotechnology diagnosis modes, follow-up, and oral cavity cancer therapy are given in Table 6.4.

6.3.7 IN TREATING COMPLEX MALADY LIKE ORAL CANCER

The uses of nanotechnology are the detection of oral cancer and it's better treatment. It has the ability to diagnose even a single cancerous vesicle *in vivo*. Most toxic drugs are delivered immediately to the cancerous vesicles. Materials used in cancer diagnosis are QDs, nanoshells, carbon nanotubes, nanowires, super magnetic NPs, dendrimers, and nanosponges. Besides, many kinds of NPs are used for drug and gene delivery. They are liposomes, solid lipid particles, polymeric NPs (nanospheres and nanocapsules), nanocrystals, polymer therapeutics, and inorganic NPs. Unlike other traditional methods, nanotechnology has the potential to destroy cancer cells without harming the normal cells. NPs can enhance the stability of drugs and control their aimed drug delivery vehicles, which can cross chemotherapeutic agents or therapeutic genes through malignant cells without harming the healthy cells. This could contain a very little portion of toxic substances due to immediate delivery of the drugs to the target tissue. Silica-coated micelles, ceramic NPs and cross-linked liposomes are some nanoscale delivery devices that can destroy cancerous cells. The permeability of drugs can be improved by the surface modification of NPs. It also provides a way to synthesize high-permeable NPs-based cancer therapeutics. It is for this reason that NPs are so therapeutically efficient in drug delivery. Besides, combinatorial chemotherapy added with nanomedicine is proved to be more useful in therapeutics in treating complex disorders like oral cancer [13].

TABLE 6.3 A List of Some Drug Delivery Systems Composed of Nanoparticles and Their Influence in Periodontal Disease Therapy

System of Delivery	Influence	References
Nanoparticles	Supplies essential material to the periodontal pocket	[14]
	Gingival inflammation diminishes by the influence of triclosan-nanoparticles	[15]
	Minocycline-loaded nanoparticles could significantly decrease periodontitis signs	[16]
	Potential materials for therapeutic purposes	[17]
Nanocomposites	2-methacryloyloxyethyl phosphorylcholine and dimethylamino hexadecyl methacrylate are very effective in Class V restorations for preventing periodontal pathogens and shielding the periodontium	[18]
Nanofibers	Introduces a balanced drug release tendency of drug enriched hyaluronic acid-polyvinyl alcohol nanofiber	[19]

TABLE 6.4 Noteworthy Depiction of Different Authors [20]

Author	Work	Mode used	References
Gharat et al.	They highlighted the etiology, line of treatment and chemopreventive measures related to OSCC, focusing on data available in the research carried out worldwide in the past 15 years.	Nanotechnology-based carrier systems	[7]
Gupta et al.	Ag5-ZnO NCs can destroy oral carcinoma (K.B.) cells in the visible light irradiation	Ag5-ZnO NCs	[8]
Lollo et al.	Investigated the capability of nanocarriers of peptide anticancer drugs for oral delivery purposes.	Nanocarriers	[12]
Rahimi et al.	Used MEDLINE/ScienceDirect/OVID for bibliographical searches until February 2015.	Nano-formulated curcumin	[19]

6.4 SHORTCOMINGS OF NANOMATERIALS

Nanomaterials are undoubtedly regarded as the future of dental material. But these are not invincible. Nanomaterials can show toxic consequences on health and the environment. The toxicity of nanoparticles depends upon various factors like concentration, types, interaction among different sites, particle size, etc. [11]. Nanomaterials vastly used in dental applications like titanium dioxide, silica, zinc oxide, cobalt, etc., can also create human problems. Their application in periodontal dressing pastes, root canal pastes, as well as implant surface coatings can easily contaminate nanoparticles with blood. As a result, they can be easily carried to the central nervous system [25]. While polishing the dental nanomaterials, nanoparticles can also spread through the air ending up reaching the human lungs through inhalation. Their exposure to the oral mucosa resulting from the dissolving and abrasion process of nanoparticles used in nanofillers, coatings, mouthwashes, and toothpaste can also cause serious health issues [5]. Different nanomaterials show different toxicity levels in different species. They are extremely small in size and are much more toxic than their bulk counterpart.

For this reason, if they penetrate through biological membranes of sensitive sites, toxicity spreads through them [26]. The main reason for genotoxicity is generating a huge amount of intracellular reactive oxygen species and a lower level of antioxidants [27]. These reactive oxygen species ultimately start the carcinogenesis process by reacting badly with DNA molecules.

Nanoparticles can also cause genetic alterations. The extremely small size of nanoparticles helps them to penetrate cell membranes very easily. After interacting with the nucleus, they transform the daughter cell through mitosis cell division. Silica and titanium dioxide nanoparticles form intranuclear protein aggregates that don't let transcription, replication, and cell. Besides, QDs, after passing through the nuclear membrane, interact with the histone protein [27]. In this way, nanomaterials can hamper our genetic activity [11].

6.5 SUMMARY

Nanomaterials have changed fiction to reality. Our teeth are more used than most other organs of our body, yet we don't treat them very well. A few decades back, we couldn't even think about tooth replacement or root canal. If any tooth was infected, in most cases, the only solution was saying bye to that to terminate the infection. Today Nanodentistry has opened a door of full possibilities of treatment to almost every kind of oral problems. Compared to the previous years, nanomaterials have developed a lot in medical science, especially in oral health care. This sector's research activities are evolving gradually, and many nanomaterials, such as zinc oxide, calcium hydroxide, Ti, etc., are widely used these days. Some nanofiber and nanocomposites are not suitable for the treatment. In that case, more comprehensive research on specific nanoparticles is mandatory. The chapter has shown what exactly a nanomaterial means and its applications in dentistry, along with the shortcomings.

ACKNOWLEDGMENTS

The authors express their heartiest gratitude to the Department of Glass and Ceramic Engineering, Rajshahi University of Engineering and Technology (RUET), Bangladesh, for the cordial assistance.

KEYWORDS

- **dentistry**
- **endodontic treatment**
- **nanomaterials**
- **periodontics**
- **prosthodontics**
- **root canal**
- **shortcomings**

REFERENCES

1. Aminu, N., Baboota, S., Pramod, K., Singh, M., Dang, S., Ansari, S. H., Sahni, J. K., & Ali, J., (2013). Development and evaluation of triclosan loaded poly-E-caprolactone nanoparticulate system for the treatment of periodontal infections. *Journal of Nanoparticle Research, 15*(11), 2075. https://doi.org/10.1007/s11051-013-2075-6.
2. Aminu, N., Chan, S. Y., & Toh, S. M., (2017). Roles of nanotechnological approaches in periodontal disease therapy. *Journal of Applied Pharmaceutical Sciences, 7*(7), 234–242.
3. Desai, S., & Chandler, N., (2009). Calcium hydroxide-based root canal sealers: A review. *Journal of Endodontics, 35*(4), 475–480. https://doi.org/10.1016/j.joen.2008.11.026.
4. Feng, X., Chen, A., Zhang, Y., Wang, J., Shao, L., & Wei, L., (2015). Application of dental nanomaterials: Potential toxicity to the central nervous system. *International Journal of Nanomedicine, 10*, 3547–3565. https://doi.org/10.2147/IJN.S79892.
5. García-Godoy, F., & Hicks, M. J., (2008). Maintaining the integrity of the enamel surface: The role of dental biofilm, saliva and preventive agents in enamel demineralization and

remineralization. *Journal of the American Dental Association, 139*(5), 25S–34S. https://doi.org/10.14219/jada.archive.2008.0352.
6. Gavrilescu, C. M., Paraschiv, C., Horjinec, P., Sotropa, D. M., & Barbu, R. M., (2018). The advantages and disadvantages of nanotechnology. *Romanian Journal of Oral Rehabilitation, 10*(2), 153–159.
7. Gharat, S. A., Momin, M., & Bhavsar, C., (2016). Oral squamous cell carcinoma: Current treatment strategies and nanotechnology-based approaches for prevention and therapy. *Critical Reviews in Therapeutic Drug Carrier Systems, 33*(4), 363–400. https://doi.org/10.1615/CritRevTherDrugCarrierSyst.2016016272.
8. Gupta, J., Mohapatra, J., & Bahadur, D., (2017). Visible light driven mesoporous ag-embedded ZnO nanocomposites: Reactive oxygen species enhanced photocatalysis, bacterial inhibition and photodynamic therapy. *Dalton Transactions, 46*(3), 685–696. https://doi.org/10.1039/c6dt03713e.
9. Joshi, D., Garg, T., Goyal, A. K., & Rath, G., (2015). Development and characterization of novel medicated nanofibers against periodontitis. *Current Drug Delivery, 12*(5), 564–577. https://doi.org/10.2174/1567201812666141205131331.
10. Katz, A., Redlich, M., Rapoport, L., Wagner, H. D., & Tenne, R., (2006). Self-lubricating coatings containing fullerene-like WS_2 nanoparticles for orthodontic wires and other possible medical applications. *Tribology Letters, 21*(2), 135–139. https://doi.org/10.1007/s11249-006-9029-4.
11. Krishnamurthy, S., & Vijayasarathy, S., (2016). Chapter 9; Role of nanomaterials in clinical dentistry. In: Grumezescu, A. M., (ed.), *Nanobiomaterials in Dentistry: Applications of Nanobiomaterials* (Vol. 11, pp. 211–240). Elsevier, Inc.
12. Lollo, G., Gonzalez-Paredes, A., Garcia-Fuentes, M., Calvo, P., Torres, D., & Alonso, M. J., (2017). Polyarginine nanocapsules as a potential oral peptide delivery carrier. *Journal of Pharmaceutical Sciences, 106*(2), 611–618. https://doi.org/10.1016/j.xphs.2016.09.029.
13. Manjunath, R. G., & Rana, A., (2015). Nanotechnology in periodontal management. *Journal of Advanced Oral Research, 6*(1), 1–8.
14. Mitthra, S., Sujatha, V., & Mahalaxmi, S., (2017). Nanodentistry-problems and challenges in research. *International Journal of Current Research, 9*(10), 58855–58858.
15. Osorio, R., Alfonso-Rodríguez, C. A., Medina-Castillo, A. L., Alaminos, M., & Toledano, M., (2016). Bioactive polymeric nanoparticles for periodontal therapy. *PLOS One, 11*(11), e0166217. https://doi.org/10.1371/journal.pone.0166217.
16. Piñón-Segundo, E., Ganem-Quintanar, A., Alonso-Pérez, V., & Quintanar-Guerrero, D., (2005). Preparation and characterization of triclosan nanoparticles for periodontal treatment. *International Journal of Pharmaceutics, 294*(1, 2), 217–232. https://doi.org/10.1016/j.ijpharm.2004.11.010.
17. Pokrowiecki, R., Pałka, K., & Mielczarek, A., (2018). Nanomaterials in dentistry: A cornerstone or a black box? *Nanomedicine, 13*(6), 639–667. https://doi.org/10.2217/nnm-2017-0329.
18. Poonia, M., Ramalingam, K., Goyal, S., & Sidhu, S. K., (2017). Nanotechnology in oral cancer: A comprehensive review. *Journal of Oral and Maxillofacial Pathology, 21*(3), 407–414. https://doi.org/10.4103/jomfp.JOMFP_29_17.
19. Rahimi, H. R., Nedaeinia, R., Sepehri, S. A., Nikdoust, S., & Kazemi, O. R., (2016). Novel delivery system for natural products: Nano-curcumin formulations. *Avicenna Journal of Phytomedicine, 6*(4), 383–398.

20. Redlich, M., Katz, A., Rapoport, L., Wagner, H. D., Feldman, Y., & Tenne, R., (2008). Improved orthodontic stainless steel wires coated with inorganic fullerene-like nanoparticles of WS(2) impregnated in electroless nickel-phosphorus film. *Dental Materials, 24*(12), 1640–1646. https://doi.org/10.1016/j.dental.2008.03.030.
21. Sasalawad, S. S., Naik, S. N., Shashibhushan, K. K., & Poornima, P., (2014). Nanodentistry: The next big thing is small. *International Journal of Contemporary Dental and Medical Reviews*, 1–6. https://doi.org/10.15713/ins.ijcdmr.12.
22. Song, W., & Ge, S., (2019). Application of antimicrobial nanoparticles in dentistry. *Molecules, 24*(6), 1–15. https://doi.org/10.3390/molecules24061033.
23. Wang, L., Xie, X., Imazato, S., Weir, M. D., Reynolds, M. A., & Xu, H. H. K., (2016). A protein-repellent and antibacterial nanocomposite for class-V restorations to inhibit periodontitis-related pathogens. *Materials Science and Engineering: C, Materials for Biological Applications, 67*, 702–710. https://doi.org/10.1016/j.msec.2016.05.080.
24. Wang, W., Liao, S., Zhu, Y., Liu, M., Zhao, Q., & Fu, Y., (2015). Recent applications of nanomaterials in prosthodontics. *Journal of Nanomaterials, 2015*, 1–11. https://doi.org/10.1155/2015/408643, PubMed: 408643.
25. Xu, H. H. K., Moreau, J. L., Sun, L., & Chow, L. C., (2011). Nanocomposite containing amorphous calcium phosphate nanoparticles for caries inhibition. *Dental Materials, 27*(8), 762–769. https://doi.org/10.1016/j.dental.2011.03.016.
26. Yao, W., Xu, P., Pang, Z., Zhao, J., Chai, Z., Li, X., Li, H., Jiang, M., Cheng, H., Zhang, B., & Cheng, N., (2014). Local delivery of minocycline-loaded PEG-PLA nanoparticles for the enhanced treatment of periodontitis in dogs. *International Journal of Nanomedicine, 9*(1), 3963–3970. https://doi.org/10.2147/IJN.S67521.
27. Zafar, M. S., Khurshid, Z., Najeeb, S., Zohaib, S., & Rehman, I. U., (2017). Chapter 26; Therapeutic applications of nanotechnology in dentistry. In: Andronescu, E., & Grumezescu, A., (eds.), *Nanostructures for Oral Medicine* (pp. 833–862). Elsevier, Inc.

PART II

NANOMATERIALS FOR DRUG DELIVERY AND THERAPY

CHAPTER 7

ADVANCES IN NANOMATERIALS: FABRICATION OF TARGETED DRUG DELIVERY SYSTEM

SHABNAM KAWOOSA, ZUBAID-UL-KHAZIR RATHER, RABIAH BASHIR, and NISAR AHMAD KHAN

ABSTRACT

The combined endeavors of nanoscience and pharmaceutical science have surged the prominence of nanomedicine that has resolved several shortcomings of conventional drug delivery system, for instance, poor pharmacokinetics and biopharmaceutical properties, non-specific biodistribution, low therapeutic indexm and selective penetrability of drug molecules. A diverse range of nano-based drug delivery vehicles such as polymeric micelles, solid-lipid liposomes, magnetic nanoparticles, liposomes, nanotubes, virus, protein, and gene-based therapies have been extensively researched in the perspective of specifically delivering the therapeutic agent to the diseased organs. Several remarkable formulations were approved by FDA after successfully clearing the pre-clinical and clinical trials, which are marketed presently viz. Abraxane®, Myocet®, Rapamune® and many more. The therapy has emerged more advantageous for treating cancers and chronic diseases in terms of precisely acting on the specific diseased site and overcoming the side effects that usually encounter the non-specific healthy organs.

Sustainable Nanomaterials for Biomedical Engineering: Impacts, Challenges, and Future Prospects.
Junaid Ahmad Malik, Megh R. Goyal & Mohamed Jaffer M. Sadiq (Eds.)

7.1 INTRODUCTION

Nano and pharmaceutical sciences have grown rapidly in recent times, offering potential combined applications and have turned out to be one of the utmost vigorous research fields in the areas of biomedicine, biotechnology, and chemistry [10]. The science of nanostructure technology has been growing extensively worldwide and has exposed wide expansions in interdisciplinary research. It has instigated revolutionary developments in the pharmaceutical field for delivering biologically active compounds in site-specific manner. Commercially available conventional dosage forms suffer from the inherent drawback of lacking the targeted drug delivery application and possess many formulation problems such as poor physicochemical properties, erratic absorption profile, affecting non-diseased tissues and stability issues. Researchers have made incessant efforts now and again to remodel drug delivery technology that can surmount the challenges confronted due to the present conventional formulations like tablets, capsules, and many other liquid dosage forms. Scientists tried hard to include multidimensional technical methodologies in improving the drug's bioavailability issues, physicochemical limitations, non-targeted drug distribution, stability of drug candidate in the microenvironment, and ameliorating the toxicity profile of therapeutic agents. There have been advances in the domain of drug delivery approach of therapeutic moieties or natural-based active composites to the intended location for management of numerous disorders. Nanocarrier-based nanomedicines have been developed and investigated employing numerous drug candidates with natural and synthetic polymers, e.g., nanoparticles (solid lipid/metal/polymer based), liposomes, niosomes, virosomes, cochleate, crystalline nanoparticles, nano-emulsions, and QDs [31, 58, 61, 69, 77]. Nano-based drug delivery system possibly will (i) enhance the physicochemical and pharmacokinetic profiles of active molecules; (ii) improve drug absorption and distribution; (iii) reduce drug degradation in various physiological systems; and (iv) superior site-specific targeting compared to conventional dosage forms.

This chapter explores the functionally active nanomaterial/biomaterials employed, categories of fabricated nano-formulations, the targeting pathways ensued are highlighted. Furthermore, the promising perspectives in this emerging field and currently marketed nano-formulations are also discussed.

7.2 NANOMATERIALS

7.2.1 TYPES OF NANOMATERIALS USED FOR TARGETING

Nanomedicine is an emergent arena that implicates the use of nanosized substances for the diagnostics and treatment of several diseases, including cancer [73, 76]. Nanoscale-type materials or substances may occur naturally but may be engineered as well, that may be purposeful in numerous profitable produces and procedures. Nanomaterials are useful in making several products like cosmetics, sunblock creams, clothing that are stain resistant, electronic engineering as well as many other daily things. The size of nanomaterials is very small, having at least one of the measurements in the nanoscale range, which is from 1 nm to 100 nm or less. Based on Siegels phenomenon of classifying nanomaterials, they are categorized based on the number of dimensions that are in nanoscale range, e.g., nanomaterials having one dimension in the nanoscale range (surface films), nanoscale in two dimensions (strands of nanofibers) or nanoscale in three dimensions (nanoparticles). Furthermore, nano-based materials can occur in various forms (single, fused, aggregated or agglomerated) with different shapes (spherical, tubular or irregular). QDs, nanotubes, dendrimers, and fullerenes are common types of nanomaterials. They have uses in the arena of nanotechnology, and show diverse physicochemical features from ordinary chemicals. A perfect drug transporting system necessarily should be not harmful to other tissues, should be biocompatible with the host environment, non-immunogenic, recyclable [71], and must escape the recognition mechanism surfaced by the host's defense. There exists the diversity of nanomaterials that are used as drug targeting vehicles such as liposomes, dendrimers, and polymeric micelles, drug carriers based on lipoproteins, different types of nanoparticle based drug carriers like solid lipid nanoparticles, gold nanoparticles (AuNPs), polymer-based nanoparticles, etc. (Figure 7.1).

7.2.1.1 LIPOSOMES

Liposomes are the distinguished drug delivery system currently used for drug targeting in various diseases. They are comprised of simple microscopic vesicles possessing the least toxic, hemolytic, and immunogenic properties. Liposomes are made from various amphiphilic molecules. The aqueous volume of liposomes is wholly enclosed by a sheath of lipid molecule typically cholesterol or non-toxic phospholipids. The ligand-coated nanocarriers may be retained in the lipophilic shell or the aqueous interior depending upon the physicochemical characteristics of the drug being encapsulated in liposome [71]. The therapeutic moiety in liposomes can either be intercalated into the lipid bilayers or encapsulated in aqueous space. They are biocompatible and biodegradable even upon repeated injections and can be manufactured using special techniques to prevent them from clearance mechanisms which include reticuloendothelial system, renal clearance, chemical or enzymatic inactivation, etc. [13, 72]. The extent of location of the drug in the liposomes is determined by the physicochemical characteristics and composition of lipids. Liposomes exhibit significant advantages over other dosage forms in consideration of drug loading capacity, size controllability, membrane permeability and deliverability of multiple drug moieties to the target site.

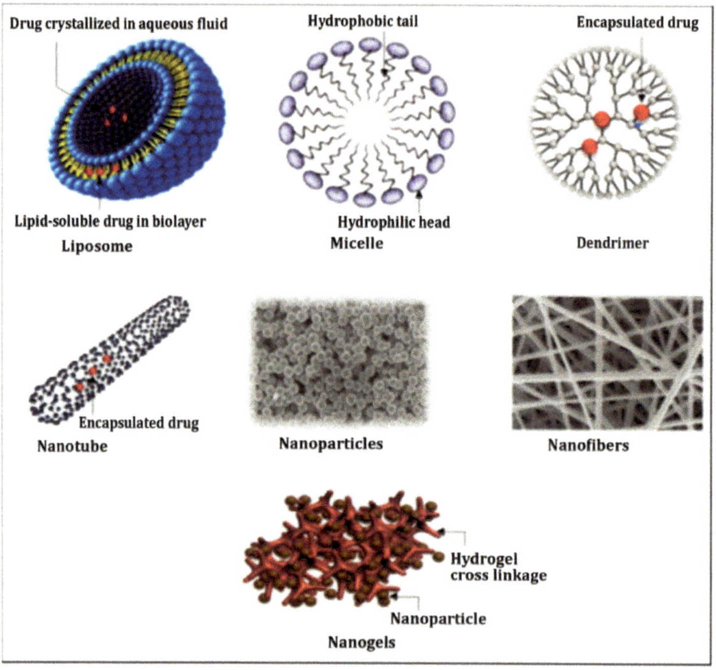

FIGURE 7.1 Various nanotechnology-based drug delivery systems.

Source: Reprinted with permission from Ref. [3].© 2020 Elsevier.

The major problem countered by liposomes is relatively lesser stability *in vitro* and clearance by the RES in the body. This enigma has been overcome by surface attachment of polyethylene glycol to the liposomes; that resulted in the prolongation of the circulation time of liposomes in blood from 200 to 1,000 min [71]. With regard to liposomes intended for anti-cancer therapy, there is the prompt release of drug in acidic environment of tumor mass due to over-reliance on glycolysis [16, 50].

7.2.1.2 MICELLES AND DENDRIMERS

Micelles and dendrimers are another type of vehicles used for targeted drug delivery and are prepared from amphiphiles comprising of both hydrophilic and hydrophobic monomer units [70]. Micelles are formed by self-assembling of bipolar copolymers in an aqueous environment as the amphiphilic concentration exceeds CMC. Drugs that have poor solubility are carried by formulating them as nano dosage forms of micelles and dendrimers; the limitations include lack of size control or functional malleability.

Dendrimers are polymer-based delivery vehicles precisely characterized as highly branched structure with bonds emerging from core, around 510 nm in diameter. They are fabricated as a central core encased by layers of polymer that branch out to form a small, spherical, and very dense nanocarrier. The therapeutic moieties may attach to different sites on the surface of dendrimers. They are useful in gene transfection and medical imaging.

7.2.1.3 NANOPARTICLES

Nanoparticles are colloidal particulates with distinctive measurements in the nanoscale range (10 to 1,000 nm). In the recent era, nanoparticles have emerged as materials having multifaceted applications such as diagnostic imaging; drug delivery carriers for targeting the therapeutically active ingredient to diseased organ/tissue [22, 42]. They have been comprehensively researched for targeting critical ailments like cancer, atherosclerosis, Alzheimer's, psoriasis, diabetes, and neurodegenerative diseases [22, 65]. Miniature dimensions, surface charge utilization of penetration enhancers, minimal side effects, and modification to target the site of action using polymeric nanoparticles are the evidenced strategies to effectively treat these diseases employing nanoparticles [23]. Polymers employed in nanoparticle formulation may be natural or synthetic polymers [19]. Naturally occurring biopolymers via cellulose, chitosan, alginate, gums, etc., have various benefits compared to synthetic materials on account of biodegradability, biocompatibility, and low immunogenicity [46]. Polymer-based nanoparticles have several shortcomings also, e.g., ambiguity of process scale-up, reproducibility, poor drug loading, and instability during storage. Biodegradable polymers possess the capability to target diseased and infectious tissue as well as operate as controlled-release therapy [70].

7.2.1.4 ARTIFICIAL DNA NANOSTRUCTURES

Based upon directly sensing the environment of infectious tissues, various nucleic acid nano-devices can be employed for targeted drug delivery of therapeutic molecules. These techniques exclusively employ DNA as specifically programed nanostructure rather than its role as the carrier of genetic information. The most widely employed technique is the DNA origami method, assembling a DNA package with a controllable lid that unseals to release the drug in response to a peripheral stimulus [14].

7.2.1.5 QUANTUM DOTS (QDS)

A quantum dot exists as a semiconductor miniscule nanostructure serving as sign poles of definite categories of compartments or molecules in the body that confines the motion of free and bound pair of electrons within conduction and valence bands/holes in all three spatial dimensions. The ability of the QDs to accustom the size is beneficial for various applications, and it is considered as one of the utmost favorable candidates that could track the drug passage within the body for diagnostic purposes and targeted drug delivery [2].

7.2.1.6 NANOCRYSTALS

Nanocrystals occur as pure solid drug particles having size within 1,000 nm range. They are chiefly manufactured by two techniques viz top down and bottom up methodologies. The top-down approaches employed for fabrication of nanocrystals are sono-crystallization, high gravity controlled precipitation, multi-inlet vortex mixing and limited impinging liquid jet precipitation techniques. But these processes are quite expensive when a carbon based solvent is used because its removal at the end is difficult. The techniques involving grinding and homogenization at higher pressure are included in bottom up scheme. Nanocrystals promote the availability of drug in the systemic circulation by improving suspension rate, drug solubility, and competence to firmly attach to the intestinal wall [41].

7.2.1.7 NANOEMULSIONS

Nanoemulsion may be defined as thermodynamically stable isotopic systems of nanosized particles with the droplet size falling characteristically in the range of 20–200 nm. Nanoemulsions comprise of two immiscible phases, mixed to form a single phase by adding an emulsifying agents – surfactant and co-surfactant. The addition of emulsifier is crucial for the stability of phases in nanoemulsion, for the reason that it reduces the interfacial tension between the immiscible phases. The basic dissimilarity between nanoemulsion and emulsion lies in the shape and size of components spread in the endless phase. Nanoemulsions may be categorized as oil in water, water in oil and bicontinuous type of nanoemulsion. They are beneficial in augmenting the therapeutic efficacy of the drug and decrease antagonistic consequences and fatal responses.

7.2.1.8 NANOSUSPENSION

Nanosuspension is the colloidal dispersion of fine particles in the range of 1 to 100 nm, stabilized by surfactants for delivering the API through oral, topical, parenteral, ocular, and pulmonary routes. The innovation of nanosuspension technology can be beneficial to surmount the hindrance of poor aqueous solubility and bioavailability of drugs. Numerous technologies like homogenization, precipitation, solvent evaporation, supercritical fluid method and nanojet techniques can be employed for the preparation of nanosuspension. They are formulated by using stabilizers, surfactants, co-surfactants, organic solvents and other additives such as buffers, polyols, osmogent, and cryoprotectant. They can function as targeted drug delivery systems when incorporated in the ocular inserts and mucoadhesive hydrogels.

7.2.1.9 NANOSPHERES

Nanospheres are homogenously structured, that exist in crystalline or amorphous form. Mostly the drug in nanospheres are encapsulated, dissolved, or attached to the polymer matrix where in the drug is physically and uniformly dispersed. They have the potential to target the concentrated dose of drug to tumor site by active or passive mechanism and keep the drug safe from enzymatic and chemical degradation as well [9, 38, 44, 47, 51]. It can restrict the drug exposure to healthy tissue by selectively dispersing the drug to the target organ. Nanospheres can attain greater concentrations in the spleen, liver, and lungs than in other parts of the body. Polymeric nanospheres facilitate entrapment of biological molecules (peptides, proteins, polysaccharides) and protect them against enzymatic degradation [47]. They selectively deliver the therapeutic agent to the brain cells by interacting with specific receptor-mediated transport system in blood-brain barrier.

7.2.1.10 NANOGELS

Nanogels are the non-fluid polymeric networks with a diameter of less than 100 nm [60]. The advantages of nanogels compared to other nanocarrier systems are minimized premature drug leak, encapsulating numerous therapeutic drug substances and biomolecules in the single dosage form, and ease of administration through diverse routes viz parenteral, oral, nasal, rectal passages. Nanogels have various applications in the arena of biochemical separation, biosensors, bio-catalysis, cell culture, antitumor therapy, etc. However, the transport of therapeutics, such as vaccines, cytokines, and nucleic acids are the vast accomplished applications of nanogels [1, 62, 74].

7.2.1.11 NANOCAPSULES

Nanocapsules are comprised of an aqueous or oily core enclosed with a polymeric shell providing a distinct nanostructure. Currently, they have enticed more attention in drug delivery applications profiting from their core-shell structure composition. The lipophilic core of nanocapsules can efficiently enhance drug loading capacity while reducing the content of polymeric vehicle in comparison to nanospheres [64]. On account of polymeric encapsulation of the payload, the abrupt eruption/degradation of drug due to local tissue environment (pH, enzymes, temperature) is avoided. They are used for targeted drug delivery because of the ability that the polymeric shell can be actuated by some responsive molecules to be able to interact with targeted biomolecules [34, 81, 85].

7.2.1.12 NANOFIBRES

Nanofibers are definite fibers with diameters less than 100 nanometers, produced by bottom up approach employing the electrospinning technique. Several polymer melts or solutions are used in the development of nanofibers, such as from synthetic polymers, e.g., polyvinyl alcohol, polylactic acid or their mixtures and processed raw materials such as cellulose, chitosan, keratin, starch or their mixtures. Nanofibers are a new emerging class of nanomaterials utilized in various fields such as health, purification, wipes, private, and personal care, nanocomposites, outfits, protection, and energy storing. There are numerous advantages of nanofibers like the possibility of releasing drug at a slow constant or fast rate, accurate dosing, increased shelf life of drugs, improved solubility and bioavailability. They possess high surface-to-volume ratio and are highly porous. More than one drug can be encapsulated directly into the nanofibers. Bioactive growth factors can be directly incorporated into nanofiber scaffolds providing an insight into the newer drug delivery systems. Delivery of drug from polymeric nanofibers depends on the opinion that dissolution rate of a drug substance rises with greater surface area of both the drug and the corresponding carrier if necessary [24, 80].

7.3 CONCEPT OF ACTIVE AND PASSIVE TARGETING

The actuality of drug targeting is reliant upon the fact that the site of application and site of action of an active pharmaceutical agent are dissimilar. The diseased/affected tissues may be localized to a specific area (as in neurological disorders associated with brain – Alzheimer's disease, Huntington's chorea, etc.), or delocalized across the body (in case of malignant tumors). As proposed by Paul Ehrlich 20th century, drug targeting was assumed to be a "magic bullet" that was proposed to be comprised of two components – the primary fraction recognizes and binds to the target, whilst the secondary one induces a therapeutic action at the site. There are various ways to exclusively target drug loaded carrier moiety to the desired affected/diseased site in the body, but the two methods are broadly accepted: Passive targeting, which depends on the prolonged residence of drug-carrier in the systemic circulation and thereby accumulating in diseased tissue due to altered vasculature; and Active targeting exploits the specificity of biological ligands that exactly recognizes the specific receptors on the diseased organs and subsequently bind pathological cells (Figure 7.2).

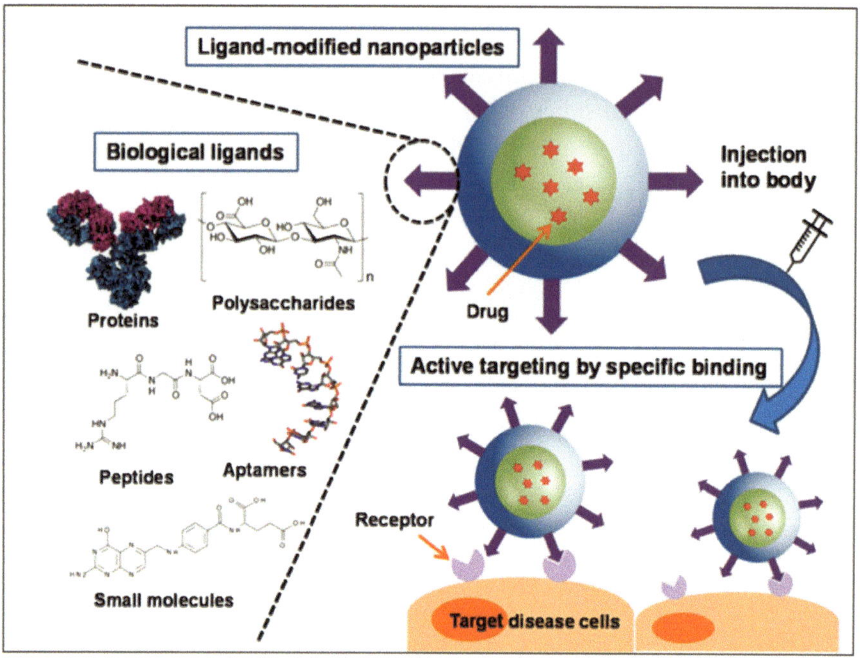

FIGURE 7.2 Scheme representing the addition of biological ligands to nanoparticles for active targeting.

Source: Reprinted from Ref. [89]. Copyright © 2019 by the authors. (http://creativecommons.org/licenses/by/4.0/).

Formerly, the cancer research done at cellular levels evidently demonstrated that the endothelial lining of the blood vessel wall of tumorous and infarcted tissues become more permeable in comparison to the normal tissues [36, 40, 63, 83]. The tumor microenvironment is composed of various cancer cells, macrophages, T cells, MDSCs, and dendritic cells entangled with collagen fibers and glycosaminoglycans (GAG) that aides tumor growth and instigates irregular structure along with altered vascularity and permeability of the lymphatic vasculature [21]. As a result, in such areas, molecules ranging 10 to 500 nm in dimensions leave out the blood vessels and get accumulated the interstitial space of tumor tissues. Now by presuming these polymeric bodies are embedded with pharmaceutical agent, where they can reach to the area with enhanced vascular permeability and finally release

API from the carrier. As the permeabilized vasculature cut-off size differs from event to event [36, 90], the efficiency of such impulsive passive drug delivery can be controlled by the size of a drug-carrying particle [55, 56]. In order to provide an adequate concentration of agglomerated entities at the target site, the drug delivery system needs to be in circulation for a prolonged period of time. By masking the drug carriers with grafting modification (by surface modification with hydrophilic polymers with flexible chain like PEG), the drug carriers can be kept in the blood for a long enough period [43, 84]. This technique is frequently developed for nanocarriers that prevent their opsonization and reticuloendothelial clearance [48, 49]. The drug doxorubicin has been loaded into the PEG-coated liposomes, resulted in extensive circulation, enhanced efficiency and reduced side effects [29, 30]. Polymeric micelles have also been used as drug carriers for anti-tumor therapy in mice bearing lung carcinoma [82, 87]. The advantages of extended presence of drug delivery system in the systemic circulation includes the maintenance of effective concentration of an API for long time even after single dosage and ability to enhance the accumulation of drug in the areas with significantly permeable vasculature and limited blood supply. However, in some diseased conditions, the EPR mechanism remains unexploited because vasculature of endothelium is unaltered. Many methods of drug targeting described so far are not general. So, it is technically difficult to directly administer the drug to implicated organ or tissue, or the disease may be delocalized. The normal tissues and the affected tissue or organ doesn't diverge much in terms of vascular permeability, local pH value or temperature.

The most extensive approach to impart affinity to a drug molecule towards its target is attachment of API with targeting or vector component, which is competent of specific recognition and thereby binding to its target site, is referred as Active targeting. There are many substances that can be used as targeting moieties: peptides, lipoproteins, hormones, antibodies (fragment/whole), polysaccharides, and some low molecular weight ligands (folate, anisamide) (Table 7.1).

TABLE 7.1 Ligands for Active Targeting of Nanoparticles

Ligand Type	Chemical Moiety Attached	Targeted Receptors	Advantages	References
Proteins	Affibodies	HER-2 receptors	Highly specific and large sized	[18, 67, 78]
	Transferrin	Tf receptors on blood-brain barrier.		

TABLE 7.1 *(Continued)*

Ligand Type	Chemical Moiety Attached	Targeted Receptors	Advantages	References
Polysaccharides	Hyaluronic acid	Binds CD44, present on the surface of cancer cells	Backbone of polymeric nanoparticles, Binds to specific over-expressed receptors on liver tissue	[17]
Peptides	IL-4R-binding peptide-1	Binds interleukin-4 receptor in lung cancer cells	Easily fabricated, small sized and cleaved by peptidase	[15, 53, 79]
	Arginylglycylaspartic acid peptide	Binds integrins in tumor affected vascular endothelial cells		
Aptamers	AS-1411	Targets nucleolin of tumor cells	Highly specific, small sized, cleaved by nuclease and costlier	[25, 35]
	CX-5461	Inhibitor of rRNA synthesis		
	GBI-10	Tenascin-C, a protein overexpressed in the extracellular matrix of pancreatic ductal adenocarcinoma.		
Small molecules	Folic acid	Target folate receptor overexpressed in cancer cells	Small sized, very low cost	[37, 54]
	Anisamide	Target sigmina 1 receptors for delivery of sunitinib to melanoma tumors		

Pairing of a drug to high affinity molecule is the simplest way of drug targeting. The most intense example in this approach is Immunotoxins [86]. The toxin is cleaved into two parts: the active fragment (the toxin one) which is coupled with an antibody; and the recognizing fragment that is cut off from the former part. In this way toxic part can be delivered to cells expressing the distinctive antigen (usually tumor cells); whereas immunotoxin remains incompetent in recognition of antigen-free cells. However, in this case, to each antibody molecule, merely a single active moiety (immunotoxin) could be attached. As immunotoxins are enormously active, they can abolish thousands of ribosomes if reached inside cell; and are primarily exploited for cancer treatment [32, 86].

Another example is regarding inhibition of plasminogen activator by specifically targeting it using urokinase-type activator and bi-specific monoclonal antibody; thereby developing fivefold greater thrombolytic action [39]. The drug potency in the murine model was sharply enhanced when antibody against proliferative component of squamous carcinomas (of mammalian) was coupled with anti-neoplastic drug daunomycin [20]. Mitomycin C was conjugated with Murine monoclonal antibody (NCC-LU-243) and was used for targeted therapy of Small-cell lung cancer (SCLC) [45]. Accordingly, various hydrophilic/hydrophobic carries could be paired with numerous active moieties and subsequently coupled with targeting unit [66].

7.4 MARKETED NANO-FORMULATIONS

Once a novel nano product is designed and produced at laboratory scale, it reaches to industrial manufacture pipeline considering to various quality regulations and finally enters the commercial market after successful clinical trials. Post market surveillance also plays a pivotal role in ascertaining the sustenance of the product in the market. The generalized factors that determine the commercialization of the novel nano-product may possibly be categorized into – technical, economic, and regulatory aspects. At the outset, it needs to be cleared that developing a targeted nano-formulation is complex [28].

The technical facet that hinders the development of nanoformulation implicates their interactions with environment, enigma regarding synthesis, sterilization, and scaling up of manufacturing process [7, 11, 88]. The results achieved on *in vitro* and *in vivo* correlation are also discordant; hence selection of cellular and animal models in the study should be done meticulously that will precisely represent the expected outcome in humans [27, 91]. Secondly, the setting up of regulatory policies for nano-formulations has been quite gradual hitherto [6–8]. Not many nano-medicines have reached the commercial market; numerous of them are in the doldrums at various stages of pharmaceutical development. Because of this reason, the market for nano-formulations is still in the emerging stage. The regulatory reforms intended for nano-products regarding manufacture, quality testing, sterilization, *in vitro/vivo* modeling and safety requires to be conscripted independently, distinct to the guidelines for conventional formulations. Latterly, in terms of economic perspective, the manufacture of nano-medicines exploits more monetary resources in comparison to small molecule synthesis, which influences the investment policy of pharmaceutical companies in the commercialization of nano-products for medical uses [12].

The majority of the nano-medicines that have been justifiably marketed primarily comprise the previously approved drug molecules; that were toxic and poorly soluble. The technology improved their toxicity and pharmacokinetic profiles and simultaneously reduced the dose of the drugs (Table 7.2).

TABLE 7.2 Exemplary Nano-Medicines at the Clinical Stage or in Market

Drug Name	Active Pharmaceutical Ingredient	Vehicle	Targeted Organ/ Disease	References
NBTXR3® (Nanobiotix)	Hafnium oxide	Radiation stimulated nanoparticle	Squamous cell carcinoma	[57]
Doxil® (Janssen-Cilag International N.V.)	Doxorubicin	PEGylated liposome	Breast cancer, Kaposi's sarcoma, ovarian cancer and other solid tumors	[5]
Myocet® (Zeneus Pharmaceuticals)	Doxorubicin	Non-PEGylated liposome	Metastatic breast cancer	[52]
Onivyde® (Ipsen Biopharmaceuticals, Inc.)	Irinotecan	Non-PEGylated liposome	Pancreatic cancer	[33]
Lipusu® (Luye Pharmaceuticals)	Paclitaxel	Non-PEGylated liposome	Gastric cancer, esophageal cancer	[26]
Mepact® (Takeda)	Mifamurtide	Non-PEGylated liposome	Osteosarcoma	[4]
DaunoXome® (Gilead Sciences, Inc.)	Daunorubicin	Non-PEGylated liposome	Kaposi's sarcoma	[68]
DepoCyte® (Pacira Pharmaceuticals Inc.)	Cytarabine	Non-PEGylated liposome	Neoplastic meningitis	[59]

TABLE 7.2 *(Continued)*

Drug Name	Active Pharmaceutical Ingredient	Vehicle	Targeted Organ/ Disease	References
Marqibo® (Spectrum Pharmaceuticals)	Vincristine	Sphingomyelin/ cholesterol liposomes	Leukemias	[75]
Visudyne® (Bausch and Lomb)	Verteporfin	Dimyristoyl-phosphatidylcholine and egg phos-phatidylglycerol (negatively charged) liposome	Age-related macular degenerations	[3]
Genexol-PM® (Zhongsheng Pharmaceuticals)	Paclitaxel	Poly(ethylene glycol)–poly(D, L-lactide) micelles	Metastatic breast cancer, pancreatic, and lung cancers	[3]
Ontak® (Eisai)	Denileukin diftitox	Recombinant fusion protein of fragment 'A' of diphtheria toxin and subunit binding to the interleukin-2 receptor	Cutaneous T-cell lymphoma, CD25-positive	[3]
Kadcyla® (Genentech USA, Inc.)	Ado-trastuzumab emtansine	Monoclonal antibody conjugate	HER-2 positive metastatic breast cancer	[3]
Mircera® (Vifor)	Epoetin beta	PEGylated aptamer	Anemia associ-ated with chronic kidney disease	[3]

7.5 SUMMARY

Drug delivery to specific sites is itself a difficult task. Tablets, capsules, powders, syrups, elixirs, suspensions, and emulsions are conventional and earlier drug delivery systems; however, they have many drawbacks like inability to reach specific sites; interactions with the local physiological environment in the gastrointestinal tract; poor bioavailability; low solubility; low permeability. Drawbacks of conventional dosage forms can be over-come by formulating a targeted drug delivery system that would deliver the drug to the affected part of the body. The advantages of using the dosage form that deliver the medicament to a intended location are reduction of dose and adverse effects related to its pharmacological actions at non-diseased sites; subsequent enhancement of efficacy of the drug. Regardless of the fact that the development of nanoparticulate systems is complex, investment by the pharmaceutical companies is up surging. There are unceasing innovative developments happening over the last decade. This unremitting growth requires the proper set up of regulatory guidelines and toxicity database that can further safeguard their use.

KEYWORDS

- **active targeting**
- **liposomes**
- **marketed nanoformulation**
- **nanomaterials**
- **passive targeting**
- **polymeric nanoparticles**
- **targeted drug delivery**

REFERENCES

1. Akiyoshi, K., Deguchi, S., Moriguchi, N., Yamaguchi, S., & Sunamoto, J., (1993). Self-aggregates of hydrophobized polysaccharides in water. Formation and characteristics of nanoparticles. *Macromolecules, 26*(12), 3062–3068. https://doi.org/10.1021/ma00064a011.
2. Allen, T. M., & Cullis, P. R., (2004). Drug delivery systems: Entering the mainstream. *Science, 303*(5665), 1818–1822. https://doi.org/10.1126/science.1095833.
3. Aminu, N., Bello, I., Umar, N. M., Tanko, N., Aminu, A., & Audu, M. M., (2020). The influence of nanoparticulate drug delivery systems in drug therapy. *Journal of Drug Delivery Science and Technology, 60*. https://doi.org/10.1016/j.jddst.2020.101961, PubMed: 101961.
4. Anselmo, A. C., & Mitragotri, S., (2019). Nanoparticles in the clinic: An update. *Bioengineering and Translational Medicine, 4*(3), e10143. https://doi.org/10.1002/btm2.10143.
5. Barenholz, Y., & Doxil, R., (2012). Doxil®--the first FDA-approved nano-drug: lessons learned. *Journal of Controlled Release, 160*(2), 117–134. https://doi.org/10.1016/j.jconrel.2012.03.020.
6. Bawa, R., Audette, G. F., & Reese, B. E., (eds.). (2016). *Clinical nanomedicine* (1st edn., p. 1502). CRC Press.

7. Bazile, D. V., (2014). Nanotechnologies in drug delivery-an industrial perspective. *Journal of Drug Delivery Science and Technology, 24*(1), 12–21. https://doi.org/10.1016/S1773-2247(14)50002-0.
8. Bergamaschi, E., Murphy, F., Poland, C. A., Mullins, M., Costa, A. L., Mcalea, E., Tran, L., & Tofail, S. A. M., (2015). Impact and effectiveness of risk mitigation strategies on the insurability of nanomaterial production: Evidences from industrial case studies. *Wiley Interdisciplinary Reviews. Nanomedicine and Nanobiotechnology, 7*(6), 839–855. https://doi.org/10.1002/wnan.1340.
9. Bhadra, D., Bhadra, S., Jain, P., & Jain, N. K., (2002). Pegnology: A review of PEG-ylated systems. *Pharmazie, 57*(1), 5–29.
10. Biju, V., (2014). Chemical modifications and bioconjugate reactions of nanomaterials for sensing, imaging, drug delivery and therapy. *Chemical Society Reviews, 43*(3), 744–764. https://doi.org/10.1039/c3cs60273g.
11. Bosetti, R., (2015). Cost-effectiveness of nanomedicine: The path to a future successful and dominant market? *Nanomedicine, 10*(12), 1851–1853. https://doi.org/10.2217/nnm.15.74.
12. Brown, P. D., & Patel, P. R., (2015). Nanomedicine: A pharma perspective. *Wiley Interdisciplinary Reviews. Nanomedicine and Nanobiotechnology, 7*(2), 125–130. https://doi.org/10.1002/wnan.1288.
13. Brufsky, A., (2010). Trastuzumab-based therapy for patients with HER2-positive breast cancer: From early scientific development to foundation of care. *American Journal of Clinical Oncology, 33*(2), 186–195. https://doi.org/10.1097/COC.0b013e318191bfb0.
14. Bujold, K. E., Lacroix, A., & Sleiman, H. F. D. N. A., (2018). DNA nanostructures at the interface with biology. *Chem, 4*(3), 495–521. https://doi.org/10.1016/j.chempr.2018.02.005.
15. Chi, L., Na, M. H., Jung, H. K., Vadevoo, S. M. P., Kim, C. W., Padmanaban, G., Park, T. I., et al., (2015). Enhanced delivery of liposomes to lung tumor through targeting interleukin-4 receptor on both tumor cells and tumor endothelial cells. *Journal of Controlled Release, 209*, 327–336. https://doi.org/10.1016/j.jconrel.2015.05.260.
16. Cho, K., Wang, X., Nie, S., Chen, Z. G., & Shin, D. M., (2008). Therapeutic nanoparticles for drug delivery in cancer. *Clinical Cancer Research, 14*(5), 1310–1316. https://doi.org/10.1158/1078-0432.CCR-07-1441.
17. Choi, K. Y., Yoon, H. Y., Kim, J. H., Bae, S. M., Park, R. W., Kang, Y. M., Kim, I. S., et al., (2011). Smart nanocarrier based on pegylated hyaluronic acid for cancer therapy. *ACS Nano, 5*(11), 8591–8599. https://doi.org/10.1021/nn202070n.
18. Clark, A. J., & Davis, M. E., (2015). Increased brain uptake of targeted nanoparticles by adding an acid-cleavable linkage between transferrin and the nanoparticle core. *Proceedings of the National Academy of Sciences of the United States of America, 112*(40), 12486–12491. https://doi.org/10.1073/pnas.1517048112.
19. Couvreur, P., (2013). Nanoparticles in drug delivery: Past, present and future. *Advanced Drug Delivery Reviews, 65*(1), 21–23. https://doi.org/10.1016/j.addr.2012.04.010.
20. Ding, L., Samuel, J., MacLean, G. D., Noujaim, A. A., Diener, E., & Longenecker, B. M., (1990). Effective drug-antibody targeting using a novel monoclonal antibody against the proliferative compartment of mammalian squamous carcinomas. *Cancer Immunology, Immunotherapy, 32*(2), 105–109. https://doi.org/10.1007/BF01754206.

21. Donahue, N. D., Acar, H., & Wilhelm, S., (2019). Concepts of nanoparticle cellular uptake, intracellular trafficking, and kinetics in nanomedicine. *Advanced Drug Delivery Reviews, 143*, 68–96. https://doi.org/10.1016/j.addr.2019.04.008.
22. Dua, K., Chellappan, D. K., Singhvi, G., De Jesus, A. P. T., Gupta, G., & Hansbro, P. M., (2018). Targeting microRNAs using nanotechnology in pulmonary diseases. *Panminerva Medica, 60*(4), 230, 231. https://doi.org/10.23736/S0031-0808.18.03459-6.
23. Dua, K., Malyla, V., Singhvi, G., Wadhwa, R., Krishna, R. V., Shukla, S. D., Shastri, M. D., et al., (2019). Increasing complexity and interactions of oxidative stress in chronic respiratory diseases: An emerging need for novel drug delivery systems. *Chemico-Biological Interactions, 299*, 168–178. https://doi.org/10.1016/j.cbi.2018.12.009.
24. Duggal, D., (2011). Role of nanotechnology in new drug delivery systems. *Drug Development Research, 3*(4), 4–8.
25. Duo, Y., Yang, M., Du, Z., Feng, C., Xing, C., Wu, Y., Xie, Z., et al., (2018). CX-5461-Loaded nucleolus-targeting nanoplatform for cancer therapy through induction of pro-death autophagy. *Acta Biomaterialia, 79*, 317–330. https://doi.org/10.1016/j.actbio.2018.08.035.
26. Farjadian, F., Ghasemi, A., Gohari, O., Roointan, A., Karimi, M., & Hamblin, M. R., (2019). Nanopharmaceuticals and nanomedicines currently on the market: Challenges and opportunities. *Nanomedicine, 14*(1), 93–126. https://doi.org/10.2217/nnm-2018-0120.
27. Finch, G., Havel, H., Analoui, M., Barton, R. W., Diwan, A. R., Hennessy, M., Reddy, V., et al., (2014). Nanomedicine drug development: A scientific symposium entitled "charting a roadmap to commercialization". *AAPS Journal, 16*(4), 698–704. https://doi.org/10.1208/s12248-014-9608-5.
28. Fornaguera, C., & García-Celma, M. J., (2017). Personalized nanomedicine: A revolution at the nanoscale. *Journal of Personalized Medicine, 7*(4), 12. https://doi.org/10.3390/jpm7040012.
29. Gabizon, A. A., (1995). Liposome circulation time and tumor targeting: Implications for cancer chemotherapy. *Advanced Drug Delivery Reviews, 16*(2, 3), 285–294. https://doi.org/10.1016/0169-409X(95)00030-B.
30. Gabizon, A. A., (2001). Pegylated liposomal doxorubicin: Metamorphosis of an old drug into a new form of chemotherapy. *Cancer Investigation, 19*(4), 424–436. https://doi.org/10.1081/cnv-100103136.
31. Girdhar, V., Patil, S., Banerjee, S., & Singhvi, G., (2018). Nanocarriers for drug delivery: Mini review. *Current Nanomedicine, 8*(2), 88–99. https://doi.org/10.2174/2468187308666180501092519.
32. Goldmacher, V. S., Blatter, W. A., Lambert, J. M., & Chari, R. V. J., (2002). Chapter 15; Immunotoxins and antibody-drug conjugates for cancer treatment. In: Muzykantov, V., & Torchilin, V. P., (eds.), *Biomedical Aspects of Drug Targeting* (pp. 291–309). Springer.
33. Grimaldi, N., Andrade, F., Segovia, N., Ferrer-Tasies, L., Sala, S., Veciana, J., & Ventosa, N., (2016). Lipid-based nanovesicles for nanomedicine. *Chemical Society Reviews, 45*(23), 6520–6545. https://doi.org/10.1039/c6cs00409a.
34. Guterres, S. S., Alves, M. P., & Pohlmann, A. R., (2007). Polymeric nanoparticles, nanospheres and nanocapsules, for cutaneous applications. *Drug Target Insights, 2*, 147–157. https://doi.org/10.1177/117739280700200002.
35. He, X., Chen, X., Liu, L., Zhang, Y., Lu, Y., Zhang, Y., Chen, Q., et al., (2018). Sequentially triggered nanoparticles with tumor penetration and intelligent drug release

for pancreatic cancer therapy. *Advanced Science, 5*(5), 1701070. https://doi.org/10.1002/advs.201701070.
36. Hobbs, S. K., Monsky, W. L., Yuan, F., Roberts, W. G., Griffith, L., Torchilin, V. P., & Jain, R. K., (1998). Regulation of transport pathways in tumor vessels: Role of tumor type and microenvironment. *Proceedings of the National Academy of Sciences of the United States of America, 95*(8), 4607–4612. https://doi.org/10.1073/pnas.95.8.4607.
37. Huo, M., Zhao, Y., Satterlee, A. B., Wang, Y., Xu, Y., & Huang, L., (2017). Tumor-targeted delivery of sunitinib base enhances vaccine therapy for advanced melanoma by remodeling the tumor microenvironment. *Journal of Controlled Release, 245*, 81–94. https://doi.org/10.1016/j.jconrel.2016.11.013.
38. Illum, L., & Davis, S. S., (1984). The organ uptake of intravenously administered colloidal particles can be altered using a non-ionic surfactant (poloxamer 338). *FEBS Letters, 167*(1), 79–82. https://doi.org/10.1016/0014-5793(84)80836-4.
39. Imura, Y., Stassen, J. M., Kurokawa, T., Iwasa, S., Lijnen, H. R., & Collen, D., (1992). Thrombolytic and pharmacokinetic properties of an immunoconjugate of single-chain urokinase-type plasminogen activator (U-PA) and a bispecific monoclonal antibody against fibrin and against U-PA in baboons. *Blood, 79*(9), 2322–2329.
40. Jain, R. K., (1999). Transport of molecules, particles, and cells in solid tumors. *Annual Review of Biomedical Engineering, 1*, 241–263. https://doi.org/10.1146/annurev.bioeng.1.1.241.
41. Junyaprasert, V. B., & Morakul, B., (2015). Nanocrystals for enhancement of oral bioavailability of poorly water-soluble drugs. *Asian Journal of Pharmaceutical Sciences, 10*(1), 13–23. https://doi.org/10.1016/j.ajps.2014.08.005.
42. Khosa, A., Singhvi, G., Saha, R. N., & Gupta, G., (2018). Drug delivery to the CNS. *Panminerva Medica, 60*(4), 226. https://doi.org/10.23736/S0031-0808.18.03471-7.
43. Klibanov, A. L., Maruyama, K., Torchilin, V. P., & Huang, L., (1990). Amphipathic polyethyleneglycols effectively prolong the circulation time of liposomes. *FEBS Letters, 268*(1), 235–237. https://doi.org/10.1016/0014-5793(90)81016-h.
44. Kommareddy, S., Tiwari, S. B., & Amiji, M. M., (2005). Long-circulating polymeric nanovectors for tumor-selective gene delivery. *Technology in Cancer Research and Treatment, 4*(6), 615–625. https://doi.org/10.1177/153303460500400605.
45. Kubota, T., Yamamoto, T., Takahara, T., Furukawa, T., Ishibiki, K., Kitajima, M., Shida, Y., & Nakatsubo, H., (1992). Targeting cancer chemotherapy using a monoclonal antibody (NCC-LU-243) conjugated with mitomycin C. *Journal of Surgical Oncology, 51*(2), 75–80. https://doi.org/10.1002/jso.2930510203.
46. Kumar, S., Kaur, P., Bernela, M., Rani, R., & Thakur, R., (2016). Ketoconazole encapsulated in chitosan-gellan gum nanocomplexes exhibits prolonged antifungal activity. *International Journal of Biological Macromolecules, 93*(A), 988–994. https://doi.org/10.1016/j.ijbiomac.2016.09.042.
47. Langer, R., (2000). Biomaterials in drug delivery and tissue engineering: One laboratory's experience. *Accounts of Chemical Research, 33*(2), 94–101. https://doi.org/10.1021/ar9800993.
48. Lasic, D. D., & Martin, F. J., (eds.). (1995). *Stealth Liposomes* (1st edn., p. 320). CRC Press.
49. Lasic, D. D., & Papahadjopoulos, D., (1998). *Medical Applications of Liposomes* (1st edn., p. 779). Elsevier Science.

50. Lee, J. H., & Yeo, Y., (2015). Controlled drug release from pharmaceutical nanocarriers. *Chemical Engineering Science, 125*, 75–84. https://doi.org/10.1016/j.ces.2014.08.046.
51. Lee, M., & Kim, S. W., (2005). Polyethylene glycol-conjugated copolymers for plasmid DNA delivery. *Pharmaceutical Research, 22*(1), 1–10. https://doi.org/10.1007/s11095-004-9003-5.
52. Leonard, R. C. F., Williams, S., Tulpule, A., Levine, A. M., & Oliveros, S., (2009). Improving the therapeutic index of anthracycline chemotherapy: Focus on liposomal doxorubicin (Myocet). *Breast, 18*(4), 218–224. https://doi.org/10.1016/j.breast.2009.05.004.
53. Lu, Z., Long, Y., Cun, X., Wang, X., Li, J., Mei, L., Yang, Y., et al., (2018). A size-shrinkable nanoparticle-based combined anti-tumor and anti-inflammatory strategy for enhanced cancer therapy. *Nanoscale, 10*(21), 9957–9970. https://doi.org/10.1039/c8nr01184b.
54. Lv, Y., Cao, Y., Li, P., Liu, J., Chen, H., Hu, W., & Zhang, L., (2017). Ultrasound-triggered destruction of folate-functionalized mesoporous silica nanoparticle-loaded microbubble for targeted tumor therapy. *Advanced Healthcare Materials, 6*(18). https://doi.org/10.1002/adhm.201700354.
55. Maeda, H., (2003). Chapter 11; Enhanced permeability and retention (EPR) effect: Basis for drug targeting to tumor. In: Muzykantov, V., & Torchilin, V. P., (eds.), *Biomedical Aspects of Drug Targeting* (pp. 211–228). Springer.
56. Maeda, H., Wu, J., Sawa, T., Matsumura, Y., & Hori, K., (2000). Tumor vascular permeability and the EPR effect in macromolecular therapeutics: A review. *Journal of Controlled Release, 65*(1, 2), 271–284. https://doi.org/10.1016/s0168-3659(99)00248-5.
57. Marill, J., Mohamed, A. N. M., & Paris, S. D. N. A., (2019). DNA Damage enhancement by radiotherapy-activated hafnium oxide nanoparticles improves cGAS-STING pathway activation in human colorectal cancer cells. *Radiotherapy and Oncology, 141*, 262–266. https://doi.org/10.1016/j.radonc.2019.07.029.
58. Martinho, N., Damgé, C., & Reis, C. P., (2011). Recent advances in drug delivery systems. *Journal of Biomaterials and Nanobiotechnology, 02*(5), 510–526. https://doi.org/10.4236/jbnb.2011.225062.
59. Mayer, A. M. S., Glaser, K. B., Cuevas, C., Jacobs, R. S., Kem, W., Little, R. D., McIntosh, J. M., et al., (2010). The odyssey of marine pharmaceuticals: A current pipeline perspective. *Trends in Pharmacological Sciences, 31*(6), 255–265. https://doi.org/10.1016/j.tips.2010.02.005.
60. Monthioux, M., & Kuznetsov, V. L., (2006). Who should be given the credit for the discovery of carbon nanotubes? *Carbon, 44*(9), 1621–1623. https://doi.org/10.1016/j.carbon.2006.03.019.
61. Muller, R. H., (1991). *Colloidal Carriers for Controlled Drug Delivery and Targeting: Modification, Characterization, and in Vivo Distribution* (1st edn., p. 379). Wissenschaftliche Verlagsgesellschaft Stuttgart: CRC Press Boca Raton.
62. Neamtu, I., Rusu, A. G., Diaconu, A., Nita, L. E., & Chiriac, A. P., (2017). Basic concepts and recent advances in nanogels as carriers for medical applications. *Drug Delivery, 24*(1), 539–557. https://doi.org/10.1080/10717544.2016.1276232.
63. Palmer, T. N., Caride, V. J., Caldecourt, M. A., Twickler, J., & Abdullah, V., (1984). The mechanism of liposome accumulation in infarction. *Biochimica et Biophysica Acta, 797*(3), 363–368. https://doi.org/10.1016/0304-4165(84)90258-7.

64. Raffin, P. A., Weiss, V., Mertins, O., Pesce Da, S. N., & Stanisçuaski, G. S., (2002). Spray-dried indomethacin-loaded polyester nanocapsules and nanospheres: Development, stability evaluation and nanostructure models. *European Journal of Pharmaceutical Sciences, 16*(4, 5), 305–312. https://doi.org/10.1016/s0928-0987(02)00127-6.
65. Rapalli, V. K., Singhvi, G., Dubey, S. K., Gupta, G., Chellappan, D. K., & Dua, K., (2018). Emerging landscape in psoriasis management: From topical application to targeting biomolecules. *Biomedicine and Pharmacotherapy = Biomedicine and Pharmacotherapie, 106*, 707–713. https://doi.org/10.1016/j.biopha.2018.06.136.
66. Ringsdorf, H., (1975). Structure and properties of pharmacologically active polymers. *Journal of Polymer Science: Polymer Symposia, 51*(1), 135–153. https://doi.org/10.1002/polc.5070510111.
67. Roncato, F., Rruga, F., Porcù, E., Casarin, E., Ronca, R., Maccarinelli, F., Realdon, N., et al., (2018). Improvement and extension of anti-EGFR targeting in breast cancer therapy by integration with the avidin-nucleic-acid-nano-assemblies. *Nature Communications, 9*(1), 4070. https://doi.org/10.1038/s41467-018-06602-6.
68. Rosenthal, E., Poizot-Martin, I., Saint-Marc, T., Spano, J. P., Cacoub, P., & DNX Study Group, (2002). Phase IV study of liposomal daunorubicin (DaunoXome) in AIDS-related Kaposi sarcoma. *American Journal of Clinical Oncology, 25*(1), 57–59. https://doi.org/10.1097/00000421-200202000-00012.
69. Safari, J., & Zarnegar, Z., (2014). Advanced drug delivery systems: Nanotechnology of health design: A review. *Journal of Saudi Chemical Society, 18*(2), 85–99. https://doi.org/10.1016/j.jscs.2012.12.009.
70. Kahan, M., Gil, B., Adar, R., & Shapiro, E., (2008). Towards molecular computers that operate in a biological environment. *Physica D: Nonlinear Phenomena, 237*(9), 1165–1172. https://doi.org/10.1016/j.physd.2008.01.027.
71. Scott, R. C., Crabbe, D., Krynska, B., Ansari, R., & Kiani, M. F., (2008). Aiming for the heart: Targeted delivery of drugs to diseased cardiac tissue. *Expert Opinion on Drug Delivery, 5*(4), 459–470. https://doi.org/10.1517/17425247.5.4.459.
72. Seidman, A., Hudis, C., Pierri, M. K., Shak, S., Paton, V., Ashby, M., Murphy, M., et al., (2002). Cardiac dysfunction in the trastuzumab clinical trials experience. *Journal of Clinical Oncology, 20*(5), 1215–1221. https://doi.org/10.1200/JCO.2002.20.5.1215.
73. Senapati, S., Thakur, R., Verma, S. P., Duggal, S., Mishra, D. P., Das, P., Shripathi, T., et al., (2016). Layered double hydroxides as effective carrier for anticancer drugs and tailoring of release rate through interlayer anions. *Journal of Controlled Release, 224*, 186–198. https://doi.org/10.1016/j.jconrel.2016.01.016.
74. Sharma, A., Garg, T., Aman, A., Panchal, K., Sharma, R., Kumar, S., & Markandeywar, T., (2016). Nanogel--and advanced drug delivery tool: Current and future. *Artificial Cells, Nanomedicine, and Biotechnology, 44*(1), 165–177. https://doi.org/10.3109/21691401.2014.930745.
75. Silverman, J. A., & Deitcher, S. R., (2013). Marqibo® (vincristine sulfate liposome injection) improves the pharmacokinetics and pharmacodynamics of vincristine. *Cancer Chemotherapy and Pharmacology, 71*(3), 555–564. https://doi.org/10.1007/s00280-012-2042-4.
76. Singh, N. K., Singh, S. K., Dash, D., Das, P. B. P., Roy, J. K., & Maiti, P., (2012). Nanostructure controlled anti-cancer drug delivery using poly(E-caprolactone) based nanohybrids. *Journal of Materials Chemistry, 22*(34), 17853–17863. https://doi.org/10.1039/c2jm32340k.

77. Singhvi, G., Banerjee, S., & Khosa, A., (2018). Chapter 11; Lyotropic liquid crystal nanoparticles: A novel improved lipidic drug delivery system. In: Grumezescu, A. M., (ed.), *Org Mater as Smart Nanocarriers Drug Delivery* (pp. 471–517). William Andrew Publishing.
78. Smith, B., Lyakhov, I., Loomis, K., Needle, D., Baxa, U., Yavlovich, A., Capala, J., et al., (2011). Hyperthermia-triggered intracellular delivery of anticancer agent to HER2(+) cells by HER2-specific affibody (ZHER2-GS-Cys)-conjugated thermosensitive liposomes (HER2(+) affisomes). *Journal of Controlled Release, 153*(2), 187–194. https://doi.org/10.1016/j.jconrel.2011.04.005.
79. Sugahara, K. N., Teesalu, T., Karmali, P. P., Kotamraju, V. R., Agemy, L., Greenwald, D. R., & Ruoslahti, E., (2010). Coadministration of a tumor-penetrating peptide enhances the efficacy of cancer drugs. *Science, 328*(5981), 1031–1035. https://doi.org/10.1126/science.1183057.
80. Suri, S. S., Fenniri, H., & Singh, B., (2007). Nanotechnology-based drug delivery systems. *Journal of Occupational Medicine and Toxicology, 2*, 16. https://doi.org/10.1186/1745-6673-2-16.
81. Teixeira, M., Alonso, M. J., Pinto, M. M. M., & Barbosa, C. M., (2005). Development and characterization of PLGA nanospheres and nanocapsules containing xanthone and 3-Methoxyxanthone. *European Journal of Pharmaceutics and Biopharmaceutics, 59*(3), 491–500. https://doi.org/10.1016/j.ejpb.2004.09.002.
82. Torchilin, V. P., (2001). Structure and design of polymeric surfactant-based drug delivery systems. *Journal of Controlled Release, 73*(2, 3), 137–172. https://doi.org/10.1016/s0168-3659(01)00299-1.
83. Torchilin, V. P., Klibanov, A. L., Huang, L., O'Donnell, S., Nossiff, N. D., & Khaw, B. A., (1992). Targeted accumulation of polyethylene glycol-coated immunoliposomes in infarcted rabbit myocardium. *FASEB Journal, 6*(9), 2716–2719. https://doi.org/10.1096/fasebj.6.9.1612296.
84. Torchilin, V. P., & Trubetskoy, V. S., (1995). Which polymers can make nanoparticulate drug carriers long-circulating? *Advanced Drug Delivery Reviews, 16*(2, 3), 141–155. https://doi.org/10.1016/0169-409X(95)00022-Y.
85. Trindade, I. C., Pound-Lana, G., Pereira, D. G. S., De Oliveira, L. A. M., Andrade, M. S., Vilela, J. M. C., Postacchini, B. B., & Mosqueira, V. C. F., (2018). Mechanisms of interaction of biodegradable polyester nanocapsules with non-phagocytic cells. *European Journal of Pharmaceutical Sciences, 124*, 89–104. https://doi.org/10.1016/j.ejps.2018.08.024.
86. Vitetta, E. S., Krolick, K. A., Miyama-Inaba, M., Cushley, W., & Uhr, J. W., (1983). Immunotoxins: A new approach to cancer therapy. *Science, 219*(4585), 644–650. https://doi.org/10.1126/science.6218613.
87. Weissig, V., Whiteman, K. R., & Torchilin, V. P., (1998). Accumulation of protein-loaded long-circulating micelles and liposomes in subcutaneous Lewis lung carcinoma in mice. *Pharmaceutical Research, 15*(10), 1552–1556. https://doi.org/10.1023/a:1011951016118.
88. Wicki, A., Witzigmann, D., Balasubramanian, V., & Huwyler, J., (2015). Nanomedicine in cancer therapy: Challenges, opportunities, and clinical applications. *Journal of Controlled Release, 200*, 138–157. https://doi.org/10.1016/j.jconrel.2014.12.030.

89. Yoo, J., Park, C., Yi, G., Lee, D., & Koo, H., (2019). Active targeting strategies using biological ligands for nanoparticle drug delivery systems. *Cancers, 11*(5), 640. https://doi.org/10.3390/cancers11050640.
90. Yuan, F., Dellian, M., Fukumura, D., Leunig, M., Berk, D. A., Torchilin, V. P., & Jain, R. K., (1995). Vascular permeability in a human tumor xenograft: Molecular size dependence and cutoff size. *Cancer Research, 55*(17), 3752–3756.
91. Zhang, X. Q., Xu, X., Bertrand, N., Pridgen, E., Swami, A., & Farokhzad, O. C., (2012). Interactions of nanomaterials and biological systems: Implications to personalized nanomedicine. *Advanced Drug Delivery Reviews, 64*(13), 1363–1384. https://doi.org/10.1016/j.addr.2012.08.005.

CHAPTER 8

SIZE AND MORPHOLOGY OF NANOFERRITES FOR DRUG DELIVERY, THERMAL HEATING, AND IMAGING IN MEDICINE

AJAY SINGH and MANJU ARORA

ABSTRACT

Ferrites as soft and hard magnetic materials have been the center of attraction for researchers and technologists for the last many decades owing to their wide range of practical applications in communication, electronics, magnetic recording, microwave absorption-based devices, etc. In nanoferrites, the structural, morphological, and synthesis procedure decides the availability of large surface area to volume ratio for surface activity, electrical, and magnetic properties for a particular application. The superparamagnetic nature of ferrites nanoparticles (NPs) opened the door for new advanced high-tech applications in the field of biotechnology and biomedical sciences, sensors, nanoelectronics, microwave devices, food processing, catalytic performance, and environmental remediation. In the present scenario, large numbers of pure and co-doped ferrites NPs have been explored for their usage as effective magnetic probes in magnetic resonance imaging (MRI) and multimodal imaging techniques, drug delivery, i.e., transport of pharmaceutical compounds to the infected part of body and the medical therapies for hyperthermia treatment, etc. For biomedical application, the high magnetocrystalline anisotropy, moderate saturation magnetization, a high coercive field, mechanical hardness, resistivity, and low loss behavior

Sustainable Nanomaterials for Biomedical Engineering: Impacts, Challenges, and Future Prospects.
Junaid Ahmad Malik, Megh R. Goyal & Mohamed Jaffer M. Sadiq (Eds.)

properties of ferrite nanoparticles are stabilized by coating them with the biocompatible surfactant. NPs exhibit core-shell structure and its shell prevents the agglomeration of ferrite NPs and provides the very large active surface for the biomolecular conjugation as desired for biomedical applications. The present chapter reports the development in nanoferrites field pertaining to biomedical applications from 2000 onwards. The investigations are discussed in terms of the biocompatibility and permissible toxicity level of ferrite nanoparticles by appropriate selection of chemicals for magnetic cores and outer shell compositions with a view to enhance stability and swift delivery of NPs, optimization of size and morphological shapes like dot, spherical, rod, wire, cubical, tetrapods, needle, and polyhedron, etc. pH and ionic strength sensitive fluid properties and the probable interactions between ferrite NPs with cells, proteins, enzymes, drugs, contrast agents and molecules which are involved in the oxidation/reduction activity undergoing in the human body or by taking orally or directly injecting into the bloodstream. These are some key factors which influences the stability, *in vivo* activity and restriction of NPs as effective contrast agent for imaging purpose, drug-delivery, and cancer medication. This chapter summarizes all the major developments that took place in this field in terms of different ferrite nanoparticles, synthesis routes, and their characterization by different analytical techniques to confirm their formation and better understanding of the science behind tentative mechanisms with all possible advantages and shortcomings. We will try to unfold some new concepts for future exploration with the help of earlier reported studies and new recent reported ideas.

8.1 INTRODUCTION

Ferrites are widely used in a variety of electronics, communication, magnetic recording, and microwave absorption-based devices. The concept of nanoscience and nanotechnology in ferrites have drawn considerable attention of researchers due to the sensitivity of their structural, surface reactivity, electrical, and magnetic with respect to the synthesis route, size, and morphology. In ferrite nanoparticles (FNPs) very large volume fraction of atoms are available around the grain boundaries as compared to bulk ferrite particles. The decrease in magnetic particles size leads to unusual disordering in surface spin known as spin canting [24, 83, 124, 168, 209], surface anisotropy and the superparamagnetic behavior [21, 32]. The size and morphology of FNPs can be easily tailored as per application demand.

It means that FNPs have core-shell structure and magnetic property is the contribution from these two factors, i.e., (i) the magnetic behavior from the finite size of particle core and (ii) atoms present at the surface of the particles. In NMP (nanomagnetic particles), the spin is ordered in core below a critical temperature and behaves as bulk analog, but in the outer shell, the spins are disordered even at the lowest temperature. That is why the magnetic behavior of the core and shell is very different and the competition between them results in a final magnetic state/property of a nanoparticle. As the size of nanoparticles decrease, magnetic properties pertaining to surface effects dominate due to the availability of a larger number of magnetic spins at the surface and on increasing FNPs size, the magnetic behavior arising from core competes. When the size of ferrite nanoparticles is below 20 nm size then these particles exhibit exceptionally high magnetization values and superparamagnetism. In superparamagnetism state, a Ferro/ferri-magnetic material behaves as paramagnetic materials even below the Curie or Neel temperature. As the NMPs size is around 10 nm, the energy barrier for magnetization reversal becomes too low that it is easily overcome by each superparamagnetic particle with no corecivity or retentivity during their magnetization investigations by vibrational sample magnetometer (VSM). It shows that superparamagnetism is a size-dependent phenomenon. When superparamagnetic (SP) particles are placed in an external magnetic field, then their magnetic moments align themselves along ~~in~~ the direction of the applied magnetic field. This disrupts on removing magnetic field, particles relax back by the random orientation of magnetic moments/spins alignment via Brownian motion and Néel rotation mechanisms, respectively. Brownian relaxation arises from the bulk rotation of the particles in colloidal dispersion. The SP particles completely randomize with the help of thermal energy available at ambient temperature in which anisotropic energy (KV) is always less than thermal energy (kT).

This superparamagnetic (SP) property of FNPs has opened a new window for researchers to use in the new field of biotechnology and biomedical applications. The SP ferrites can be used in early diagnosis and effective treatment of some diseases such as cancer [141]. Laurent et al. [96] have reviewed the preparation of superparamagnetic iron oxide particles for biomedical applications. The magnetic properties of FNPs are now exploited in various biomedical applications such as biosensors, targeted drug delivery, and contrast agents in magnetic resonance imaging (MRI) [63, 111, 131, 150, 192, 195]. This new application has created interest in the exploration of new economic strategies to develop FNPs of desired stoichiometric compositions, high purity without any other phases/impurities, homogeneity, uniform size, narrow size distribution, morphology with tunable magnetic and electrical properties. The variety of physical and chemical preparation methods [46, 63, 96, 111, 136, 150, 192, 195] is resulted.

Ferrites are basically ferromagnetic materials, in terms of their magnetic properties, they are categorized into two classes: soft and hard magnetic materials. The word "soft" refers to temporary, in these ferrites ferromagnetism emerges only on applying magnetic field. While in hard ferrites, ferromagnetism still persists even after removing the external magnetic field. Fe_3O_4, Cobalt ferrite, nickel ferrite, zinc ferrite, Mn–Zn ferrite and Ni–Zn ferrite are some of the examples of soft ferrites and $BaFe_{12}O_{19}$ and $SrFe_{12}O_{19}$ come under hard ferrites. The characteristic magnetic properties of ferrites are influenced by the interactions between metallic ions with oxygen ions present at octahedral/tetrahedral sites in the ferrite lattice [111].

The ferrites acquire four different types of crystal structures as shown in the flowchart in Figure 8.1: (i) spinel ferrites; (ii) hexagonal ferrites; (iii) garnet ferrites; and (iv) orthoferrites. The structural details are described into the following section:

1. **Spinel Ferrites:** The cubic structured spinel ferrites are generally represented by AB_2O_4 chemical formula, where A is a present divalent metal ion (e.g., Mn, Fe, Co, Ni, Cu, or Zn, etc.), with an ionic radius from ~ 0.6 to 1 Å range and B is a trivalent metal ion in present case Fe^{3+} ions. In spinel cubic unit cell 64 tetrahedral sites and 32 octahedral sites are present, out of which metal ions occupy only 8 and 16 sites, respectively (referred to as A and B sites, respectively). In AB_2O_4 structure, out of three metal ions, one occupies a tetrahedral site and two occupy the octahedral site. In a normal spinel the divalent M ion occupies the tetrahedral (A) site while the trivalent Fe ion occupies the octahedral (B) site and degree of inversion x=0. In an inverse spinel structure, the divalent A ions occupy one of the B sites, and trivalent Fe ions occupy both B and A sites [111] and x =1. In a spinel ferrite, the metal ions are separated by oxygen ions, and the ions in the A sites are antiparallel to those in B sites. In the majority of ferrites, the two sublattices are different in number and in the types of ions, so there is a resultant magnetization. The doping/substitution of two or more metal ions is also possible in spinel ferrites. The third is mixed spinel structure in which x lies between $0<x<1$.

2. **Hexagonal Ferrites:** These have general formula $MFe_{12}O_{19}$, composed of complex hexagonal unit cell with magnetoplumbite structures. Magnetoplumbite structures have chemical formula $A^{2+}O\cdot 6(B^{3+}_2 O_3)$. Around 12 Fe^{3+} ions in the unit cell of the hexagonal structure are located at three different sites: (i) nine ions at octahedral sites with six nearest O^{2-} ions, (ii) two ions at the tetrahedral sites with four nearest O^{2-} ions, and (iii) one ion in the hexagonal site with five nearest O^{2-} ions. These ferrites exhibit strong uniaxial magnetic direction and available in six different types (i) M-type ferrites ($BaFe_{12}O_{19}$, $SrFe_{12}O_{19}$), (ii) Z-type ferrites ($Ba_2Co_2Fe_{24}O_{41}$), (iii) Y-type ferrites ($Ba_2Co_2Fe_{12}O_{22}$), (iv) W-type ferrites ($BaCo_2Fe_{16}O_{27}$), (v) X-type ferrites ($Ba_2Co_2Fe_{28}O_{46}$) and (vi) U-type ferrites ($Ba_2Co_4Fe_{36}O_{60}$) [143]. These ferrites are used in permanent magnets [56].
3. **Garnet Ferrites:** These have general formula either $3M_2O_3 \cdot 5Fe_2O_3$ or two units of $Me_3Fe_2(FeO_4)_3$ where Me is trivalent rare earth or a magnetic rare earth, i.e., ytterbium. In the orthorhombic crystal structured garnets, trivalent rare earth and Fe^{3+} cations are present at tetrahedral, octahedral, or dodecahedral sites. As the interaction between tetrahedral and octahedral sites is antiparallel to each other. This causes the rare earth ions present at dodecahedral sites to have antiparallel net magnetic moment. The garnet structure is very complex due to the combination of 4 units of $3M_2O_3\cdot 5Fe_2O_3$ [63].
4. **Orthoferrites:** These are weak ferromagnetic materials, represented by chemical formula $MFeO_3$, where M is one or more rare earth metal ions with ionic radii bigger then Fe^{3+} ions and acquire orthorhombic structure with a space group Pbnm [64, 133], i.e., distorted perovskite structure (ABO_3). Fe^{3+} ions are coordinated to six oxygen anions to form FeO_6 octahedron. The cationic M (rare earth diamagnetic ions) are present in the interstitial area between the octahedral structures and coordinated to 12 oxygen anions. They generally exhibit antiferromagnetic ordering, when the Néel temperature is below 750 K. Dzyaloshinskii-Moriya (DM) interaction in some materials having iron in its structure results in the spin-lattice canting, a weak ferromagnetic ordering in these ferrites [18, 153]. The magnetic transition temperature of orthoferrites depends upon the ionic radius of the A site element. The transition temperature increases with the increase in ionic radius [33], e.g., transition from antiferromagnetic

to paramagnetic state in $LaFeO_3$ and $YFeO_3$ takes place at 740 K and 655 K, respectively [167, 171]. Some groups also reported the attainment of ferroelectricity in few centrosymmetric orthoferrites.

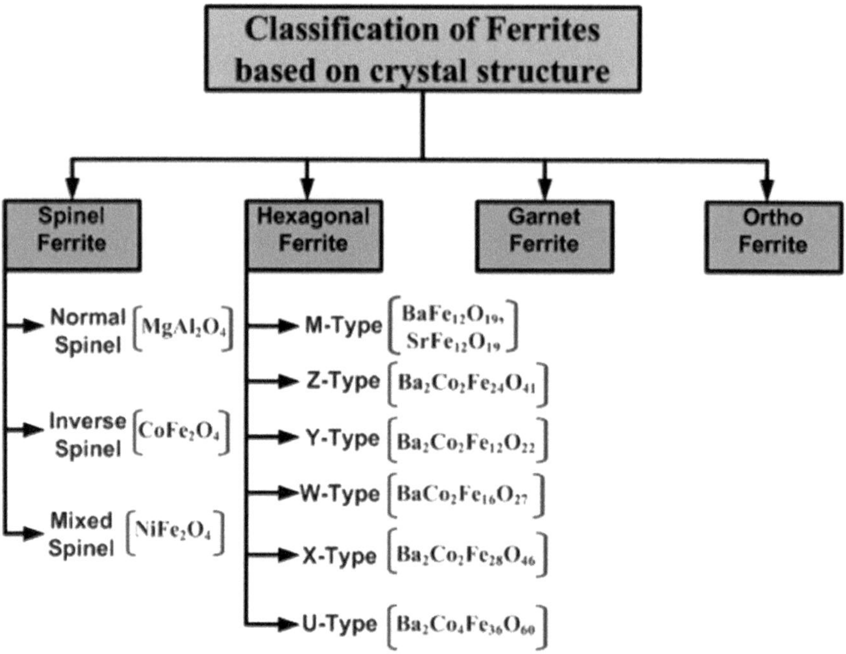

FIGURE 8.1 Classification of ferrites based on crystal structure.

This chapter explored the potential biomedical applications, e.g., disease therapy, diagnosis, and biosensors of ferrite nanoparticles derived by different chemical and physical routes having optimized physical and chemical characteristic properties. These nanoparticles are characterized by different analytical techniques to confirm their formation, desired structure, morphology, particle size and magnetic properties.

8.2 SYNTHESIS OF FERRITES

Ferrite bulk, thin films, and nanoparticles are prepared by different methods such as conventional solid-state reaction, gas phase reaction (chemical vapor deposition (CVD), metal-organic chemical vapor deposition (MOCVD), and a sputtering laser ablation technique) and wet chemical co-precipitation, sol-gel, hydrothermal, and a precursor based method to obtain desired specific size, shape, and morphology of as-prepared particles/thin films for a particular application. The brief details of these methods are described in subsections.

8.2.1 SOLID-STATE REACTION METHOD

Bulk iron oxide or ferrites are synthesized by classical high-temperature solid-state method, which requires mechanical mixing/ball milling of oxides, hydroxide or carbonates of respective metals in stoichiometric ratio followed by calcination at high temperatures ~1,000°C [12, 77, 154, 164, 175, 179, 213]. The calcined powder was ground into fine powder and then pressed into a pellet form for again sintering at different temperatures in the furnace from 1,200–1,350°C in air for about 4–5 hrs. to obtain desired composition of ferrites. This method suffers with many limitations due to high operating temperatures and long heating schedules, coarsening of grains, and volatilization of low melting components, which restricts the formation of desired stoichiometry and nano-form ferrites.

8.2.2 GAS PHASE METHODS

These methods are generally employed for thin films deposition and nanoparticles which include chemical vapor deposition (CVD), metal-organic chemical vapor deposition (MOCVD), and a sputtering laser ablation technique.

8.2.2.1 CHEMICAL VAPOR DEPOSITION (CVD) TECHNIQUE

In the CVD process, a carrier gas with precursors is continuously passed into a reaction chamber maintained under vacuum at high temperature (> 900°C) by gas delivery setup to obtain products in the form of clusters or nanoparticles. Then as obtained nanopowders were successively heated in ultra-pure gas atmosphere to achieve desired composition, structure, and morphology of the nanoparticles [19, 140, 191]. For ferrites nanoparticles deposition, iron trichloride precursor has been used.

8.2.2.2 METAL ORGANIC CHEMICAL VAPOR DEPOSITION (MOCVD)

In the MOCVD process, Metallo-organic precursors have been used to undertake reactions in 300–800°C temperature range and at pressures from 1 Torr to ambient for thin film deposition. Tris (2,2,6,6-tetramethyl-3,5-heptadionato) Fe(III), tris(tbutyl-3-oxo-butanoato) Fe(III), and Fe(II) dihydride complexes $H_2Fe[P(CH_3)_3]_4$ are used as precursors for the synthesis of ferrite thin film [137, 170]. The depositions of thin films of many other ferrites are also reported in literature [37, 65].

8.2.2.3 LASER ABLATION METHOD

This is one of the most used methods for the deposition of thin films. It involves the stoichiometric removal of constituent species from the target during the ablation with lesser controlling parameters like substrate temperature, pressure, ablation yield, and deposition rate, influence the chemical composition as compared to the other physical deposition techniques [28, 144]. The uniform size hematite and maghemite nanoparticles are prepared using Iron pentacarbonyl as a precursor and ethylene as the carrier gas [14] by one step laser pyrolysis [2]. A series of Ni–Zn ferrite and Zn ferrite thin films are deposited on Si substrate by the laser ablation technique [181]. In addition to these, many researchers have also reported the deposition of ferrite thin film using PLD technique [5, 31, 107, 186]. Like solid-state reaction, the gas phase techniques produce high-quality products but also have some drawbacks like low yields, small-scale production, purity of products used, difficult to maintain stoichiometric composition, expensive instruments and their maintenance.

8.2.3 WET (SOLUTION PHASE) METHODS

In this chapter, the brief details on high quality, uniform particle size with narrow size distribution and morphology of ferrite nanopowders synthesized by wet synthetic routes via chemical co-precipitation, sol-gel, hydrothermal, precursor-based methods for various biomedical applications are presented as in subsections.

8.2.3.1 CO-PRECIPITATION METHOD

This method is extensively used for the synthesis of FNPs with several different stoichiometric compositions [7, 51, 119, 122, 126, 142, 149, 185, 224]. Co-precipitation reactions comprise of the following four characteristic steps: (i) The precipitation reactions occur at high supersaturation state. (ii) In the nucleation process, large numbers of fine small particles separate out from the solution. (iii) Under Secondary processes, the growth in particle size and aggregation of particles takes place. (iv) Supersaturation state is a basic necessary conditions for the precipitation of reaction product. The size, morphology, and particle size distribution of the produced nanoparticles during chemical reaction are influenced by the solutions mixing process, rate of reactant addition, stirring rate, temperature, and pH of solutions. The precipitation step of chemical reaction does not always give nano size product particles. The nucleation and growth of product particles steps optimize particle size and morphology. During precipitation reaction, at first numerous very fine small crystallites are formed which act as nucleation center and further grows to bigger size with time or rapidly to form bigger stable particles ensembles. For Fe_3O_4 and other FNPs chemical synthesis, one molar aqueous solutions of ferrous, ferric, and other chloride/nitrate/carbonate/sulfate salts are prepared at 60°C for precipitation hydrous ferrous ferric oxide by drop-wise addition of alkaline salts like ammonia/sodium hydroxide solution at about 8–10 pH of solution. The fine metals hydroxides particles were first obtained in the co-precipitation reaction performed in an alkaline medium at low temperatures. The precipitates were washed many times with distilled water to remove the residual salts and impurities. The precipitated particles (metal hydroxides) solid solution in alkaline medium were then again heated at 80°C to obtain the Fe_3O_4 nanopowder. This is a very slow process and takes a few hours' time for transformation to ferrite nanoparticles. Due to small size of precipitated hydrous metal oxides, they try to agglomerates simultaneously. In order to restrict particles agglomeration process, metal salts concentration ratio, pH of the solution, precipitation temperature, speed of adding the salt mixture into the base solution, and stirring speed parameters are critically controlled for optimizing nucleation rate and particles growth [22]. The particles are stabilized by forming a uniform monolayer coating with suitable surfactant.

8.2.3.2 SOL-GEL METHOD

In this method, the alkoxide, epoxide, or ligand compounds of different metals are used as starting materials. The hydrolysis of acid-or base-catalyzed is followed by the condensation reaction of metal alkoxides precursors which result in dispersion of nano-sized oxide particles in a "sol." The sol then forms a "gel" via solvent (ethanol or water/alcohol mixture) removal or by further condensation or inorganic polymerization in the form of a three-dimensional metal oxide network. The basic catalysis forms a colloidal gel, whereas acid catalysis produces polymeric gel [84, 94, 109]. The gel on calcination at ~300–800°C temperature in an oxygen atmosphere, produces the desired ferrites. The structure and properties of the gel are influenced by the nature and concentration of starting precursor materials, solvent, temperature, pH of solution, etc.

8.2.3.3 OXIDATION METHOD

Oxidation method for the synthesis of ferrites involves oxidation of ferrous ions to ferric ions in the presence of air or KNO_3 [71, 72]. The size-controlled nanoparticles synthesis is described in detail by LaMer and Dinegar [95]. While in modified oxidation method is utilized for the production of large size ferrites particles by using optimal concentration of Fe^{3+} ion with the precursor salts by discriminating nucleation and growth stages [25].

8.2.3.4 HYDROTHERMAL METHOD

In this method, uniform shape, small sized ferrite nanoparticles ~10 nm or below are obtained. The metal salts or metal hydroxides solutions in water are mixed with NaOH or Ammonia solution. The mixed solution is transferred into Teflon-lined stainless steel autoclave cell for hydrothermal treatment at about 150–200°C temperature. The size and morphology of as grown nanoparticles depends upon the pressure, temperature, and reaction time. These parameters are optimized to control the nucleation rate and their growth [15, 20, 92]. As this method execute reaction at low-temperature, so it forms homogeneous mono/polycomponent ferrites of desired chemical compositions. Since the hydrothermal process is performed at low temperatures and preferred because it forms small and uniform size ferrite nanoparticles as compared to earlier high temperature reaction processes. Chen et al. [22] have reported synthesis of Ni, Co, Mn, and Zn ferrite-nanoparticles by the direct hydrothermal decomposition of the trinuclear heterometallic oxocentered acetate cluster of $[MFe_2O(CH_3CO_2.6(H_2O)_3]_nH_2O$ in solution. This method has been used for the synthesis of both hard and soft ferrites [29, 59, 69, 86, 104–106, 108, 163, 174, 188, 196, 201, 204, 212, 221, 223]. Although the hydrothermal method is very versatile, but it has disadvantages of slow reaction kinetics at any given temperature and the need for a high-pressure reactor that restricts the large scale production of nanoparticles.

8.2.3.5 COMBUSTION METHOD

In the combustion method, the decomposition reaction between metal nitrates and organic compounds resulted from exothermic reaction at 300°C, which acts as fuel to form oxides within few minutes [90, 156, 172, 197]. The reaction is carried out in a Pyrex dish by dissolving metal nitrates and fuel in a very small amount of water, and then the mixture is heated to evaporate water to form a viscous liquid foam which ignites to initiate a self-sustained combustion reaction to form the desired product. The ignition of organic precursor is a gas-producing self-propagating exothermic reaction which produces voluminous fine oxides in a very small time. The controlling factor for the combustion method is the exothermic decomposition, which depends upon the nature of the fuel used and the ratio oxidizer/fuel. The method suffers from limitations of controlling exothermic reaction for large-scale production, particle size control and impurity phases.

8.2.3.6 REVERSE MICELLE AND MICROEMULSION METHOD

The method produces monodispersed nanoparticles of different morphologies. In water-in-oil microemulsions, the aqueous phase is dispersed as in the form of microdroplets with a monolayer of organic surfactant molecules in the hydrophobic medium. These microdroplets provide a unique microenvironment for the nanoparticles formation with restricted size growth. The droplet size of reverse micelles is easily modulated in the nanometer range by optimizing water surfactant molar ratios. The microemulsion method can be used by the following two ways to synthesize nanoparticles: (i) precursor salts are dissolved in the aqueous phases of two identical water/oil microemulsions and precipitation takes place upon mixing. In this process, the precipitated particles are confined within the droplets, of controlled size and shape of nanoparticles. (ii) The reducing or precipitating agent addition to a microemulsion for nanoparticles precipitation [135, 189]. This method is also used for the synthesis of ferrites [16, 17, 114, 128, 166, 211, 217, 231]. The large amount of solvent consumption is one of the major drawbacks of this method, and that limits its usage for large amount production of nanoparticles.

8.2.3.7 PRECURSOR-BASED METHOD

In this process, a precursor compound, i.e., complex ion, a combination of metal ions and a chelating agent is prepared which is then calcined at different temperatures to obtain nanoparticles of the desired mixed oxide system. The organometallic complexes, metal hydroxides/oxalates citrates/hydrazine carboxylates, etc., are some of the commonly used precursor compounds [60, 151, 165, 200]. Mallapur [115] have reported the preparation of a spinel ferrite system of using a precursor-based method in which PVA acts as a chelating agent and sucrose as a fuel. The synthesis of $Ni_xZn_{1-x}Fe_2O_4$ and $-Fe_2O_3$ using a PVA precursor-based method and an oxalate precursor-based method [47, 157, 160, 162]. Mohamed et al. [123] have synthesized cobalt ferrite using an organic acid precursor method where oxalic acid was used as a chelating agent. Zinc ferrite nanopowder was synthesized using a tartrate precursor method [116]. There are some other precursor-based methods also available for the synthesis of nanoferrites by using citrate precursor [80, 177, 199] and polymeric precursor route, etc. [44, 67].

8.2.3.8 OTHER SYNTHESIS METHODS

Mechanochemical synthesis is the combination of chemical and mechanical alloying synthesis method in which precursors (ferrites) are first prepared by a chemical route, e.g., sol-gel or co-precipitation method. These as-prepared precursors are milled under different conditions by using a planetary ball mill and finally calcined at 600–900°C. Cobalt ferrite nanopowder using a mechanochemical route where cobalt ferrite precursor was obtained by a co-precipitation method, and then milled using a Fritsch Planetary mill and calcined at 500°C to form cobalt ferrite nanopowder [117]. Maensiri et al. [112] synthesized ~~of~~ $NiFe_2O_4$ nanoparticles from nickel nitrate, ferric nitrates aqueous solutions and egg white (ovalbumin). Sonochemical synthesis technique has been used for the synthesis of $NiFe_2O_4$, $SrFe_{12}O_{19}$ ferrites [138, 169, 178]. Although the wet chemical methods produce a variety of nanostructured ferrite powders, they are also suffered from the following limitations like: (i) costly metal alkoxides or metal complexes, (ii) usage of delicate reagents, (iii) concentrated acids/bases and organic solvents, and (iv) the formation of unwanted heterogeneous crystalline phases. To overcome these drawbacks, now a very simple and cost-effective ethylenediaminetetraacetic acid (EDTA) precursor-based synthesis method has been employed for the preparation of several types of ferrite nanoparticles [47, 54, 55, 145, 157–159, 161]. In this method, an aqueous solution of EDTA is mixed with the appropriate quantity of metal nitrates. The nanosized ferrite powders are obtained by drying the reaction mixture, followed by calcination of the precursor powder at various temperatures. A variety of soft ferrites ($NiFe_2O_4$, Ni–Zn ferrite, Mn–Ni–Zn ferrite, etc.), and the hard ferrite ($CoFe_2O_4$, $BaFe_{12}O_{19}$, $SrFe_{12}O_{19}$, etc.), with particle size ranging from 20–40 nm can be prepared by employing this method. In this method, the chelating agent (EDTA) helps the formation of voluminous, fluffy, porous, carbon-rich precursor. Apart from this, the chelating agent (EDTA) also controls the agglomeration of metal ions. A small atomic clusters of nascent metal oxides with proper chemical homogeneity embedded in the precursor is formed during decomposition of the precursor. These embedded metal oxides finally produce ferrite nanoparticles. In the process of thermal decomposition, the heat of combustion is dissipated by the production of gases like CO_2 and NO_X. This dissipation of heat prevents the sintering of nanoferrites [158]. The advantages of this method are (i) low calcination temperature, (ii) no impurity phases formation, (iii) use of economic metal nitrates and EDTA as reactants and water as solvent, (iv) uses non-corrosive reagents (NaOH), (v) cost-effective experimental set up, and (vi) ease of controlling final products composition. The facile, simple, cost-effective, and its capability of developing several types of ferrite nanoparticles make it a very attractive option for researchers. This method has also been employed for the preparation of hard-soft ferrite nanocomposites ($NiFe_2O_4$–$BaFe_{12}O_{19}$). Figure 8.2 shows the flow chart depicting various chemical synthesis techniques of ferrite.

To restrict the agglomeration of as prepared FNPs, stabilization of FNPs is necessary during synthesis or post-synthesis process. The details of FNPs stabilization are discussed in detail in Section 8.4.

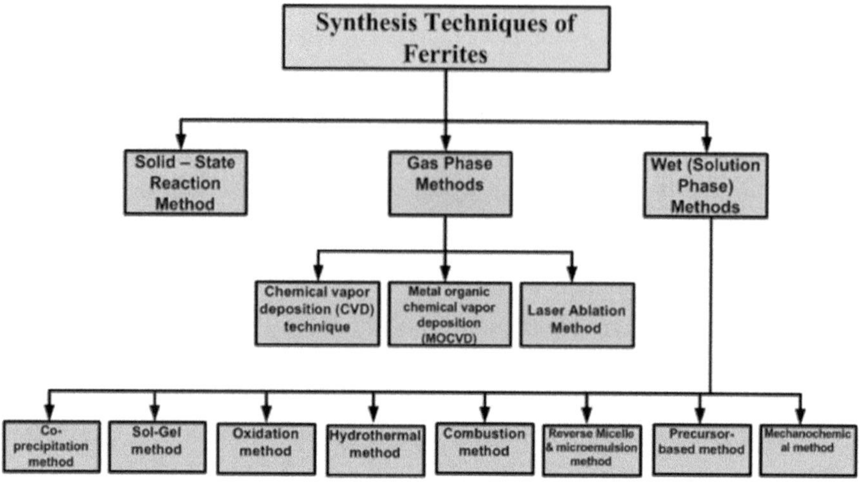

FIGURE 8.2 Flow chart showing various synthesis techniques of ferrite.

8.3 ANALYTICAL TECHNIQUES USED FOR CHARACTERIZATION OF FERRITE NANOPARTICLES

The structural characteristics of ferrite nanoparticles are carried out by employing powder X-ray diffraction (XRD)/XRF, Fourier transform infrared (FTIR), Raman, and scanning electron microscopy (SEM). In order to have the morphological information of the prepared ferrite energy dispersive X-ray electron microscopy (EDX), Transmission electron microscopy (TEM)/high resolution transmission electron microscopy (HRTEM), atomic force microscopy (AFM) techniques are used. The magnetic properties such as saturation magnetization, remanent magnetization, presence of paramagnetic center, g-value, peak to peak line width are studied using vibrational sample magnetometry (VSM) and electron paramagnetic resonance (EPR) techniques. The information regarding crystal structure, crystallite shape and size and calculation of lattice parameters of prepared ferrite nanoparticles are derived from their XRD patterns. IR transmittance and Raman absorption spectra throws light on the vibrational peaks pertaining to different molecular groups of tentative symmetry present in the ferrite nanoparticles to be used for different biomedical applications. SEM/EDX, TEM/HRTEM, AFM images reveal the morphology details of prepared nanoparticles in terms of their shape, size, degree of agglomeration and particle size distribution and elemental compositions. Magnetization vs. applied magnetic field related parameters are obtained from the hysteresis loop recorded by VSM of ferrite nanoparticles. Hysteresis loop shows that the magnetic moment of ferrite nanoparticles aligned in the direction of external applied magnetic field and the point where maximum magnetization occurs is called saturation magnetization (M_S). On removal of magnetic field, the magnetization of the ferrite nanoparticles reverse direction for attaining previous state and this magnetization is called remanent magnetization (M_r). On reversing the direction of the external applied magnetic field, the ferromagnetic domains acquire a compensated state in which effective magnetization appears to be zero. The field required to demagnetize the ferrite nanoparticles is called Coercivity (H_C). The unique superparamagnetic feature of ferrite nanoparticles is zero coercivity. It means each ferrite nanoparticle behaves as a single domain magnetic particle which at blocking temperature has thermal energy which is sufficient enough to overcome the anisotropy barrier and randomize the magnetization. Therefore, on removal of external magnetic fields, superparamagnetic ferrite nanoparticles exhibit negligible M_r. The FNPs behavior is similar to paramagnetic materials on the application of external magnetic field [13, 28].

Electron spin-related parameters like g-value, peak-to-peak linewidth, spin concentration of paramagnetic centers, spin-spin lattice relaxation time constant is evaluated from their EPR spectra. These are some of the basic techniques generally used for characterization of ferrite nanoparticles. In addition to these, sometimes microwave, adsorption, and electrical studies are also done as per application demand.

8.4 STABILIZATION OF FNPS THROUGH FUNCTIONALIZATION FOR CONJUGATION TO BIOLOGICAL ENTITIES

The stability of magnetic FNPs without/with minimum agglomeration is an important issue for using in a particular application due to the attractive van der Waals and magnetic inter-particle forces. The properties of agglomerated particles are different from monodispersed nanoparticles owing to lesser surface area. The aggregation of FNPs during synthesis by wet chemical routes is a natural phenomenon. It is therefore necessary to stabilize magnetic nanoparticles for technological as well as biomedical applications. For this, the surfaces of FNPs are functionalised with ligands which are charged as it provides electrostatic and steric repulsive force for their stabilization in solution by anionic surfactants. These coatings protect the FNPs magnetic cores, minimize toxicity effects and provide function/reactive group present in ligands for conjugation to biomolecules. For this purpose, ligands such as inorganic and organic are available to stabilize and functionalize FNPs. The functionalization of FNPs with organic ligands is carried out either *in-situ* or post-synthesis. *In-situ* functionalization process ligands/surfactant are added during FNPs synthesis, and the obtained FNPs can be directly used for further conjugation [13]. The terminal hydroxyl (–OH), amino ($–NH_2$), and carboxyl (-COOH) functional groups present in the surfactant like dextran, chitosan, poly (acrylic acid), interact with FNPs during their precipitation process. But this method is not applicable for all types of organic surfactants. The post-synthesis method has been used for functionalization of FNPs by ligands. The two primary techniques applied during post-synthesis functionalization of FNPs are ligand addition and ligand exchange. Double-layer structure is formed by ligand addition which involves adsorption of amphiphilic molecules having both a hydrophobic and a hydrophilic component. The metallic atoms on FNPs surface are electron deficient and the surface have natural affinity for electron-rich functional groups. In ligand-exchange procedure, the original ligands/surfactant molecules are directly replaced by the new functional ligands/molecules whose functional group binds by chemical bonding or electrostatic attraction on FNPs surface and another functional group for bio-functionalization. Dispersibility in water can be enhanced by ligand exchange approach in which hydrophobic capping ligands are replaced by hydrophilic ligands. For example, in Fe_3O_4 nanoparticles the hydrophobic ligands were replaced with organosilane terminated with $–NH_2$, –SH, and –COOH groups for further conjugation with fluorescent dyes. Another study reported the use of silane ligands with different terminal functional groups like amine, aldehyde, acrylate, isocyanate, thiol, PEG, cyano, or carboxylate groups for exchanging oleic acid as surfactant present on $CoFe_2O_4$ nanoparticles. The ligands can be classified as "small monomeric" and "polymeric" as per their chain structure. Small monomeric ligands help in enhancing ionic stability and chemical modification on FNPs surfaces having a small hydrodynamic radius. Such FNPs capped ligands are desired for endocytosis application. For MRI T2 imaging contrast agents and to kill HER2+ cell selectively, thiol-based functionalization of FNPs have been used. The nature of the counter ions present in the solution, pH, and ionic strength play crucial role in the stabilization of FNPs by the interactions between electrical double layers. In the core-shell structure, the shell protects the core from the surrounding environment. This is a very promising approach for effectively reducing the agglomeration of nanoparticles.

Normally, two types of coatings are used to form the core-shell structured magnetic nanoparticle: (i) organic coating shells, including surfactant, polymers, and proteins, and (ii) Inorganic coating shells which includes silica and precious noble metals Au and Ag. The organic surfactants or polymers are attached on the surface of magnetic nanoparticles either chemically or physically to form a single or double layer core-shell structures in which as generated repulsive forces tries to balance inter-particle forces acting on ferrite nanoparticles. Dextran, carboxymethylated dextran, carboxydextran, glycosaminoglycan, arabinogalactan, sulfonated styrene-divinylbenzene, polyethylene glycol (PEG), polyvinyl alcohol (PVA), chitosan, polyethylene glycol, polyaniline, are some polymers used and oleylamine, oleic acid, stearic acid, reduced graphene oxide, etc., as surfactants [30, 79, 111, 116, 155, 190, 192, 210, 228]. Polymeric functionalization causes a steric barrier to inhibit agglomeration of FNPs and this provides the surface functionality. This increases of hydrodynamic radius and restricts their use in biological applications. The characteristic features of polymers such as length and width are cautiously selected as per requirement. The polymeric chains interact on FNPs surface via physical adsorption and/or through chemical covalent bonding. For the functionalization of FNPs, both natural and synthetic polymeric ligands have been assessed. Natural polymeric ligands like starch, dextran, and chitosan are widely used because they have characteristic large water solubility and extremely small toxicity. Dextran-coated FNPs are used for intracellular labeling in μNMR biosensors. Iron oxide nanoparticles modified with poly(L-lysine), a biocompatible polymer are used for cancer stem cell labeling.

Inorganic silica, gold, silver or titanate encapsulates FNPs to form a core-shell structure which prevents the leakage of metal ions as this encapsulation restricts degradation or chemical oxidation of the magnetic core. The shell also helps in modifying the surface of particles with various ligands to act as a linker for conjugation to different biomolecules. The studies on magnetic nanoparticles coated with inert silica/gold have already been made for inhibiting their aggregation in liquid, improving chemical stability and reduce toxicity for biomedical applications. Silica core-shell structured ferrite nanoparticles are formed either by *in situ* hydrolysis of sol-gel precursor, e.g., tetraethyl orthosilicate (TEOS) followed by condensation [9, 48, 58, 62, 73] or by using a silicic acid [110] solution and a microemulsion procedure. To control the silica coating, micelles or inverse micelles process have been employed [232]. Silane-functionalized FNPs to conjugate with protease for the enzymatic activity evaluation. While multifunctional mesoporous silica

FNPs dispersion in aqueous solution are in use for delivery of drug, MRI, fluorescence imaging, and cell targeting. A silanol groups present on the surface of ferrite nanoparticles coated with silica are easily functionalized by reacting with different coupling agents and encouraged usage in various biomedical applications like bio-labeling, target drug delivery, etc. Gold-coated FNPs have been used in MRI and fluorescence imaging applications [208]. While silver-coated iron oxide nanoparticles integrated with carbon paste are used in immunosensor.

8.5 FERRITE NANOPARTICLES FOR BIOMEDICAL APPLICATION

Ferrite nanoparticles (FNPs) having characteristic superparamagnetic property are desired for various biomedical and bioanalytical applications [3, 27, 29, 57, 68, 85, 88, 218, 226] because at ambient temperature and on the application of magnetic field, these particles exhibit magnetic property that rapidly randomize on removing magnetic field. Depending on the magnetic behavior on the application and removal of external applied magnetic field, FNPs are categorized as paramagnetic, ferromagnetic, ferromagnetic, anti-ferromagnetic or superparamagnetic [29, 68, 85].

The detection of FNPs is very simple and economic due to their extremely small background interference as compared to optical and electrical techniques [27]. The easy functionalization and integration of FNPs have made them a suitable for usage as imaging agents, heat generators, drug carriers [88, 218] and for designing biosensors. FNPs interact with biomolecules to magnetically labeled target and behaves like ligand-receptor complex. These complex molecules are explored for designing their application in different diagnostic techniques. This process involves several steps, e.g., (i) preparation of desirable size FNPs, (ii) Modification of FNPs by a suitable linker or ligands which are biocompatible, (iii) Optimization of FNPs and ligands size for the desired magnetic properties as per application requirement, and (iv) Selection of a suitable technique for magnetic detection of FNPs. Earlier for biomedical imaging and drug delivery, FNPs less than 100 nm are used for endocytosis of affected organs and cells [226] and in biosensors, bigger sized FNPs are used for strong signals arising from higher magnetization [218]. The magnetic core is protected by surface functionalization with appropriate surfactant, so that it can easily attach to target molecules. Now a day, the different biocompatible ferrite superparamagnetic particles are also used for preparing their stable colloidal dispersion also known as ferrofluid in a suitable carrier medium. The small size of ferrite nanoparticles used for ferrofluids preparation inhibits the precipitation of nanoparticles due to gravitational effect and agglomeration of nanoparticles into bigger size ensembles. For biomedical applications, these ferrofluids should have pH 7, i.e., they should be stable in water.

8.5.1 USE OF FERRITE NANOPARTICLES FOR HYPERTHERMIA TREATMENT

FNPs are used in hyperthermia treatment [38, 50, 52, 66, 85, 97, 118, 125, 127, 146, 173, 184, 219]. The single domain FNPs have size in ~10–50 nm range which directly affects their intrinsic properties, e.g., saturation magnetization, magnetic anisotropy constant, and the exchange stiffness, etc. [88]. FNPs become superparamagnetic when their size is less than one single domain. Two relaxation mechanism occurs in FNPs: (i) Néel relaxation and (ii) Brownian relaxation. Néel relaxation dominates when the size of FNPs is $<$ 15 nm for iron oxide and it arises due to spin-spin relaxation because of rotation of magnetic moment to overcome the anisotropy energy barrier. However, the Brownian relaxation is due to mechanical rotation of FNPs. Both these relaxation phenomena are utilized for hyperthermia therapy. In this therapy, heat is generated up to 45°C due to repeated Néel and Brownian relaxation when magnetic particles (MPs) are exposed to externally applied alternating magnetic field. This heat is transferred to the surrounding which treat/destroy the unwanted target tumor tissue/cancer cell. The heat generation is dependent on magnetic properties like Ms, Hc, and magnetic anisotropy (K) [38, 88, 218]. Mazario et al. [118] have synthesized biocompatible cobalt ferrite nanoparticles for hyperthermia application. Multifunction FNPs are now used for cancer treatment [97, 127, 173, 184].

8.5.2 FERRITE NANOPARTICLES FOR DRUG DELIVERY

In this application, magnetometer imaging is used to trace FNPs and is guided by an external magnetic field. To initiate drug release or for tissue ablation heat is used. As distance increases, the external magnetic field decays and high-saturation magnetization (M_S) FNPs are desirable because of their easy magnetization in weak field [168]. NMR technique is used to detect the relaxation of nuclear spins induced by FNPs. In the presence of an external magnetic field, the FNPs generate a small local magnetic field which induces faster spin-spin relaxation time (T_2) in the surrounding protons. The important parameter for imaging is the transverse relaxation rate (R_2) which is inversely proportional to T2, i.e., $R_2 = 1/T_2$. It has been demonstrated that FNPs can induce high T_2 relaxation time constant [113, 218]. Higher relaxation time constant is obtained in the bigger size FNPs with higher M_S values [74]. In the promising functionalized and conjugated FNPs, drug molecules are attached and guided to a chosen infected/diseased site in the human body by applying an external magnetic field.

It continues to remain there till the drug is completely discharged at the site and then the magnetic field is removed. Nigam et al. [129] have used citrate-stabilized 8–10 nm sized Fe_3O_4 nanoparticles aqueous ferrofluid for doxorubicin hydrochloride drug delivery. Mitoxantrone loaded on magnetic nanoparticles was studied for its anti-tumor activity [102]. Recently, researchers have used polymeric nanomaterials embedded with superparamagnetic iron oxide nanoparticles hybrids for targeted drug delivery, cell imaging, etc. [6, 43, 78, 87, 130, 132, 148, 215].

8.5.3 FERRITE NANOPARTICLES FOR MAGNETIC RESONANCE IMAGING (MRI)

MRI is the most powerful non-invasive imaging technique based on spin relaxation of water protons and is widely employed in clinical imaging. MRI requires no ionizing irradiation and have spatial resolution at the cellular level ~ 10 μm. By the use of exogenous contrast agents, the process of relaxation of water protons by causing local magnetic fields is accelerated which increase the sensitivity of MRI and signal-to-noise ratio for precise and accurate molecular imaging [205, 206].

Clinically, more than 30% of all MRI scans are done using MRI contrast agents which are either paramagnetic or superparamagnetic. The contrast agents are categorized into two different groups depending upon their working mechanism. T_2 contrast agents are paramagnetic FNPs and are called negative contrast agents [193]. They generate negative (dark) image by shortening T_2 of water protein [229]. Earlier, ~~the~~ studies made on the pure and doped FNPs of composition MFe_2O_4 (M = Co, Ni, Mn) showed a noticeable effect on their superparamagnetic properties and T_2 image contrast capacity with respect to size, magnetic dopants, particle morphology and chelating agents. Zhao and coworkers prepared octopod iron oxide NPs and showed that the material produces 5 times higher ultrahigh transverse relaxivity of 679.3±30

$mM^{-1}s^{-1}$ than spherical iron oxide NPs of similar geometric volume [227]. Yang et al. [216] prepared six different morphological shapes like sphere, cube, plates, tetrahedral, rhombohedra, and octopod of manganese doped iron oxide NPs and concluded that the T_2 relaxation rates of nearby protons depends upon the effective radii of the nanomaterials. The transverse relaxation time which is due to the inhomogeneity of induced local magnetic field of FNPs are affected by chelating agent [220]. T_1 contrast agents consists of paramagnetic FNPs works by reducing the longitudinal relaxation time (T_1) of water protons and is capable of producing positive (bright) image and are called positive contrast agents. Gadolinium (Gd) complexes are used *in vivo* circulation time. This disadvantage restricts their clinical applications. To overcome this T_1 MRI agents based on Mn and Fe are in use [23, 203, 230]. The relaxivity ratio (i.e., r_2/r_1) decides whether the contrast agent is T_1 or T_2. An ideal T_1 contrast agent should have high r_1/r_2 ratio. Iron oxide NPs with diameter > 5 nm have high r_2 value (i.e., large r_2/r_1 ratio). Therefore, they are good for T_2 imaging. T_1-MRI contrast agent based on Mn^{2+} ions developed by Chen and coworkers is used for acidic tumor microenvironment [23]. Ultra small iron oxide NPs having size 3.5 nm developed recently by Wang et al. [207]. These NPS agglomerates in acidic tumor environment to enhance the T_2 signal. Another factor apart from pH which initiate (dis)aggregation of magnetic NPs is enzyme activity. Glutathione (GSH) – responsive MRI agent developed by Gao et al. [42] works by crosslinking of adjacent Fe_2O_4 NPs [42] enhances image contrast. The thiol groups and melamine moieties on the NPs aggregates in the presence of GSH within the tumor microenvironment. These aggregated particles showed enhanced T_2 contrast. So magnetic nanoparticles and fluorescent magnetic nanocomposites are used in magnetic resonance imaging (MRI) contrast agents of molecular and cell imaging [1, 4, 8, 35, 41, 45, 61, 82, 89, 98, 101, 103, 118, 139, 176, 180, 183, 194, 198, 202, 222, 225]. Fan et al. [35] have studied the biological activity and MRI of adipose-derived stem cells and compared them with bone marrow mesenchymal stem cells.

8.5.4 FERRITE NANOPARTICLES FOR MAGNETIC SEPARATION OF BIOMOLECULES

Magnetic separation is a quick and simple method used to capture specific proteins or biomolecules. This purification process is done in one vessel without using expensive liquid chromatography or other techniques [49,

152]. Recently Lai et al. [93] have successfully demonstrated the binding of Concanavalin A (Con A) onto Fe_3O_4 magnetic nanoparticles via carbodiimide activation. This material acts as a magnetic nano-adsorbent for glycoprotein separation.

8.5.5 FERRITE NANOPARTICLES FOR BIOSENSORS

FNPs are widely used in various types of biosensors application due to their unique advantages like economic synthesis and environment safe. Due to the absence of magnetic background of biological entities, accurate measurements are done in turbid visually obscured samples with no processing. While in optical techniques, the scattering, absorption, and autofluorescence within the sample interferes. The large numbers of procedures have been devised using magnetic labels to sense biomolecules [120]. The biosensor basically consists of main three components (i) bioreceptor, (ii) transducer, and (iii) the detector. A bioreceptor act as a template for the material to be detected. The function of transducer is to convert interaction of bioanalyte and bioreceptor into an electrical signal. The detector receives the electrical signal and amplifies it so that the response can be read and studied properly. Another essential requirement of the nanobiosensors is the immobilization process of the bioreceptor which facilitate fast, more feasible and efficient reaction with bioanalyte. Biosensor's performance is sensitive to temperature change, pH, interfering contaminants and other physicochemical variations [81]. Several FNPs with nanotubes, nanowires, nanorods, nanoparticles morphologies in the powder and thin films form studied showed enhanced biological signaling and transduction mechanism [70]. FNPs-based biosensors for clinical use possess high figures of merit, sensitivity, high signal-to-noise ratio with low detection limit and analytical time as compared to non-FNPs-based strategies [75, 76].

The important parameters in magnetic biosensing are magnetic susceptibility and saturation magnetization (M_S). A sensor based on FNPs with high M_S is more sensitive as high M_S induces stronger translational attractive force on exposure to larger magnetic gradient [219]. FNPs play a dual role both as force transducer and signal provider by using different multimodal techniques in the development of biosensors. Till today, a large number of biosensors for different applications are available by the integration of magnetic sensors with different techniques. But still, it is a challenging problem for researchers working in the field of magnetic nanotechnology due to their increasing demand in the bioanalytical and biotherapy fields with high detection sensitivity, molecular specificity and fast recovery time. Magnetic biosensors owing to their remarkable properties like high sensitivity, compact instrumentation, and flexible integration emerged as efficient detection device for diagnostics. A number of sensitive magnetic detection devices such as magnetoresistive sensors, spin valves, superconducting quantum interference devices (SQUIDs), Hall sensors, giant magneto-impedance-based sensors, and micro-NMR sensors have been developed in recent years [10, 11, 34, 36, 53, 91, 100, 121]. Nowadays in clinical laboratories with the help of biosensors glucose in diabetic patients, bacterial infection in the urinary tract, HIV-AIDS, and cancer are easily diagnosed.

8.6 SUMMARY

The functionalized FNPs in pure, doped hybrids, nanocomposite, and multifunctional analogs are used in with imaging, bio-sensing, therapeutic, and theranostics applications. This encouraged the development of synthesis routes for their usage in many precise, accurate, advanced imaging techniques like MRI/PET and PET/CT for better understanding of cancers and diseases. The nanohybrids have shown efficient with better contrast and resolution imaging, targeting, and chemotherapeutic efficacy on a single platform. FNPs owing to low toxicity, high colloidal stability, magnetic behavior, and easy functionalization and proved to be a better option in clinical practice. The FDA has approved the usage of ferumoxytol, GastroMARK, and Feridex in MNP-based products for clinical trials. Such treated FNPs are used for the treatment of anemic patients with chronic renal diseases and as imaging agents in clinical trials. The future needs the development of new advanced hybrid FNPs with low toxicity, good biocompatibility, and fast kinetics. For imaging purposes, new nano contrast FNPs-based agents will provide high-resolution, more accurate pictures of infected organs/tissues/cells. So by considering these facts, the future of FNPs is very attractive and challenging. The more detailed information on FNPs for radiologists and physicians will drastically improve their knowledge and increase clinical output as their demand is huge from the consumer market for a long span of time.

KEYWORD

- **biomolecules**
- **biosensor**
- **drug delivery**
- **ferrite**
- **ferrite nanoparticles**
- **hyperthermia**
- **wet methods**

REFERENCES

1. Ahmed, N., Jaafar-Maalej, C., Eissa, M. M., Fessi, H., & Elaissari, A., (2013). New oil-in-water magnetic emulsion as contrast agent for *in vivo* magnetic resonance imaging (MRI). *Journal of Biomedical Nanotechnology, 9*(9), 1579–1585. https://doi.org/10.1166/jbn.2013.1644.
2. Alexandrescu, R., Morjan, I., Voicu, I., Dumitrache, F., Albu, L., Soare, I., & Prodan, G., (2005). Combining resonant/non-resonant processes: Nanometer-scale iron-based material preparation via CO_2 laser pyrolysis. *Applied Surface Science, 248*(1–4), 138–146. https://doi.org/10.1016/j.apsusc.2005.03.016.
3. Ansari, M. A., Baykal, A., Asiri, S., & Rehman, S., (2018). Synthesis and characterization of antibacterial activity of spinel chromium-substituted copper ferrite nanoparticles for biomedical application. *Journal of Inorganic and Organometallic Polymers and Materials, 28*(6), 2316–2327. https://doi.org/10.1007/s10904-018-0889-5.
4. Anzai, Y., Mclachlan, S., Morris, M., Saxton, R., & Lufkin, R. B., (1994). Dextran-coated superparamagnetic iron oxide, an MR contrast agent for assessing lymph nodes in the head and neck. *AJNR. American Journal of Neuroradiology, 15*(1), 87–94.
5. Axelsson, A. K., Valant, M., Fenner, L., Wills, A. S., & Alford, N. M., (2009). Chemistry of post-annealing of epitaxial $CoFe_2O_4$ thin films. *Thin Solid Films, 517*(13), 3742–3747. https://doi.org/10.1016/j.tsf.2009.01.142.
6. Babic, M., Horák, D., Jendelová, P., Glogarová, K., Herynek, V., Trchova, M., Likavanová, K., et al., (2009). Poly(N,N-Dimethylacrylamide)-coated maghemite nanoparticles for stem cell labeling. *Bioconjugate Chemistry, 20*(2), 283–294. https://doi.org/10.1021/bc800373x.
7. Banerjee, M., Verma, N., & Prasad, R., (2007). Structural and catalytic properties of $Zn_{1-X}Cu_xFe_2O_4$ nanoparticles. *Journal of Materials Science, 42*(5), 1833–1837. https://doi.org/10.1007/s10853-006-0821-1.
8. Bárcena, C., Sra, A. K., Chaubey, G. S., Khemtong, C., Liu, J. P., & Gao, J., (2008). Zinc ferrite nanoparticles as MRI contrast agents. *Chemical Communications, 19*(19), 2224–2226. https://doi.org/10.1039/b801041b.

9. Barnakov, Y. A., Yu, M. H., & Rosenzweig, Z., (2005). Manipulation of the magnetic properties of magnetite–silica nanocomposite materials by controlled stober synthesis. *Langmuir, 21*(16), 7524–7527. https://doi.org/10.1021/la0508893.
10. Baselt, D. R., Lee, G. U., Natesan, M., Metzger, S. W., Sheehan, P. E., & Colton, R. J., (1998). A biosensor based on magnetoresistance technology. *Biosensors and Bioelectronics, 13*(7–8), 731–739. https://doi.org/10.1016/s0956-5663(98)00037-2.
11. Besse, P. A., Boero, G., Demierre, M., Pott, V., & Popovic, R., (2002). Detection of a single magnetic microbead using a miniaturized silicon hall sensor. *Applied Physics Letters, 80*(22), 4199–4201. https://doi.org/10.1063/1.1483909.
12. Bhame, S. D., & Joy, P. A., (2008). Effect of sintering conditions and microstructure on the magnetostrictive properties of cobalt ferrite. *Journal of the American Ceramic Society, 91*(6), 1976–1980. https://doi.org/10.1111/j.1551-2916.2008.02367.x.
13. Bohara, R. A., Thorat, N. D., & Pawar, S. H., (2016). Role of functionalization: Strategies to explore potential nano-bio applications of magnetic nanoparticles. *RSC Advances, 6*(50), 43989–44012. https://doi.org/10.1039/C6RA02129H.
14. Bomatí-Miguel, O., Mazeina, L., Navrotsky, A., & Veintemillas-Verdaguer, S., (2008). Calorimetric study of maghemite nanoparticles synthesized by laser-induced pyrolysis. *Chemistry of Materials, 20*(2), 591–598. https://doi.org/10.1021/cm071178o.
15. Burda, C., Chen, X., Narayanan, R., & El-Sayed, M. A., (2005). Chemistry and properties of nanocrystals of different shapes. *Chemical Reviews, 105*(4), 1025–1102. https://doi.org/10.1021/cr030063a.
16. Cao, X., & Gu, L., (2005). Spindly cobalt ferrite nanocrystals: Preparation, characterization and magnetic properties. *Nanotechnology, 16*(2), 180–185. https://doi.org/10.1088/0957-4484/16/2/002.
17. Capek, I., (2004). Preparation of metal nanoparticles in water-in-oil (W/O) microemulsions. *Advances in Colloid and Interface Science, 110*(1, 2), 49–74. https://doi.org/10.1016/j.cis.2004.02.003.
18. Chakraborty, T., Yadav, R., Elizabeth, S., & Bhat, H. L., (2016). Evolution of Jahn-Teller distortion, transport and dielectric properties with doping in perovskite $NdFe_{1-X}Mn_xO_3$ ($0 \leq X \leq 1$) compounds. *Physical Chemistry Chemical Physics, 18*(7), 5316–5323. https://doi.org/10.1039/c5cp07083j.
19. Chang, W., Skandan, G., Danforth, S. C., Kear, B. H., & Hahn, H., (1994). Chemical vapor processing and applications for nanostructured ceramic powders and whiskers. *Nanostructured Materials, 4*(5), 507–520. https://doi.org/10.1016/0965-9773(94)90058-2.
20. Chen, D., & Xu, R., (1998). Hydrothermal synthesis and characterization of nanocrystalline Fe_3O_4 powders. *Materials Research Bulletin, 33*(7), 1015–1021. https://doi.org/10.1016/S0025–5408(98)00073–7.
21. Chen, J. P., Sorensen, C. M., Klabunde, K. J., & Hadjipanayis, G. C., (1995). Enhanced magnetization of nanoscale colloidal cobalt particles. *Physical Review: B, Condensed Matter, 51*(17), 11527–11532. https://doi.org/10.1103/physrevb.51.11527.
22. Chen, L., Shen, Y., & Bai, J., (2009). Large-scale synthesis of uniform spinel ferrite nanoparticles from hydrothermal decomposition of trinuclear heterometallic oxo-centered acetate clusters. *Materials Letters, 63*(12), 1099–1101. https://doi.org/10.1016/j.matlet.2009.02.034.
23. Chen, Y., Yin, Q., Ji, X., Zhang, S., Chen, H., Zheng, Y., Sun, Y., et al., (2012). Manganese oxide-based multifunctionalized mesoporous silica nanoparticles for pH-responsive

MRI, ultrasonography and circumvention of MDR in cancer cells. *Biomaterials, 33*(29), 7126–7137. https://doi.org/10.1016/j.biomaterials.2012.06.059.

24. Chinnasamy, C. N., Narayanasamy, A., Ponpandian, N., Chattopadhyay, K., Shinoda, K., Jeyadevan, B., Tohji, K., et al., (2001). Mixed spinel structure in nanocrystalline $NiFe_2O_4$. *Physical Review. Part B, 63*(18). https://doi.org/10.1103/PhysRevB.63.184108, PubMed: 184108.
25. Chinnasamy, C. N., Senoue, M., Jeyadevan, B., Perales-Perez, O., Shinoda, K., & Tohji, K., (2003). Synthesis of size-controlled cobalt ferrite particles with high coercivity and squareness ratio. *Journal of Colloid and Interface Science, 263*(1), 80–83. https://doi.org/10.1016/s0021-9797(03)00258-3.
26. Chrisey, D. B., & Hubler, G. K., (1994). *Pulsed laser deposition of thin films* (1st ed) p. 707. Wiley.
27. Colombo, M., Carregal-Romero, S., Casula, M. F., Gutiérrez, L., Morales, M. P., Böhm, I. B., Heverhagen, J. T., et al., (2012). Biological applications of magnetic nanoparticles. *Chemical Society Reviews, 41*(11), 4306–4334. https://doi.org/10.1039/c2cs15337h.
28. Cullity, B. D., & Graham, C. D., (2009). *Introduction to Magnetic Materials* (2nd edn., p. 550). IEEE Publications.
29. Deng, H., Li, X., Peng, Q., Wang, X., Chen, J., & Li, Y., (2005). Monodisperse magnetic single-crystal ferrite microspheres. *Angewandte Chemie, 44*(18), 2782–2785. https://doi.org/10.1002/anie.200462551.
30. Deng, J., Ding, X., Zhang, W., Peng, Y., Wang, J., Long, X., Li, P., & Chan, A. S. C., (2002). Magnetic and conducting Fe_3O_4–cross-linked polyaniline nanoparticles with core–shell structure. *Polymer, 43*(8), 2179–2184. https://doi.org/10.1016/S0032-3861(02)00046-0.
31. Dixit, G., Singh, J. P., Srivastava, R. C., Agrawal, H. M., Choudhary, R. J., & Gupta, A., (2010). Structural and magnetic behavior of $NiFe_2O_4$ thin film grown by pulsed laser deposition. *Indian Journal of Pure and Applied Physics, 48*, 287–291.
32. Dunlop, D. J., (1972). Magnetite: Behavior near the single-domain threshold. *Science, 176*(4030), 41–43. https://doi.org/10.1126/science.176.4030.41.
33. Eibschutz, M., Shtrikman, S., & Treves, D., (1967). Mössbauer studies of Fe^{57} in orthoferrites. *Physical Review Journals Archive, 156*, 562–577.
34. Enpuku, K., Minotani, T., Gima, T., Kuroki, Y., Itoh, Y., Yamashita, M., Katakura, Y., & Kuhara, S., (1999). Detection of magnetic nanoparticles with superconducting quantum interference device (squid) magnetometer and application to immunoassays. *Japanese Journal of Applied Physics, 38*(2, No. 10A), L1102–L1105. https://doi.org/10.1143/JJAP.38.L1102.
35. Fan, J., Tan, Y., Jie, L., Wu, X., Yu, R., & Zhang, M., (2013). Biological activity and magnetic resonance imaging of superparamagnetic iron oxide nanoparticles-labeled adipose-derived stem cells. *Stem Cell Research and Therapy, 4*(2), 44–54. https://doi.org/10.1186/scrt191.
36. Ferreira, H. A., Graham, D. L., Freitas, P. P., & Cabral, J. M. S., (2003). Biodetection using magnetically labeled biomolecules and arrays of spin valve sensors (invited). *Journal of Applied Physics, 93*(10), 7281–7286. https://doi.org/10.1063/1.1544449.
37. Fitzgerald, A. G., & Engin, R., (1974). Defects in nickel ferrite thin films grown by chemical vapor deposition. *Thin Solid Films, 20*(2), 317–327. https://doi.org/10.1016/0040-6090(74)90067-4.

38. Fortin, J. P., Wilhelm, C., Servais, J., Ménager, C., Bacri, J. C., & Gazeau, F., (2007). Size-sorted anionic iron oxide nanomagnets as colloidal mediators for magnetic hyperthermia. *Journal of the American Chemical Society, 129*(9), 2628–2635. https://doi.org/10.1021/ja067457e.
39. Frey, N. A., Peng, S., Cheng, K., & Sun, S., (2009). Magnetic nanoparticles: Synthesis, functionalization, and applications in bioimaging and magnetic energy storage. *Chemical Society Reviews, 38*(9), 2532–2542. https://doi.org/10.1039/b815548h.
40. Gallo, J., Kamaly, N., Lavdas, I., Stevens, E., Nguyen, Q. D., Wylezinska-Arridge, M., Aboagye, E. O., & Long, N. J. C., (2014). $_{x}Cr^{4-}$ targeted and MMP-responsive iron oxide nanoparticles for enhanced magnetic resonance imaging. *Journal of German Chemical Society, 53*, 9550–9554.
41. Gao, H., Goriacheva, O. A., Tarakina, N. V., & Sukhorukov, G. B., (2016). Intracellularly biodegradable polyelectrolyte/silica composite microcapsules as carriers for small molecules. *ACS Applied Materials and Interfaces, 8*(15), 9651–9661. https://doi.org/10.1021/acsami.6b01921.
42. Gao, Z., Hou, Y., Zeng, J., Chen, L., Liu, C., Yang, W., & Gao, M., (2017). Tumor microenvironment-triggered aggregation of antiphagocytosis ^{99m}Tc-labeled Fe_3O_4 nanoprobes for enhanced tumor imaging *in vivo*. *Journal of Advanced Materials, 29*(24). https://doi.org/10.1002/adma.201701095, PubMed: 1701095.
43. Ge, J., Zhang, Y., Dong, Z., Jia, J., Zhu, J., Miao, X., & Yan, B., (2017). Initiation of targeted nanodrug delivery *in vivo* by a multifunctional magnetic implant. *ACS Applied Materials and Interfaces, 9*(24), 20771–20778. https://doi.org/10.1021/acsami.7b05009.
44. Gharagozlou, M., (2009). Synthesis, characterization and influence of calcination temperature on magnetic properties of nanocrystalline spinel co-ferrite prepared by polymeric precursor method. *Journal of Alloys and Compounds, 486*(1, 2), 660–665. https://doi.org/10.1016/j.jallcom.2009.07.025.
45. Ghasemian, Z., Shahbazi-Gahrouei, D., & Manouchehri, S., (2015). Cobalt zinc ferrite nanoparticles as a potential magnetic resonance imaging agent: An *in vitro* study. *Avicenna Journal of Medical Biotechnology, 7*(2), 64–68.
46. Ghosh, N. N., (2011). Chapter 7; A commentary on nanostructured ferrite materials: Synthesis, properties and applications. In: Barker, J. M., (ed.), *Powder Engineering, Technology and Applications* (pp. 197–206). Nova Science Publishers.
47. Ghosh, N. N., & Sarangi, P. P., (2010). Chapter 10; Development of novel but simple aqueous solution based chemical methodologies for synthesis of pure nanostructured A-Fe_2O_3 powders and the effect of nanostructure on the electrical and magnetic properties. In: Cabral, V., & Silva, R., (eds.), *Nanomaterials: Properties, Preparation and Processes* (pp. 367–379). Nova Science Publishers.
48. Girginova, P. I., Daniel-Da-Silva, A. L., Lopes, C. B., Figueira, P., Otero, M., Amaral, V. S., Pereira, E., & Trindade, T., (2010). Silica coated magnetite particles for magnetic removal of Hg^{2+} from water. *Journal of Colloid and Interface Science, 345*(2), 234–240. https://doi.org/10.1016/j.jcis.2010.01.087.
49. Giri, S., Trewyn, B. G., Stellmaker, M. P., & Lin, V. S. Y., (2005). Stimuli-responsive controlled-release delivery system based on mesoporous silica nanorods capped with magnetic nanoparticles. *Angewandte Chemie, 44*(32), 5038–5044. https://doi.org/10.1002/anie.200501819.
50. Gonzales-Weimuller, M., Zeisberger, M., & Krishnan, K. M., (2009). Size-dependant heating rates of iron oxide nanoparticles for magnetic fluid hyperthermia. *Journal of*

Magnetism and Magnetic Materials, 321(13), 1947–1950. https://doi.org/10.1016/j.jmmm.2008.12.017.
51. Gul, I. H., Ahmed, W., & Maqsood, A., (2008). Electrical and magnetic characterization of nanocrystalline Ni-Zn ferrite synthesis by co-precipitation route. *Journal of Magnetism and Magnetic Materials, 320*(3, 4), 270–275. https://doi.org/10.1016/j.jmmm.2007.05.032.
52. Gupta, A. K., & Gupta, M., (2005). Synthesis and surface engineering of iron oxide nanoparticles for biomedical applications. *Biomaterials, 26*(18), 3995–4021. https://doi.org/10.1016/j.biomaterials.2004.10.012.
53. Haun, J. B., Yoon, T. J., Lee, H., & Weissleder, R., (2010). Magnetic nanoparticle biosensors. *Wiley Interdisciplinary Reviews. Nanomedicine and Nanobiotechnology, 2*(3), 291–304. https://doi.org/10.1002/wnan.84.
54. Hazra, S., & Ghosh, N. N., (2014). Preparation of nanoferrites and their applications. *Journal of Nanoscience and Nanotechnology, 14*(2), 1983–2000. https://doi.org/10.1166/jnn.2014.8745.
55. Hazra, S., Patra, M. K., Vadera, S. R., & Ghosh, N. N., (2012). Preparation of $Mn_{0.2}Zn_{(0.8-X)}$ Fe_2O_4 nanopowders by using a novel EDTA precursor method and their microstructure, DC resistivity and magnetic properties. *Optoelectronic and advanced materials—Rapid communication, 6*, 451–455.
56. Hessien, M. M., Radwan, M., & Rashad, M. M., (2007). Enhancement of magnetic properties for the barium hexaferrite prepared through ceramic route. *Journal of Analytical and Applied Pyrolysis, 78*(2), 282–287. https://doi.org/10.1016/j.jaap.2006.08.005.
57. Hou, L., Udangawa, W. M. R. N., Pochiraju, A., Dong, W., Zheng, Y., Linhardt, R. J., & Simmons, T. J., (2016). Synthesis of heparin-immobilized, magnetically addressable cellulose nanofibers for biomedical applications. *ACS Biomaterials Science and Engineering, 2*(11), 1905–1913. https://doi.org/10.1021/acsbiomaterials.6b00273.
58. Hurley, K. R., Lin, Y. S., Zhang, J., Egger, S. M., & Haynes, C. L., (2013). Effects of mesoporous silica coating and post-synthetic treatment on the transverse relaxivity of iron oxide nanoparticles. *Chemistry of Materials, 25*(9), 1968–1978. https://doi.org/10.1021/cm400711h.
59. Hwang, C. S., & Wang, N. C., (2004). Preparation and characteristics of ferrite catalysts for reduction of CO_2. *Materials Chemistry and Physics, 88*(2, 3), 258–263. https://doi.org/10.1016/j.matchemphys.2004.02.028.
60. Iijima, M., Sato, K., Kurashima, K., Ishigaki, T., & Kamiya, H., (2008). Low-temperature synthesis of redispersible iron oxide nanoparticles under atmospheric pressure and ultra dense reagent concentration. *Powder Technology, 181*(1), 45–50. https://doi.org/10.1016/j.powtec.2007.06.019.
61. Im, G. H., Kim, S. M., Lee, D. G., Lee, W. J., Lee, J. H., & Lee, I. S., (2013). $Fe3O_4$/mno hybrid nanocrystals as a dual contrast agent for both T1- and T2-weighted liver MRI. *Biomaterials, 34*(8), 2069–2076. https://doi.org/10.1016/j.biomaterials.2012.11.054.
62. Lee, Y. T., Im, S. H., Wiley, B., & Xia, Y., (2005). Quick formation of single-crystal nanocubes of silver through dual functions of hydrogen gas in polyol synthesis. *Chemical Physics Letters, 411*(4–6), 479–483. https://doi.org/10.1016/j.cplett.2005.06.080.
63. Indira, T. K., & Lakshmi, P. K., (2010). Magnetic nanoparticles-A review. *International Journal of Pharmaceutical Science and Nanotechnology, 3*, 1085–1091.

64. Ita, B., Murugavel, P., Ponnambalam, V., & Raju, A. R., (2003). Magnetic properties of lanthanum orthoferrite fine powders prepared by different chemical routes. *Journal of Chemical Sciences, 115*(5–6), 519–524. https://doi.org/10.1007/BF02708243.
65. Itoh, H., Uemura, T., Yamaguchi, H., & Naka, S., (1989). Chemical vapor deposition of epitaxial Ni-Zn ferrite films by thermal decomposition of acetylacetonato complexes. *Journal of Materials Science, 24*(10), 3549–3552. https://doi.org/10.1007/BF02385738.
66. Jang, D. H., Lee, Y. I., Kim, K. S., Park, E. S., Kang, S. C., Yoon, T. J., & Choa, Y. H., (2013). Induced heat property of polyethyleneglycol-coated iron oxide nanoparticles with dispersion stability for hyperthermia. *Journal of Nanoscience and Nanotechnology, 13*(9), 6098–6102. https://doi.org/10.1166/jnn.2013.7635.
67. Jang, J. S., Borse, P. H., Lee, J. S., Jung, O. S., Cho, C. R., Jeong, E. D., Ha, M. G., et al., (2009). Synthesis of nanocrystalline $ZnFe_2O_4$ by polymerized complex method for its visible light photocatalytic application: An efficient photo-oxidant. *Bulletin of the Korean Chemical Society, 30*(8), 1738–1742. https://doi.org/10.5012/bkcs.2009.30.8.1738.
68. Jeong, U., Teng, X., Wang, Y., Yang, H., & Xia, Y., (2007). Superparamagnetic colloids: Controlled synthesis and niche applications. *Advanced Materials, 19*(1), 33–60. https://doi.org/10.1002/adma.200600674.
69. Jiang, W., Cao, Z., Gu, R., Ye, X., Jiang, C., & Gong, X., (2009). A simple route to synthesize $ZnFe_2O_4$ hollow spheres and their magneto rheological characteristics. *Smart Materials and Structures, 18*(12). https://doi.org/10.1088/0964-1726/18/12/125013, PubMed: 125013.
70. Jianrong, C., Yuqing, M., Nongyue, H., Xiaohua, W., & Sijiao, L., (2004). Nanotechnology and biosensors. *Biotechnology Advances, 22*(7), 505–518. https://doi.org/10.1016/j.biotechadv.2004.03.004.
71. Joseyphus, R. J., Kodama, D., Matsumoto, T., Sato, Y., Jeyadevan, B., & Tohji, K., (2007). Role of polyol in the synthesis of Fe particles. *Journal of Magnetism and Magnetic Materials, 310*(2), 2393–2395. https://doi.org/10.1016/j.jmmm.2006.10.1132.
72. Justin, J. R. J., Narayanasamy, A., Shinoda, K., Jeyadevan, B., & Tohji, K., (2006). Synthesis and magnetic properties of the size-controlled Mn-Zn ferrite nanoparticles by oxidation method. *Journal of Physics and Chemistry of Solids, 67*(7), 1510–1517. https://doi.org/10.1016/j.jpcs.2005.11.015.
73. Joshi, H. M., De, M., Richter, F., He, J., Prasad, P. V., & Dravid, V. P., (2013). Effect of silica shell thickness of Fe_3O_4-SiO_x core-shell nanostructures on MRI contrast. *Journal of Nanoparticle Research, 15*(3), 1448–1456. https://doi.org/10.1007/s11051-013-1448-1.
74. Jun, Y. W., Lee, J. H., & Cheon, J., (2008). Chemical design of nanoparticle probes for high-performance magnetic resonance imaging. *Angewandte Chemie, 47*(28), 5122–5135. https://doi.org/10.1002/anie.200701674.
75. Justino, C. I. L., Rocha-Santos, T. A. P., Cardoso, S., & Duarte, A. C., (2013). Strategies for enhancing the analytical performance of nanomaterial-based sensors. *Trends in Analytical Chemistry, 47*, 27–36. https://doi.org/10.1016/j.trac.2013.02.004.
76. Justino, C. I. L., Rocha-Santos, T. A. P., Duarte, A. C., & Rocha-Santos, T. A., (2010). Review of analytical figures of merit of sensors and biosensors in clinical applications. *TrAC Trends in Analytical Chemistry, 29*(10), 1172–1183. https://doi.org/10.1016/j.trac.2010.07.008.
77. Kaiser, M., (2009). Effect of nickel substitutions on some properties of Cu-Zn ferrites. *Journal of Alloys and Compounds, 468*(1, 2), 15–21. https://doi.org/10.1016/j.jallcom.2008.01.070.

78. Khandare, J., & Minko, T., (2006). Polymer-drug conjugates: Progress in polymeric prodrugs. *Progress in Polymer Science, 31*(4), 359–397. https://doi.org/10.1016/j.progpolymsci.2005.09.004.
79. Kim, B. J., Cho, H., Park, J. H., Mano, J. F., & Choi, I. S., (2018). Strategic advances in formation of cell-in-shell structures: From syntheses to applications. *Journal of Advanced Materials, 30*(14), e1706063. https://doi.org/10.1002/adma.201706063, PubMed: 29441678.
80. Kim, S. G., Wang, W. N., Iwaki, T., Yabuki, A., & Okuyama, K., (2007). Low-temperature crystallization of barium ferrite nanoparticles by a sodium citrate-aided synthetic process. *Journal of Physical Chemistry C, 111*(28), 10175–10180. https://doi.org/10.1021/jp068249b.
81. Kissinger, P. T., (2005). Biosensors: A perspective. *Biosensors and Bioelectronics, 20*(12), 2512–2516. https://doi.org/10.1016/j.bios.2004.10.004.
82. Kleinauskas, A., Kim, J. K., Choi, G. H., Kim, H. T., Røe, K., & Juzenas, P., (2012). Superparamagnetic magnetite nanoparticles for cancer theranostics. *Reviews in Nanoscience and Nanotechnology, 1*(4), 271–283. https://doi.org/10.1166/rnn.2012.1018.
83. Kodama, R. H., & Berkowitz, A. E., (1999). Atomic-scale magnetic modeling of oxide nanoparticles. *Physical Review. Part B, 59*(9), 6321–6336. https://doi.org/10.1103/PhysRevB.59.6321.
84. Kojima, K., Miyazaki, M., Mizukami, F., & Maeda, K., (1997). Selective formation of spinel iron oxide in thin films by complexing agent-assisted sol–gel processing. *Journal of Sol-Gel Science and Technology, 8*(1–3), 77–81. https://doi.org/10.1007/BF02436821.
85. Kolhatkar, A. G., Jamison, A. C., Litvinov, D., Willson, R. C., & Lee, T. R., (2013). Tuning the magnetic properties of nanoparticles. *International Journal of Molecular Sciences, 14*(8), 15977–16009. https://doi.org/10.3390/ijms140815977.
86. Komarneni, S., D'Arrigo, M. C., Leonelli, C., Pellacani, G. C., & Katsuki, H., (1998). Microwave-hydrothermal synthesis of nanophase ferrites. *Journal of the American Ceramic Society, 81*(11), 3041–3043. https://doi.org/10.1111/j.1151-2916.1998.tb02738.x.
87. Kooti, M., Sedeh, A. N., Motamedi, H., & Rezatofighi, S. E., (2018). Magnetic graphene oxide inlaid with silver nanoparticles as antibacterial and drug delivery composite. *Applied Microbiology and Biotechnology, 102*(8), 3607–3621. https://doi.org/10.1007/s00253-018-8880-1.
88. Krishnan, K. M., (2010). Biomedical nanomagnetics: A spin through possibilities in imaging, diagnostics, and therapy. *IEEE Transactions on Magnetics, 46*(7), 2523–2558. https://doi.org/10.1109/TMAG.2010.2046907.
89. Kucheryavy, P., He, J., John, V. T., Maharjan, P., Spinu, L., Goloverda, G. Z., & Kolesnichenko, V. L., (2013). Superparamagnetic iron oxide nanoparticles with variable size and an iron oxidation state as prospective imaging agents. *Langmuir, 29*(2), 710–716. https://doi.org/10.1021/la3037007.
90. Kumar, G. S., Raguram, T., & Rajni, K. S., (2019). Synthesis and characterization of nickel-substituted cobalt ferrite nanoparticles using sol–gel auto-combustion method. *Journal of Superconductivity and Novel Magnetism, 32*(6), 1715–1723. https://doi.org/10.1007/s10948-018-4867-5.

91. Kurlyandskaya, G., & Levit, V., (2005). Magnetic dynabeads detection by sensitive element based on giant magnetoimpedance. *Biosensors and Bioelectronics, 20*(8), 1611–1616. https://doi.org/10.1016/j.bios.2004.04.027.
92. Kushnir, S. E., Gavrilov, A. I., Kazin, P. E., Grigorieva, A. V., Tretyakov, Y. D., & Jansen, M., (2012). Synthesis of colloidal solutions of $SrFe_{12}O_{19}$ plate-like nanoparticles featuring extraordinary magnetic-field-dependent optical transmission. *Journal of Materials Chemistry, 22*(36), 18893–18901. https://doi.org/10.1039/c2jm33874b.
93. Lai, B. H., Chang, C. H., Yeh, C. C., & Chen, D. H., (2013). Direct binding of concanvalin a onto iron oxide nanoparticles for fast magnetic selective separation of lactoferrin. *Separation and Purification Technology, 108*, 83–88. https://doi.org/10.1016/j.seppur.2013.02.020.
94. Lam, U. T., Mammucari, R., Suzuki, K., & Foster, N. R., (2008). Processing of iron oxide nanoparticles by supercritical fluids. *Journal of Industrial and Engineering Chemistry Research, 47*, 599–614.
95. Lamer, V. K., & Dinegar, R. H., (1950). Theory, production and mechanism of formation of monodispersed hydrosols. *Journal of the American Chemical Society, 72*(11), 4847–4854. https://doi.org/10.1021/ja01167a001.
96. Laurent, S., Forge, D., Port, M., Roch, A., Robic, C., Vander, E. L. V., & Muller, R. N., (2008). Magnetic iron oxide nanoparticles: Synthesis, stabilization, vectorization, physicochemical characterizations, and biological applications. *Chemical Reviews, 108*(6), 2064–2110. https://doi.org/10.1021/cr068445e.
97. Laurent, S., & Mahmoudi, M., (2011). Superparamagnetic iron oxide nanoparticles: Promises for diagnosis and treatment of cancer. *International Journal of Molecular Epidemiology and Genetics, 2*(4), 367–390.
98. Lee, N., & Hyeon, T., (2012). Designed synthesis of uniformly sized iron oxide nanoparticles for efficient magnetic resonance imaging contrast agents. *Chemical Society Reviews, 41*(7), 2575–2589. https://doi.org/10.1039/c1cs15248c.
99. Lee, H., Shin, T. H., Cheon, J., & Weissleder, R., (2015). Recent developments in magnetic diagnostic systems. *Chemical Reviews, 115*(19), 10690–10724. https://doi.org/10.1021/cr500698d.
100. Lee, H., Yoon, T. J., Figueiredo, J. L., Swirski, F. K., & Weissleder, R., (2009). Rapid detection and profiling of cancer cells in fine-needle aspirates. *Proceedings of the National Academy of Sciences of the United States of America, 106*(30), 12459–12464. https://doi.org/10.1073/pnas.0902365106.
101. Lee, J. H., Jung, M. J., Hwang, Y. H., Lee, Y. J., Lee, S., Lee, D. Y., & Shin, H., (2012). Heparin-coated superparamagnetic iron oxide for *in vivo* MR imaging of human MSCs. *Biomaterials, 33*(19), 4861–4871. https://doi.org/10.1016/j.biomaterials.2012.03.035.
102. Lee, K. J., An, J. H., Chun, J. R., Chung, K. H., Park, W. Y., Shin, J. S., Kim, D. H., & Bahk, Y. Y., (2013). *In vitro* analysis of the anti-cancer activity of mitoxantrone loaded on magnetic nanoparticles. *Journal of Biomedical Nanotechnology, 9*(6), 1071–1075. https://doi.org/10.1166/jbn.2013.1530.
103. Lee, N., Cho, H. R., Oh, M. H., Lee, S. H., Kim, K., Kim, B. H., Shin, K., et al., (2012). Multifunctional Fe_3O_4/TaO_x core/shell nanoparticles for simultaneous magnetic resonance imaging and x-ray computed tomography. *Journal of the American Chemical Society, 134*(25), 10309–10312. https://doi.org/10.1021/ja3016582.

104. Li, X., & Wang, G., (2009). Low-temperature synthesis and growth of superparamagnetic $Zn_{0.5}Ni_{0.5}Fe_2O_4$ nanosized particles. *Journal of Magnetism and Magnetic Materials, 321*(9), 1276–1279. https://doi.org/10.1016/j.jmmm.2008.11.006.
105. Li, X., Li, Q., Xia, Z., & Yan, W., (2008). Effects on direct synthesis of large scale mono-disperse $Ni_{0.5}Zn_{0.5}Fe_2O_4$ nanosized particles. *Journal of Alloys and Compounds, 458*(1, 2), 558–563. https://doi.org/10.1016/j.jallcom.2007.04.214.
106. Li, Y., Yi, R., Yan, A., Deng, L., Zhou, K., & Liu, X., (2009). Facile synthesis and properties of $ZnFe_2O_4$ and $ZnFe_2O_4$/polypyrrole core-shell nanoparticles. *Solid State Sciences, 11*(8), 1319–1324. https://doi.org/10.1016/j.solidstatesciences.2009.04.014.
107. Lisjak, D., Bobzin, K., Richardt, K., Bégard, M., Bolelli, G., Lusvarghi, L., Hujanen, A., et al., (2009). Preparation of barium hexaferrite coatings using atmospheric plasma spraying. *Journal of the European Ceramic Society, 29*(11), 2333–2341. https://doi.org/10.1016/j.jeurceramsoc.2009.01.028.
108. Liu, Q., Huang, H., Lai, L., Sun, J., Shang, T., Zhou, Q., & Xu, Z., (2009). Hydrothermal synthesis and magnetic properties of $NiFe_2O_4$ nanoparticles and nanorods. *Journal of Materials Science, 44*(5), 1187–1191. https://doi.org/10.1007/s10853-009-3268-3.
109. Liu, X. Q., Tao, S. W., & Shen, Y. S., (1997). Preparation and characterization of nanocrystalline α-Fe_2O_3 by a sol–gel process. *Sensors and Actuators. Part B, 40*(2, 3), 161–165. https://doi.org/10.1016/S0925-4005(97)80256-0.
110. Liu, X., Ma, Z., Xing, J., & Liu, H., (2004). Preparation and characterization of amino-silane modified superparamagnetic silica nanospheres. *Journal of Magnetism and Magnetic Materials, 270*(1, 2), 1–6. https://doi.org/10.1016/j.jmmm.2003.07.006.
111. Lu, A. H., Salabas, E. L., & Schüth, F., (2007). Magnetic nanoparticles: Synthesis, protection, functionalization, and application. *Angewandte Chemie, 46*(8), 1222–1244. https://doi.org/10.1002/anie.200602866.
112. Maensiri, S., Masingboon, C., Boonchom, B., & Seraphin, S., (2007). A simple route to synthesize nickel ferrite ($NiFe_2O_4$) nanoparticles using egg white. *Scripta Materialia, 56*(9), 797–800. https://doi.org/10.1016/j.scriptamat.2006.09.033.
113. Mahmoudi, M., Hosseinkhani, H., Hosseinkhani, M., Boutry, S., Simchi, A., Journeay, W. S., Subramani, K., & Laurent, S., (2011). Magnetic resonance imaging tracking of stem cells *in vivo* using iron oxide nanoparticles as a tool for the advancement of clinical regenerative medicine. *Chemical Reviews, 111*(2), 253–280. https://doi.org/10.1021/cr1001832.
114. Makovec, D., Kodre, A., Arčon, I., & Drofenik, M., (2009). Structure of manganese zinc ferrite spinel nanoparticles prepared with co-precipitation in reversed microemulsions. *Journal of Nanoparticle Research, 11*(5), 1145–1158. https://doi.org/10.1007/s11051-008-9510-0.
115. Mallapur, M. M., Shaikh, P. A., Kambale, R. C., Jamadar, H. V., Mahamuni, P. U., & Chougule, B. K., (2009). Structural and electrical properties of nanocrystalline cobalt substituted nickel zinc ferrite. *Journal of Alloys and Compounds, 479*(1, 2), 797–802. https://doi.org/10.1016/j.jallcom.2009.01.142.
116. Malyutin, A. G., Cheng, H., Sanchez-Felix, O. R., Carlson, K., Stein, B. D., Konarev, P. V., Svergun, D. I., Dragnea, B., & Bronstein, L. M., (2015). Coat protein-dependent behavior of poly(ethylene glycol) tails in iron oxide core virus-like nanoparticles. *ACS Applied Materials and Interfaces, 7*(22), 12089–12098. https://doi.org/10.1021/acsami.5b02278.

117. Manova, E., Paneva, D., Kunev, B., Estournès, C., Rivière, E., Tenchev, K., Léaustic, A., & Mitov, I., (2009). Mechanochemical synthesis and characterization of nanodimensional iron–cobalt spinel oxides. *Journal of Alloys and Compounds, 485*(1, 2), 356–361. https://doi.org/10.1016/j.jallcom.2009.05.107.
118. Mazario, E., Menéndez, N., Herrasti, P., Cañete, M., Connord, V., & Carrey, J., (2013). Magnetic hyperthermia properties of electrosynthesized cobalt ferrite nanoparticles. *Journal of Physical Chemistry C, 117*(21), 11405–11411. https://doi.org/10.1021/jp4023025.
119. Mendoza, M. E., Donado, F., Silva, R., Pérez, M. A., & Carrillo, J. L., (2005). Magnetite microcrystals for magneto-rheological fluids. *Journal of Physics and Chemistry of Solids, 66*(6), 927–931. https://doi.org/10.1016/j.jpcs.2004.06.019.
120. Merkoçi, A., Aldavert, M., Marın, S., & Alegret, S., (2005). New materials for electrochemical sensing V: Nanoparticles for DNA labeling. *TrAC Trends in Analytical Chemistry, 24*(4), 341–349.
121. Miller, M. M., Prinz, G. A., Cheng, S. F., & Bounnak, S., (2002). Detection of a micron-sized magnetic sphere using a ring-shaped anisotropic magnetoresistance-based sensor: A model for a magnetoresistance-based biosensor. *Applied Physics Letters, 81*(12), 2211–2213. https://doi.org/10.1063/1.1507832.
122. Modak, S., Ammar, M., Mazaleyrat, F., Das, S., & Chakrabarti, P. K., (2009). XRD, HRTEM and magnetic properties of mixed spinel nanocrystalline Ni-Zn-Cu-ferrite. *Journal of Alloys and Compounds, 473*(1, 2), 15–19. https://doi.org/10.1016/j.jallcom.2008.06.020.
123. Mohamed, R. M., Rashad, M. M., Haraz, F. A., & Sigmund, W., (2010). Structure and magnetic properties of nanocrystalline cobalt ferrite powders synthesized using organic acid precursor method. *Journal of Magnetism and Magnetic Materials, 322*(14), 2058–2064. https://doi.org/10.1016/j.jmmm.2010.01.034.
124. Morales, M. P., Serna, C. J., Bødker, F., & Mørup, S., (1997). Spin canting due to structural disorder in maghemite. *Journal of Physics: Condensed Matter, 9*(25), 5461–5467. https://doi.org/10.1088/0953–8984/9/25/013.
125. Mornet, S., Vasseur, S., Grasset, F., & Duguet, E., (2004). Magnetic nanoparticle design for medical diagnosis and therapy. *Journal of Materials Chemistry, 14*(14), 2116–2175. https://doi.org/10.1039/b402025a.
126. Nair, S. S., Rajesh, S., Abraham, V. S., Anantharaman, M. R., & Nampoori, V. P. N., (2006). Magnetic field-induced cluster formation and variation of magneto-optical signals in zinc-substituted ferrofluids. *Journal of Magnetism and Magnetic Materials, 305*(1), 28–34. https://doi.org/10.1016/j.jmmm.2005.11.036.
127. Nan, X., Zhang, X., Liu, Y., Zhou, M., Chen, X., & Zhang, X., (2017). Dual-targeted multifunctional nanoparticles for magnetic resonance imaging guided cancer diagnosis and therapy. *ACS Applied Materials and Interfaces, 9*(11), 9986–9995. https://doi.org/10.1021/acsami.6b16486.
128. Ngo, A. T., & Pileni, M. P., (2000). Nanoparticles of cobalt ferrite: Influence of the applied field on the organization of the nanocrystals on a substrate and on their magnetic properties. *Advanced Materials, 12*(4), 276–279. https://doi.org/10.1002/(SICI)1521-4095(200002)12:4<276::AID-ADMA276>3.0.CO;2-D.
129. Nigam, S., Barick, K. C., & Bahadur, D., (2011). Development of citrate-stabilized Fe_3O_4 nanoparticles: Conjugation and release of doxorubicin for therapeutic

applications. *Journal of Magnetism and Magnetic Materials, 323*(2), 237–243. https://doi.org/10.1016/j.jmmm.2010.09.009.

130. Nishiyama, N., & Kataoka, K., (2006). Nanostructured device based on block copolymer assemblies for drug delivery: Designing structures for enhanced drug function. *Advances in Polymer Science, 193*, 67–101.
131. Oh, J. K., & Park, J. M., (2011). Iron oxide-based superparamagnetic polymeric nanomaterials: Design, preparation, and biomedical application. *Progress in Polymer Science, 36*(1), 168–189. https://doi.org/10.1016/j.progpolymsci.2010.08.005.
132. Oh, J. K., Lee, D. I., & Park, J. M., (2009). Biopolymer-based microgels/nanogels for drug delivery applications. *Progress in Polymer Science, 34*(12), 1261–1282. https://doi.org/10.1016/j.progpolymsci.2009.08.001.
133. Didosyan, Y. S., Hauser, H., Reider, G. A., & Nicolics, J., (2004). Sensors and actuators on orthoferrites. *Proceedings of the IEEE Sensors, 1032–1035*. https://doi.org/10.1109/icsens.2004.1426350.
134. Osborne, E. A., Jarrett, B. R., Tu, C., & Louie, A. Y., (2010). Modulation of T2 relaxation time by light-induced, reversible aggregation of magnetic nanoparticles. *Journal of the American Chemical Society, 132*(17), 5934–5935. https://doi.org/10.1021/ja100254m.
135. Pang, Y. X., & Bao, X. J., (2002). Aluminium oxide nanoparticles prepared by water-in-oil microemulsions. *Journal of Materials Chemistry, 12*(12), 3699–3704. https://doi.org/10.1039/b206501k.
136. Ghosh, N., Pant, P., & Bhuvaneswari, S., (2008). Chemical methodologies for preparation of micron and nanometer scale ferrites: A mini review of patents. *Recent Patents on Nanotechnology, 2*(1), 8–18. https://doi.org/10.2174/187221008783478653.
137. Park, S., Lim, S., & Choi, H., (2006). Chemical vapor deposition of iron and iron oxide thin films from Fe(II) dihydride complexes. *Chemistry of Materials, 18*(22), 5150–5152. https://doi.org/10.1021/cm0601990.
138. Perelshtein, I., Perkas, N., Magdassi, S., Zioni, T., Royz, M., Maor, Z., & Gedanken, A., (2008). Ultrasound-assisted dispersion of $SrFe_{12}O_{19}$ nanoparticles in organic solvents and the use of the dispersion as magnetic cosmetics. *Journal of Nanoparticle Research, 10*(1), 191–195. https://doi.org/10.1007/s11051-007-9303-x.
139. Pernia, L. M., Rivera-Fernández, S., Franco, J. M., Pozo, D., De La Fuente, J. M., & GarcíA-MartíN, M. L., (2015). Long-circulating pegylated manganese ferrite nanoparticles for MRI-based molecular imaging. *Nanoscale, 7*(5), 2050–2059. https://doi.org/10.1039/c4nr05781c.
140. Pierson, H. O., (1999). *Handbook of Chemical Vapor Deposition: Principles, Technology, and Applications* (2nd edn., p. 506). William Andrew Publishing Ltd.
141. Pita, M., Abad, J. M., Vaz-Dominguez, C., Briones, C., Mateo-Martí, E., Martín-Gago, J. A., Morales, M. D. P., & Fernández, V. M., (2008). Synthesis of cobalt ferrite core/metallic shell nanoparticles for the development of a specific PNA/DNA biosensor. *Journal of Colloid and Interface Science, 321*(2), 484–492. https://doi.org/10.1016/j.jcis.2008.02.010.
142. Predoi, D., Kuneser, V., & Filoti, G., (2004). Magnetic behavior of maghemite nanoparticles studied by Mössbauer spectroscopy. *Romanian Reports in Physics, 56*, 373–378.
143. Pullar, R. C., (2012). Hexagonal ferrites: A review of the synthesis, properties and applications of hexaferrite ceramics. *Progress in Materials Science, 57*(7), 1191–1334. https://doi.org/10.1016/j.pmatsci.2012.04.001.

144. Rafique, M. S., Khaleeq-Ur-Rahman, M., Saif-ur-Rehman, A., Anjum, S., Anwar, M. S., Bhatti, K. A., Saeed, S., & Awan, M. S., (2008). Effect of external magnetic field on the deposition of $BaFe_{12}O_{19}$. *Vacuum, 82*(11), 1233–1237. https://doi.org/10.1016/j.vacuum.2008.01.052.
145. Rajput, A. B., Hazra, S., & Ghosh, N. N., (2013). Synthesis and characterization of pure single-phase $CoFe_2O_4$ nanopowder via a simple aqueous solution-based EDTA-precursor route. *Journal of Experimental Nanoscience, 8*(4), 629–639. https://doi.org/10.1080/17458080.2011.582170.
146. Ramesh, R., Ponnusamy, S., & Muthamizhchelvan, C., (2013). Synthesis, properties, and heating characteristics of bovine serum albumin coated Fe₃O₄ magnetic fluid for magnetic fluid hyperthermia application. *Science of Advanced Materials, 5*(9), 1250–1255. https://doi.org/10.1166/sam.2013.1579.
147. Raming, T. P., Winnubst, A. J. A., Van, K. C. M., & Philipse, A. P., (2002). The synthesis and magnetic properties of nanosized hematite (A-Fe_2O_3) particles. *Journal of Colloid and Interface Science, 249*(2), 346–350. https://doi.org/10.1006/jcis.2001.8194.
148. Rana, S., Gallo, A., Srivastava, R. S., & Misra, R. D. K., (2007). On the suitability of nanocrystalline ferrites as a magnetic carrier for drug delivery: Functionalization, conjugation and drug release kinetics. *Acta Biomaterialia, 3*(2), 233–242. https://doi.org/10.1016/j.actbio.2006.10.006.
149. Rao, B. P., Caltun, O., Cho, W. S., Kim, C. O., & Kim, C. G., (2007). Synthesis and characterization of mixed ferrite nanoparticles. *Journal of Magnetism and Magnetic Materials, 310*(2), e812–e814. https://doi.org/10.1016/j.jmmm.2006.10.771.
150. Roca, A. G., Costo, R., Rebolledo, A. F., Veintemillas-Verdaguer, S., Tartaj, P., González-Carreño, T., et al., (2009). Progress in the preparation of magnetic nanoparticles for applications in biomedicine. *Journal of Physics. Part D, 42*(22). https://doi.org/10.1088/0022-3727/42/22/224002, PubMed: 224002.
151. Rubinger, C. P. L., Gouveia, D. X., Nunes, J. F., Salgueiro, C. C. M., Paiva, J. A. C., Graça, M. P. F., André, P., & Costa, L. C., (2007). Microwave dielectric properties of $NiFe_2O_4$ nanoparticles ferrites. *Microwave and Optical Technology Letters, 49*(6), 1341–1343. https://doi.org/10.1002/mop.22402.
152. Safarik, I., & Safarikova, M., (2004). Magnetic techniques for the isolation and purification of proteins and peptides. *Biomagnetic Research and Technology, 2*(1), 7. https://doi.org/10.1186/1477-044X-2-7.
153. Saha, R., Sundaresan, A., & Rao, C. N. R., (2014). Novel features of multiferroic and magnetoelectric ferrites and chromites exhibiting magnetically driven ferroelectricity. *Mater. Horiz., 1*(1), 20–31. https://doi.org/10.1039/C3MH00073G.
154. Sakellari, D., Tsakaloudi, V., Polychroniadis, E. K., & Zaspalis, V., (2008). Microstructural phenomena controlling losses in nicuzn-ferrites as studied by transmission electron microscopy. *Journal of the American Ceramic Society, 91*(2), 366–371. https://doi.org/10.1111/j.1551-2916.2007.02030.x.
155. Sakho, E. H. M., Jose, J., Thomas, S., Kalarikkal, N., & Oluwafemi, O. S., (2019). Antimicrobial properties of MFe_2O_4 (M= Mn, Mg)/reduced graphene oxide composites synthesized via solvothermal method. *Materials Science and Engineering C, Materials for Biological Applications, 95*, 43–48. https://doi.org/10.1016/j.msec.2018.10.067.
156. Salunkhe, A. B., Khot, V. M., Phadatare, M. R., & Pawar, S. H., (2012). Combustion synthesis of cobalt ferrite nanoparticles—Influence of fuel to oxidizer ratio. *Journal of Alloys and Compounds, 514*, 91–96. https://doi.org/10.1016/j.jallcom.2011.10.094.

157. Sarangi, P. P., Naik, B., & Ghosh, N. N., (2009). Low temperature synthesis of single-phase α-Fe_2O_3 nano-powders by using simple but novel chemical methods. *Powder Technology, 192*(3), 245–249. https://doi.org/10.1016/j.powtec.2009.01.002.
158. Sarangi, P. P., Naik, B., & Ghosh, N. N., (2008). Synthesis of single-phase α-Fe_2O_3 nanopowders by using a novel low temperature chemical synthesis route. *Journal of the American Ceramic Society, 91*(12), 4145–4147. https://doi.org/10.1111/j.1551-2916.2008.02794.x.
159. Sarangi, P. P., Naik, B., Vadera, S. R., Patra, M. K., Prakash, C., & Ghosh, N. N., (2010). Preparation of polybenzoxazine-Ni-Zn ferrite magnetic nanocomposite and its magnetic property. *Materials Technology, 25*(5), 271–275. https://doi.org/10.1179/106678509X12519671504700.
160. Sarangi, P. P., Vadera, S. R., Patra, M. K., Prakash, C., Selvin, R., & Ghosh, N. N., (2010). Development of a novel aqueous solution based chemical methodology for synthesis of $Ni_{(1-X)}Zn_xFe_2O_4$ nanopowders and their electrical and magnetic property. *Integrated Ferroelectrics, 116*(1), 1–15. https://doi.org/10.1080/10584587.2010.503513.
161. Sarangi, P. P., Vadera, S. R., Patra, M. K., & Ghosh, N. N., (2010). Microstructure Dc resistivity and magnetic property of Ni-Zn ferrite nano-powders synthesized by EDTA precursor based method. *Journal of Optoelectronics and Advanced Materials, 12*, 1279–1285.
162. Sarangi, P. P., Vadera, S. R., Patra, M. K., & Ghosh, N. N., (2010). Synthesis and characterization of pure single phase Ni-Zn ferrite nanopowders by oxalate based precursor method. *Powder Technology, 203*(2), 348–353. https://doi.org/10.1016/j.powtec.2010.05.027.
163. Sato, T., Sue, K., Suzuki, W., Suzuki, M., Matsui, K., Hakuta, Y., Hayashi, H., et al., (2008). Rapid and continuous production of ferrite nanoparticles by hydrothermal synthesis at 673 K and 30 Mpa. *Industrial and Engineering Chemistry Research, 47*(6), 1855–1860. https://doi.org/10.1021/ie071168x.
164. Sattar, A. A., El-Sayed, H. M., El-Shokrofy, K. M., & El-Tabey, M. M., (2007). Study of the dc resistivity and thermoelectric power in Mn-substituted Ni–Zn ferrites. *Journal of Materials Science, 42*(1), 149–155. https://doi.org/10.1007/s10853-006-1087-3.
165. Satyanarayana, L., Reddy, K. M., & Manorama, S. V., (2003). Nanosized spinel $NiFe_2O_4$: A novel material for the detection of liquefied petroleum gas in air. *Materials Chemistry and Physics, 82*(1), 21–26. https://doi.org/10.1016/S0254-0584(03)00170-6.
166. Scano, A., Ennas, G., Frongia, F., La Barbera, A. L., López-Quintela, M. A., Marongiu, G., Paschina, G., et al., (2011). Mn-ferrite nanoparticles via reverse microemulsions: Synthesis and characterization. *Journal of Nanoparticle Research, 13*(7), 3063–3073. https://doi.org/10.1007/s11051-010-0205-y.
167. Seo, J. W., Fullerton, E. E., Nolting, F., Scholl, A., Fompeyrine, J., & Locquet, J. P., (2008). Antiferromagnetic $LaFeO_3$ thin films and their effect on exchange bias. *Journal of Physics–Condensed Matter, 20*(26). PubMed: 264014-264023.
168. Šepelák, V., Baabe, D., Mienert, D., Litterst, F. J., & Becker, K. D., (2003). Enhanced magnetization in nanocrystalline high-energy milled $MgFe_2O_4$. *Scripta Materialia, 48*(7), 961–966. https://doi.org/10.1016/S1359-6462(02)00600-0.
169. Shafi, K. V. P. M., Koltypin, Y., Gedanken, A., Prozorov, R., Balogh, J., Lendvai, J., & Felner, I., (1997). Sonochemical preparation of nanosized amorphous $NiFe_2O_4$ particles.

Journal of Physical Chemistry B, 101(33), 6409–6414. https://doi.org/10.1021/jp970893q.
170. Shalini, K., Subbanna, G. N., Chandrasekaran, S., & Shivashankar, S. A., (2003). Thin films of iron oxide by low pressure MoCVD using a novel precursor: Tris(*T*-butyl-3-Oxo-butanoato) iron(III). *Thin Solid Films, 424*(1), 56–60. https://doi.org/10.1016/S0040-6090(02)00903-3.
171. Shang, M., Zhang, C., Zhang, T., Yuan, L., Ge, L., Yuan, H., & Feng, S., (2013). The multiferroic perovskite $YFeO_3$. *Applied Physics Letters, 102*(6). https://doi.org/10.1063/1.4791697, PubMed: 062903.
172. Shanmugavel, T., Raj, S. G., Rajarajan, G., Kumar, G. R., & Boopathi, G., (2015). Single step synthesis and characterization of nanocrystalline cobalt ferrite by auto-combustion method. *Materials Today: Proceedings, 2*(4, 5), 3605–3609. https://doi.org/10.1016/j.matpr.2015.07.106.
173. Shen, B., Ma, Y., Yu, S., & Ji, C., (2016). Smart multifunctional magnetic nanoparticle-based drug delivery system for cancer thermo-chemotherapy and intracellular imaging. *ACS Applied Materials and Interfaces, 8*(37), 24502–24508. https://doi.org/10.1021/acsami.6b09772.
174. Shen, X., Wang, Y., Yang, X., Xia, Y., Zhuang, J., & Tang, P., (2009). Megahertz magneto-dielectric properties of nanosized NiZnCo ferrite from CTAB-assisted hydrothermal process. *Transactions of Nonferrous Metals Society of China, 19*(6), 1588–1592. https://doi.org/10.1016/S1003-6326(09)60075-3.
175. Shin, J. Y., & Oh, J. H., (1993). The microwave absorbing phenomena of ferrite microwave absorbers. *IEEE Transactions on Magnetics, 29*(6), 3437–3439. https://doi.org/10.1109/20.281188.
176. Silva, A. K. A., Wilhelm, C., Kolosnjaj-Tabi, J., Luciani, N., & Gazeau, F., (2012). Cellular transfer of magnetic nanoparticles via cell microvesicles: Impact on cell tracking by magnetic resonance imaging. *Pharmaceutical Research, 29*(5), 1392–1403. https://doi.org/10.1007/s11095-012-0680-1.
177. Singh, A. K., Goel, T. C., & Mendiratta, R. G., (2003). Effect of cation distribution on the properties of $Mn_{0.2}Zn_xNi_{0.8-X}Fe_2O_4$. *Solid State Communications, 125*(2), 121–125. https://doi.org/10.1016/S0038-1098(02)00626-9.
178. Sivakumar, M., Gedanken, A., Zhong, W., Du, Y. W., Bhattacharya, D., Yeshurun, Y., & Felner, I., (2004). Nanophase formation of strontium hexaferrite fine powder by the sonochemical method using $Fe(CO)_5$. *Journal of Magnetism and Magnetic Materials, 268*(1, 2), 95–104. https://doi.org/10.1016/S0304-8853(03)00479-7.
179. Sivakumar, N., Narayanasamy, A., Ponpandian, N., & Govindaraj, G., (2007). Grain size effect on the dielectric behavior of nanostructured $Ni_{0.5}Zn_{0.5}Fe_2O_4$. *Journal of Applied Physics, 101*(8). https://doi.org/10.1063/1.2721379, PubMed: 084116.
180. Smitha, S., Haseena, V. S., Narayanan, T. N., Reena, M. A. P., Ajayan, P. M., Puthumana, J., Aziz, A. A., et al. (2012). Fluorescent superparamagnetic iron oxide core–shell nanoprobes for multimodal cellular imaging. *Materials Express, 2*(4), 265–274. https://doi.org/10.1166/mex.2012.1083.
181. Sorescu, M., Diamandescu, L., Swaminathan, R., Mchenry, M. E., & Feder, M., (2005). Structural and magnetic properties of NiZn and Zn ferrite thin films obtained by laser ablation deposition. *Journal of Applied Physics, 97*(10). https://doi.org/10.1063/1.1854416.

182. Sun, C., Lee, J. S. H., & Zhang, M., (2008). Magnetic nanoparticles in MR imaging and drug delivery. *Advanced Drug Delivery Reviews, 60*(11), 1252–1265. https://doi.org/10.1016/j.addr.2008.03.018.
183. Sun, L., Li, X., Wei, X., Luo, Q., Guan, P., Wu, M., Zhu, H., et al., (2016). Stimuli-responsive biodegradable hyperbranched polymer-gadolinium conjugates as efficient and biocompatible nanoscale magnetic resonance imaging contrast agents. *ACS Applied Materials and Interfaces, 8*(16), 10499–10512. https://doi.org/10.1021/acsami.6b00980.
184. Sun, L., Wan, J., Schaefer, C. G., Zhang, Z., Tan, J., Guo, J., Wu, L., & Wang, C., (2017). Specific on-site assembly of multifunctional magnetic nanocargos based on highly efficient and parallelized bioconjugation: Toward personalized cancer targeting therapy. *ACS Biomaterials Science and Engineering, 3*(3), 381–391. https://doi.org/10.1021/acsbiomaterials.6b00773.
185. Suwalka, O., Sharma, R. K., Sebastian, V., Lakshmi, N., & Venugopalan, K., (2007). A study of nanosized Ni substituted Co-Zn ferrite prepared by coprecipitation. *Journal of Magnetism and Magnetic Materials, 313*(1), 198–203. https://doi.org/10.1016/j.jmmm.2006.12.026.
186. Swaminathan, R., Woods, J., Calvin, S., Huth, J., & Mchenry, M. E., (2006). Microstructural evolution model of the sintering behavior and magnetic properties of NiZn ferrite nanoparticles. *Journal of Advances in Science and Technology, 45*, 2337–2344.
187. Prithviraj, S. P. M., Basavaraja, S., Lagashetty, A., Srinivas, R. N. V., Nijagunappa, R., & Venkataraman, A., (2011). Synthesis and characterization of zinc ferrite nanoparticles obtained by self-propagating low-temperature combustion method. *Bulletin of Materials Science, 34*(7), 1325–1330. https://doi.org/10.1007/s12034-011-0323-x.
188. Tang, X., Hong, R. Y., Feng, W. G., & Badami, D., (2013). Ethylene glycol assisted hydrothermal synthesis of strontium hexaferrite nanoparticles as precursor of magnetic fluid. *Journal of Alloys and Compounds, 562*, 211–218. https://doi.org/10.1016/j.jallcom.2013.02.049.
189. Tartaj, P., & De Jonghe, L. C., (2000). Preparation of nanospherical amorphous zircon powders by a microemulsion-mediated process. *Journal of Materials Chemistry, 10*(12), 2786–2790. https://doi.org/10.1039/b002720k.
190. Tassa, C., Shaw, S. Y., & Weissleder, R., (2011). Dextran-coated iron oxide nanoparticles: A versatile platform for targeted molecular imaging, molecular diagnostics, and therapy. *Accounts of Chemical Research, 44*(10), 842–852. https://doi.org/10.1021/ar200084x.
191. Tayakoli, A., Sohrabi, M., & Kargari, A., (2007). A review of methods for synthesis of nanostructured metals with emphasis on iron compounds. *Chemical Papers, 61*, 151–170.
192. Teja, A. S., & Koh, P. Y., (2009). Synthesis, properties, and applications of magnetic iron oxide nanoparticles. *Progress in Crystal Growth and Characterization of Materials, 55*(1, 2), 22–45. https://doi.org/10.1016/j.pcrysgrow.2008.08.003.
193. Tong, S., Hou, S., Zheng, Z., Zhou, J., & Bao, G., (2010). Coating optimization of superparamagnetic iron oxide nanoparticles for high T2 relaxivity. *Nano Letters, 10*(11), 4607–4613. https://doi.org/10.1021/nl102623x.
194. Tsai, Z. T., Wang, J. F., Kuo, H. Y., Shen, C. R., Wang, J. J., & Yen, T. C., (2010). *In situ* preparation of high relaxivity iron oxide nanoparticles by coating with chitosan: A potential MRI contrast agent useful for cell tracking. *Journal of Magnetism and Magnetic Materials, 322*(2), 208–213. https://doi.org/10.1016/j.jmmm.2009.08.049.

195. Umut, E., (2013). Chapter 8; Surface modification of nanoparticles used in biomedical applications. In: Aliofkhazraei, M., (ed.), *Modern Surface Engineering Treatments* (pp. 185–208) Intech, Rijeka.
196. Upadhyay, C., Mishra, D., Verma, H. C., Anand, S., & Das, R. P., (2003). Effect of preparation conditions on formation of nanophase Ni-Zn ferrites through hydrothermal technique. *Journal of Magnetism and Magnetic Materials, 260*(1, 2), 188–194. https://doi.org/10.1016/S0304-8853(02)01320-3.
197. Velinov, N., Petrova, T., Tsoncheva, T., Genova, I., Koleva, K., Kovacheva, D., & Mitov, I., (2016). Auto-combustion synthesis, mössbauer study and catalytic properties of copper-manganese ferrites. *Hyperfine Interactions, 237*(1), 1–11. https://doi.org/10.1007/s10751-016-1222-8.
198. Venkatesha, N., Poojar, P., Qurishi, Y., Geethanath, S., & Srivastava, C., (2015). Graphene oxide-Fe_3O_4 nanoparticle composite with high transverse proton relaxivity value for magnetic resonance imaging. *Journal of Applied Physics, 117*(15). https://doi.org/10.1063/1.4918605, PubMed: 154702.
199. Verma, A., Alam, M. I., Chatterjee, R., Goel, T. C., & Mendiratta, R. G., (2006). Development of a new soft ferrite core for power applications. *Journal of Magnetism and Magnetic Materials, 300*(2), 500–505. https://doi.org/10.1016/j.jmmm.2005.05.040.
200. Verma, A., Thakur, O. P., Prakash, C., Goel, T. C., & Mendiratta, R. G., (2005). Temperature dependence of electrical properties of nickel-zinc ferrites processed by the citrate precursor technique. *Materials Science and Engineering, 116*(1), 1–6. https://doi.org/10.1016/j.mseb.2004.08.011.
201. Verma, S., Joy, P. A., Khollam, Y. B., Potdar, H. S., & Deshpande, S. B., (2004). Synthesis of nanosized $MgFe_2O_4$ powders by microwave hydrothermal method. *Materials Letters, 58*(6), 1092–1095. https://doi.org/10.1016/j.matlet.2003.08.025.
202. Wadajkar, A. S., Menon, J. U., & Nguyen, K. T., (2012). Polymer-coated magnetic nanoparticles for cancer diagnosis and therapy. *Reviews in Nanoscience and Nanotechnology, 1*(4), 284–297. https://doi.org/10.1166/rnn.2012.1019.
203. Walter, A., Billotey, C., Garofalo, A., Ulhaq-Bouillet, C., Lefèvre, C., Taleb, J., Laurent, S., et al., (2014). Mastering the shape and composition of dendronized iron oxide nanoparticles to tailor magnetic resonance imaging and hyperthermia. *Chemistry of Materials, 26*(18), 5252–5264. https://doi.org/10.1021/cm5019025.
204. Wang, H. W., & Kung, S. C., (2004). Crystallization of nanosized Ni-Zn ferrite powders prepared by hydrothermal method. *Journal of Magnetism and Magnetic Materials, 270*(1, 2), 230–236. https://doi.org/10.1016/j.jmmm.2003.09.019.
205. Wang, Y. J., Hussain, S. M., & Krestin, G. P., (2001). Superparamagnetic iron oxide contrast agents: Physicochemical characteristics and applications in MR imaging. *European Radiology, 11*(11), 2319–2331. https://doi.org/10.1007/s003300100908.
206. Wáng, Y. X., & Idée, J. M., (2017). A comprehensive literatures update of clinical researches of superparamagnetic resonance iron oxide nanoparticles for magnetic resonance imaging. *Quantitative Imaging in Medicine and Surgery, 7*(1), 88–122. https://doi.org/10.21037/qims.2017.02.09.
207. Wang, L., Huang, J., Chen, H., Wu, H., Xu, Y., Li, Y., Yi, H., et al., (2017). Exerting enhanced permeability and retention effect driven delivery by ultrafine iron oxide nanoparticles with T1-T2 switchable magnetic resonance imaging contrast. *ACS Nano, 11*(5), 4582–4592. https://doi.org/10.1021/acsnano.7b00038.

208. Wang, L., Luo, J., Fan, Q., Suzuki, M., Suzuki, I. S., Engelhard, M. H., Lin, Y., et al., (2005). Monodispersed core-shell Fe_3O_4@Au nanoparticles. *Journal of Physical Chemistry. B, 109*(46), 21593–21601. https://doi.org/10.1021/jp0543429.
209. Wang, Z. L., (2001). Chapter 10; Walt, characterization of nanophase materials. In: De Heer, W. A., (ed.), *Nanomagnetism,* (Vol. 18, pp. 142–165). Wiley-VCH Verlag GmbH.
210. Wu, Y., Guo, J., Yang, W., Wang, C., & Fu, S., (2006). Preparation and characterization of chitosan-poly(acrylic acid) polymer magnetic microspheres. *Polymer, 47*(15), 5287–5294. https://doi.org/10.1016/j.polymer.2006.05.017.
211. Xu, P., Han, X., & Wang, M., (2007). Synthesis and magnetic properties of $BaFe_{12}O_{19}$ hexaferrite nanoparticles by a reverse microemulsion technique. *Journal of Physical Chemistry C, 111*(16), 5866–5870. https://doi.org/10.1021/jp068955c.
212. Xu, X. J., Zhou, L. H., Zhai, Q. G., & Lu, C. Z., (2007). Synthesis, properties, and formation mechanism of zinc ferrite hollow spheres. *Journal of the American Ceramic Society, 90*(6), 1959–1962. https://doi.org/10.1111/j.1551-2916.2007.01731.x.
213. Yadoji, P., Peelamedu, R., Agrawal, D., & Roy, R., (2003). Microwave sintering of Ni–Zn ferrites: Comparison with conventional sintering. *Materials Science and Engineering, 98*(3), 269–278. https://doi.org/10.1016/S0921-5107(03)00063-1.
214. Yang, J. M., & Yen, F. S., (2008). Evolution of intermediate phases in the synthesis of zinc ferrite nanopowders prepared by the tartrate precursor method. *Journal of Alloys and Compounds, 450*(1, 2), 387–394. https://doi.org/10.1016/j.jallcom.2006.10.139.
215. Yang, J., Park, S. B., Yoon, H. G., Huh, Y. M., & Haam, S., (2006). Preparation of poly E-caprolactone nanoparticles containing magnetite for magnetic drug carrier. *International Journal of Pharmaceutics, 324*(2), 185–190. https://doi.org/10.1016/j.ijpharm.2006.06.029.
216. Yang, L., Wang, Z., Ma, L., Li, A., Xin, J., Wei, R., Lin, H., et al., (2018). The roles of morphology on the relaxation rates of magnetic nanoparticles. *ACS Nano, 12*(5), 4605–4614. https://doi.org/10.1021/acsnano.8b01048.
217. Yener, D. O., & Giesche, H., (2004). Synthesis of pure and manganese-, nickel-, and zinc-doped ferrite particles in water-in-oil microemulsions. *Journal of the American Ceramic Society, 84*(9), 1987–1995. https://doi.org/10.1111/j.1151-2916.2001.tb00947.x.
218. Yoo, D., Lee, J. H., Shin, T. H., & Cheon, J., (2011). Theranostic magnetic nanoparticles. *Accounts of Chemical Research, 44*(10), 863–874. https://doi.org/10.1021/ar200085c.
219. Zélis, P. M., Pasquevich, G. A., Stewart, S. J., Raap, M. B. F., Aphesteguy, J., Bruvera, I. J., Laborde, C., Pianciola, B., Jacobo, S., & Sánchez, F. H., (2013). Structural and magnetic study of zinc-doped magnetite nanoparticles and ferrofluids for hyperthermia applications. *Journal of Physics: Part D, 46*(12). https://doi.org/10.1088/0022-3727/46/12/125006, PubMed: 125006.
220. Zeng, J., Jing, L., Hou, Y., Jiao, M., Qiao, R., Jia, Q., Liu, C., et al., (2014). Anchoring group effects of surface ligands on magnetic properties of Fe_3O_4 nanoparticles: Towards high performance MRI contrast agents. *Advanced Materials, 26*(17), 2694–8, 2016. https://doi.org/10.1002/adma.201304744.
221. Zhang, X., & Li, Q., (2008). Microwave assisted hydrothermal synthesis and magnetic property of hematite nanorods. *Materials Letters, 62*(6, 7), 988–990. https://doi.org/10.1016/j.matlet.2007.07.029.
222. Zhang, B., Yang, B., Zhai, C., Jiang, B., & Wu, Y., (2013). The role of exendin-4-conjugated superparamagnetic iron oxide nanoparticles in beta-cell-targeted MRI. *Biomaterials, 34*(23), 5843–5852. https://doi.org/10.1016/j.biomaterials.2013.04.021.

223. Zhang, D. E., Zhang, X., Ni, X. M., Song, J. M., & Zheng, H. G., (2006). Synthesis and characterization of $CoFe_2O_4$ octahedrons via an EDTA-assisted route. *Journal of Magnetism and Magnetic Materials, 305*(1), 68–70. https://doi.org/10.1016/j.jmmm.2005.11.030.
224. Zhang, H., Wang, W. W., Li, H., Meng, S., & Li, D., (2008). A strategy to prepare ultrafine dispersed Fe_2O_3 nanoparticles. *Materials Letters, 62*(8, 9), 1230–1233. https://doi.org/10.1016/j.matlet.2007.08.026.
225. Zhang, L., Liu, Y., Zhang, Q., Li, T., Yang, M., Yao, Q., Xie, X., & Hu, H., (2018). Gadolinium-labeled aminoglycoside and its potential application as a bacteria-targeting magnetic resonance imaging contrast agent. *Analytical Chemistry, 90*(3), 1934–1940. https://doi.org/10.1021/acs.analchem.7b04029.
226. Zhao, F., Zhao, Y., Liu, Y., Chang, X., Chen, C., & Zhao, Y., (2011). Cellular uptake, intracellular trafficking, and cytotoxicity of nanomaterials. *Small, 7*(10), 1322–1337. https://doi.org/10.1002/smll.201100001.
227. Zhao, Z., Zhou, Z., Bao, J., Wang, Z., Hu, J., Chi, X., Ni, K., et al., (2013). Octapod iron oxide nanoparticles as high-performance T2 contrast agents for magnetic resonance imaging. *Nature Communications, 4*, 2266. https://doi.org/10.1038/ncomms3266.
228. Zhou, L., Yuan, J., & Wei, Y., (2011). Core-shell structural iron oxide hybrid nanoparticles: From controlled synthesis to biomedical applications. *J. Mater. Chem., 21*(9), 2823–2840. https://doi.org/10.1039/C0JM02172E.
229. Zhou, Z., Huang, D., Bao, J., Chen, Q., Liu, G., Chen, Z., Chen, X., & Gao, J., (2012). A synergistically enhanced T(1) -T(2) dual-modal contrast agent. *Advanced Materials, 24*(46), 6223–6228. https://doi.org/10.1002/adma.201203169.
230. Zhou, Z., Wang, L., Chi, X., Bao, J., Yang, L., Zhao, W., Chen, Z., et al., (2013). Engineered iron–oxide-based nanoparticles as enhanced T1 contrast agents for efficient tumor imaging. *ACS Nano, 7*(4), 3287–3296. https://doi.org/10.1021/nn305991e.
231. Zhu, H., Gu, X., Zuo, D., Wang, Z., Wang, N., & Yao, K., (2008). Microemulsion-based synthesis of porous zinc ferrite nanorods and its application in a room-temperature ethanol sensor. *Journal of Nanotechnology, 19*(40), 405503. https://doi.org/10.1088/0957-4484/19/40/405503.
232. Zhu, Y., Jiang, F. Y., Chen, K., Kang, F., & Tang, Z. K., (2013). Modified reverse microemulsion synthesis for iron oxide/silica core-shell colloidal particles. *Journal of Sol-Gel Science and Technology, 66*(1), 180–186. -https://doi.org/10.1007/s10971–013-2985-x.

CHAPTER 9

ROLE OF NANOPARTICLES IN CHEMOTHERAPY IN CANCER AND DRUG DELIVERY: CURRENT SCENARIO AND FUTURE CHALLENGES

SHUBHJEET MANDAL, MOHD. ANEES, HARPAL SINGH, and AZIZ UNNISA

ABSTRACT

Chemotherapy is widely utilized and has proven an effective treatment method for all forms of cancer. However, its conventional approach shows many limitations, including poor pharmaceutical properties of drugs, higher off-target cytotoxicity, and multidrug resistance (MDR) against the therapy, making conventional chemotherapy a less preferable treatment option. Many potent drugs have lost their value due to these limitations and, were ignored for a long time in cancer chemotherapy. Drug delivery systems propose a promising way to overcome almost all the limitations associated with conventional chemotherapy. Nanotechnology as a therapeutic transport mechanism for cancer therapy has received a lot of interest in recent years. Nanoparticle-based drug delivery approach begins with the encapsulation or attachment of the chemotherapeutic drugs within/on the nanocarrier, followed by their successful delivery to the target site to achieve numerous advantages, including protection, stabilization, and sustained release of drug molecules. These offered benefits provide a chance to re-think the possible drugs before being ignored because of their toxicity and poor pharmacokinetics. The first

Sustainable Nanomaterials for Biomedical Engineering: Impacts, Challenges, and Future Prospects.
Junaid Ahmad Malik, Megh R. Goyal & Mohamed Jaffer M. Sadiq (Eds.)

successful liposome nanoparticle-based formulation, "Doxil" was permitted by the Food and Drug Administration (FDA) for cancer management in 1995. Since then, the field has developed exponentially and drawn the interest of a variety of investigators. Currently, several nanoparticle-based cancer chemotherapy drugs have been licensed, although many are at different levels of clinical or pre-clinical growth.

9.1 INTRODUCTION

Cancer is the worst of all disorders with unregulated metastatic tissue growth. The WHO estimates that about 7.8 million people die from cancer each year, the second most significant cause of death worldwide. There are various cancers, the cervical, breast, lung, stomach, and colorectal cancer being the most common type among females. The most common type among men is prostate, lung, stomach, colorectal, and pancreatic cancer. For more than half a century, scientific research has concentrated on cancer, with significant developments in the essential biology of cancer, but nothing has been incorporated into clinics. This difference is always attributed to the heterogeneity of cancer.

In cancer care, there are many difficulties, including diagnosis. Early diagnosis of neoplastic lesions remains an elusive task, pending scientific advancement in diagnostic radiology. In some instances, only after metastasis is detected, leaving a severe risk of recovery and eventual survival. Cancer Research UK evidence shows that early-stage diagnosis leads to 3X the existence rate of cancer. Diagnostic techniques may also have a notable effect on cancer morbidity and death.

After diagnosis, the second big problem for cancer therapy is to identify the optimal regime for cancer cell removal. The three treatment choices are based mainly on oncologists – chemotherapy, radiation, and surgical removal. Chemical treatment is the primary clinical method for both local and metastatic cancer, whereas operation and radiation are primarily used in native and non-metastatic cancer [19]. Chemotherapy is a single or multifunctional cytotoxic treatment to suppress malignant cell growth and differentiation. A range of chemical-therapeutic drugs is the form alkylative, neoplastic like doxorubicin, daunorubicin, etc., plant alkaloids, antimetabolite, and other miscellaneous antitumor agents (vincristine, vinblastine, paclitaxel, etc.). The fundamental theory behind these chemotherapeutic drugs is that they affect the most harmful cells. Apart from cancer cells, the body contains other rapidly proliferating cells such as bone marrow, intestinal epithelial cells, and hair follicles. Thus the adverse effects on these issues are also high. In the treatment of cancer, this unspecific biodistribution of current therapeutic agents is crucial. Furthermore, other inefficiencies in traditional chemotherapy include low solubility, poor bioavailability, and inconsequential stability, failure to penetrating cancer cells, a small therapeutic window, multidrug resistance growth, and fast reticular clearance [94]. These suggest that new techniques that can solve the above limits and target cancer cells, in particular, ensures maximum reducing the toxicity of normal tissues.

Many of the above-mentioned cross-cutting problems are being solved and rectified by nanotechnologies made with proven cancer science. It provides particular diagnostic, medicinal, and predictive approaches. In diagnostics, it is used in imagination, which is essential for clinical cancer treatment. All existing imaging methods involve contrasting agents in their molecular expression profiles to determine malignant conditions. External contrast agents frequently employed in facilities are either unspecific or resistant to cancer detection in the early stages. Nanocarriers have been researched to recognize early-stage cancers, track molecular expressions, and microenvironmental neoplasms to produce selective external agents [39]. The capacity to function cancer-focused agents on the surface, biodegradability, and biocompatibility of magnet nanoparticles, including iron oxide nanoparticles, indicate effective contrasting agents of MRI [76].

One example is the use of reserve organs to diagnose and characterize early-stage liver lesions caused by a specific superparamagnetic iron oxide-carboxy dextran-coated nanoparticle. Cross-linked iron oxide nanoparticles and annexin-V were investigated in the MRI detection of camptothecin-mediated Jurkat T cells *in vitro*. Because of their accuracy and responsiveness, these nanoparticles have the potential for future therapeutic uses [37].

Nanotechnology has prepared the path for therapeutic systems as a drug delivery mechanism with a tremendous capacity for producing targeted formulations and the opportunity to solve bio barriers [79]. Liposomes, micelles, carbon nanotubes, dendrimers, and polymer and magnet nanoparticles are some nanocarriers utilized in nanomedicines [69]. These nanocarriers are ideal supplies because they allow several anti-cancer drugs to be administered simultaneously, protecting the drug from degradation, increasing water solubility, lowering renal clearance, increasing blood distribution, and using controlled drug release [67]. Nanotechnology also provides a new forum to build tumor-targeted vehicles through either passive or active targeted mechanisms [74]. The EPR effect in cancer tissue is used in passive targeting to accrue nanoscale portion in the tumor interstitial preferentially. The tumor vasculature, unlike usual vessels, is organizational and structural pathological, and heterogeneous. They are fitted with vascular permeability and angiogenesis mediators to provide hyper-permeability and leakage to these vessels. To facilitate this, they have a fast growth rate.

In comparison, because of the loss of functioning lymph vessels, cancer is improperly drained, resulting in longer retention. These architectural defects contribute to the gold standard EPR effect while firing nanoparticles at cancer [50]. Therefore, this technique contributes to the position but does not facilitate their uptake in cancer cells of the nanoparticles of the tumor microenvironment. Therefore, the pathways for aggressive targeting are being researched. For this reason, nanoparticles' surface functionalization is achieved by various binding receptor and surface-specific ligands to a cancer cell, thereby facilitating their interconnection by receptor-mediated endocytosis [77]. A range of targeting ligands was widely used in this function, including albumins, biotins, transferrin, folates, antibodies, peptides, and aptamers. Successful targeting increases cytotoxic medication therapeutic effectiveness and decreases drug resistance incidents.

In the latest domain of customized cancer treatment, such as nanotheranostics, nanotechnology will also find therapeutic and diagnostic feature outgrowth of Nano preparations. Patient stratification, early diagnosis, and disease staging are just a few of the applications for this technology, including patient subgroup testing, therapeutic response management, and early detection. It is critical in dealing with the tumor's heterogeneity, which necessitates a tailored treatment. We must also abandon a one-for-all approach and implement innovations that better cater to the hope of adapting to cancer treatment therapy. Each tumor is distinct, including those of the exact origin. Nanotheranostics are an emerging area with scope for clinical oncology to be transfigured [17].

Polymeric and inorganic nanoparticles are different nanocarriers currently used for drug delivery. Recently, the latest contribution to delivery vehicles is viral proteins. This chapter will describe the different methods and processes associated with the nanoparticle-based drug delivery on/to the cancer site.

9.2 CHEMOTHERAPY AND LIMITATIONS

The use of chemicals to destroy or inhibit cancer cells is chemotherapy. Chemotherapy is not local and involves the whole body as compared to surgery and radiation therapy. The critical goal is quickly to separate cells in chemotherapy, as the capacity of most cancer cells to break efficiently. Hair follicle cells, stem cells, mucous, and Intestinal cells, and bone marrow cells differentiate readily into the body and are also affected by chemotherapy [85]. All cancer cells, on the other hand, do not easily separate. Tumors might have a duplication rate of more than 70 days or less than 30 days. To compensate for this deficiency, selective, and targeted medicines are required to increase chemotherapeutic effectiveness. By inhibiting proliferation and transcription by targeting DNA, cell growth, and division through suppression of protein synthesis, orthodox chemotherapy aims to stop cancer cells from multiplying and dividing. Furthermore, specific protein synthesis routes, mitotic motors, specific antigens, and protein kinase inhibitors are being investigated to treat medicines that can impact mitotic spindle formation at various phases of cell division.

In chemotherapy, toxicity is the primary concern. The chemical agents are highly volatile, and high doses for treating different forms of cancer are also seen to be toxic. These results can be categorized as long and short-term toxicity under two headings. In tissues that have fast splitting capacities, short-term toxicity typically evolves rapidly. The most common side effects of short-term toxicity are nausea and dizziness [79]. Long-term exposure typically contributes to cancer and infertility. In addition to these issues, cancer cells may acquire resistance to chemotherapy drugs, reducing the efficacy of the treatment. Drug resistance can result from multidrug expression, a reduction in membrane absorption, and increased cytoplasm tolerance, DNA reparation, and DNA damage. New active agents need to be continuously treated for cancer in this case [99].

Both of these therapies can be mixed or used individually. The argument is; however that mixed and prolonged treatment will enhance the damage to healthy cells as they are not explicitly discussed. In recent years, new therapeutic models have been used to reduce average cell exposure concentrations. The most frequent methods are immunotherapy and hormone treatment [27]. The immune system of the patient battles cancer and eliminates the side effects of immunotherapy linked to medication. Cancer cells expand more slowly by stopping hormones that can aid cancer growth or increasing hormone activity to combat carcinoma cells.

The following complications are addressed by existing cancer chemotherapy [7]:

- Traditional anti-cancer medications can intravenously deliver the whole body, distribute narcotics over the bloodstream, and hurt both malignant and human cells.
- A high hydrostatic pressure defines tumor interstation in contrast to normal tissues, which results in a convective external interstitial flow that can remove medication from malignancy.
- When effectively given to the interstitial tumor, cancer cells that developed multidrug resistance (MDR) impair the medication's effectiveness. MDR is characterized by overexpression of the P-glycoprotein plasma (P-gp) membrane, pushing medicines away from the cell. To get around P-gp-mediated MDR, researchers used various methods, including co-administration of P-gp inhibitors and attaching anticancer drugs inside nanoparticles. To guarantee that the medication does not accept P-gp in the plasma membrane before being given to the cell's cytoplasm or nucleus [80].

9.3 A NEW SOLUTION OF DRUG DELIVERY: NANOTECHNOLOGY

The distribution of therapeutic compounds to their intended recipients is crucial in the cure of many diseases. Small efficacy, weak biodistribution, and lack of selectivity characterize the traditional application of drugs [70]. The management of drug distribution will solve these drawbacks and disadvantages. In the controlled drug delivery system (DDS), the medication is administered to the place of action, limiting its effects on critical issues and unwanted side effects.

DDS also prevents the chemical from degrading and being cleared as quickly and increases the drug's concentration in the target area. This new treatment method is critical where a dosage or concentration of a drug is different from the therapeutic consequences or adverse effects. Adding medications to customized carriers can help to concentrate cells in a specific way. Latest Nanotechnological advances have shown the great potential of nanoparticles (structures below 100 nm in at least one dimension) as drug transporters.

Nanostructures offer unique physicochemical and biological characteristics due to their small sizes, such as increased reactive surface and the ability to cross barriers between tissues, making them an attractive material for biomedical applications. Nanotechnology is now a booming and multidisciplinary field of nanoscale engineering and production [42]. It significantly influences the pharmaceuticals and medical sectors, where techniques from micro and micro-making to nanometer scale have changed [34]. The production and deployment of nm-sized materials have also been enabled in drug-supply applications. In the era of genetic engineering, the use of polymers and colloids for various applications modifies the transfection mechanism of nanoparticles. Developing artificial cells or exploring physiological aspects using nanomaterials can help us gain a better understanding of natural physiological processes. These milestones are just two of the latest accomplishments of nanotechnology. Drug supply has been highly affected by nanotechnology, and it is a significant medical field.

9.3.1 HISTORY OF NANOTECHNOLOGY

A famous quote from Richard Feynman, inspired by a modern research area called nanotechnologies, "There is plenty of room at the bottom." Eric Drexler later suggested a nanoscale creator in 1980 that could produce a clone of itself and other structures and ultimately developed into a sophisticated system. Fullers were discovered in 1985 when the scan tunneling microscope was invented, followed by carbon nanotubes.

The first well-researched nano made is nanotubes belonging to the structural fullerene family. The term comes from the car diabetes' long hollow tube structure, consisting of one dense atomic sheet. These sheets are often rolling at many angles, and these characteristics dictate the nature of the nanotubes. Cough particles have a diameter of 2,500 to 10,000 nanometers, whereas fine particles have a diameter of 100 to 2,500 nanometers. Tiny by 1 to 100 nanometers, ultrafine particles called nanoparticles. Scientists are fascinated by nanoparticles because they act as a connection between bulk and atomic structures. The fullerene nanotubes, which belong to the fullerene family, are the first well-studied nano mate in the experiment. The name derives from their long hollow tunnel, a structure of one thick atomic carbon shell.

9.3.2 NANOCARRIERS USED AS DRUG DELIVERY SYSTEM

Nanoparticles as range of 1 nm to 100 nm are objects of at least one dimension in compliance with the concepts of the National Nanotechnology Initiative. The "nano" prefix, on the other hand, is usually reserved for particles smaller than a few 100 nanometers. Cells are more conveniently used than large molecules by nanocarriers with customized biological and physicochemical properties, meaning that they can carry commercially available bioactive substances [96]. Examples of drug deliver tested nanocarriers are liposomes, dendrimers, polymers, stable lipid nanoparticles, silicon or carbon, and magnetic nanoparticles.

For a particular therapy, how the medication is paired with the nanocarrier and the targeted approach. The surface of nanocarriers can be adsorbed or covalently attached to a material. Covalent binding has an advantage over other forms of connection. It allows for complete control of the amount of pharmaceutical substance supplied and the number of drug molecules bound to the nanocarrier. Active or passive pathways may be used to hit cells with nanocarriers. The first approach is based on the attractiveness of the therapeutic substance. Nanocarriers are used to identify ligands bound to the surface of conjugates, peptides, and low molecular ligands like folic acids, among other things. Modulation of physical stimuli can also be used to achieve a practical approach (e.g., temperature, pH, magnetism). Passive targeting occurs due to increased vascular permeability and preservation, which is common in cancerous tumors with leaky tissues [70].

After the medication-nanocarrier conjugates reach the sick tissue, the therapeutic agents are released. Temperature, pH, osmolality, and enzymes can all be used to control nanocarrier release.

Biocompatible (that is, they must be able to integrate with the biological system without triggering immunological responses or creating negative effects) and non-toxic nanocarriers are required for medical applications. Nanoparticles have an undesirable effect, relying heavily on their hydrodynamic scale, structure, quantity, surface chemistry, delivery route, and immune response (mainly the macrophage and granulocyte route of absorption) and the period they live in their bloodstream. Since a certain number of variables will influence the toxicity of nanoparticles, it isn't easy to quantify them, and toxicological tests are required with each new formulation. However, some generalizations can be made concerning their size – smaller particles have a wider region, so they are more active and thus toxic [1]. Nanoparticles of 10–100 nm hydrodynamic diameter are widely recognized as having ideal pharmacokinetics for *in vivo* applications. Larger nanoparticles cause the tissue extravasations and renal clarification of smaller nanoparticles, and they are opsonized and separated by the reticuloendothelial macrophages from blood [22].

9.3.3 *NANOTECHNOLOGY FOR CANCER TREATMENT*

The type of medication currently used in cancer patients has saved many lives. Still, the treatment's side effects are severe and affect the entire body because chemotherapy agents are not specific. Cancer is a very complex biological condition, and certain illnesses may be called a disorder. Cancer cells are distinguished by their fast and out-of-reach division and proliferation. The fundamental goal of modern chemotherapy is to destruct all cells easily. The disadvantage of this treatment is that it destroys other rapidly growing cells in the body, such as intestinal epithelium and hair follicles, leading the individual to experience side effects [10]. The creation of nanoparticles has brought chemotherapy a different route. The targeted distribution of drugs to the tumor site or a definite cell group larger than any other natural tissue and toxic organ effects were prevented by intelligently engineered nanoparticles [92]. Multiple structures have been studied for this method of procedure.

Micelles and liposomes are other methods for delivering chemotherapeutic drugs. With their hydrophobic center and hydrophilic shell, micelles are also an excellent approach to overcome insoluble medicines. As the outside of the micelle is more PEGylated, the capacity of nanocarriers to travel through the local vasculature of tumors and inflamed tissue via passive transport increases, resulting in higher drug concentration in malignancies. A few polymer micelles containing cancer medicines, SP1049C, NK105, NC-6004, NK911, and NK012, have been confirmed in clinical trials as of this date [108], and one such device is licensed for patients with breast cancer with paclitaxel Genexol-PM [71].

Dendrimers are highly ramified macromolecules that provide multiple functional groups for drug, target, and imaging agent attachment. Their profiles focus on different structural features [53] for absorption, delivery, metabolism, and elimination (ADME). A polyfunctional dendrimer device was used to record active localization of Folic acid, fluorescein imaging, and the supply of anticancer *in-vitro* methotrexate [84]. The use of biocompatible components and surface derivative treatments with PEGylation, acetylation, glycosylation, and several amino acids can boost the therapy index for cytotoxic medications employing dendrimers [18].

While many other nanoparticles are promising in cancer therapy, carbon nanotubes are one of the latest structures. Carbon nanotubes (CNTs) are a allotropic type of carbon that may be separated into single-walled carbon nanotubes (SWCNTs) and multi-walled carbon nanotubes (MWCNTs) on a multitude of sheets in concentration cylinders [88].

As carbon-hydrophobic nanotubes are very hydrophobic inside, they can quickly charge with water-insoluble drugs. The large surface area enables the functioning of the outer surfaces and allows for the accurate application of a specific cancer receptor and contrast agent [30]. Ultimately, the study of Buckminsterfullerene C60 (spherical molecule) and its derivatives is used to cure cancer [16]. Fullerene C60 can scavenge up to six electrons from reactive oxygen (ROS) molecules, making it an effective scavenger [81].

Nano-C60 fullerene nanocrystals have been shown to enhance the cytotoxicity of chemical therapeutic agents, allowing for future testing of Nano-C60 chemotherapy [107].

Prylutska et al. [82] investigated the Fullerene C60 complex with doxorubicin, finding that tumor volumes in the treated rats (C60+Dox) were 1.4 times lower than in the untreated rats (control group) [82]. Additionally, the C60+Dox complex is thought to be the mechanism of action on tumor cells and immunomodulating effects.

9.4 TARGETED DRUG DELIVERY MECHANISM

9.4.1 DELIVERY TO THE TUMOR SITE

Nanoparticles can be delivered to the tumor location in two ways: passively or actively. Nanoparticles penetrate tumor tissue via passive delivery due to an enhanced permeability and retention (EPR) effect. Angiogenesis that is both rapid and deficient is marked by increased permeability of blood vessels in the tumor. Inactive distribution, the modified nanoparticles bind selectively to express antigens or receptors in the cancer cells with molecules (ligands, antigens). Because of the solid ligand-receptor interactions expressed on cancer cells, the spread of nanoparticles to cancer cells may be decreased when healthy cells are circumvented [33].

9.4.2 DRUG LOADING

Based on nanoparticles morphology and chemistry, chemotherapeutic drugs can either be conjugated, entrapped, adsorbed, or encapsulated to them. Drug solubility in the nanoparticle polymeric matrix, molar mass, drug-delivery interaction, and the presence of end functional groups in either the drug or the matrix all have an impact on drug loading and entrapment [21].

9.4.3 *DRUG RELEASE*

To practice its pharmacological effects, one must extract an encapsulated substance from the nanoparticle. The nature of this product, its nanomaterial structure, and drug location inside nanoparticles was strongly determined by the kinetics of drug releases (i.e., if the medication is encapsulated in the nanoparticle matrix, adsorbed to the surface of the nanoparticle, or covalently attached to the surface of the particle, increased permeability of the tumor's blood vessels is indicative of rapid and unstable angiogenesis) [103]. Any nanoparticles can exhibit incremental or controlled release properties that provide various advantages as far as dosing frequency is concerned. Drug discharge pathways from the nanoparticles include diffusion, oxidation of polymers, and triggers. In the event of endogenous or environmental temperature, electromagnetic radiation, light, pH, oxidative stress ion power, echolocation, biochemical or chemical compounds, enzymes, and redox potential, [26, 89] the amount of drug released will increase stimulus sensitive nanoparticles [11, 89]. For polymer nanoparticles, the initial freezing of smaller particles is typically more resistant than larger particles; however, the faster *in vivo* degradation of larger particles appears to occur. The deterioration of polymers relies on the size and form of the polymer within the body and the polymer matrix's polymer MW. The degradation of high MW polymers is more likely than that of low MW polymers [48].

A physiologically suitable well-stirred medium, such as phosphate saline buffered at pH 7.4 and 37°C, is used to determine drug release kinetics. Sink requirements need to be preserved, indicating an appropriate amount of material used not to reach 20% of the drug's saturated concentration when 100% of the drug release has occurred in the medium [91]. Nanoparticles shall be applied to the media at zero, and medium samples were taken for the analysis duration at specified times. Depending on the drug release kinetics for the nanoformulation in question, a release analysis can take anywhere from hours to weeks. At the start of the experiment, a sample is taken at zero time (or, practically speaking, a few seconds after the nanoparticles have been added to the medium). When the proportion of prescriptions released is set to zero, the percentage of unencapsulated free medicines calculated by the encapsulation quality calculation should be quite close. The initial burst is usually released after 30 to 60 minutes (Unless the release kinetics are steady over many weeks, in which case a sample taken after one day could be a better indicator of burst release). Individual samples of the release analysis can be treated as previously indicated for encapsulation quality evaluation; for example, nanoparticles can be separated at each time by high-speed centrifugation or filtration. The drug concentration released by the nanoparticles at each stage would be calculated using a validated HPLC technique (or other suitable methods applicable to the drug of interest). To research opioid release, dialysis bags may also be used. The suspension of the nanopharmaceuticals shall, in this case, be mixed up with the release medium if the suspension is put inside the dialysis tube or a floating dialysis system with an adequate pore size to avoid that the nanoparticles will be able to leave the dialysis tubing or device. Still, the released molecules will pass via the pores at the prescribed periods from the external media. The pores are also identified with molecular weight cut-offs (MWCO). The 12–14 kDa MWCO, for example, is equal to a 2.4 nm pore size [86]. Data for drug launches are also appropriate for a descriptive model.

The Higuchi formula, which compares the proportion of medication released to the square root of time (as indicated by the equation below), is most common.

$$\text{MMtMM halide} = \text{KK halide}$$

where; *Mt* is the mass of the drug released at time *t*, *M* halide is the total accumulated mass of the drug released at time *t*, and constant *K* is dependent on experimental variables [36].

When nanoparticles are attached to the surface of cancer cells, the medication is released by receptor-mediated endocytosis inside the cell. The release of drugs is based on nanoparticle size, drug solubility, surface corrosion, and nanoparticle degradation [100]. Another main factor in drug release is the micro-environment of the tumor. Several studies have documented that "smart" nanoparticles are engineered and produced to release loaded medicines only on the tumor and not in healthy cells to avoid target cytotoxicity [21].

9.5 TRANSPORT OF NANOPARTICLES

Externally administered NP can bind to cellular plasma membranes and enter the cell via passive diffusion or endocytosis. Extracellular components interact with invaginated cell membranes during endocytosis, which is a multi-stage process. These invaginations are subsequently squeezed into endosomes or phagosomes (depending on the transport pathway) and delivered to the cell's appropriate compartments.

9.5.1 PASSIVE TRANSPORT

Where a lipophilic NP crosses the lipid membrane, the process of passive transport will occur. Many other factors related to NP characteristics will rely on diffusion. The amount of passive dispersion of molecules through the lipophilic plasma membrane of various cells [105] might be influenced by NP surface charge. NPs were shown to promote a 3- to 5-fold increase in absorption across a lipid bilayer containing cholesterol. A study focused on the effects of NPs on tight junctions has clarified the diffusion of NPs into cells. The effect of NP formulation on the opening of tight junctions [32].

NP treatment has been shown to result in significantly expanded apical membrane space, promoting paracellular insulin transport. These findings imply that using NPs to transport medicines across cellular membranes via diffusion processes may benefit existing drug delivery methods.

9.5.2 ACTIVE TRANSPORT

The medication preferentially accumulates in the target tissue, organ, or cells due to the chemical bonding of a targeting substance that intimately interacts with the anti-genetic agents (or receptors) present in the target tissue. The use of a targeted motor not only minimizes harmful side effects by delivering the drug to a specific location of the action, but it also promotes cellular drug absorption via receptor-mediated endocytosis, an active mechanism with a slightly lower plasma membrane concentration than basic endocytosis. Monoclonal antibodies are also used for aggressive targeting. Today, however, it's preferable to goal folate. Folate is a low vitamin of molecular weight that is needed for eukaryotic cells, with its conjugates able to provide pathological cell pathologies with several drugs or imaging agents without harming normal tissue [101].

9.5.3 ENDOCYTOSIS

Pinocytosis (macropinocytosis and receptor-mediated endocytosis) and phagocytosis are the two main types of endocytosis [31]. Receptor-mediated endocytosis involves swallowing the receptor into a coated pit in combination with the corresponding ligand. The most frequently associated receptor-mediated endocytic pathway is clathrin-protein-coated pits, commonly known as the 'classical' kind of receptor-mediated endocytosis. The clathrin-coated pits are primarily responsible for absorbing vital nutrients such as cholesterol and iron [28].

Adipocytes, muscle cells, and epithelial cells all have caveolae, caveolin structural proteins that serve as membrane proteins [25]. Caveolae-induced endocytosis works by engulfing molecules that stick to the caveolae's surfaces. Unlike clathrin-dependent endocytosis, caveolae proteins do not dissociate from the vesicle after endocytosis. Processes of macropinocytosis cause larger solute macromolecules to be engulfed and can measure up to 5 μm in diameter.

Phagocytosis typically refers to the cellular absorption of large nanoparticles absorbed and transported by phagosomes, such as microorganisms and dead cells. It's crucial to note that phagocytosis is often limited to cultured mammalian cells like macrophages, which play a critical role in detecting and removing foreign substances.

9.6 APPLICATIONS OF NANOPARTICLES IN CANCER THERAPY

One of the most critical outcomes of nanoparticle-based drug delivery systems promises to be cancer-focused therapy. New distribution technologies allow the passive aggregation of nanoparticles intravenously injected from the leaky vasculature (20–150 nm). Nanostructured vehicles can enter tumors due to the interruptive and pooling character of the cancer microvasculature, which has larger pores (100–1,000 nm in diameter). E.g., nanoparticles have minimal diffusion distances from the vasculature to adjoining parenchyma [63].

Additionally, nanoparticles more significant than the intercellular gap of healthy tissue but smaller than cancer cell pores can be targeted. While abnormal and improperly perfused cancer cells can, to any degree, hinder the efficacy of nanoparticles, long-term formulations and the location of cancers are feasible [3]. In comparison to medicine alone, many formulations have a 300-fold increase in the region under the curve and a lower clearance and delivery volume [40].

Liposomes and polymeric micelle are other effective ways to combat cancer by nanostructured materials. Medication release can be caused by liposomes accumulated in enzyme-mediated liposome destabilization found in an interstitial position [24, 52]. Thus vaghilinha such nanostructures can be administered by passive delivery procedures to cancer cells and associated extracellular components. One of the most interesting tumor-associated tumor-specific identification and imagery methods is metalloproteinase 7, with a fluorogenic substratum dependent on dendrimer [4]. Successful targeting has been shown in some research. On the surface of long-circulating macrophage-evading nanoparticles, folic acid was attached to the poly(ethylene glycol) chains [2, 9, 102]. In comparison to monoclonal antibodies, folate-based targeting offers significant advantages. Folate is nonimmunogenic nanoparticles, and folate is quickly internalized dormant cells that circumvent multi-drug-efflux pumps in cancer cells.

Subjectively, targeting cancer blood arteries with customized medicines would help with early identification and treatment of metastatic cancers, as well as resistant tumor therapy and diagnostics [49, 68]. The idea is also intuitively appealing. Molecular biological advances have recently revealed possible targets in cancer vasculature, including integrins with cancer angiogenesis functions [8]. Integrins are bound to the sequence of RGD motifs (Arg-Gly-Asp) and built into cyclic peptides, including RGD-4C [6]. Compared to doxorubicin alone, the combination of RGD-4C and doxorubicin increases chemotherapeutic effectiveness while reducing cardiac toxicity and toxicity [5]. The HWGF peptide motif and the NGR hexapeptide are two examples of specific recognition sequences that combat human and mouse tumor cells [23]. Another anti-angiogenic technique for treating solid tumors is cationic nanoparticles complex with specific therapeutic individuals [47]. Similar approaches to the characterization of early angiogenesis [90] of paramagnetic nanoparticles with site-specific magnetic resonance (MR) have been extended. Additional techniques involve utilizing ornamental liposome nanoparticles to slow tumor blood flow and target peptide-coated quantum dot blood and lymphatic channels in cancer [56, 75].

Nanoparticle-based therapy was widely utilized in animal models to treat a variety of cancers. In the case of a PLGA nanoparticle armed with an HER-2-positive toxin in the breast, Chen et al. [17] researched therapeutic efficiency. Crosslinked serum albumins [95] and quantum-dot-laden chitosan nanoparticles containing absorbed siRNA [97] targeting breast cancer also revealed positive HER-2 results.

Some experimental targeting approaches have recently been proposed for localized and metastatic colorectal cancer treatment [38]. In addition to the concomitant bacterial heat-stable enterotoxins, gold nanoparticles (AuNPs) absorbed near-infrared light were used by investigators to heat-based targeted cancer cells [106]. The use of nanoscale polymeric micelles with chemotherapeutic medicines has been studied for the passive targeting of colorectal cancer [58].

Besides, nanostructured polymer conjugates are used to enhance colorectal cancer treatment [51]. Nanoparticles containing targeted aptamers to destroy small-cell lung cancer cells were recently produced [14]. Targeted nanoparticles of lung cancer are used with EGF ligand [104]. Anti-sensory oligodeoxynucleotides or siRNAs have also been studied as targeted nano-scaling carriers for lung cancer therapy [61].

One of the most studied targets for prostate cancer therapy is prostate-specific membrane antigen (PSMA). The objective potential of pharmaceutical/aptamer conjugates and polymer stealth nanoparticles to treat chemotherapeutic drugs for prostate cancer has been studied at Langer's group at MIT [35]. The *in vivo* trials of aptamer-compound nanoparticles showed positive findings for prostate cancer treatment [20]. Folateligands is another path to a targeted prostate tumor by PSMA [46].

Efficient nanoparticles transmission to cancer has resulted in more methods for treating this disorder (Figure 9.1). Therapeutic devices focused on Nanoparticles are capable of attacking multiple parts of the tumor with various targeting moieties to prevent such drug resistance issues.

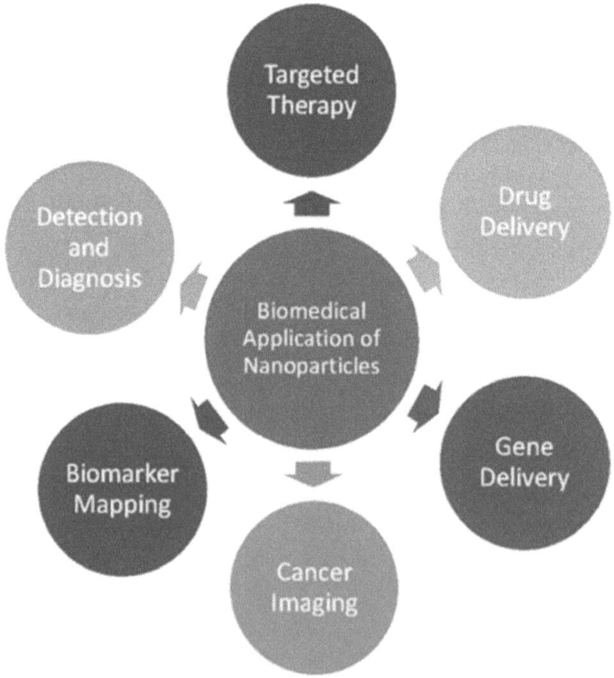

FIGURE 9.1 Biomedical application of nanoparticles.

9.7 ADVANTAGE OF USING NANOPARTICLE FOR DRUG DELIVERY

Nanoparticle-based drug delivery platforms have many benefits over traditional drug delivery platforms, making them a good candidate for DDS. Encapsulation of therapeutics into nanocarriers protects them against degradation and prolongs their circulation. The high surface area of nanoparticles provides an opportunity to load higher drug amounts into them [41].

Similarly, the ability for controlled release activity and active targeting allows attention drugs selectively at the cancer site and thereby reduces cytotoxicity of the off-target. In this way, a nanoparticle improves drug efficacy, reduces unwanted cytotoxicity, and reduces the cost of dose formulation [12]. Some of the other advantages are shown in Figure 9.2.

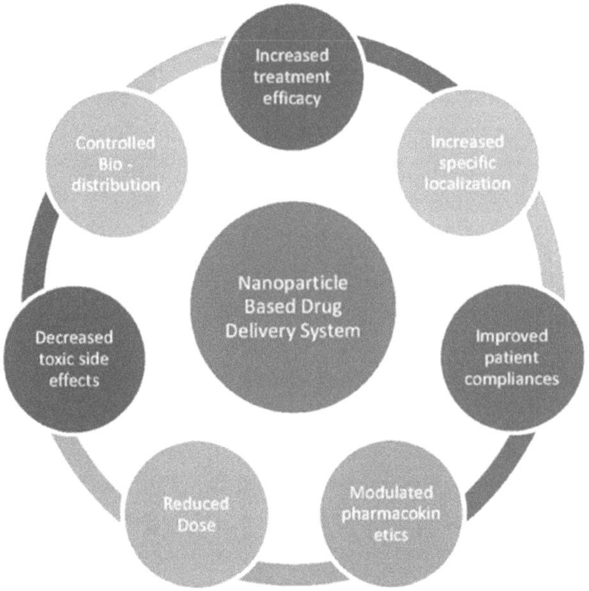

FIGURE 9.2 Advantages of using nanoparticle-based drug delivery.

Apart from this, nanoparticles have the following advantages:

- Better/Improved cancer diagnosis and treatment application;
- The entry into tissue at the molecular level;
- Feasibility to programed nanoparticles for recognizing cancerous cells;

- Increased drug localization and cellular uptakes of drug;
- Direct/selective targeting of the drug by active and passive targeting to a cancerous cell;
- The larger the surface area of changeable electrical, magnetic, optical, and biological characteristics;
- Drug delivery that is selective and effective while avoiding interactions with healthy cells;
- Assisting therapeutic drugs in crossing biological barriers, mediating molecular interactions, and detecting molecular changes.

9.8 COMMERCIALIZED CANCER THERAPEUTIC NANOPARTICLES PRODUCTS AVAILABLE IN THE MARKET

The number of commercial medicinal drugs based on nanoparticles has grown dramatically over the last 20 years. There are currently over 150 companies around the world that have developed nanoscale therapy [66]. Thus far, gross revenue of more than $5.4 billion has been licensed for clinical purposes for over 20 pharmaceutical drugs based on nanoparticles [98]. Liposome pharmaceutical products and polymer-drug conjugates constitute the two major classes, which make up about 80% of the total.

Liposomes are spherical, bilayered lipid complexes composed of amphiphilic molecules which are natural or synthetic [72]. The FDA of the United States authorized the first liposomal formula for AIDS linked with Kaposi's sarcoma in 1995 for nanoparticle-related medicines. Another commonly explored medium for the provision of nanoparticles on the market today is the provision of polymer-drug conjugates [101]. Small molecule therapeutic medicines, particularly anticancer chemotherapeutic drugs, have a limited circulation length and non-specific targeting. On the other hand, it would minimize unwanted side effects and increase circulation times by conjugating small molecular drugs to polymeric nanostructures. This technique improves the passive supply of leaky blood vessel drugs to the tissue, including tumors and atherosclerotic plaques [31].

Although several polymeric nanoparticles have been proposed as drug transporters, only a few have been authorized for clinical trials with linear design. The first introduction into clinical use was the introduction in the 1990s of PEG combined nanoparticles. The technique will increase plasma stability, solubility, and immunogenicity of the medicinal substance.

Additional polymer medication conjugates are 5–200 nm hydrodynamic in size and known as drug carriers.

Abraxane, which contains 130 nm of packed paclitaxel albumin, was approved by the FDA in 2005 to treat breast cancer. Growing the length of the condition and raising the mean survival rate of women with breast cancer by about double the therapeutic response rate [28].

In addition to already approved nanoparticle-based drugs, several other platforms in the field of clinical application for therapeutic drug enhancement, such as various liposomes [25], polymer conjugates [41], AuNPs [66], calcium nanoparticles [98], dendrimers [12], and QDs are currently in different stages of clinical and preclinical development. Ostim® is a hydroxyapatite pulp commercially available in the bone tissue field [72]. More and more medicinal drugs based on nanoparticles will become commercially available on the market with ongoing efforts in research and development. Here are some nanoparticle-based cancer therapeutics available in the market (Table 9.1).

9.9 FUTURE OF NANOCARRIER-BASED CANCER THERAPEUTIC DRUG DELIVERY SYSTEM

One of today's most exciting research areas in nanomedicine, Over the previous two decades, extensive research has resulted in the filing of 1,500 patents and the completion of hundreds of clinical trials [73]. As discussed in the above section, cancer seems to be the best case of diseases that have benefited from non-medical technology for their diagnosis and treatment. Nanomedicine and the nano-drug delivery mechanism will continue to be a trend that will be the subject of research and development for decades to come, with the use of numerous nanoparticles to distribute the exact amount of drug to affect cells such as cancer/tumor cells without disrupting the physiology of normal cells.

Since some nanoparticles are measured in nanometers and others are measured in sub-micrometers, the representations of nanoparticles in this relationship are not standardized (over 100 nm). It will conduct more research on materials with more standardized homogeneity and loading and release capabilities in the future. This analysis also discussed essential improvements in the analytical application of metal-based nanoparticles. The diagnostic and medicinal use of these metals, like gold and silver, could lead to more extensive use of nanomedicine in the future. One of the critical

TABLE 9.1 Commercially Available Nanoparticle Used for Cancer Drug Delivery

Type of Nanoparticle	Name of the Drug	Company/Institute	Active Ingredient	Admiration Route	Type of Targeted Cancer	Year of FDA Approval
PEGylated Liposomes	Doxil	Johnson and Johnson, USA	Doxorubicin hydrochloride	Injection I.V.	Ovarian cancer, AIDS related Kaposi's sarcoma	1995
Albumin bound paclitaxel nanospheres	Abraxane	AbraxisBiosciencs	Paclitaxel	Injectable suspension (infusion)	Breast cancer, lung cancer, pancreatic cancer	2005
Liposome encapsulated doxorubicin	Myocet	Teva	Doxorubicin	Powder, dispersion for infusion	Breast cancer	2000
Liposome encapsulated daunorubicin	DaunoXome	Galen, UK	Daunorubicin	Injection	AIDS related Kaposi's sarcoma	1996
Liposome encapsulated vincristine	Marqulbo	Talon therapeutics	Vincristine	Injection I.V.	Acute lymphoblastic leukemia	2012
Polymer micelle	Genexol-PM	Celgene	Paclitaxel	Injection I.V.	Breast cancer, pancreatic cancer	2005, 2012, 2013
PEGylated liposomes	Caelyx	Janssen	Doxorubicin hydrochloride	Injection I.V.	Multiple myeloma, AIDS-related Kaposi's sarcoma, ovarian cancer,	2005, 1995, 2008
Liposome encapsulated cytarabine	DepoCyt	Sigma-Tau	Cytarabine	Injection	lymphomatous meningitis	1996
Liposome encapsulated vincristine	Marqibo	Onco TCS	Vincristine sulfate	Injection	Acute lymphoblastic leukemia	2012
Liposome encapsulated irinotecan	Onivyde	Merrimack	Irinotecan	Injection	Pancreatic cancer	2015

TABLE 9.1 *(Continued)*

Type of Nanoparticle	Name of the Drug	Company/Institute	Active Ingredient	Admiration Route	Type of Targeted Cancer	Year of FDA Approval
Albumin bound paclitaxel nanoparticles	Abraxan	Celgene	Paclitaxel	Injection	Breast cancer, pancreatic cancer, NSCLC,	2005, 2013, 2012
Polymer-protein conjugate	Oncaspar	Enzon pharmaceuticals	PEGylated L-Asparaginase	Injection	Acute lymphoblastic leukemia	1994
Polymer-protein conjugate	Neulasta	Amgen	PEGylated granulocyte colony-stimulating factor (GCSF) protein	Injection	Neutropenia chemotherapy-induced	2002
Leuprolide acetate polymer	Eligard	Tolmar	Poly(DL-lactide-coglycolide (PLGH)	Injection	Prostate cancer	2002
Random copolymer	Copaxone	Teva	GlatiramerAcetate (L-tyrosine, L-lysine, L-alanine, and L-glutamate)	Injection	Multiple sclerosis	1996

enthusiasts for this is AuNPs, absorbed in soft tumor tissues and make the tumor susceptible to radiation.

While the potential opportunities for nanomedicine and nano-drug distribution are overwhelmingly understandable, their actual effect in the healthcare sector remains incredibly minimal, even in cancer treatment/diagnosis. It is a recent branch of science with two decades of research and several essential characteristics that are not known. Fundamental indications of sick tissues, including primary biological markers that allow total concentration without altering the normal cell phase, will be one of the significant study fields in the future. Finally, with our growing understanding of the molecular disease, the application of nanomedicine progresses or represents a similar marker nanomaterial-subcellular scale to create avenues for new diagnosis/therapeutic applications. Therefore, the potential understanding of disease's molecular signatures will lead to advancement in applications of nanomedicine. In addition to what we detailed in that assessment utilizing known nanoprobes and nanotheranostics devices, more research will be required for the broader use of nanomedicine.

The idea of controlled release of specific medicines at threatened sites and technology for assessing these events, the influence of therapeutic goods on tissues/cells, and theoretical mathematical prediction models are all still being worked out. Many stories in the fields of nanomedicine concentrate on biomaterials and formulation trials, which seem to be the early stages in the application of biomedicine. Future helpful evidence such as opioid treatment and medical trials, animal studies, and multidisciplinary studies need considerable time and testing funding.

With the growing movement to search for more reliable drugs and diagnostics, it seems evident in the future that nanomedicine and nano-drug delivery technologies are multi-centered and more intelligent. The fundamental perspective of creating nanorobot (and nanodevice) tissue diagnostic and healing mechanisms with comprehensive external control mechanisms has been quite positive. It was not yet a reality and appears to be a futuristic science that humans could soon do. However, as with their advantages, the possible risk for human and environmental nanomedicines still needs to be studied on a long-term basis.

It can also effectively analyze people and the environment for modern nanomaterials' potential acute or chronic toxicity. As nanomedicines become more common, they will be an inexpensive field of research that would require further research, finally, along with developments in nanomedicine technologies, the regulation of nanomedicines.

9.10 CHALLENGES ASSOCIATED WITH NANOCARRIER-BASED DRUG DELIVERY

While nanotechnology in the drug supply has proved to be popular on the market, not all techniques have experienced the same popularity as other nano-drug products. The creation of modern nanomaterials entails problems that must solve. However, some of the challenges faced have, and still do, to enhance properties like extensive circulation in the blood, improved functional region, drug safety against degradation, site-specific targeting, and crossings of bio-barrier barriers by changing physicochemical characteristics of nanomaterials.

The large-scale processing of nanomaterials is yet another task of R&D for drug entry. Laboratory or pilot innovations still need to be extended to subsequent marketing. Several nano-drug delivery methods may not be scalable due to the shape and process of processing and the high cost of materials employed. The problems of size are low nanomaterials concentration, aggregation, and the chemical method – nanomaterials are more straightforward to change on a lab-scale to increase their quality than on a wide scale. It is also a struggle to preserve the large scale and structure of nanomaterials.

Despite the number of nano-drug-supply patents, marketing is only in its early stages. It is in part because university scientists perform most nano-drug distribution analysis trials. Therefore, it must be step up collaborations with pharmaceutical firms for these technologies to enter the market. Sadly, most large pharmaceutical companies currently have no goals for nanotech due to the absence of regulatory standards and the difficulties involved with scaling up. However, more pharmaceutical companies are to take nano-medicinal drugs to contend favorably with the expiry of more patents and a drop on the market.

Advances in nano-drug distribution technologies are also emerging regulatory obstacles. Legislation that considers the pharmacokinetic and physicochemical characteristics of nano-drug products, which differ from regular drug products, is becoming more critical.

The USFDA and EMEA have taken the lead in recognizing some potential issues in terms of research and policy. Also, a TC 229 scientific committee has now been established to establish guidelines for terminology and vocabulary in the sector of nanotechnology, steps, and characterization, and, among other standards, for health, protection, and the ecosystem. The standards are still being established.

9.10.1 SAFETY ISSUES

Concerns regarding the protection of nanotech in human beings are emerging with the increased R&D on nano-drug distribution. Because some nanomaterials degrade while others do not, the by-products have severe side effects.

Materials healthy at the macroscale might not be nanoscale because they may modify physicochemical properties at the nanoscale. These nanomaterials cannot be entirely extracted from the body and can have many potential effects [44]. It should consider nanomaterials' protection and future effects for patients and the whole production and disposal process.

The manufacture and manufacturing of nanomaterials will not be subject to standard safety precautions in pharmaceutical factories. It should also take additional steps to avoid the increased harmful impacts of nanomaterials on the environment. Although one of the stated advantages of nanotechnology is that it requires fewer items to be created than bulk production, it is uncertain whether this will be the case because effective marketing would be costly. The public is still hesitant to accept nanotechnology because documented safety standards are not available. Yet nano-drug supplies are a trend that cannot neglect in terms of these difficulties to resolve challenges over time.

Apart from these, there are some disadvantages of using nanocarrier based drug delivery:

- Dynamic changes in cancer cells;
- Lack of drug solubility;
- Lack of selectivity;
- Nonspecific targeting of conventional delivery;
- Drug resistance;
- The small amount of drug reaches the targeted cancer cells;
- Poor targeting of heterogenic tumors;
- Serious side effects;
- The inability of the medication to reach the tumor's core, leading to less effective treatment with a lower dose and a worse survival rate.

9.11 ETHICAL ISSUES RELATED TO NANOPARTICLE-BASED CANCER DRUG DELIVERY

If their efficacy is established, the usage of NPs in medicine and bioscience will increase. Medical researchers may utilize NPs to analyze and explain therapy efficacy and drug administration observations using various techniques. The use of NPs is also investigated by researchers in several medical ways, from imaging to skin grafting [55]. The potential for NP-based medical products is nearly infinite, and our present understanding of the mechanical and molecular characteristics that make Nanoparticle function can only be tested. For a variety of medical, pharmaceutical products, NP medication distribution must be included over and above existing prescription and medical products. The pharmaceutical industry is now researching changed methods of supply of innovative and currently licensed medications to offset exponential costs in creating and preventing new drugs [59]. The potency and clinical use of different medications have been successfully improved with product design and distribution advances. The current state of knowledge about NP drug delivery designs is intended to promote commercial use, boost economic growth, and compensate for the economic loss associated with the development of new drugs. Companies will alter the efficacy and functionality of existing common medicines used in many disease-based therapies thanks to these new supply systems. As a result, the use of NP technology should enhance existing medication development, promote targeted new drug research, and boost economic growth. It should greatly expand the use of NP technology in manufacturing sectors.

While seldom discussed in such papers, it is crucial to note that the expense of manufacture and model scale-up for reformulated NP designs is a significant impediment to any progression of NP drugs.

Although examples of the use of NPs as new treatment options in the future are encouraging, most formulations have struggled to advance beyond preclinical trials based on the high cost of development and reproducibility of the NP formulation. Industrial production capabilities have thankfully improved due to the launch of new, novel FDA-regulated NP-based medicines. The NP drug products currently on the market have opened the way to a more straightforward model size transformation for several potential NP drug products. Unfortunately, not all NP systems have been approved on a mass-market basis, leaving scale-up and mass production paths capable of selecting NP systems.

For a complete understanding of the most successful administrative pathways, it must illustrate further study into various avenues of NP distribution. Presently, the primary subject of research of NP-reformulated medications is oral, I.V. or distribution transdermal. Researches on the efficient methods to treat the NP formulated medication must be considered to accurately interpret the consequences of the current drug delivery mechanism of NPs. NPs can modify different pharmaceutical properties that influence the conventional method of supply. To calculate the impact of NP drug reformulation on availability correctly, we must illustrate different modes of administration. For instance, PLGA is known to provide NP-encapsulated medication with hydrophobic properties depending on the lactic acid content of the medicines [15]. Changes in the lipophilic properties of drugs will potentially affect the best way of delivery of narcotics.

As a result, the effects of NP encapsulated medications and the influence of NP encapsulated drugs in various modes of distribution must be highlighted, dependent on the researchers chosen to disseminate them. Additional clarity in the best possible way to ensure the application of NPs in newly formulated medicinal drugs is improved.

The benefits of NP formulated approaches to renal problems shown in this analysis help establish effective alternate drug distribution to quench and eliminate significant side effects of medication. Additional experiments are also required to explain their advantages and the general best practice for NP technology. Exact pathways are currently not well known for minimizing kidney side effects observed with NP-trapped medications. Ultimately, recognizing Pharmacodynamics and pharmacokinetic implications relational to nephrotoxicity and onsets of renal side effects is warranted for further investigations of the mechanical impact of NP architecture on drug excretion, retinal aggregation, and cell absorption. Renal problems are also common in NSAID use. Surprisingly, some research examines the formulations of NSAID NP and the effect on renal function has been carried out. More experiments on NSAID-induced nephrotoxicity are strongly warranted to understand the successful impact of NP formulations.

9.12 SUMMARY

Recent advancements in nanomaterial and nanotechnology have created a new baseline in life science research, including cancer research or molecular medicine. Every segment of biological sciences is dependent on nanomaterials, either its general molecular laboratory practice or baseline surgery. Nanotechnology has prepared the path for therapeutic systems as a new drug delivery mechanism. Nanomedicines include liposomes, micelles, carbon nanotubes, dendrimers, polymer, and magnet nanoparticles. Nanocarriers have been researched to detect early-stage cancers and track molecular expressions. Drug supply has been highly affected by nanotechnology, and it is a significant medical field. The medication is paired with the nanocarrier, and the targeted approach is critical for particular therapy, says Dr. Andrew Boulden. Nanotechnology provides a new forum to build tumor-targeted drug carriers through modern passive or active targeted mechanisms to deliver therapeutic drugs to tumor sites.

ACKNOWLEDGMENTS

I'd like to express my gratitude to IIT Delhi for supplying papers and research journals and support and helpful suggestions in completing this chapter. Also, I would want to Thank Mr. Rahul Sharma, Mr. Sachin Kumar, Ms. Anchal (University of Delhi, New Delhi), Ms. Akansha Masooma, Ms. Salini Nair, Ms. Rama Singh (Dr. Reddy's Laboratories, Hyderabad), Mrs. Haritha Sivadas (Hyde E+C, Hyderabad), Mr. Arun Kumar A. (Sartorius Stedim India Limited, Bengaluru), Mr. Vipul Saxena (Stelis Biopharma, Bengaluru) and Ms. Kanupriya R. Daga (University of Georgia, USA) for their love, encouragement, and support. I express my sincere gratitude to Dr. M. Nataraj (Sardar Patel University, Gujarat, India), Prof. R. B. Subramanian (Sardar Patel University, Gujarat, India), Prof. B. Rainer (MHH, Hanover, Germany), Prof. Klaus Futterer (University of Birmingham, Birmingham, UK) and Dr. Sourav Singh Deo (University of Delhi, New Delhi, India) for their intellectual guidance and inspiration for the successful completion of this chapter. With heartfelt privilege, For their love and motivation, I am grateful to my parents, grandparents, brother, and cousins.

KEYWORDS

- **cancer biology**
- **cancer therapy**
- **chemotherapy**
- **drug delivery**
- **drug targeting**
- **future vaccine**
- **nanocarrier**
- **nanoparticles**
- **tumor targeting**

REFERENCES

1. Ai, J., Biazar, E., Jafarpour, M., Montazeri, M., Majdi, A., Aminifard, S., Zafari, M., et al., (2011). Nanotoxicology and nanoparticle safety in biomedical designs. *International Journal of Nanomedicine, 6*, 1117–1127.
2. Allen, T. M., (2002). Ligand-targeted therapeutics in anticancer therapy. *Nature Reviews Cancer, 2*(10), 750–763.
3. Allen, T. M., & Cullis, P. R., (2004). Drug delivery systems: Entering the mainstream. *Science, 303*(5665), 1818–1822.
4. Andresen, T. L., Davidsen, J., Begtrup, M., Mouritsen, O. G., & Jørgensen, K., (2004). Enzymatic release of antitumor ether lipids by specific phospholipase A2 activation of liposome-forming prodrugs. *Journal of Medicinal Chemistry, 47*(7), 1694–1703.
5. Arap, W., Pasqualini, R., & Ruoslahti, E., (1998). Cancer treatment by targeted drug delivery to tumor vasculature in a mouse model. *Science, 279*(5349), 377–380.
6. Assa-Munt, N., Jia, X., Laakkonen, P., & Ruoslahti, E., (2001). Solution structures and integrin binding activities of an RGD peptide with two isomers. *Biochemistry, 40*(8), 2373–2378.
7. Baltimore, M. D., (1950). Dendritic cell-based xenoantigen vaccination for prostate cancer immunotherapy. *Journal of Immunology, 167*(12), 7150–7156.
8. Baluk, P.,Hashizume, H., & McDonald, D., (2005). Cellular abnormalities of blood vessels as targets in cancer. *Current Opinion in Genetics & Development, 15*, 102–111.
9. Barenholz, Y., (2001). Liposome application: Problems and prospects. *Current Opinion in Colloid & Interface Science, 6*, 66–77.
10. Baudino, T. A., (2015). Targeted cancer therapy: The next generation of cancer treatment. *Current Drug Discovery of Technology, 12*(1), 3–20.
11. Bhadra, D., Bhadra, S., Jain, S., & Jain, N. K., (2003). A PEGylated dendritic nanoparticulate carrier of fluorouracil. *International Journal of Pharmaceutics, 257*(1, 2), 111–124.
12. Bissett, D., Cassidy, J., De Bono, J.,Muirhead, F., Main, M., Robson, L., Fraier, D., Magne, M. L., Pellizzoni, C., & Porro, M. G., (2004). Phase I and pharmacokinetic (PK) study of MAG-CPT (PNU 166148): A polymeric derivative of camptothecin (CPT). *British Journal of Cancer, 91*, 50–55.
13. Chen, H.,Gao, J., Lu, Y., Kou, G., Zhang, H., Fan, L.,Guo, Y., & Zhong, Y., (2008). Preparation and characterization of PE38KDEL-loaded anti-HER2 nanoparticles for targeted cancer therapy. *Journal of Controlled Release, 128*, 209–216.
14. Chen, H., Medley, C.,Sefah, K.,Shangguan, D., Tang, Z.,Meng, L., & Tan, W., (2008). Molecular recognition of small-cell lung cancer cells using aptamers. *ChemMedChem, 3*, 991–1001.
15. Chen, S., Pieper, R., Webster, D. C., & Singh, J., (2005). Triblock copolymers: Synthesis, characterization, and delivery of a model protein. *International of Journal Pharmacy, 288*(2), 207–218.
16. Chen, Z., Mao, R., & Liu, Y., (2012). Fullerenes for cancer diagnosis and therapy: Preparation, biological and clinical perspectives. *Current Drug Metabolism, 13*(8), 1035–1045.
17. Chen, Guizhi, Jin, & Xiaoyuan, (2017). Rethinking cancer nanotheranostics. *Nature Reviews Materials, 2*, 1–18.

18. Cheng, Y., & Zhao, L., (2011). Design of biocompatible dendrimers for cancer diagnosis and therapy: Current status and future perspectives. *Chemical Society Reviews, 40*(5), 2673–2703.
19. Chidambaram, M., Manavalan, R., & Kathiresan, K., (2011). Nanotherapeutics to overcome conventional cancer chemotherapy limitations. *Journal of Pharmacy & Pharmaceutical Sciences, 14*(1), 67–77.
20. Chu, Marks, J., Lavery, L., Faulkner, S., Rosenblum, M., Ellington, A. D., & Levy, M., (2006). Aptamer: Toxin conjugates that specifically target prostate tumor cells. *Cancer Research, 66*, 5989–5992.
21. Chu, K. S., Schorzman, A. N., Finniss, M. C., Bowerman, C. J., Peng, L., Luft, J. C., & DeSimone, J. M., (2013). Nanoparticle drug loading as a design parameter to improve docetaxel pharmacokinetics and efficacy. *Biomaterials, 34*(33), 742–749.
22. Cole, A. J., Yang, V. C., & David, A. E., (2011). Cancer theranostics: The rise of targeted magnetic nanoparticles. *Trends Biotechnology, 29*, 323–332.
23. Curnis, F., Sacchi, A., & Corti, A., (2002). Improving chemotherapeutic drug penetration in tumors by vascular targeting and barrier alteration. *Journal of Clinical Investigation, 110*, 475–482.
24. Davidsen, J., Jørgensen, K., Andresen, T., & Mouritsen, O., (2003). Secreted phospholipase A2 as a new enzymatic trigger mechanism for localized liposomal drug release and absorption in diseased tissue. *Biochimica et Biophysica Acta (BBA)-Biomembranes, 1609*, 95–101.
25. Davis, F., (2002). The origin of pegnology. *Advanced Drug Delivery Reviews, 54*, 457–458.
26. Davis, M. E., Chen, Z., & Shin, D. M., (2010). Nanoparticle therapeutics: An emerging treatment modality for cancer. *Nanoscience and Technology: A Collection of Reviews from Nature Journals, 7*(9), 239–250.
27. De Schryver, A., Huys, J., & Vakaet, L., (1992). Systemic treatment of early breast cancer by hormonal, cytotoxic, or immune therapy: 133 randomized trials involving 31000 recurrences and 24000 deaths among 75000 women: 1. *Lancet, 339*(8784), 1–15.
28. Deguchi, J., Aikawa, M., Tung, C., Aikawa, E., Kim, D., Ntziachristos, V., Weissleder, R., & Libby, P., (2006). Inflammation in atherosclerosis: Visualizing matrix metalloproteinase action in macrophages *in vivo*. *Circulation, 114*, 55–62.
29. Des Rieux, A., Fievez, V., Garinot, M., Schneider, Y. J., & Preat, V., (2006). Nanoparticles as potential oral delivery systems of proteins and vaccines: A mechanistic approach. *Journal of Controlled Release, 116*(1), 1–27.
30. Dinesh, B., Bianco, A., & Ménard-Moyon, C., (2016). Designing multimodal carbon nanotubes by covalent multi-functionalization. *Nanoscale, 8*(44), 18596–18611.
31. Duncan, R., (2006). Polymer conjugates as anticancer nanomedicines. *Nature Reviews Cancer, 6*, 688–701.
32. Duncan, R., & Gaspar, R., (2011). Nanomedicine(s) under the microscope. *Molecular Pharmaceutics, 8*, 2101–2141.
33. Duncan, R., & Richardson, S. C., (2012). Endocytosis and intracellular trafficking as gateway for nanomedicine delivery: Opportunities and challenges. *Molecular Pharmaceutics, 9*(9), 2380–2402.
34. Emerich, D., & Thanos, C., (2003). Nanotechnology and medicine: The here and now and future prospects. *Expert Opinion on Biological Therapy, 3*, 655–663.

35. Farokhzad, O., Jon, S., Khademhosseini, A., Tran, T., LaVan, D., & Langer, R., (2004). Nanoparticle-aptamer bioconjugates: A new approach for targeting prostate cancer cells. *Cancer Research, 64*, 7668–7672.
36. Fenart, L., Casanova, A., Casanova, A., Dehouck, B., Duhem, C., Slupek, S., Cecchelli, R., & Betbeder, D., (1999). Evaluation of effect of charge and lipid coating on ability of 60-nm nanoparticles to cross an *in vitro* model of the blood brain barrier. *Journal of Pharmacology and Experimental Therapeutics, 291*(3), 1017–1022.
37. Ferrari, M., (2005). Cancer nanotechnology: Opportunities and challenges. *Nature Reviews Cancer, 5*(3), 161–171.
38. Fortina, P., Kricka, L., Graves, D., Park, J., Hyslop, T., Halas, N., Surrey, S., & Waldman, S. A., (2007). Applications of nanoparticles to diagnostics and therapeutics in colorectal cancer. *Trends in Biotechnology, 25*, 145–152.
39. Frangioni, J. V., (2008). New technologies for human cancer imaging. *Journal of Clinical Oncology, 26*(24), 4012–4021.
40. Gabizon, A., Shmeeda, H., Horowitz, A., & Zalipsky, S., (2004). Tumor cell targeting of liposome-entrapped drugs with phospholipid-anchored folic acid-PEG conjugates. *Advanced Drug Delivery Reviews, 56*, 1177–1192.
41. Gordon, A., Fleagle, J., Guthrie, D., Parkin, D., Gore, M., & Lacave, A., (2001). Recurrent epithelial ovarian carcinoma: A randomized phase III study of pegylated liposomal doxorubicin versus topotecan. *Journal of Clinical Oncology, 19*, 3312–3322.
42. Gourley, P., (2008). Brief overview of biomicronano technologies. *Biotechnology Progress, 21*, 2–10.
43. Hanahan, D., & Weinberg, R. A., (2011). Hallmarks of cancer: The next generation. *Cell, 144*, 646–674.
44. Handy, R. D., & Shaw, B. J., (2007). Toxic effects of nanoparticles and nanomaterials: Implications for public health, risk assessment and the public perception of nanotechnology. *Health, Risk and Society, 9*(2), 125–144.
45. Hang, S., Gao, H., & Bao, G., (2015). Physical principles of nanoparticle cellular endocytosis. *ACS Nano, 9*(9), 8655–8671.
46. Mandal, S., Piyush, K. T., Anees, M., & Anchal, D., (2020). Cancer gene therapy and its techniques in cancer biology. *International Journal of Cell Biology and Cellular Processes, 6*(1), 38–54.
47. Mandal, S., & Piyush, K. T., (2021). Gene therapy and gene editing for cancer therapeutics. In: *Handbook of Research on Advancements in Cancer Therapeutics* (pp. 116–204). IGI Global.
48. Iversen, T. G., Skotland, T., & Sandvig, K., (2011). Endocytosis and intracellular transport of nanoparticles: Present knowledge and need for future studies. *Nano Today, 6*(2), 176–185.
49. Jain, K., (2009). Role of nanobiotechnology in the development of personalized medicine. *Nanomedicine, 4*, 249–252.
50. Jain, R. K., & Stylianopoulos, T., (2010). Delivering nanomedicine to solid tumors. *Nature Reviews Clinical Oncology, 7* (11), 653–664.
51. Jelínková, M., Strohalm, J., Etrych, T., Ulbrich, K., & íhová, B., (2003). Starlike vs. classic macromolecular prodrugs: Two different antibody-targeted HPMA copolymers of doxorubicin studied *in vitro* and *in vivo* as potential anticancer drugs. *Pharmaceutical Research, 20*, 1558–1564.

52. Jørgensen, K., Davidsen, J., & Mouritsen, O., (2002). Biophysical mechanisms of phospholipase A2 activation and their use in liposome-based drug delivery. *FEBS Letters, 531*, 23–27.
53. Kaminskas, L. M., & Boyd, B. J., (2011). Dendrimer pharmacokinetics: The effect of size, structure and surface characteristics on ADME properties. *Nanomedicine, 6*(6), 1063–1084.
54. Karode, S. K., Kulkarni, S. S., Suresh, A. K., & Mashelkar, R. A., (1998). New insights into kinetics and thermodynamics of interfacial polymerization. *Chemical Engineering Science, 53*, 2649–2663.
55. Kawai, K., Larson, B. J., Ishise, H., Carre, A. L., Nishimoto, S., Longaker, M., & Lorenz, H. P., (2011). Calcium-based nanoparticles accelerate skin wound healing. *PLoS One, 6*(11), e27106-e27106.
56. Kerman, M., Chan, W., Laakkonen, P., Bhatia, S., & Ruoslahti, E., (2002). Nanocrystal targeting *in vivo*. *Proceedings of the National Academy of Sciences of the United States of America, 99*, 12617–12621.
57. Khoury-Fallouh, A. N., Roblot-Treupel, L., Fessi, H., Devissaguet, J. P., & Puisieux, F., (1986). Development of a new process for the manufacture of poly isobutyl cyanoacrylate nanocapsules. *International Journal of Pharmacy, 28*, 125–136.
58. Koizumi, F., Kitagawa, M., Negishi, T., Onda, T., Matsumoto, S. I., Hamaguchi, T., & Matsumura, Y., (2006). Novel SN-38-incorporating polymeric micelles, NK012, eradicate vascular endothelial growth factor-secreting bulky tumors. *Cancer Research, 66*, 10048–10056.
59. Kola, I., & Landis, J., (2004). Can the pharmaceutical industry reduce attrition rates?. *Nature Reviews Drug Discovery, 3*(8), 711–715.
60. Lenggoro, I. W., Okuyama, K., Fernández De La, M. J., & Tohge, N., (2000). Preparation of ZnS nanoparticles by electrospray pyrolysis. *Journal of Aerosol Science, 31*(1), 121–136.
61. Li, S., & Huang, L., (2006). Targeted delivery of antisense oligodeoxynucleotide and small interference RNA into lung cancer cells. *Molecular Pharmacy, 3*, 579–588.
62. Li, D., & Kaner, R. B., (2006). Shape and aggregation control of nanoparticles: Not shaken, not stirred. *Journal of American Chemical Society, 128*(3), 968–975.
63. Lockman, P., Mumper, R., Khan, M., & Allen, D., (2002). Nanoparticle technology for drug delivery across the blood-brain barrier. *Drug Development and Industrial Pharmacy, 28*, 1–13.
64. Low, P., & Antony, A., (2004). Folate receptor-targeted drugs for cancer and inflammatory diseases. *Advanced Drug Delivery Reviews, 56*, 1055–1058.
65. Lu, Y., & Low, P., (2002). Folate-mediated delivery of macromolecular anticancer therapeutic agents. *Advanced Drug Delivery Reviews, 54*, 675–693.
66. McCarthy, T., Karellas, P., Henderson, S., Giannis, M., O'Keefe, D., & Heery, G., (2005). Dendrimers as drugs: Discovery and preclinical and clinical development of dendrimer-based microbicides for HIV and STI prevention. *Molecular Pharmacy, 2*, 312–318.
67. Misra, R., Acharya, S., & Sahoo, S. K., (2010). Cancer nanotechnology: Application of nanotechnology in cancer therapy. *Drug Discovery Today, 15*(19, 20), 842–850.
68. Moghimi, S., Hunter, A., & Murray, J., (2005). Nanomedicine: Current status and future prospects. *The FASEB Journal, 19*, 311–330.

69. Mohanty, C., Das, M., & Sahoo, S. K., (2012). Emerging role of nanocarriers to increase the solubility and bioavailability of curcumin. *Expert Opinion on Drug Delivery, 9*(11), 1347–1364.
70. Nevozhay, D., Kañska, U., Budzyñska, R., & Boratyñski, J., (2007). Current status of research on conjugates and related drug delivery systems in the treatment of cancer and other diseases (polish). *Postepy Hig. Med. Dosw., 61*, 350–360.
71. Oerlemans, C., & Bult, W., (2010). Polymeric micelles in anticancer therapy: Targeting, imaging and triggered release. *Pharmaceutical Research, 27*, 2569–2589.
72. Paciotti, G., Myer, L., Weinreich, D., Goia, D., Pavel, N., McLaughlin, R. E., & Tamarkin, L., (2004). Colloidal gold: A novel nanoparticle vector for tumor directed drug delivery. *Drug Delivery, 11*, 169–183.
73. Pandit, A., & Zeugolis, D. I., (2016). Twenty-five years of nano-bio-materials: Have we revolutionized healthcare?. *Future Medicine, 11*(9), 985–987.
74. Parveen, S., & Sahoo, S. K., (2008). Polymeric nanoparticles for cancer therapy. *Journal of Drug Targeting, 16*(2), 108–123.
75. Pastorino, F., Brignole, C., Marimpietri, D., Cilli, G. C., Ribatti, L. R., Allen, T. M., Corti, A., & Ponzoni, M., (2003). Vascular damage and anti-angiogenic effects of tumor vessel-targeted liposomal chemotherapy. *Cancer Research, 63*, 7400–7409.
76. Peng, X. H., Qian, X., Mao, H., Wang, A. Y., Chen, Z. G., & Nie, S., (2008). Targeted magnetic iron oxide nanoparticles for tumor imaging and therapy. *International Journal of Nanomedicine, 3*(3), 311–321.
77. Perez-Herrero, E., & Fernandez-Medarde, A., (2015). Advanced targeted therapies in cancer: Drug nanocarriers, the future of chemotherapy. *European Journal of Pharmaceutics and Biopharmaceutics, 93*, 52–79.
78. Perez-Soler, R., Lopez-Berestein, G., Lautersztain, J., Al-Baker, S., Francis, K., Macias-Kiger, D., Raber, M. N., & Khokhar, A. R., (1990). Phase I clinical and pharmacological study of liposome-entrapped cis-bis-neodecanoato-trans-R, R,-1, 2-diaminocyclohexane platinum (II). *Cancer Research, 50*, 4254–4259.
79. Poli-Bigelli, S., Rodrigues-Pereira, J., Carides, A. D., Julie Ma, G., Eldridge, K., Hipple, A., Evans, J. K., Horgan, K. J., & Lawson, F., (2003). Addition of the neurokinin 1 receptor antagonist aprepitant to standard antiemetic therapy improves control of chemotherapy-induced nausea and vomiting. *Cancer, 97*(12), 3090–3098.
80. Provenge, R., (2004). Immunotherapy (APC8015) targeting prostatic acid phosphatase can induce durable remission of metastatic androgen-independent prostate cancer: A phase 2 trial. *The Prostate, 60*(3), 197–204.
81. Prylutska, S. V., Grynyuk, I. I., Matyshevska, O. P., Prylutskyy, Y. I., Ritter, U., & Scharff, P., (2008). Anti-oxidant properties of C60 fullerenes *in vitro*. *Fuller Nanotube Carbon Nanostructure, 16*, 698–705.
82. Prylutska, S. V., Skivka, L. M., Didenko, G. V., Prylutskyy, Y. I., & Evstigneev, M. P., (2015). Complex of C60 fullerene with doxorubicin as a promising agent in antitumor therapy. *Nanoscale Research Letters, 10*(1), 499. doi: 10.1186/s11671-015-1206-7.
83. Qi, L., & Gao, X., (2008). Emerging application of quantum dots for drug delivery and therapy. *Expert Opinion on Drug Delivery, 5*(3), 263–267.
84. Quintana, A., Raczka, E., & Piehler, L., (2002). Design and function of a dendrimer based therapeutic nanodevice targeted to tumor cells through the folate receptor. *Pharmaceutical Research, 19*, 1310–1316.

85. Rani, D., Somasundaram, V. H., Nair, S., & Koyakutty, M., (2012). Advances in cancer nanomedicine. *Journal of the Indian Institute of Science, 92*(2), 187–218.
86. Sahay, G., Alakhova, D. Y., & Kabanov, A. V., (2010). Endocytosis of nanomedicines. *Journal of Controlled Release, 145*(3), 182–195.
87. Sahoo, S. K., & Labhasetwar, V., (2003). Nanotech approaches to drug delivery and imaging. *Drug Discovery Today, 8*(24), 1112–1120.
88. Sanginario, A., Miccoli, B., & Demarchi, D., (2017). Carbon nanotubes as an effective opportunity for cancer diagnosis and treatment. *Biosensors, 7*(1), 2079–6374.
89. Sawant, R. R., & Torchilin, V. P., (2012). Multifunctional nanocarriers and intracellular drug delivery. *Current Opinion in Solid State and Materials Science, 16*(6), 269–275.
90. Schmieder, A., Winter, P., Caruthers, S., Harris, T. D., Williams, T. A., Allen, J. S., Lacy, E. K., et al., (2005). Molecular MR imaging of melanoma angiogenesis with 3-targeted paramagnetic nanoparticles. *Magnetic Resonance in Medicine, 53*, 621–627.
91. Seleci, M., Ag, S., Didem, B., Rebecca, S., Frank, B., & Cornelia, S. T., (2016). Smart multifunctional nanoparticles in nanomedicine. *BioNanoMaterials, 52*, 101–114.
92. Shen, B., & Ma, Y., (2016). Smart multifunctional magnetic nanoparticle-based drug delivery system for cancer thermo-chemotherapy and intracellular imaging. *ACS Applied Materials and Interfaces, 8*(37), 24502–24508.
93. Singh, A., &Sahoo, S. K., (2014). Magnetic nanoparticles: A novel platform for cancer theranostics. *Drug Discovery Today, 19*(4), 474–481.
94. Singh, D., Minz, A. P., & Sahoo, S. K., (2017). Nanomedicine-mediated drug targeting of cancer stem cells. *Drug Discovery Today, 22*(6), 952–959.
95. Steinhauser, I., Spänkuch, B., Strebhardt, K., & Langer, K., (2006). Trastuzumab-modified nanoparticles: Optimization of preparation and uptake in cancer cells. *Biomaterials, 27*, 4975–4983.
96. Suri, S. S., Fenniri, H., & Singh, B., (2007). Nanotechnology-based drug delivery systems. *Journal of Occupational Medicine and Toxicology, 2*, 1–6.
97. Tan, W., Jiang, S., & Zhang, Y., (2007). Quantum-dot based nanoparticles for targeted silencing of HER2/neu gene via RNA interference. *Biomaterials, 28*, 1565–1571.
98. Tiwari, S., & Amiji, M., (2006). Improved oral delivery of paclitaxel following administration in nanoemulsion formulations. *Journal of Nanoscience and Nanotechnology, 6*(9), 3215–3221.
99. Tiwari, A. K., Sodani, K., Dai, C. L., Ashby, C. R., Chen, Z. S., & Chen, Z. S., (2011). Revisiting the ABCs of multidrug resistance in cancer chemotherapy. *Current Pharmaceutical Biotechnology, 12*(4), 570–594.
100. Tiwari, G., Tiwari, R., Sriwastawa, B., Bhati, L., Pandey, S., Pandey, P., & Bannerjee, S. K., (2012). Drug delivery systems: An updated review. *International Journal of Pharmaceutical Investigation, 2*(1), 2–11.
101. Torchilin, V., (2005). Recent advances with liposomes as pharmaceutical carriers. *Nature Reviews Drug Discovery, 4*, 145–160.
102. Torchilin, V., Lukyanov, A., Gao, Z., & Papahadjopoulos-Sternberg, B., (2003). Immunomicelles: Targeted pharmaceutical carriers for poorly soluble drugs. *Proceedings of the National Academy of Sciences of the United States of America, 100*, 6039–6044.
103. Tsai, Y. M., Chang-Liao, W. L., Chien, C. F., Lin, L. C., & Tsai, T. H., (2012). Effects of polymer molecular weight on relative oral bioavailability of curcumin. *International Journal of Nanomedicine, 7*, 2957–2966.

104. Tseng, C., Wang, T., Dong, G. C., Wu, S. Y. H., Young, T. H., Shieh, M. J., Lou, P. J., & Lin, F. H., (2007). Development of gelatin nanoparticles with biotinylated EGF conjugation for lung cancer targeting. *Biomaterials, 28*, 3996–4005.
105. Wagner, V., Dullaart, A., Bock, A., & Zweck, A., (2006). The emerging nanomedicine landscape. *Nature biotechnology, 24*, 1211–1218.
106. Waldman, S., Fortina, P., Surrey, S., Hyslop, T., Kricka, L., & Graves, D., (2006). Opportunities for near-infrared thermal ablation of colorectal metastases by guanylyl cyclase C-targeted gold nanoshells. *Future Oncology, 2*, 705–716.
107. Zhang, Q., Yang, W., Man, N., Zheng, F., Shen, Y., Sun, K., Li, Y., & Wen, L. P., (2009). Autophagy-mediated chemosensitization in cancer cells by fullerene C60 nanocrystal. *Autophagy, 5*(8), 1107–1117.
108. Zhang, X., & Huang, Y., (2014). Nanomicellar carriers for targeted delivery of anticancer agents. *Therapeutic Delivery, 5*(1), 53–68.
109. Zhao, X., & Li, H., (2008). Targeted drug delivery via folate receptors. *Expert Opinion on Biological Therapy, 5*(3), 309–319.

CHAPTER 10

APPLICATIONS OF NANOMATERIALS IN DIAGNOSTICS AND TREATMENT OF CANCER

TARUN KUMAR KUMAWAT, VARSHA KUMAWAT, VISHNU SHARMA, ANJALI PANDIT, and MANISH BIYANI

ABSTRACT

The second most significant cause of death in the world is cancer. High mortality occurs in underdeveloped nations due to a lack of adequate and cheap treatments. Many cancer therapies, including surgery, radiation, and chemotherapy, have been explored, but they are all restricted. Nanotechnology may greatly enhance bench-to-clinical treatment systems. Biotechnology has shown the potential of direct evolution in developing industrially significant molecules like nanoparticles and bionanomaterials for over three decades. Bionanomaterials are made up of biological molecules that may be formed into molecular structures of nanoscale dimensions. Bionanomaterials' tiny size, high surface area, stability, and biocompatibility may allow *in situ* disease diagnosis. With the increase in cancer incidence, effective clinical diagnostic and therapeutic methods are more important in the early or late stages. Bionanomaterials are used to detect cancer at the molecular level. The development of science and technology and economic affluence has increased people's concern about health. This chapter's main aim is to emphasize the role of advanced bionanomaterials in cancer detection and treatment. We will briefly discuss the use of electrochemical nano-bio sensors based on advanced functional nanomaterials for point-of-care diagnostics.

Sustainable Nanomaterials for Biomedical Engineering: Impacts, Challenges, and Future Prospects.
Junaid Ahmad Malik, Megh R. Goyal & Mohamed Jaffer M. Sadiq (Eds.)

10.1 INTRODUCTION

Nanotechnology is a rapidly growing area of science [84]. The development of nanoparticles (NPs) can increase medication penetration, refocus treatment, and precisely target cancer cells [39]. Applied nanotechnology is a multidisciplinary area that includes devices created in engineering, biology, physics, and chemistry [30]. The scientific literature defines nanoparticles (NP) and nanomaterials (NM) as tiny particles with new characteristics and functionalities or the capacity to control or modify matter on an atomic scale. Particles derived from biological sources are referred to as "bio nanoparticles" [31]. Bioinspiration through biological nanomachines and their molecular assemblies provides guiding principles for the development of functional nanoscale devices. In recent decades, significant advancements in the biosynthesis of nanomaterials, including hybrid nanomaterials, as well as their applications, have been studied [11]. The most challenging aspect is synthesizing nanomaterials with precise geometries. This obstacle has motivated material scientists to envision biosystems as bio templates for producing bio nanoparticles (BNPs), which vary in their characteristics and functions from their counterparts due to their majority of nanoscale constituent pieces [80].

Nanostructure materials have been used in antitumor pharmaceuticals [93]. Rapid advancements in the science of nanoparticles over the past several decades have resulted in their increasing use in medication delivery strategies. Nanomedicine has the potential to revolutionize world health via its potential uses in prevention, diagnostics, and treatments [81]. The benefits of nanotechnology-based medication delivery systems include increased therapeutic effectiveness and decreased adverse effects [52]. The chapter discusses the many methods for synthesizing GNPs utilizing plant extracts, plant wastes, bacteria, and fungus. It also discusses the biomedical applications, particularly cancer in which they have been used, and the rationale for their adaptability to science and technology.

10.2 CANCER: AN INSIDIOUS GLOBAL THREAT

Cancer ranks second in global mortality [37]. Surgery, radiation, and chemotherapy are used to treat cancer cells, but they are cytotoxic to normal cells and therefore ineffective [38]. Chemotherapy agents have low bioavailability. Hepatocellular carcinoma is common liver cancer and a significant cause of cancer mortality globally [41]. Blood cancer occurs when malignant alterations accumulate in cells of the primary or secondary lymphoid organs. Multiple myeloma, lymphoma, and leukemia are the three most common blood malignancies. Chemotherapy, immunotherapy, and bone marrow transplantation may all treat blood cancer [32].

An early cancer diagnosis is critical since it affects treatment options and outcomes, ultimately affecting the patient's quality of life. Optical, magnetic resonance, and computed tomography are today's diagnostic methods. These technologies need radioactive materials, fluorescent dyes, and contrast agents [42]. Consequently, the primary difficulties in anticancer nanotechnology treatment have been the design and production of multifunctional nanomaterials. Anticancer therapy using multifunctional nanomaterials can target cancer or tumor, administer therapeutic medicines, and monitor tumor tissues [49]. Nanotechnology has flourished in recent years and has many uses in healthcare. Bio-nanoparticles are cheap, safe, and biocompatible, attracting global scientific attention. Theranostic nanomedicine for cancer detection and treatment has recently gained popularity [96].

Liposomes, nanoparticles, and dendrimers are improved drug delivery methods with better pharmacokinetic characteristics for anticancer treatments [32]. Drug delivery techniques for cancer therapy are used to enhance effectiveness while simultaneously reducing side effects. Most chemotherapy drugs cause cytotoxicity in both standard and malignant cells. Indeed, researchers should commit significant resources to develop more effective immunotherapy medicines and treatment regimens for cancer.

10.3 ADVANCED BIONANOMATERIALS: AN OVERVIEW

Bionanomaterials are essential for developing novel customized medication delivery methods and the diagnosis and treatment of a wide range of life-threatening illnesses. Nanobiomaterials can be synthesized quickly, ecofriendly, and non-toxic utilizing biological processes. These materials are entirely biocompatible in their natural state and are thus ideal for biological applications. Numerous metals, including Ag, Au, Pt, Cu, and Se, have been utilized to synthesize the equivalent nanomaterials from various biological sources [61]. The lack of information regarding nanoparticle's health effects is a significant barrier to nanotechnology implementation. It is possible to minimize these negative consequences by using bionanomaterials instead of synthetic nanomaterials generated via biosynthesis or green methods

[23]. Dynamic light scattering, differential scanning calorimetry, electron microscopy, etc., are all employed to assess these nanobiomaterials (Figure 10.1) [10].

Bionanomaterials (BNMs) are biomaterials that have been enhanced by nanotechnology. The development of functional nanostructured biological materials, i.e., bionanomaterials designed biological beings, has been made possible by advances in nanotechnology (e.g., liposomes, proteins, enzymes, nucleic acids, oligonucleotides, receptor, antibodies, and macromolecules of peptides, polysaccharides, biological cofactors, organelles, cell, tissue, microorganism, and ligands) [57]. Bionanomaterials based on molecularly imprinted polymers (MIPs) have generated an appealing vision for biomedical applications by creating specialized cavities for targets in the presence of a target. Furthermore, this is a low-cost, fast, and simple-to-use sensing method that is extremely sensitive and selective, making it particularly well suited for theranostics, diagnostic, and screening applications [4]. Bionanotechnology is a rapidly growing area as advances in nanotechnology are applied to unmet requirements in biology, biomedicine, and other fields, including different sensor and drug delivery technologies. Among the diverse array of produced bionanomaterials, stimuli-sensitive bionanomaterials are of particular interest [79].

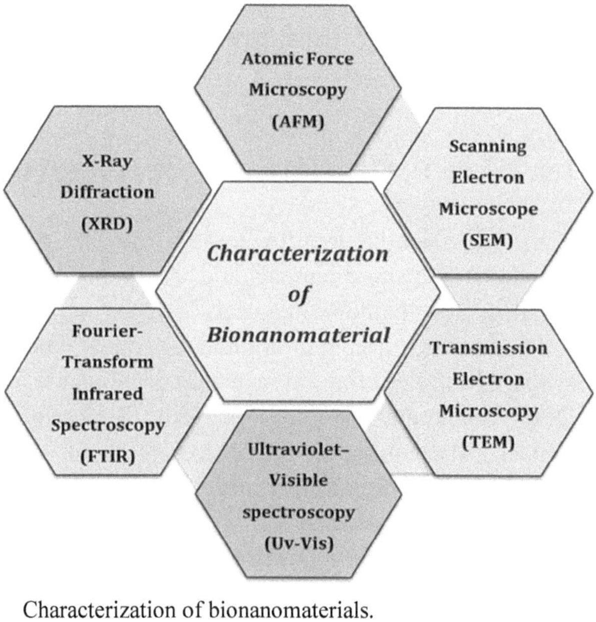

FIGURE 10.1 Characterization of bionanomaterials.

10.4 BIOSYNTHESIS OF BIONANOMATERIALS

Microbes like bacteria, actinomycetes, yeast, viruses, and marine algae have lately acquired an interest in green nanotechnology [84]. Toxic byproducts from traditional physical and chemical nanoparticle production pose environmental and health risks. Bionanomaterials are materials with nanoscale dimensions that are biocompatible, such as silver and AuNPs, QDs, liposomes, and other similar materials [55]. Several kinds of biomaterials derived from plants, animals, and microorganisms have been used to cure cancer (Figure 10.2).

The biological production of NPs has several advantages, including cheap cost, high safety, stability, and rapidity. This reduces possible health and environmental hazards while improving the nanomaterial's size, shape, biocompatibility, and stability [87]. At the moment, an unprecedented level of interest in herbal medicine is surfacing worldwide. The western world began comparing allopathic (many-sided) treatments to traditional ones and recognizing the latter as safer options [51]. Natural products have been researched to discover novel therapies for human dysfunctions due to their many therapeutic characteristics [18]. Instead of chemicals, scientists have tried to minimize the dangers by utilizing novel eco-friendly agents. A non-toxic technique of manufacturing NPs is the biological method, which uses microorganisms and plant extracts to create "bio nanoparticles" [50].

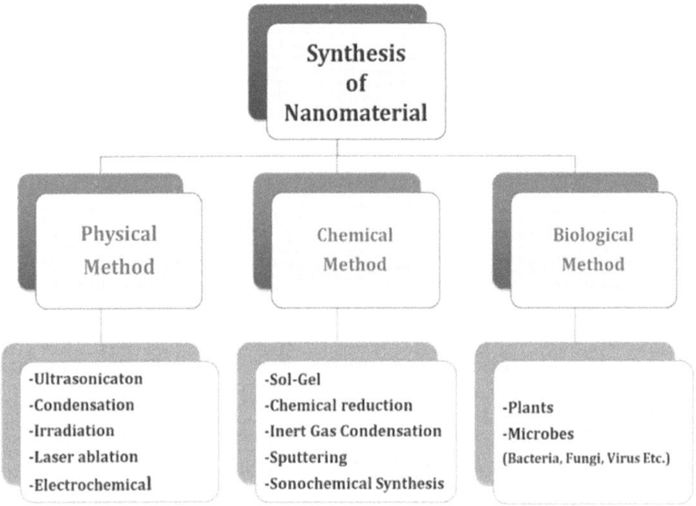

FIGURE 10.2 Methods for synthesis of nanomaterial.

10.4.1 PLANT-MEDIATE BIOSYNTHESIS OF BIONANOMATERIAL

Even while plant-mediated synthesis produces smaller-sized bionanomaterials in less time, their stability does not match well with conventional techniques [43]. Around the globe, a diverse range of plants are collected, and the various components of these plants may be transformed into a diverse range of bionanomaterials [12] (Table 10.1). Biomedical uses of silver nanoparticles (Ag NPs) utilizing plant resources have been approved [8]. Ahmed et al. [3] made AgNPs from *Jurinea dolomiaea* and tested their anticancer potential using the MTT assay on cervical cancer (HeLa), breast cancer (MCF-7), and NIH-3 T3 cells. Majeed et al. [56] produced Ag NPs using *Artocarpus integer* leaves. AgNPs demonstrated superior antitumor activity against the MCF-7 and MG-63 cancer cell lines. While displaying a lower affinity for normal 3T3 cells. Ag NPs have many bioactivities such as antibacterial and antifungal activities with a vast potential for cancer diagnostics and cancer treatments, anti-inflammation, anti-angiogenic activity, and anticancer activities [39].

TABLE 10.1 Plant Employed in the Synthesis of Nanoparticle

Name of Nanoparticle	Name of Plant	Size (in nm)	Application	References
Ag NPs	*Sesamum indicum*	6.6–14.8	Antitumor activity	[7]
Au NPs	*Tasmannia lanceolata*	7.10 ± 0.66	Anticancer activity	[46]
Au NPs	*Alternanthera bettzickiana*	80–120	Anticancer activity	[64]
CuO	*Acalypha indica*	26–30	Anticancer activity	[86]
ZnO	*Delonix regia*	65–184	Anticancer activity	[1]

10.4.2 BACTERIA-MEDIATE BIOSYNTHESIS OF BIONANOMATERIAL

Nanoparticle manufacturing techniques that are nontoxic and environmentally acceptable are becoming more critical since chemical nanoparticle synthesis comes with many environmental concerns and problems. Due to their rapid development rate, biological systems, particularly bacteria, are ideal for accomplishing this objective [33]. Bacterial Ag NPs are also known to have anticancer and antioxidant activity [85]. Al-Zubaidi et al.

[9] used *Bacillus thuringiensis* israelensis extract to manufacture silver bio nanoparticles (biological methods). Bustos et al. [16] showed that *Pseudomonas aeruginosa* culture supernatant biosynthesized AgNPs. Manivasagan et al. [58] biosynthesized gold bio nanoparticles using *Nocardiopsis* sp. MBRC-48.

10.4.3 ACTINOMYCETES-MEDIATE BIOSYNTHESIS OF BIONANOMATERIAL

Actinomycetes are regarded as a significant source of novel medicinal and industrial goods [62]. Most research has been conducted on *Streptomyces* species owing to their intrinsic capacity to produce redox-active macromolecules/secondary metabolites [40]. Abd-Elnaby et al. [2] produced Ag NPs isolated from *Streptomyces rochei* MHM13 from Suez Gulf sediments in the Red Sea. Ranjitha and Ravishankar Rai [73] used *Saccharomonas glauca* to help in the production of AuNPs. The produced GNPs were cytotoxic to HT-29 and Hep 2 cancer cells. A catalyst (*S. laurentii* R-1) for silver nanoparticle-assisted biosynthesis was obtained from the roots of the medicinal plant *Achillea fragrantissima* by Eid et al. [27]. Malignant Caco-2 cells were susceptible to the Ag-NPs that were synthesized. Metal nanoparticles made from actinomycetes are praised for their naturalness [54].

10.4.4 FUNGI-MEDIATE BIOSYNTHESIS OF BIONANOMATERIAL

However, the apparent drawbacks of many physical and chemical techniques have prompted researchers to concentrate on creating bio nanoparticles. Fungal systems are eco-friendly, biodegradable, and simple to grow and scale [59]. *Fomes fomentarius* is a wild fungus whose intracellular extract was used to produce titanium oxide and AgNPs combined with aqueous titanium isopropoxide and silver nitrate solutions [76]. Saravanakumar et al. [78] synthesized and characterized copper oxide nanoparticles from *Trichoderma asperellum*. Majeed et al. [56] examined the filamentous fungus *Penicillium italicum* for the extracellular generation of AgNPs that exhibited good anticancer activity. Fungi are capable of secreting significant quantities of secondary metabolites and reducing metal ions to metallic nanoparticles [5]. Fungi have a larger secretome than bacteria, including enzymes, active chemicals, and proteins involved in capping and reducing Au NPs [83]. The enormous and varied fungal flora on our planet has piqued scientists worldwide in exploring it for NP synthesis. The production of NPs by fungi is a novel platform in nanobiotechnology [65].

10.4.5 YEAST-MEDIATE BIOSYNTHESIS OF BIONANOMATERIAL

Biomedical uses of nanoparticles (NPs) having at least one dimension less than 100 nm. The delivery of drugs through yeast is a potential platform for treating a variety of medical problems [77]. Distinct mechanisms used by yeast strains from various genera produce nanoparticles with variable size, location, monodispersity, and other properties [35]. Gold and AgNPs are produced by silver-tolerant *Saccharomyces cerevisiae* [53]. In recent years, interest in biosynthesized AgNPs has grown [74]. A431 epidermoid carcinoma and MCF-7 breast cancer were both destroyed by platinum nanoparticles produced by *Saccharomyces boulardii* [15].

10.4.6 VIRUS-MEDIATE BIOSYNTHESIS OF BIONANOMATERIAL

The virus is an infectious microbe that only lives within living cells [45]. The virus-assisted production of nanomaterials is reliable, simple, non-toxic, and environmentally friendly [47]. Viral nanoparticles (VNPs) are viruses that include mammal, plant, and bacteriophage viruses. They have been utilized as building blocks and supramolecular templates in nanotechnology [95]. In the future, plant virus nanoparticles may be used to deliver cancer therapy medications [48]. Many virus-metallic hybrid nanomaterials have been described as model systems for manufacturing [34]. Cao et al. [19] created nanoparticles for doxorubicin administration using Red Clover Necrotic Mosaic Virus (RCNMV). Hollow nanotubes with a polyanionic inner surface delivered phenanthriplatin [25]. Thangavelu et al. [91] used the plant pathogenic squash leaf curl coronavirus (SLCCNV) to make Au-Ag composite semiconductor nanoparticles (NPs) (Table 10.2).

TABLE 10.2 Microbes Employed in the Synthesis of Nanoparticle

Name of Nanoparticle	Name of Microbe	Size (in nm)	Application	References
ZnO	*Aspergillus terreus*	28–63	Anticancer activity	[13]
Gold	*Fusarium solani* ATLOY-8	40–45	Anticancer activity	[24]
Gold	*Pleurotus ostreatus*	10–30	Anticancer activity	[88]
Silver	*Deinococcus radiodurans*	4–50	Anticancer activity	[26]
Silver	*Streptacidiphilus durhamensis*	8–48	Anticancer activity	[17]
FA-CMV-Dox assemblies	Cucumber mosaic virus (CMV)	~29	Anticancer drug delivery	[99]
HEVNP	Hepatitis E virus	27–34	Cancer therapy	[20]

10.4.7 ALGAE-MEDIATE BIOSYNTHESIS OF BIONANOMATERIAL

The production of nanoparticles (NPs) by algae has been assigned many essential functions in developing nano-based medicines for human health improvement [82]. Algae include bioactive chemicals and secondary metabolites that reduce, cap, and stabilize nanoparticles. Algae extracts are great for making AgNPs in a greenway. Brown seaweed extract helps make AgNPs [71]. Algal silver oxides and AuNPs are anticancer agents [29]. Venkatesan et al. [94] made Ag NPs using *Ecklonia cava* extract that inhibits apoptosis in human cervical cancer cells. Using copper sulfate ($CuSO_4$) as a precursor, researchers have discovered that "green" copper nanoparticles synthesized from *Sargassum polycystum* may effectively suppress the development of breast cancer cells (MCF-7) by over 93% [72]. Supraja et al. [89] made Ag NPs from the extract of *Sargassum muticum*. They found that Ag NPs generated had anticancer effects on a breast cancer cell line.

10.5 NANOTECHNOLOGY AND CANCER

Cancer nanotechnology has numerous uses in molecular imaging, molecular diagnostics, and customized therapy [66]. Nanotechnology's use in cancer treatment has garnered significant interest in recent years

[60]. Multifunctional nanoparticles can target cancer cells, transport, and release medicines in controlled ways [6]. Bionanosynthesis is a critical component of the rapidly expanding area of nanotechnology because it produces high-quality functional materials with biological applications [73].

Nanotechnology may revolutionize cancer detection and treatment. Protein engineering and materials science nanoscale targeting methods may hope for people living with cancer [67]. Nanomedicine currently provides new approaches to cancer therapy by combining several modalities of treatment into a single formulation. Additionally, nanoparticles function as nanocarriers, altering therapeutic molecules' solubility, biodistribution, and efficacy, resulting in more effective therapies with fewer adverse effects [92]. New nanoparticle-biomolecule interactions on the surface and within cells may revolutionize cancer detection and treatment [28].

10.6 ROLE OF BIONANOMATERIALS IN CANCER DIAGNOSTICS: A MEDICAL APPROACH

Cancers are degenerative illnesses that pose a significant danger to global public health at any point in time [90]. This is a group of diseases that significantly impair cellular metabolism due to aberrant signaling pathways such as cell proliferation, angiogenesis, and metastasis. Lifestyle changes, chemical exposure, pollution, and genetic factors all lead to cancer. The high death rate in developing countries is related to late diagnosis and inadequate treatment [70]. Cancer nanotechnology may substantially improve current cancer detection, diagnostic, imaging, and therapy techniques while reducing side effects [36].

Several nanotechnology research and advances have been made, and numerous nanomaterials have been used to diagnose cancer in its earliest stages. Early detection, imaging, and molecular imaging techniques have been improved by developing new bio-nanotools invented by researchers [75]. Nanomaterials offer a unique set of physical, optical, and electrical characteristics that make them valuable for sensing. Over the years, AuNPs, QDs, magnetic nanoparticles, gold nanowires, and carbon nanotubes have been created [22]. Bionanomaterials play an increasingly significant role in engineering and medicinal research because of their environmentally friendly and sustainable characteristics.

10.7 APPLICATION OF ADVANCED BIONANOMATERIALS IN CANCER TREATMENT: A SCIENTIFIC APPROACH

Numerous cancer treatments have made significant strides during the last decade. Surgery, chemotherapy, and radiation treatment are now used to treat cancer, but they all have side effects on noncancerous cells. As a result, the quest for novel therapies is a high priority today, and nanoparticles are one such promising anticancer agent [97]. Researchers are developing more ecologically friendly and sustainable methods to produce nanoparticles (NPs) [63]. A new approach to cancer immunotherapy improves the host immune system's capacity to detect and destroy specific cancer cells. Recent years have seen a surge in interest in nanoparticle-based cancer immunotherapies [98].

Recent advancements have resulted in the creation of multifunctional bionanomaterials capable of delivering therapeutic medicines or genes to a bone malignancy. Bionanomaterial-based bone cancer treatment offers hope to patients and opens up exciting new therapeutic possibilities [14]. As with other chemotherapeutic agents, nanomaterial-based treatments often need a relatively high degree of accumulation inside cancer cells to be practical. Although nanoparticles may passively target cancer cells and act as "nanocarriers" for chemotherapeutics, their random distribution restricts their utility as "nanocarriers" [21]. Nanoparticles (NPs) have shown efficacy in delivering anticancer medicines to the brain [68]. Metal nanoparticles (MNPs) are used in several nanomedicine applications, notably cancer nanotheranostics [44]. The use of nanoparticles produced in a green way to treat cancer is a relatively new development [69].

10.8 CHALLENGES AND FUTURE PERSPECTIVES

A constantly changing global environment necessitates the accessibility and translation of diagnostics and treatments to clinical settings. Researchers have been investigating bionanomaterials for over centuries. The development of ecologically safe and reliable nanoparticle manufacturing methods is essential in nanotechnology. In the last 15 years, remarkable progress has been achieved in the area of nanobiotechnology. The increasing need for analytical evidence on nanoscale biomaterials necessitates modifying existing instruments and methods or the development of new ones.

Targeted delivery attempts to increase cellular absorption and treatment success while maintaining a greater degree of disease specificity. Self-assembled peptides and proteins have enormous promise as tailored drug delivery systems, with significant applications in anticancer treatment. The use of microbes for novel NP synthesis and their application in biomedicine and cancer treatment has recently been demonstrated. Still, the methods for commercial production of novel NPs must be modified, and the NP's synthesis must be scaled up from current techniques, as recent studies have demonstrated. As a result of the above discussion, it is anticipated that bionanomaterials will be used soon to develop cancer treatments and diagnostic agents for cancer patients.

10.9 SUMMARY

Over the years, nanotechnology has shown a great deal of promise in cancer diagnosis and treatment. With the advancement of cutting-edge technology, scientists are attempting to create new methods for cancer. The introduction and subsequent development of nanotechnology-based bionanomaterials have shown to be beneficial to the healthcare industry. Cancer is a global epidemic and the leading cause of mortality, necessitating the development of tailored bionanosystems. The product of bionanomaterials is a medical revolution and changed the way medicines are administered and treated for many diseases. The numerous advantages of green nanoparticle synthesis over conventional synthesis have attracted the attention of many academic and industrial experts. Cancer-fighting smart nanostructure weapons are being developed using nanotechnology. This chapter discussed the critical significance of bridging nanotechnology and cancer biology for the effective creation of bionanomaterials for cancer nanodiagnostic and nanotherapy.

ACKNOWLEDGMENTS

The authors want to thank the Chairman of the Biyani Group of Colleges, Jaipur, India, for guidance and support.

KEYWORDS

- **bionanomaterials**
- **biosensor**
- **cancer**
- **diagnosis**
- **health**
- **molecularly imprinted polymers**
- **nanotechnology**

REFERENCES

1. Abbasi, T., Anuradha, J., Ganaie, S. U., & Abbasi, S. A., (2015). Biomimetic synthesis of nanoparticles using aqueous extracts of plants (botanical species). *Journal of Nano Research, 31*, 138–202.
2. Abd-Elnaby, H. M., Abo-Elala, G. M., Abdel-Raouf, U. M., & Hamed, M. M., (2016). Antibacterial and anticancer activity of extracellular synthesized silver nanoparticles from marine *Streptomyces rochei* MHM13. *The Egyptian Journal of Aquatic Research, 42*(3), 301–312.
3. Ahmed, M. J., Murtaza, G., Rashid, F., & Iqbal, J., (2019). Eco-friendly green synthesis of silver nanoparticles and their potential applications as antioxidant and anticancer agents. *Drug Development and Industrial Pharmacy, 45*(10), 1682–1694.
4. Akgonullu, S., Bakhshpour, M., & Denizli, A., (2021). Molecularly imprinted bionanomaterials and their biomedical applications. In: Singh, R. P., & Singh, K. R. B., (eds.), *Bionanomaterials: Fundamentals and Biomedical Applications* (pp. 1–27). IOP Publishing Ltd.
5. Akther, T., Vabeiryureilai, M., Nachimuthu, S. K., Davoodbasha, M., & Srinivasan, H., (2019). Fungal-mediated synthesis of pharmaceutically active silver nanoparticles and anticancer property against A549 cells through apoptosis. *Environmental Science and Pollution Research, 26*(13), 13649–13657.
6. Alexis, F., Rhee, J. W., Richie, J. P., Radovic-Moreno, A. F., Langer, R., & Farokhzad, O. C., (2008). New frontiers in nanotechnology for cancer treatment. *Urologic Oncology: Seminars and Original Investigations, 26*(1), 74–85.
7. Alfuraydi, A. A., Devanesan, S., Al-Ansari, M., AlSalhi, M. S., & Ranjitsingh, A. J., (2019). Eco-friendly green synthesis of silver nanoparticles from the sesame oil cake and its potential anticancer and antimicrobial activities. *Journal of Photochemistry and Photobiology B: Biology, 192*, 83–89.
8. Almatroudi, A., Khadri, H., Azam, M., Rahmani, A. H., Al Khaleefah, F. K., Khateef, R., Ansari, M. A., & Allemailem, K. S., (2020). Antibacterial, antibiofilm and anticancer activity of biologically synthesized silver nanoparticles using seed extract of *Nigella sativa*. *Processes, 8*(4), 388. doi: 10.3390/pr8040388.

9. Al-Zubaidi, L. A., Alnuaimi, M. T., Aljanabi, Z. Z., & Adel, M. M., (2021). Biodegradation of chlorpyrifos pesticide using silver bio-nanoparticles *bacillus thuringiensis* israelensis extracts. *IOP Conference Series: Earth and Environmental Science, 779*(1), 012113. doi: 10.1088/1755-1315/779/1/012113.
10. Arunachalam, P., Alaswad, S. O., & Karuppiah, P., (2021). Analytical methods for the characterization of bionanomaterials. In: Krishnan, A., Muthupandian, S., Balakumar, C., Suvardhan, K., Sarojini, J. P., & Quansheng, C., (eds.), *Handbook on Nanobiomaterials for Therapeutics and Diagnostic Applications* (pp. 117–131). Elsevier.
11. Balaji, S., Mukunthan, K. S., & Kannan, N., (2014). Bio-nanomaterials: Structure and assembly. *Reviews in Advanced Sciences and Engineering, 3*(3), 250–260.
12. Barhoum, A., Jeevanandam, J., Rastogi, A., Samyn, P., Boluk, Y., Dufresne, A., Danquah, M. K., & Bechelany, M., (2020). Plant celluloses, hemicelluloses, lignins, and volatile oils for the synthesis of nanoparticles and nanostructured materials. *Nanoscale, 12*(45), 22845–22890.
13. Baskar, G., Chandhuru, J., Sheraz, F. K., Praveen, A. S., Chamundeeswari, M., & Muthukumar, T., (2015). Anticancer activity of fungal L-asparaginase conjugated with zinc oxide nanoparticles. *Journal of Materials Science: Materials in Medicine, 26*(1), 5380. doi: 10.1007/s10856-015-5380-z.
14. Blackburn, G., Scott, T. G., Bayer, I. S., Ghosh, A., Biris, A. S., & Biswas, A., (2013). Bionanomaterials for bone tumor engineering and tumor destruction. *Journal of Materials Chemistry B, 1*(11), 1519–1534.
15. Borse, V., Kaler, A., & Banerjee, U. C., (2015). Microbial synthesis of platinum nanoparticles and evaluation of their anticancer activity. *International Journal of Emerging Trends in Electrical and Electronics, 11*(2), 26–31.
16. Bustos, P. S., Quinteros, De Los, Á. M., Gomez, D. S., Ortega, M. G., Páez, P. L., & Guiñazú, N. L., (2021). Silver bionanoparticles toxicity in trophoblast is mediated by nitric oxide and glutathione pathways. *Toxicology, 454*, 152741. doi: 10.1016/j.tox.2021.152741.
17. Buszewski, B., Railean-Plugaru, V., Pomastowski, P., Rafińska, K., Szultka-Mlynska, M., Golinska, P., Wypij, M., et al., (2018). Antimicrobial activity of biosilver nanoparticles produced by a novel *Streptacidiphilus durhamensis* Strain. *Journal of Microbiology, Immunology and Infection, 51*(1), 45–54.
18. Cadona, F. C., Dantas, R. F., De Mello, G. H., & Silva-Jr, F. P., (2021). Natural products targeting into cancer hallmarks: An update on caffeine, theobromine, and (+)-catechin. *Critical Reviews in Food Science and Nutrition, 23*, 1–20.
19. Cao, J., Guenther, R. H., Sit, T. L., Opperman, C. H., Lommel, S. A., & Willoughby, J. A., (2014). Loading and release mechanism of red clover necrotic mosaic virus derived plant viral nanoparticles for drug delivery of doxorubicin. *Small, 10*(24), 5126–5136.
20. Chen, C. C., Stark, M., Baikoghli, M., & Cheng, R. H., (2018). Surface functionalization of hepatitis e virus nanoparticles using chemical conjugation methods. *Journal of Visualized Experiments, 135*, 57020. doi: 10.3791/57020.
21. Chen, T., Yang, Tang, Zhong, Bai, Zhang, Li, & Zheng, (2012). Surface decoration by spirulina polysaccharide enhances the cellular uptake and anticancer efficacy of selenium nanoparticles. *International Journal of Nanomedicine, 7*, 835–844.
22. Choi, Y. E., Kwak, J. W., & Park, J. W., (2010). Nanotechnology for early cancer detection. *Sensors, 10*(1), 428–455.

23. Chowdary, A. R., & Sanivada, S. K., (2021). Impact of nanomaterials on the food chain. In: Kumar, V., Guleria, P., Ranjan, S., Dasgupta, N., & Lichtfouse, E., (eds.), *Nanotoxicology and Nanoecotoxicology Environmental Chemistry for a Sustainable World* (Vol. 2, pp. 97–117). Springer.
24. Clarance, P., Luvankar, B., Sales, J., Khusro, A., Agastian, P., Tack, J. C., Al Khulaifi, M. M., et al., (2020). Green synthesis and characterization of gold nanoparticles using endophytic fungi *fusarium solani* and its *in-vitro* anticancer and biomedical applications. *Saudi Journal of Biological Sciences, 27*(2), 706–712.
25. Czapar, A. E., Zheng, Y. R., Riddell, I. A., Shukla, S., Awuah, S. G., Lippard, S. J., & Steinmetz, N. F., (2016). Tobacco mosaic virus delivery of phenanthriplatin for cancer therapy. *ACS Nano, 10*(4), 4119–4126.
26. Deobagkar, D., Kulkarni, R., Shaiwale, N., & Deobagkar, D., (2015). Synthesis and extracellular accumulation of silver nanoparticles by employing radiation-resistant *deinococcus radiodurans*, their characterization, and determination of bioactivity. *International Journal of Nanomedicine, 10*, 963–974.
27. Eid, A. M., Fouda, A., Niedbała, G., Hassan, S. E. D., Salem, S. S., Abdo, A. M., Hetta, H. F., & Shaheen, T. I., (2020). Endophytic *streptomyces laurentii* mediated green synthesis of Ag-Nps with antibacterial and anticancer properties for developing functional textile fabric properties. *Antibiotics, 9*(10), 641. doi: 10.3390/antibiotics9100641.
28. El-Batal, A. I., & Al Tamie, M. S., (2015). Biosynthesis of gold nanoparticles using marine *streptomyces cyaneus* and their antimicrobial, antioxidant and antitumor (*in vitro*) activities. *Journal of Chemical and Pharmaceutical Research, 7*(7), 1020–1036.
29. El-Sheekh, M. M., Hassan, L. H. S., & Morsi, H. H., (2021). Assessment of the *in vitro* anticancer activities of cyanobacteria mediated silver oxide and gold nanoparticles in human colon CaCo-2 and cervical Hela cells. *Environmental Nanotechnology, Monitoring & Management, 16*, 100556. doi: 10.1016/j.enmm.2021.100556.
30. Ferrari, M., (2005). Cancer nanotechnology: Opportunities and challenges. *Nature Reviews Cancer, 5*(3), 161–171.
31. Garcia-Mayagoitia, S., Perez-Hernandez, H., Medina-Perez, G., Campos-Montiel, R. G., & Fernandez-Luqueno, F., (2020). Bio-nanomaterials in the air pollution treatment. In: Abdeltif, A., Assadi, A. A., Nguyen-Tri, P., Nguyen, T. A., & Rtimi, S., (eds.), *Nanomaterials for air Remediation* (pp. 227–248). Elsevier.
32. Garg, A., Garg, S., Mishra, N., Enaganti, S., & Shukla, A., (2021). Advanced drug delivery systems in blood cancer. In: Yadav, A. K., Gupta, U., & Sharma, R., (eds.), *Advanced Drug Delivery Systems in the Management of Cancer* (pp. 141–154). Academic Press.
33. Ghashghaei, S., & Emtiazi, G., (2015). The methods of nanoparticle synthesis using bacteria as biological nanofactories, their mechanisms and major applications. *Current Bionanotechnology, 1*(1), 3–17.
34. Ghosh, S., & Webster, T. J., (2021). Viruses and nanotechnology. In: Ghosh, S., & Webster, T. H., (eds.), *Nanobiotechnology* (pp. 133–143). Elsevier.
35. Ghosh, S., Ahmad, R., Zeyaullah, M., & Khare, S. K., (2021). Microbial nano-factories: Synthesis and biomedical applications. *Frontiers in Chemistry, 9*, 626834. doi: 10.3389/fchem.2021.626834.
36. Grobmyer, S. R., Iwakuma, N., Sharma, P., & Moudgil, B. M., (2010). What is cancer nanotechnology?. In: Grobmyer, S., & Moudgil, B., (eds.), *Methods in Molecular Biology (Clifton, N.J.)* (pp. 1–9). Springer.

37. Guo, Y., Liu, Y., Wu, W., Ling, D., Zhang, Q., Zhao, P., & Hu, X., (2021). Indoleamine 2,3-dioxygenase (IDO) inhibitors and their nanomedicines for cancer immunotherapy. *Biomaterials, 276*, 121018. doi: 10.1016/j.biomaterials.2021.121018.
38. Hu, X., Saravanakumar, K., Jin, T., & Wang, M. H., (2019). Mycosynthesis, characterization, anticancer and antibacterial activity of silver nanoparticles from endophytic fungus *Talaromyces purpureogenus*. *International Journal of Nanomedicine, 14*, 3427–3438.
39. Hussein, H. A., & Abdullah, M. A., (2020). Biosynthesis, mechanisms, and biomedical applications of silver nanoparticles. In: Thangadurai, D., Sangeetha, J., & Prasad, R., (eds.), *Nanotechnology in the Life Sciences* (pp. 313–332). Springer.
40. Hwang, K. S., Kim, H. U., Charusanti, P., Palsson, B. Ø., & Lee, S. Y., (2014). Systems biology and biotechnology of Streptomyces species for the production of secondary metabolites. *Biotechnology Advances, 32*(2), 255–268.
41. Jagwani, S., Jalalpure, S., Dhamecha, D., Jadhav, K., & Bohara, R., (2020). Pharmacokinetic and pharmacodynamic evaluation of resveratrol loaded cationic liposomes for targeting hepatocellular carcinoma. *ACS Biomaterials Science & Engineering, 6*(9), 4969–4984.
42. Jani, P., Subramanian, S., Korde, A., Rathod, L., & Sawant, K. K., (2020). Theranostic nanocarriers in cancer: Dual capabilities on a single platform. In: Thangadurai, D., Sangeetha, J., & Prasad, R., (eds.), *Nanotechnology in the Life Sciences* (pp. 293–312). Springer.
43. Jeevanandam, J., Danquah, M. K., & Pan, S., (2021). Plant-derived nanobiomaterials as a potential next generation dental implant surface modifier. *Frontiers in Materials, 8*, 1–7.
44. Khalil, M. S. M., Marouf, S., Element, D., Timbs, A., Gallienne, A., Schuh, A., & Old, J. M., (2014). A Study of δ-globin gene mutations in the UK population: Identification of three novel variants and development of a novel DNA test for Hb A′ 2, *Hemoglobin, 38*(3), 201–206.
45. Khan, S. A., & Lee, C. S., (2020). Green biological synthesis of nanoparticles and their biomedical applications. In: Inamuddin, & Asiri, A. M. (eds.), *Nanotechnology in the Life Sciences* (pp. 247–280). Springer.
46. Khandanlou, R., Murthy, V., & Wang, H., (2020). Gold nanoparticle-assisted enhancement in bioactive properties of Australian native plant extracts, *Tasmannia lanceolata* and *Backhousia litriodora*. *Materials Science and Engineering: C, 112*, 110922. doi: 10.1016/j.msec.2020.110922.
47. Krishnan, S., Bhardwaj, S. K., Liu, S., Xing, R., & Chavali, M., (2021). Virus-assisted biological methods for greener synthesis of nanomaterials. In: Kharisov, B., & Kharissova, O., (eds.), *Handbook of Greener Synthesis of Nanomaterials and Compounds* (pp. 785–806). Elsevier.
48. Le, D. H. T., Lee, K. L., Shukla, S., Commandeur, U., & Steinmetz, N. F., (2017). Potato virus X, a filamentous plant viral nanoparticle for doxorubicin delivery in cancer therapy. *Nanoscale, 9*(6), 2348–2357.
49. Lee, H. U., Park, S. Y., Park, E. S., Son, B., Lee, S. C., Lee, J. W., Lee, Y. C., Kang, K. S., Kim, M. I., Park, H. G., et al., (2014). Photoluminescent carbon nanotags from harmful cyanobacteria for drug delivery and imaging in cancer cells. *Scientific Reports, 4*(1), 4665. doi: 10.1038/srep04665.

50. Lee, Y. C., & Moon, J. Y., (2020). Bionanomaterials production. In: Lee, Y. C., & Moon, J. Y., (eds.), *Introduction to Bionanotechnology* (pp. 33–60). Springer.
51. Leonida, M. D., & Kumar, I., (2016). Bionanomaterials from plant sources. In: Leonida, M. D., & Kumar, I., (eds.), *Bionanomaterials for Skin Regeneration* (pp. 91–101). Springer.
52. Lim, E. K., Chung, B. H., & Chung, S. J., (2018). Recent advances in pH-sensitive polymeric nanoparticles for smart drug delivery in cancer therapy. *Current Drug Targets, 19*(4), 300–317.
53. Liu, L., Liu, T., Tade, M., Wang, S., Li, X., & Liu, S., (2014). Less is more, greener microbial synthesis of silver nanoparticles. *Enzyme and Microbial Technology, 67*, 53–58.
54. Hamed, M. M., & Abdelftah, L. S., (2019). Biosynthesis of gold nanoparticles using marine *Streptomyces griseus* isolate (M8) and evaluating its antimicrobial and anticancer activity. *Egyptian Journal of Aquatic Biology and Fisheries, 23*(1), 173–184.
55. Madgula, K., & Puli, V. S., (2021). Bionanomaterials: Concept of nanomedicine. In: Singh, R. P., & Singh, K. R. B., (eds.), *Bionanomaterials: Fundamentals and biomedical applications* (pp. 1–20). IOP Publishing Ltd.
56. Majeed, S., Bakhtiar, N. F. B., Danish, M., Mohamad, I., M. N., & Hashim, R., (2019). Green approach for the biosynthesis of silver nanoparticles and its antibacterial and antitumor effect against osteoblast MG-63 and breast MCF-7 cancer cell lines. *Sustainable Chemistry and Pharmacy, 12*, 100138. doi: 10.1016/j.scp.2019.100138.
57. Mallick, S., Shankar, A., & Prasad, B., (2021). Bionanomaterials utility for therapeutic applications. In: Singh, R. P., & Singh, K. R. B., (eds.), *Bionanomaterials: Fundamentals and Biomedical Applications* (pp. 1–29). IOP Publishing Ltd.
58. Manivasagan, P., Alam, M. S., Kang, K. H., Kwak, M., & Kim, S. K., (2015). Extracellular synthesis of gold bionanoparticles by *Nocardiopsis* sp. and evaluation of its antimicrobial, antioxidant and cytotoxic activities. *Bioprocess and Biosystems Engineering, 38*(6), 1167–1177.
59. Misra, M., Sachan, A., & Sachan, S. G., (2021). Role of fungal endophytes in the green synthesis of nanoparticles and the mechanism. In: Sharma, V. K., Shah, M. P., Parmar, S., & Kumar, A., (eds.), *Fungi Bio-prospects in Sustainable Agriculture, Environment and Nano-technology* (pp. 489–513). Elsevier.
60. Misra, R., Acharya, S., & Sahoo, S. K., (2010). Cancer nanotechnology: Application of nanotechnology in cancer therapy. *Drug Discovery Today, 15* (19, 20), 842–850.
61. Mittal, A. K., & Banerjee, U. C., (2016). Current status and future prospects of nanobiomaterials in drug delivery. In: Grumezescu, A. M., (ed.), *Nanobiomaterials in Drug Delivery* (pp. 147–170). Elsevier.
62. Mohamedin, A., El-Naggar, N. E. A., Shawqi, H. S., & Sherief, A. A., (2015). Green synthesis, characterization and antimicrobial activities of silver nanoparticles by *Streptomyces viridodiastaticus* SSHH-1 as a living nanofactory: Statistical optimization of process variables. *Current Nanoscience, 11*(5), 640–654.
63. Moradi, F., Sedaghat, S., Moradi, O., & Arab, S. S., (2021). Review on green nano-biosynthesis of silver nanoparticles and their biological activities: With an emphasis on medicinal plants. *Inorganic and Nano-Metal Chemistry, 51*(1), 133–142.
64. Nagalingam, N., Kalpana, V. N., Devi, R. V., & Panneerselvam, A., (2018). Biosynthesis, characterization, and evaluation of bioactivities of leaf extract-mediated biocompatible

gold nanoparticles from *Alternanthera bettzickiana. Biotechnology Reports, 19*, e00268. doi: 10.1016/j.btre.2018.e00268.
65. Netala, V., Bethu, M., Pushpalatha, B., Baki, V., Aishwarya, S., Rao, J., & Tartte, V., (2016). Biogenesis of silver nanoparticles using endophytic fungus *Pestalotiopsis microspora* and evaluation of their antioxidant and anticancer activities. *International Journal of Nanomedicine, 11*, 5683–5696.
66. Nie, S., Xing, Y., Kim, G. J., & Simons, J. W., (2007). Nanotechnology applications in cancer. *Annual Review of Biomedical Engineering, 9*(1), 257–288.
67. Peer, D., Karp, J. M., Hong, S., Farokhzad, O. C., Margalit, R., & Langer, R., (2007). Nanocarriers as an emerging platform for cancer therapy. *Nature Nanotechnology, 2*(12), 751–760.
68. Perini, G., Giulimondi, F., Palmieri, V., Augello, A., Digiacomo, L., Quagliarini, E., Pozzi, D., et al., (2021). Inhibiting the growth of 3D brain cancer models with bio-coronated liposomal temozolomide. *Pharmaceutics, 13*(3), 378. doi: 10.3390/pharmaceutics13030378.
69. Pouresmaeil, V., Haghighi, S., Raeisalsadati, A. S., Neamati, A., & Homayouni-Tabrizi, M., (2020). The anti-breast cancer effects of green-synthesized zinc oxide nanoparticles using carob extracts. *Anti-Cancer Agents in Medicinal Chemistry, 21*(3), 316–326.
70. Rai, N. K., Singh, S. K., Tiwari, M., & Sharma, L. K., (2020). Implications of nanotechnology in cancer diagnostics and therapeutics. In: Thangadurai, D., Sangeetha, J., & Prasad, R., (eds.), *Nanotechnology in the Life Sciences* (pp. 271–291). Springer.
71. Rajeshkumar, S., Malarkodi, C., Paulkumar, K., Vanaja, M., Gnanajobitha, G., & Annadurai, G., (2014). Algae mediated green fabrication of silver nanoparticles and examination of its antifungal activity against clinical pathogens. *International Journal of Metals, 2014*, 1–8.
72. Ramaswamy, S. V. P., Narendhran, S., & Sivaraj, R., (2016). Potentiating effect of eco-friendly synthesis of copper oxide nanoparticles using brown alga: Antimicrobial and anticancer activities. *Bulletin of Materials Science, 39*(2), 361–364.
73. Ranjitha, V. R., & Ravishankar, R. V., (2021). Bioassisted synthesis of gold nanoparticles from *Saccharomonospora glauca*: Toxicity and biocompatibility study. *BioNanoScience, 11*(2), 371–379.
74. Ratan, Z. A., Haidere, M. F., Nurunnabi, M., Shahriar, S. M., Ahammad, A. J. S., Shim, Y. Y., Reaney, M. J. T., & Cho, J. Y., (2020). Green chemistry synthesis of silver nanoparticles and their potential anticancer effects. *Cancers, 12*(4), 855. doi: 10.3390/cancers12040855.
75. Rawal, S., & Patel, M., (2021). Bio-nanocarriers for lung cancer management: Befriending the barriers. *Nano-Micro Letters, 13*, 142. doi: 10.1007/s40820-021-00630-6.
76. Rehman, S., Farooq, R., Jermy, R., Asiri, S. M., Ravinayagam, V., Al Jindan, R., Alsalem, Z., et al., (2020). A wild *fomes fomentarius* for bioremediation of one-pot synthesis of titanium oxide and silver nanoparticles for antibacterial and anticancer application. *Biomolecules, 10*(4), 622. doi: 10.3390/biom10040622.
77. Sabu, C., Mufeedha, P., & Pramod, K., (2019). Yeast-inspired drug delivery: Biotechnology meets bioengineering and synthetic biology. *Expert Opinion on Drug Delivery, 16*(1), 27–41.
78. Saravanakumar, K., Shanmugam, S., Varukattu, N. B., MubarakAli, D., Kathiresan, K., & Wang, M. H., (2019). Biosynthesis and characterization of copper oxide nanoparticles

from indigenous fungi and its effect of photothermolysis on human lung carcinoma. *Journal of Photochemistry and Photobiology B: Biology, 190*, 103–109.

79. Shah, R. A., Frazar, E. M., & Hilt, J. Z., (2020). Recent developments in stimuli responsive nanomaterials and their bionanotechnology applications. *Current Opinion in Chemical Engineering, 30*, 103–111.
80. Sharma, D., Rajput, J., Kaith, B. S., Kaur, M., & Sharma, S., (2010). Synthesis of ZnO nanoparticles and study of their antibacterial and antifungal properties. *Thin Solid Films, 519*(3), 1224–1229.
81. Sheridan, E., Vercellino, S., Cursi, L., Adumeau, L., Behan, J. A., & Dawson, K. A., (2021). Understanding intracellular nanoparticle trafficking fates through spatiotemporally resolved magnetic nanoparticle recovery. *Nanoscale Advances, 3*(9), 2397–2410.
82. Shukla, A. K., Upadhyay, A. K., & Singh, L., (2021). Algae-mediated biological synthesis of nanoparticles: Applications and prospects. In: Mandotra, S. K., Upadhyay, A. K., & Ahluwalia, A. S., (eds.), *Algae* (pp. 325–338). Springer.
83. Siddiqi, K. S., & Husen, A., (2016). Fabrication of metal nanoparticles from fungi and metal salts: Scope and application. *Nanoscale Research Letters, 11*(1), 98. doi: 10.1186/s11671-016-1311-2.
84. Singh, P., Kim, Y. J., Zhang, D., & Yang, D. C., (2016). Biological synthesis of nanoparticles from plants and microorganisms. *Trends in Biotechnology, 34*(7), 588–599.
85. Singh, R., Shedbalkar, U. U., Wadhwani, S. A., & Chopade, B. A., (2015). Bacteriagenic silver nanoparticles: Synthesis, mechanism, and applications. *Applied Microbiology and Biotechnology, 99*(11), 4579–4593.
86. Sivaraj, R., Rahman, P. K. S. M., Rajiv, P., Narendhran, S., & Venckatesh, R., (2014). Biosynthesis and characterization of *acalypha indica* mediated copper oxide nanoparticles and evaluation of its antimicrobial and anticancer activity. *Spectrochimica Acta-Part A: Molecular and Biomolecular Spectroscopy, 129*, 255–258.
87. Souza, G. M., De Oliveira, V. K. C., Naldi, L. V., Pereira, V. C., & Winkelstroter, L. K., (2021). Green synthesized nanoparticles as a promising strategy for controlling microbial biofilm. In: Maddela, N. R., Chakraborty, S., & Prasad, R., (eds.), *Nanotechnology for Advances in Medical Microbiology* (pp. 1–28). Springer.
88. Supandi, Yeni, & Merdekawati, F., (2018). *In silico* study of pyrazolylaminoquinazoline toxicity by lazar, protox, and admet predictor. *Journal of Applied Pharmaceutical Science, 8*(9), 119–129.
89. Supraja, N., Dhivya, J., Prasad, T. N. V. K. V., & David, E., (2018). Synthesis, characterization and dose dependent antimicrobial and anticancerous efficacy of phycogenic (*Sargassum muticum*) silver nanoparticles against breast cancer cells (MCF 7) cell line. *Advances in Nano Research, 6*(2), 183–200.
90. Tarhan, T., Dündar, A., Okumuş, V., & Çulha, M., (2021). Synthesis and characterization of bionanomaterials and evaluation of their antioxidant, antibacterial, and DNA cleavage activities. *Chemistry Select, 6*(17), 4217–4223.
91. Thangavelu, R. M., Ganapathy, R., Ramasamy, P., & Krishnan, K., (2020). Fabrication of virus metal hybrid nanomaterials: An ideal reference for bio semiconductor. *Arabian Journal of Chemistry, 13*(1), 2750–2765.
92. Trabulo, S., Aires, A., Aicher, A., Heeschen, C., & Cortajarena, A. L., (2017). Multifunctionalized iron oxide nanoparticles for selective targeting of pancreatic cancer cells. *Biochimica et Biophysica Acta (BBA)-General Subjects, 1861*(6), 1597–1605.

93. Tran, P. H. L., & Tran, T. T. D., (2019). Current designs and developments of fucoidan-based formulations for cancer therapy. *Current Drug Metabolism, 20*(12), 933–941.
94. Venkatesan, J., Kim, S. K., & Shim, M., (2016). Antimicrobial, antioxidant, and anticancer activities of biosynthesized silver nanoparticles using marine algae *Ecklonia cava*. *Nanomaterials, 6*(12), 235. doi: 10.3390/nano6120235.
95. Wu, Y., Li, J., & Shin, H. J., (2021). Self-assembled viral nanoparticles as targeted anticancer vehicles. *Biotechnology and Bioprocess Engineering, 26*(1), 25–38.
96. Xue, Y., Yu, G., Shan, Z., & Li, Z., (2018). Phyto-mediated synthesized multifunctional Zn/CuO NPs hybrid nanoparticles for enhanced activity for kidney cancer therapy: A complete physical and biological analysis. *Journal of Photochemistry and Photobiology B: Biology, 186*, 131–136.
97. Yadav, A., & Mendhulkar, V., (2018). Antiproliferative activity of *Camellia sinensis* mediated silver nanoparticles on three different human cancer cell lines. *Journal of Cancer Research and Therapeutics, 14*(6), 1316–1324.
98. Yang, L., Shi, P., Zhao, G., Xu, J., Peng, W., Zhang, J., Zhang, G., et al., (2020). Targeting cancer stem cell pathways for cancer therapy. *Signal Transduction and Targeted Therapy, 5*, 8. doi: 10.1038/s41392-020-0110-5.
99. Zeng, Q., Wen, H., Wen, Q., Chen, X., Wang, Y., Xuan, W., Liang, J., & Wan, S., (2013). Cucumber mosaic virus as drug delivery vehicle for doxorubicin. *Biomaterials, 34*(19), 4632–4642.

CHAPTER 11

ACTIVE-TARGETED NANODRUG CARRIERS FOR CANCER THERANOSTICS

SHARMILADEVI PALANI

ABSTRACT

Nanomaterials offer several advantageous features over their bulk counterparts and have emerged as a promising drug carrier in the field of nanomedicine. In the battle with cancer, the effectiveness of chemotherapy is affected by severe systemic toxicity. These side effects primarily arise due to the lack of specificity of the anticancer drugs. Nanomaterials offer promising strategies to overcome this issue. The high surface-to-volume ratio, biocompatibility, and easily modifiable surface makes nanomaterials an ideal platform to act as targeted drug carriers. The nanodrug carriers are directed to the cancer site by modifying the surface with specific moieties and functional groups that have the affinity to bind the surface receptors of cancer cells. Thus targeted drug delivery can enhance the efficacy of the therapy and minimize side effects in normal tissues. Nanomaterials also offer the possibility of conjugating fluorescent tags to trace the path of the drug carrier after administration. In addition to this, nanomaterials can also be used to provide combinatorial therapy and can effectively act as a theranostic agent. This chapter discusses the utilization of different nanostructures as active targeted drug carriers and the mechanisms adapted to target the cancer cells.

Sustainable Nanomaterials for Biomedical Engineering: Impacts, Challenges, and Future Prospects.
Junaid Ahmad Malik, Megh R. Goyal & Mohamed Jaffer M. Sadiq (Eds.)

11.1 INTRODUCTION

The incidence of cancer keeps increasing and leads to millions of deaths worldwide. According to the World Health Organization (WHO), deaths related to cancer were estimated at 9.6 million in 2018. Men most often suffer from lung, prostate, colorectal, stomach, and liver cancers, whereas women often suffer from breast, colorectal, lung, cervical, and thyroid cancer. To advance the effectiveness of cancer treatment, it is essential to understand the pathophysiology of cancer in detail. The efficacy of chemotherapeutic agents is largely limited by the poor solubility and non-specific biodistribution. A multifaceted approach between clinicians, researchers, and engineers is needed to develop effective anticancer therapeutics. Anticancer drugs have to overcome the distinct physiological feature of the tumors to reach them. Thus the complex biological system poses a significant hurdle to the administered anticancer drugs. Most of the anticancer drugs get distributed all over the system and thus affect the healthy tissues leading to severe side effects.

Cancer therapy is mainly focused on the elimination of the tumor site as the growth of cancer cells leads to the formation of new blood vessels. Thus cancer research is primarily focused on the design and formulation of chemical agents and drug delivery vehicles that can load more amounts of drugs and deliver at a specific site. The overall objective of cancer therapy is to prolong the lifespan of the patients by pain management and to avoid the recurrence of cancer. Therefore, it is essential to lessen the toxic side effects of chemotherapy to provide effective cancer therapy.

Cancer research is no longer viewed from the point of cell type and composition alone, as cancer cells have adapted to various survival strategies. Hence the thrust in research needs to be focused on understanding cancer biology to design targeted drug carriers that can specifically exhibit their therapeutic action on cancer cells. Even with intensive chemotherapies, radiotherapies, and surgical removal complete eradication of cancer is still not achievable as cancer cells show high resistance under various adverse conditions. In several cases, cancer cells develop resistance to treatment and recur. Thus a multifaceted theranostic approach is needed to understand the various adaptation mechanisms of the cancer cells and monitor their responses to cancer therapy [37].

The term "theranostics" refers to a nanoplatform that integrates both therapeutic and diagnostic agents. In recent years, theranostics has become an interesting and widely researched topic in cancer research. Initially, theranostic agents were developed by the addition of an imaging moiety to the drug carrier. Imaging moieties such as MRI, CT, and PET contrast agents were attached to the drug carrier Advances in the fabrication of nanomaterials accelerated the development of nanotheranostics.

Nanotheranostic agents were able to provide non-invasive imaging and deliver the drug carrier to the pathological site, thereby minimizing the non-selective accumulation of drugs. Hence, nanotheranostic agents can greatly improve the prognosis and the therapeutic decision of the clinician's as well. Nanotheranostic agents also enable to track of the path of drug carrier, drug release, and biodistribution *in vivo*. Targeted nanodrug carriers promise reduced toxic side effects of chemotherapeutic agents by delivering the drug specifically to the tumor site. So far several strategies have been used to target the cancer site using nanodrug carriers. The primary aim of active targeting is to protect the therapeutic agent from the biological environment and to prevent non-specific interactions.

This chapter briefly discusses the design of actively targeted theranostic nanomaterials and the type of diagnosis and therapy they provide for effective cancer therapy.

11.2 DRUG TARGETING

Depending upon the approach employed, drug targeting is classified into passive targeting and active targeting. These targeting strategies are classified based on the pathophysiology of the tumor microenvironment [11].

11.2.1 PASSIVE DRUG TARGETING

Passive targeting allows the drug or a drug carrier to get deposited in the cancer site due to the physicochemical factors. Tumor cells undergo rapid cell growth and thus leads to the formation of new blood vessels to supply them with oxygen and other nutrients. The newly formed blood vessels are often disordered and enlarged which presents them with large fenestrations. This allows the nanodrug carriers to extravasate through the tumor vessels and accumulate in the tumor site. The poor lymphatic drainage of the fast-growing tumors thus retains the accumulated nanocarriers. This effect is called the enhanced permeability and retention (EPR) effect and is mostly observed in solid tumors. The mechanism behind active and passive targeting is presented in Figure 11.1.

Even though, EPR is used as the gold standard approach to target the drugs it is a major challenge to channelize the nanocarrier to a specific target. The administered nanocarriers are distributed all over the body, to get distributed in the tumor site by passive targeting. The lack of selectivity with the reference to the EPR effect is thus a major challenge to overcome and improve the efficacy of the drug carriers. The various factors that affect the EPR effect and the mobility of the nanocarriers have been discussed in several literatures [6, 31] hence it is not elaborated in this chapter.

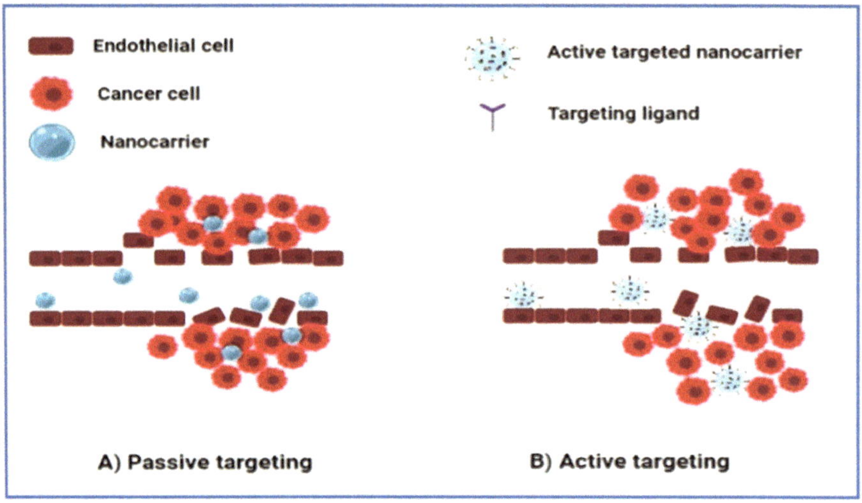

FIGURE 11.1 Schematic illustration of (A) passive targeting; and (B) active targeting.

11.2.2 ACTIVE DRUG TARGETING

While several studies have shown significant results with passive targeting, the lack of specificity led to research in active targeting methods. Active targeting refers to the attachment of a ligand that can be recognized by the over-expressed receptors on the surface of cancer cells. These ligands have the potential to bind with the over-expressed receptors and thus helps in actively targeting the drug. Thus the ligands help the nanocarrier to specifically release the therapeutic payload in the tumor site and prevent the non-specific distribution and toxic side effects.

Targeting ligands are chosen based on the high amount of particularly over-expressed receptors on the tumor cell surface or the tumor site. Targeting ligands are designed based on the larger presence of targetable components on the target cell or region. The targeting ligands often used in active targeting are antibodies, peptides, sugars, and lectins. Active targeting is aimed at enhancing the internalization of the drug carrier by binding the ligand with the over-expressed receptors. The efficacy of the antitumor response is due to the enhanced cellular uptake rather than the increased accumulation. Thus the binding affinity of the ligand is an important factor in designing actively targeted nanocarriers. The effectiveness of active targeting is independent of the tumor vasculature and enables direct cell killing and superior cytotoxic effects against cancer cells. The commonly targeted receptors in active targeting are listed in Table 11.1.

TABLE 11.1 List of Overexpressed Receptors on the Cancer Cell and in the Tumoral ndothelium

Cell Surface Receptors	Tumor Endothelial Receptors
Transferrin receptor	Vascular endothelial growth factor receptor (VEGF)
Folate receptor	$\alpha_v\beta_3$ integrin
Glycoproteins	Vascular cell adhesion molecule (VCAM-1)
Epidermal growth factor receptor (EGFR)	Matrix metalloproteinases (MMPs)

11.3 ACTIVELY TARGETED THERANOSTIC NANOCARRIERS

Theranostic nanoparticles generally comprise of a surface-modified nanocarrier loaded with a biomedical payload (Figure 11.2). In general, the biomedical payloads are imaging probes and therapeutic agents. Commonly loaded imaging agents are MR, CT, SPECT contrast agents, and fluorescent dyes, and therapeutic agents include chemotherapeutics, peptides, phototherapeutic agents, and proteins. The biomedical payloads are loaded into a nanocarrier for protecting the drug cargo under physiological conditions and to ensure the delivery of the cargo to the tumor site. Several nanocarriers have been designed and tailored specifically to target the cancer site. The surface of the nanodrug carriers is modified with modifiers to improve the biocompatibility, half-life time, and targeting ability [26].

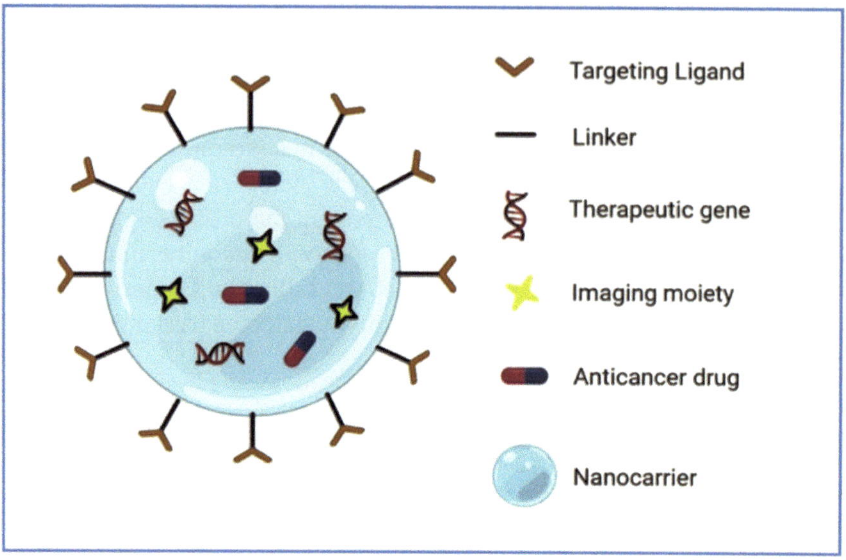

FIGURE 11.2 Design of actively targeted theranostic nanocarrier.

11.3.1 MAGNETIC NANOPARTICLES

Theranostic agents primarily require a drug carrier, a therapeutic drug, and an imaging agent. Many nanoparticles have the potential to function as a therapeutic agent and an imaging agent. Nanoparticles like carbon nanomaterials, QDs, magnetic nanoparticles, and metal nanoclusters possess the ability to act as imaging probes for MRI and optical imaging. Since these nanoparticles can act as a drug carrier and possess intrinsic imaging characteristics they are much preferred to be used as nanotheranostic agents.

Magnetic nanoparticles (MNPs) have drawn the attention of researchers in the field of cancer theranostics due to their potential to act as T2 contrast agents in magnetic resonance imaging. MNPs were initially researched for their diagnostic ability and was later explored for drug delivery. The delivery of MNPs can also be mediated by using external magnetic fields and they also have the potential to act as PTT agents. MNPs offer the advantage of negligible toxic effects and good biocompatibility. MNPs are poorly soluble in aqueous solutions and tend to aggregate hence the surface of the MNPs is often functionalized with functional groups that can stabilize and improve the solubility.

In active targeting, overexpression of the receptors is taken advantage of, and the surface of the nanoparticles are conjugated with targeting ligands. EGFR receptors are overexpressed in several tumors such as breast cancer, colorectal carcinoma, lung cancer, and ovarian cancer. Total EGF proteins are usually conjugated to the surface of the nanocarriers to target these receptors. EGF proteins using a short peptide fragment sequence called EGF fragment (EGFfr). The anticancer drug doxorubicin (DOX) and the peptide fragment was grafted to the MNPs surface to direct the nanocarrier to the cancer site and release the drug. It was reported that the nanocarrier released the drug in the cytosol by cleavage of the hydrazone bond due to the low pH of the endosomes. The attachment of the peptide fragment improved the targeting ability of the conjugate and the anticancer activity of the drug was improved by 7-fold [22].

The tumor microenvironment presents several barriers to the drug carriers as in the case of pancreatic cancer. The tumor stroma and the irregularities in the vasculature of pancreatic tumor tissues serve as a difficult barrier for the drug carriers to extravasate and deliver the drug. To overcome this barrier, Lee et al. [24] engineered a nanofabricate consisting of Iron oxide nanoparticles (FeNPs), gemcitabine, tetrapeptide (GFLG) linker, and an amino-terminal fragment (ATF) peptide. The nanoconjugate was designed to target the tumor overexpressing urokinase-type plasminogen activator receptor (uPAR) using the ATF peptide. The anticancer drug (Gem) was then linked with the carrier using GFLF linker which will undergo lysosomal cleavage to release the drug. This nanoconjugate showed significant cell killing in human pancreatic xenografts in mice. The biodistribution of the drug was assessed with T2 MRI. Thus the theranostic nanocarrier showed potential to overcome the tumor-stromal barrier and to enhance the therapeutic effect of the nanoparticle. The study showed that the anticancer drug is selectively released in the targeted site by receptor-mediated internalization [24].

Nanodrug carriers are not only used to target and ferry the drug cargo, but also to improve the solubility of the drugs. Many anticancer drugs suffer from poor solubility and thus their efficacy is limited. In a study by Luong et al. [28], a superparamagnetic iron oxide nanoparticle (SPION) based polyvalent theranostic nanoparticle was constructed using PAMAM dendrimers. The SPIONs were coated with PAMAM dendrimers decorated with folic acid (FA). Since folate receptors are expressed in higher amounts in tumor cells, modification of the drug carrier surface with FA leads to target the drug actively to the tumor site. The poorly soluble 3,4-diflurobenzylidene diferuloylmethane (CDF) drug was encapsulated into magnetic nanocarriers.

This encapsulation improved the stability, bioavailability, and half-life of the drug by 16-fold. The distribution of the drug was monitored by the T2 contrast of MRI provided by SPIONs [28].

Magnetic nanoparticles are also conjugated with other metal nanoparticles to produce multifunctional nanohybrids. These nanohybrids allow integrating dual-modal imaging agents and drugs to offer dual-modal imaging and combinatorial therapy. In one such approach, Nan et al. [29], fabricated a multifunctional nanohybrid to provide combinatorial therapy and imaging. In this interesting study, a core composed of the anticancer drug methotrexate (MTX) was encapsulated into thin gold nanoshells. The surface of the gold nanoshells was later distributed with FeNPs. This nanocomposite was then coated with polyethylene glycol to impart biocompatibility. The molecular structure of MTX is similar to that of folic acid and this enables the drug to selectively get attached to the overexpressed folate receptors. Also, the localization of the drug carrier by MRI allows to selectively radiate using a strong near-infrared light at 808 nm. This NIR radiation elevates the temperature of the nanocarrier thereby producing a photothermal effect on the cancer cells and also triggers the release of the drug MTX. Thus the nanohybrid produces synergistic chemo and photothermal therapy to ablate the cancer cells [29].

Biomimetic theranostic platforms are being increasingly researched for the treatment of cancer. One such approach is the use of living cells called Cell-based drug delivery systems (CDDSs). The CDDSs are being increasingly used for treating glioma bypassing the blood-brain barrier (BBB) and blood tumor barrier. Some living cells (e.g., mesenchymal stem cells and neural stem cells) possess certain intrinsic properties such as tumor-homing and drug-loading, hence cell-based therapies hold the potential for promising drug delivery systems. But the distribution and bio-effects of these cells need to be monitored to ensure the safety *in vivo.* Wu et al. [38], engineered inflammation-activatable neutrophils to actively target the glioma cells. A core-shell nanoparticle of FeNPs and mesoporous silica was synthesized and the shell was loaded with DOX. Then the drug-loaded nanoparticle was co-incubated with neutrophils to fabricate the biomimetic theranostic platform. This cell-based drug carrier demonstrated the potential to target the inflamed glioma site and the accumulation of the nanocarrier was monitored by MRI [38].

Prodrug-based approaches require the activation of the drug by enzymes that are specifically overexpressed in the tumor site. One such enzyme present in the tumor site is matrix metalloproteinases (MMPs) and it can cleave specific c peptide sequences. MMPs are a large family of enzymes, among them, the membrane-type MMP (MTMMP) has been identified to play a pivotal role in increasing the invasiveness of the cancer cells. Besides, it also acts as a direct cellular target for the activation of the prodrug. Ansari et al. [4], linked the prodrug azademethycolchicine to FeNPs using an MMP-14 cleavable peptide substrate to generate an enzyme activated nanotheranostic agent. The drug carrier was tracked using MRI and the drug conjugated showed improved antitumor efficacy [4].

Magnetic nanoparticles are regularly coated with biocompatible compounds and natural polymers to improve their solubility and stability. Cyclodextrin (CD) is most preferred for drug delivery applications as the central core serves as an excellent cavity for entrapping hydrophobic drugs. Further, the high amount of hydroxyl groups present on the surface of the CDs imparts them with hydrophilic nature. Gholibegloo et al. [17], synthesized CD-based nanosponges (CDNSs) to improve the porosity and surface area of the CDs to load more drugs and to maintain a sustained release of the drugs. The presence of a large amount of carbonyl and hydroxyl groups on the surface of the CDNSs facilitated the linker free linkage on the surface of FeNPs. This nanocomposite was later grafted with molecules of FA to target the tumor cells. The poorly soluble drug curcumin was entrapped in the porous matrix of the nanosponges to provide a therapeutic effect. Thus the smart nanocarrier provided active targeted controlled release of the drug as well as localization of the tumor [17].

FeNPs are well known for their T2 contrasting ability in MRI. Due to their excellent biocompatibility, several studies have also developed FeNPs for positive contrasting in MRI. Cancer therapy is not only focused on killing the cancer kills but it is also targeted at inhibiting the angiogenesis. In that aspect, a study by Groult et al. [18], investigated the effect of heparin coating over the FeNPs. The utility of heparins (unfractionated and low-molecular-weight) as anticoagulants is well-known. Heparins also possess a strong affinity to bind growth factors and can inhibit the hydrolytic enzyme heparanase over-expressed in the TME. This study showed that the bioactivity of the heparins was not affected upon functionalization on the surface of FeNPs. The heparin-coated FeNPs demonstrated good positive contrast in MRI and anti-heparanase activity. Thus this nanocomposite showed the potential to act as an actively targeted nanocarrier for cancer imaging and therapy [18].

Along with MR imaging, other imaging modalities are also combined to be used for bimodal imaging applications. Guo et al. [19], fabricated a nanocomposite that is capable of producing bimodal contrast in photoacoustic

(PA) imaging and MR imaging. The nanocomposite was also engineered to treat breast cancer with chemotherapy and photothermal effect. This combinatorial theranostic platform was designed utilizing FeNPs, perfluorohexane(PFH), carboxyl modified PEG, and poly (lactic-co-glycolic acid) (PLGA), paclitaxel (PTX), and herceptin. The antibody herceptin is used as the ligand to facilitate the active target of the HER2 receptor and accumulate the drug carrier in the TME. The encapsulated SPIONS served as a source of PA imaging probe and photothermal agent as well. Further, the NIR radiation also induced the formation of PFH gas bubbles which helped to image the tumor by ultrasound imaging. This nanosystem demonstrated active targeting of breast cancer and real-time bimodal imaging and synergistic cancer therapy [19].

Dual targeting strategies are also employed to improve the localization of the drug carrier in the TME. In an interesting study Liu et al. [27], conjugated the surface of MNPs with hyaluronic acid (HA) and folic acid (FA) to actively transport the drug carrier. Another interesting aspect of this study was the utilization of S-nitrosothiols as therapeutic moiety. Nitric Oxide (NO) is a diatomic free radical associated with the regulation of several physiological functions. Several NO-releasing materials have been researched and utilized for addressing diseases caused by the insufficiency of nitrogen. S-nitrosothiols can be stimulated by temperature/light or other reactive substances to release NO in a controlled manner. Higher concentrations of NO can induce cell apoptosis. Hence localization of the S-nitrosothiols can be utilized to kill the cancer cells [27].

FeNPs are not only used as the core component in core-shell material several studies have also utilized the FeNPs for surface modification. In a study by Shi et al. [33], FeNPs were decorated on the surface of fullerene (C60) and PEG was coated on the surface to improve biocompatibility. To impart therapeutic property a PDT agent was conjugated to the C60-FeNPs-PEG surface. Hematoporphyrin monomethyl ether (HMME) is a porphyrin photosensitizer used in PDT therapy but its use is limited due to the light absorption and less bioavailability. Hence the nano-formulated C60-FeNPs-PEG/HMME in comparison with HMME, showed enhanced photothermal ablation of cancer cells *in vitro* B16-F10 cell line and *in vivo* murine tumor model. The cellular uptake was found to be increased by 23-fold thereby improving the efficacy of the treatment. The surface decorated FeNPs provided good T2 contrasting in MRI [33].

Multifunctional nanotheranostic agents integrate multiple imaging and therapeutic moieties and target them to the tumor site by active targeting.

Yue et al. [41], designed a pH-responsive nanotheranostic agent capable of multimodal imaging and effective killing of the cancer cells. The nanosystem was fabricated using iron platinum nanoparticles (FePt NPs) and graphene oxide (GO). The acidic pH in tumor cells triggers the release of Fe ions which in turn catalyzes the decomposition of H_2O_2 into reactive oxygen species (ROS) in the cancer cells thereby inducing cell death. The nanoconjugate was coated with PEG to improve the biocompatibility and FA was used as the targeting moiety. Thus the multifunctional nanocarrier provided real-time monitoring by T2 MRI and CT along with targeted drug delivery [41].

11.3.2 QUANTUM DOTS (QDS)

Quantum dots (QDs) have gained a lot of interest in biomedical research as optical imaging probes. These colloidal semiconductor nanocrystals offer a long fluorescence lifetime and high stability due to their unique physicochemical properties. QDs offer the advantage of fine-tuning the optical properties by varying the composition, shape, and size. Core-shell type QDs are often utilized for imaging important biological events in real-time *in vivo* imaging. Multifunctional QDs are fabricated by entrapping them in a carrier along with therapeutic agents [42].

Doxorubicin, despite being a common first-line therapy for various cancers including breast, bladder, lung, and ovarian cancer it's application is limited due to the cardiotoxicity caused by the cumulative dose-dependent release. Alibolandi et al. [3], designed a QD-based theranostic agent for the treatment and diagnosis of breast cancer. In this work, the hydrophobic DOX and hydrophilic MSA-capped QDs were encapsulated into polymersomes to release the drugs in a sustained manner. The surface of the polymersomes was coated with PEG and with folate receptors to target the TME. The high photostability of the QDs provided the good fluorescent ability and the drug carrier showed sustained drug release without causing severe side effects [3].

To overcome the issue of poor solubility several nanodrug formulations have experimented. In a study by Abdelhamid et al. [1], a nanotheranostic agent was developed based on the encapsulation of the hydrophobic drugs rapamycin (RAP) and celecoxib (CXB) into the nanocapsules. The drug CXB inhibits the cyclooxygenase (COX-2) enzyme, which is over-expressed in cancer cells, and this enzyme is reported to promote the growth of cancer cells by activating procarcinogens into carcinogens. Even though CXB can selectively inhibit the COX-2 it is poorly soluble in water and also leads to serious side effects. Hence a targeted approach is needed to improve the efficacy without affecting the healthy tissues. The drug RAP also suffers from poor bioavailability due to poor solubility, gastric acid sensitivity, and first-pass metabolism. The nanocapsules were incorporated with positively charged benzalkonium chloride (BKC) followed by the deposition of anionic polysaccharide chondroitin sulfate (CS). The anionic charge of the CS-nanocapsules (CS-NCs) was used to electrostatically deposit a positively charged layer of enzyme-responsive type-A gelatin. This layer prevents the CS-NCs from non-selective internalization and gets degraded by the matrix metalloproteinases (MMPs) thereby facilitating receptor-mediated internalization into the tumor site. Finally, the surface of the NCs was coated with thiol-capped cadmium telluride (CdTe) QDs via a tumor-cleavable bond to offer imaging-guided therapy. This nanodrug carrier showed superior antitumor property along with the imaging-guide route [1].

Semiconductor QDs are doped with impurities to alter their fluorescent properties. Bwatanglang et al. [7], doped the ZnS QDs with Mn^{2+} ions to emit orange-red fluorescence around 600 nm. The doping of the impurity significantly improved the fluorescence efficiency of the QDs. In this study, a nanocomposite was fabricated by loading the drug 5-fluorouracil (5-FU) into chitosan biopolymers encapsulated with Mn-ZnS QDs. The surface of chitosan biopolymers was later coated with folic acid to target the folate receptors. The drug 5-FU is commonly utilized for treating neck, colorectal, and breast cancers but has limited bioavailability and causes severe side effects. The inclusion of folate receptors on the nanocarrier significantly improved the anti-tumor efficacy *in vitro* and *in vivo*. The cytotoxicity assays also revealed that the nanoconjugate did not show any toxic effects. Confocal laser scanning images showed high-intensity fluorescence images of the breast cancer cells MCF-7 and MDA-MB231 [7].

Nutlin-3a is a potent anticancer drug, but its efficacy is limited due to the poor solubility and non-specific biodistribution. Polymeric nanoparticles are used to encapsulate poorly soluble anticancer drugs to improve solubility. Further polymeric nanoparticles are biodegradable and can release the drug in a sustained manner. In a study by Das et al. [12], PLGA nanoparticles were synthesized and the anticancer drug nutlin-3a was encapsulated into the nanoparticles. The nanocarrier was conjugated with QDs for bioimaging and tracking the carrier *in vivo*. To actively target this nanocarrier to the target site the PLGA nanoparticles were conjugated with nucleic acid aptamers that are specific to bind with epithelial cell adhesion molecule (EpCAM). Aptamers are short single-stranded oligonucleotides based on DNA or RNA.

These nucleic acid aptamers upon folded into a three-dimensional (3D) structure holds a high affinity to bind to ligands or protein-based targets. The overexpression of EpCAM in tumor cells enables the aptamer conjugated nanocarrier to selectively deliver the drug in the TME [13].

NIR emitting QDs provides high penetration depth than the QDs emitting in the visible region. Imaging in the NIR region provides high resolution and high-intensity images by eliminating the autofluorescence of the collagen tissues. Duman et al. [15], fabricated a nanotheranostic agent using Ag_2S QDs that emit in the NIR range. The QDs were coated with PEG and Cetuximab (Cet) antibody was conjugated as the targeting moiety due to its high affinity to bind EGFR. The Cet conjugated nanocarrier was loaded with the anticancer drug, 5-fluorouracil (5FU) to exhibit a synergistic effect. The fabricated theranostic agent was targeted for the visualization and treatment of lung cancer using cell lines that express the EGFR receptors in low (H1299) and high (A549) levels. The antibody tagged QDs efficiently targeted the A549 cells and demonstrated a superior cell-killing effect than the treatment of 5FU alone [15].

Protein-inorganic hybrid nanoplatforms are gaining interest as drug carriers and biosensors. Xie et al. [39], synthesized a protein-inorganic hybrid nanoplatform using a photo-treated disulfide bond rich protein of lysozyme (LYS), NIR QDs, and the chemotherapeutic drug paclitaxel (PTX). The QDs and PTX were encapsulated into LYS and this protein-inorganic nanocarrier was modified with the cRGD peptide and the surface was coated with PEG. Arg-Gly-Asp (RGD) peptides have a good affinity to bind with the overexpressed integrins $\alpha_v\beta_3$. The prepared nanoconjugate showed good intracellular uptake in human esophageal cancer cells [ECA109] due to the active targeting of the peptides with the integrins [39].

11.3.3 CARBON-BASED NANOMATERIALS

Rapid development in the field of nanotechnology led to more fascinating ideas and applications of nanomaterials in the diagnosis and treatment of diseases. In particular, carbon nanomaterials have gained a lot of attention from researchers and have generated huge impacts in the field of materials science research. Carbon-based nanomaterials possess unique intrinsic properties thus graphene oxide (GO), carbon nanotubes (CNTs), graphene quantum dots (GQDs), and carbon dots (CDs) are extensively used for biomedical applications. Das et al. [12, 13], fabricated an active targeted multifunctional theranostic agent capable of exhibiting cytotoxic effect, monitoring of internalization, and biodistribution of the nanocarrier via fluorescence imaging and radioscintigraphy. Folic acid was used as the targeting ligand and the anticancer drug methotrexate (MTX) was conjugated on the surface of multi-walled carbon nanotubes (MWCNTs). The nanoconjugate was further labeled with a radio-tracer Technitium-99m and a fluorescent dye Alexa-Flour (AF-647/488). MWCNTs offer the advantage of conjugating all the agents through chemical linkages and provides stability *in vivo*. Interestingly, the imaging probes were linked through hydrolytically stable amide linkages, whereas the therapeutic moiety was coupled through an ester linkage that can be hydrolyzed in intracellular conditions. This study also studied the impact of functional molecules on biodistribution, cellular uptake, and antitumor efficacy. The cytotoxic efficacy of the cells was tested on lung cancer (A549) and breast cancer (MCF-7) cell lines. The multimodal imaging probes enabled the real-time visualization of treatment response [12].

Functionalized CNT-based platforms are used for the integration of multifunctional agents for theranostic applications. Al-Faraj et al. [2], synthesized polyvinylpyrrolidone (PVP) functionalized single-walled carbon nanotubes (SWCNTs) and tagged the surface with FeNPs and an Endoglin/CD105 antibody. The nanocarrier was designed to deliver the drug DOX in the tumor site, thereby improving the cell-killing effect. The coupled monoclonal antibody targets the tumor site and the FeNPs enable the non-invasive tracking of the drug carrier *in vivo*. The synthesized CNT conjugate was stable *in vivo* and in addition to the CD105 high-energy flexible magnet was used to magnetically target the drug delivery system. The CNT-based nanotheranostic agent demonstrated a superior cell-killing effect in the murine breast cancer model [2].

Carbon dots (CDs) are an interesting class of carbon-based materials for nanotheranostics. CDs are intrinsically fluorescent, highly soluble, and are highly biocompatible. The high fluorescence of the CDs is used to non-invasively image the drug carrier and the tumors as well and can also be easily functionalized with surface groups. Hence CDs are often used in the preparation of nanotheranostic agents by conjugation of anticancer drugs and targeting moieties. Gao et al. [16], fabricated an interesting theranostic agent based on a turn-on fluorescent nanoprobe. The CDs surface was modified with polyethyleneimine (PEI) and hyaluronic acid (HA)-conjugated doxorubicin (Dox). Hyaluronic acid (HA) has a high affinity to bind overexpressed CD44 receptors on many cancer cells. Hence HA acts as the active targeting moiety to penetrate cancer cells. Hyaluronidase (HAase) has been reported to be associated with the presence of malignant tumors and can degrade HA. High expression of HAase is reported in cancer cells. Hence the presence of HAase in the tumor cells is used as a fluorescent turn-on mechanism for the PEI-CDs. The fluorescence intensity of the prepared conjugate is lesser in the physiological environment and high in the cancer site due to the degradation of HA in the presence of HAase. This degradation process restores the fluorescence of the P-CDs. The nanoconjugate demonstrated good sensitivity up to 0.65 UmL^{-1}. The MTT assay performed on HeLa cells showed that the nanoconjugated DOX efficiently killed the cells [16].

Carbon dots are also conjugated with metals to be used as dual-modal imaging agents. In a study by Yao et al. [40], Gadolinium-CDs complexes were synthesized to act as nanoprobes for optical imaging and magnetic resonance imaging (MRI). These nanoprobes were later decorated on the surface of apoferritin nanocages encapsulated with a high concentration of DOX. Apo ferritin nanocages are preferred as nano-drug carriers because of their hollow structure and their ability to unfold and refold at pH 2.0 and 7.4. The surface of the nanocages was modified with folic acid to actively the drug to tumor cells. The anticancer activity of the active targeted drug was tested on MCF-7 cell lines and in mice models [40].

Graphene oxide-based nanocarriers are also designed to actively target drugs. Tian et al. [35], designed a nanoconjugate of graphene oxide (GO) conjugated with PEG and FA. The surface was further labeled with a peptide and a fluorescent dye. This nanoconjugate was designed to provide imaging, therapy along with the detection of caspase-3 activity. Higher levels of caspase-3 are associated with an increased rate of cancer recurrence and deaths. Hence detection of caspase-3 activity will help in the prognosis of therapeutic response. To achieve this aim, caspase-3-specific substrate peptide (CALNNDEVDK-FAM) was linked to the carrier. The substrate peptide was labeled with fluorescein, which can undergo cleaving in the presence of active caspase-3. The high surface area of the GO was used to conjugate aromatic and hydrophobic drugs such as camptothecin, curcumin, and evodiamine. This nanoconjugate showed selective uptake in the cancer cells and it was imaged by the high fluorescence generated by the fluorescein-conjugated to the peptide. Hela cell line was used to determine the efficacy of nano-formulated drugs. Thus this GO nanoconjugate demonstrated a potential nanotheranostic agent for the delivery and therapeutic self-monitoring [35].

11.3.4 SILICA NANOPARTICLES

Mesoporous silica nanoparticles (MSNs) are considered important nanodrug carriers because of their ability to release the drugs in a controlled manner. The well defined internal pores that are in the range of mesopores (2–50 nm) offers large pore volume and surface area. The pore size of the MSNs can be easily tuned according to the application. Besides, the easily modifiable surface of the MSNs makes them ideal candidates for designing controlled release drug delivery platforms. MSNs have been largely explored to fabricate multifunctional nanosystems by integrating therapeutic and imaging moieties. MSNs are also well tolerated both *in vitro* and *in vivo* systems and do not cause toxic side effects [36].

Chen et al. [9], designed an MSNs based multifunctional agent comprised of FeNPs in the core and a shell loaded with the anticancer drug DOX. The surface of the MSNs was conjugated with a targeting peptide adamantane-PEG_8-glycine-arginine-glycine-aspartic-serine (AD-PEG_8-GRGDS). To prevent the premature release of the drug a gatekeeper (β-cyclodextrin (β-CD)) was linked to the shell via the platinum (IV) drug that can be intracellularly reduced to platinum (II) drug. The prepared nanoconjugate results in the detachment of the gatekeeper in the TME and the activation of the platinum (II) results in the ablation of the cancer cells. The FeNPs in the core allows for the visualization of the drug accumulation in the tumor site in MR imaging. Thus the prepared conjugate exhibited a synergistic effect on the killing of cancer cells. The targeting efficacy and cellular uptake of the nanocarrier was assessed in two different cell line of COS7 (African green monkey kidney fibroblast cells) and Hela cell line using flow cytometry and confocal laser scanning microscopy [9].

The multistep engineering process is involved in the fabrication of multifunctional theranostic nanoagents. Multimodal imaging moieties are conjugated to enhance the detection of the tumor. In a study by Chen et al. [8], a multifunctional agent targeting the *in vivo* vasculature was developed using copper sulfide (CuS) nanoparticles and MSNs. The CuS nanoparticles were used as the core and MSNs was used as the shell to act as a reservoir for an anticancer drug. CuS nanoparticles exhibit strong near-infrared (NIR) optical absorption and thus this conjugate will provide synergistic photothermal and chemotherapy in cancer cell killing. The outer surface of the MSNs was functionalized with amino groups, PEG, and a chelator 1,4,7-triazacyclononanetriacetic acid (NOTA) for the labeling of copper-64. This conjugate was targeted to the cancer site by engineering the surface with IgG1 monoclonal antibody (TRC 105). This antibody has a high affinity to bind both human and murine CD105 (endoglin). CD105 is one of the endothelial cell proliferation biomarkers and is found to be overexpressed in most of the tumor neovasculature. TRC105 is also used for PET imaging of endoglin. Photothermal therapy of the tumor cells carried out in 4T1 tumor-bearing mice showed that the cells were killed efficiently and there was no recurrence for two months. Thus the nanosystem served as a promising agent for photothermally enhanced drug release and thermochemotherapy [8].

Cancer treatment requires constant monitoring of therapeutic response thus several theranostic formulations were developed. Pasha et al. [30], used a platinum(II)-based molecule BMPP-Pt which produces intracellular fluorescence by 'aggregation-induced emission' and also has the potential to kill the cancer cells. This dual functional molecule was encapsulated in MSNs surface modified with anti-EpCAM aptamer. The aptamer modified nano-conjugate enhanced the intracellular uptake and also showed high-intensity fluorescence and effective cell killing in human hepatocellular carcinoma cell line Huh7. This study showed the potential of a single compound-based actively targeted theranostic modality [30].

MSNs are attractive platforms for constructing multifunctional nanotheranostic agents due to their porous nature and the easily modifiable surface. Li et al. [25], synthesized bismuth sulfide (Bi_2S_3) and MSNs core-shell nanoparticles for targeted therapy of HER-2 positive breast cancer cells. The core material composed of rod-shaped Bi_2S_3 NPs surface modified with polyvinylpyrrolidone and mesoporous silica was used as the shell. The anticancer drug DOX was loaded into the porous shell and the surface was modified with trastuzumab, a monoclonal antibody with high affinity to target the HER-2 overexpressed cells. The *in vitro* and *in vivo* studies conducted in breast cancer cell line SkBR-3 showed that the accumulation of the nanoparticle in the targeted group was 16 times higher than in the non-targeted group. The presence of the element bismuth with strong X-ray attenuation properties facilitated the imaging of the tumor in computed tomography and photothermal therapy. Thus the nanoconjugate provided image-guided photothermal and chemotherapy [25].

11.3.5 METAL NANOPARTICLES

Nanoparticles (NPs) and noble metal NPs are attractive candidates for biomedical applications and have been proposed as nontoxic drug carriers.

These metal nanoparticles can provide simultaneous diagnosis and therapy. Colloidal gold nanoparticles (AuNPs) display localized surface plasmon resonance (LSPR) and can thus absorb a specific wavelength of light. Thus AuNPs have photoacoustic and photothermal properties which promote their use in photothermal therapies and diagnostic imaging applications. Modification of the shape and size of the AuNPs results in tuning their wavelength and other physicochemical properties. Thus AuNPs in combination with other metals and anticancer drugs offer synergistic therapeutic effects in killing the cancer cells. Hao et al. [20], designed a nanocomposite composed of docetaxel (DTX), PLGA, and AuNPs. A core-shell type nanomaterial was formed with the drug-loaded PLGA as the core and gold nanoshell. This nanocomposite was directed to the tumor site with the peptide angiopep-2 through Au-S bond. This nanocomposite used the NIR laser as the external stimulus to generate the heat and to release the drug to kill the cancer kills. The drug release showed a strong dependence on the NIR laser and the nanocomposite also showed potential in X-ray imaging. The synergistic effect of this nanosystem was demonstrated in U87MG glioma cells *in vitro* and *in vivo* [20].

AuNPs are also combined with other metal nanoparticles to enhance the targeted drug release properties of the nanoconjugate. Jenkins et al. [21], designed a plasmonically active drug conjugate system comprised of gold nanorods coated with silver nanoparticles (AgNPs). This nanosystem was loaded with DOX and the surface was modified with an EpCAM antibody. High levels of EpCAM expression is associated with triple-negative breast cancer (TNBC) and hence it is considered a viable target to develop therapeutic agents. The presence of silver coating over the surface of the nanosystem enabled the tracking of the drug carrier and identification of the tumor site by surface-enhanced Raman spectroscopy (SERS) and photoacoustic imaging. The efficacy of the drug release was demonstrated in 4T1 and MDA-MD-231 cell lines. A superior cell-killing effect was observed in 4T1 cells as they express EpCAM two times higher than the MDA-MD-231 cells. Thus this conjugate showed the potential of the nanoconjugate in efficiently targeting the overexpressed EpCAM tumors [21].

Near-infrared (NIR) imaging (700–900 nm) is considered superior to visible light (400–700 nm) as it offers high penetration depth with low tissue absorption. Researchers are also investigating the possibility of imaging in the second near-infrared window (NIR-II, 1,000–1,700 nm) to further enhance the penetration depth and absorption by the tissues. Rare earth metal nanoparticles (RE NPs) are highly stable and have a high fluorescence lifetime. Ding et al. [14] synthesized $NaNdF_4$ NPs (NNF NPs) and found that the photoluminescence and photothermal conversion efficiency differs with an increase in the size of the NPs (4.7, 5.9, 12.8, and 15.6 nm). Among these NPs 12.8 nm was further experimented based on the obtained results. The surface of the NNF NPs was coated with PEG and modified with RGD peptide to target the cancer cell. These NPs showed better photothermal effect under 808 nm laser and photoluminescence effect under 793 nm laser in both *in vitro* and *in vivo*. The fabricated NNF NPs were also labeled with the radiotracer ^{99m}TC to visualize the targeting ability of the NPs towards 4T1 cancer cells in SPECT/CT imaging [14].

11.3.6 LIPOSOMES

Among all the nanomaterials that are being experimented with for use in nanomedicine, liposomes are considered an important member of the therapeutic nanodrug carriers. Liposomal nanoformulations were the first FDA approved nanotherapeutic agent for cancer therapy. Until now liposomal formulations are largely explored in the development of nanotherapeutics. The amphiphilic phospholipid bilayer of the liposomes not only enables the encapsulation of hydrophobic and hydrophilic drugs but also protects the drug cargo from destructive conditions like pH and other enzymes. Liposomes are highly biocompatible and are negligibly toxic. Liposomal formulations also enhance the cellular uptake and the surface modifications are carried out to target them to the cancer site [5, 32].

An on-demand drug-releasing multifunctional liposomal formulation was fabricated by Dai et al. [10], based on thermosensitive liposomes. In this work, DOX, and the dye indocyanine green (ICG) were co-encapsulated into the liposomes. The surface was conjugated with folic acid and gadolinium ions to actively target the carrier and to visualize the drug carrier in MRI. The ICG dye upon NIR irradiation triggered the drug release along with photothermal and photodynamic therapy. In this work novel, fatty acids were mixed to form a phase change material (PCM) core and the encapsulation of drugs in this core enables to release of the drug upon NIR irradiation. This nanoconjugate offered multimodal imaging such as PA, MR, and fluorescence imaging *in vitro* and *in vivo*. The anticancer effect of combination therapy (chemo/PTT/PDT) was shown in HeLa cell lines [10].

Clinical translation of nanoformulation is limited due to the poor distribution and uptake of the drug in the tumor region. External stimuli triggered drug release have been experimented to overcome this limitation. Thomas et al. [34], used ultrasound (US) as an external stimulus to enhance the liposomal delivery to the tumor site. In this study, DOX was encapsulated into liposomes and the surface was modified with Indium-111 tagged epidermal growth factor to target the breast cancer cells. The tagged Indium not only helps in SPECT imaging but also induces DNA damage in cancer cells. Thus the combination of In and DOX provides synergistic chemo-radiation therapy. The combination of US and gas-filled microbubbles helps the degree of extravasation by the nanodrug carriers by increasing the permeability of the blood vessels and convective forces. The efficacy of the nanocarrier in cancer cell killing was investigated in two different cancer cell line of MDA-MB-468 and MCF-7 with high and low expression of EGFR. Selective uptake and high cell killing were observed in the MDA-MB-468 cell line in comparison with the MCF-7 cell line. The enhancement in the tumor uptake was found to be increased by 66% in MDA-MB-469 xenografts [34].

Multifunctional liposomes with dual targeting and combination therapy have shown favorable results in cancer treatment. Kono et al. [23], developed a multifunctional liposomal formulation for dual-targeted cancer. This study used the external trigger and gradient magnetic field guided approach to release and target the drug carrier to the cancer site. The multifunctional assembly was composed of fullerene (C60) decorated with FeNPs-PEG and this nanohybrid was co-encapsulated into thermosensitive liposomes along with docetaxel (DTX). The surface of the liposomes with grafted with FA to target the nanocarrier to the cancer site. Radiofrequency (RF) thermal therapy (RTT) has gained attention in cancer therapy due to its non-invasive nature, controllable features, and efficacy in killing cancer cells. In RTT, RF absorbing agents convert the RF energy into heat, thereby killing the cancer cells. Further RF radiation can penetrate the tissues without much energy loss. In this study, the exposure of C60 to RF radiation (13.56 MHz) leads to heat release capable of killing cancer cells. Thus the generated heat was used as a trigger to release the drugs from the thermosensitive liposomes. *In vitro* and *in vivo* studies carried on the MCF-7 cell line showed that the multifunctional liposomes significantly killed the cancer cells through RTT, DTX, and also showed better T2 contrasting ability in MRI in locating the tumor site [23].

11.4 SUMMARY

Active targeting of the theranostic nanosystems demonstrates superior internalization and high therapeutic effect in cancer therapy. While active targeting of the nanocarriers is impactful in determining the therapeutic efficacy of the therapeutic moieties still thorough investigation is needed in some areas. It is essential to optimize the size and physicochemical properties of the nanosystem according to the tumor density, extracellular matrix organization, and the extent of the EPR effect. Further, it is essential to develop models for the quantification of EPR along with efficacy models. The need for EPR effect can be bypassed in tumors that can be easily accessed by other administration routes such as local delivery. Several researchers have tried to improve the amount of targeting ligands on the surface of nanocarriers to maximize the targeting efficiency. It is to be noted that high targeting ligand density might lead to steric hindrance and an increase in the size of the nanosystem. Thus the density of the targeting ligands has to be optimized using imaging techniques and efficacy studies in both *in vitro* and *in vivo*. Therefore, it is important to design methods to optimize the ligand density and to further understand the receptor interaction. Along with the toxicity evaluations, it is of prime importance to consider the pharmacokinetic and pharmacodynamic properties of the nanomaterials while designing the theranostic nanomaterials. Since cancer therapy involves gene-level control, studies should also focus on developing gene-delivery vehicles that will lead to the design of personalized cancer therapy. Future theranostic approaches can consider the above-stated points to develop actively targeted nanotheranostics to meet the criteria required for clinical translation.

KEYWORDS

- **active targeting**
- **cancer therapy**
- **cell surface receptors**
- **diagnosis**
- **multimodal imaging**
- **nanodrug carriers**
- **targeted drug delivery**

REFERENCES

1. Abdelhamid, A. S., Helmy, M. W., Ebrahim, S. M., Bahey-El-Din, M., Zayed, D. G., Zein El, D. E. A., El-Gizawy, S. A., & Elzoghby, A. O., (2018). Layer-by-layer gelatin/chondroitin quantum dots-based nanotheranostics: Combined rapamycin/celecoxib delivery and cancer imaging. *Nanomedicine, 13*(14), 1707–1730. https://doi.org/10.2217/nnm-2018-0028.
2. Al Faraj, A., Shaik, A. P., & Shaik, A. S., (2015). Magnetic single-walled carbon nanotubes as efficient drug delivery nanocarriers in breast cancer murine model: Noninvasive monitoring using diffusion-weighted magnetic resonance imaging as sensitive imaging biomarker. *International Journal of Nanomedicine, 10*, 157–168. https://doi.org/10.2147/IJN.S75074.
3. Alibolandi, M., Abnous, K., Sadeghi, F., Hosseinkhani, H., Ramezani, M., & Hadizadeh, F., (2016). Folate receptor-targeted multimodal polymersomes for delivery of quantum dots and doxorubicin to breast adenocarcinoma: *In vitro* and *in vivo* evaluation. *International Journal of Pharmaceutics, 500*(1, 2), 162–178. https://doi.org/10.1016/j.ijpharm.2016.01.040.
4. Ansari, C., Tikhomirov, G. A., Hong, S. H., Falconer, R. A., Loadman, P. M., Gill, J. H., Castaneda, R., et al., (2014). Development of novel tumor-targeted theranostic nanoparticles activated by membrane-type matrix metalloproteinases for combined cancer magnetic resonance imaging and therapy. *Small, 10*(3), 566–75, 417. https://doi.org/10.1002/smll.201301456.
5. Beltrán-Gracia, E., López-Camacho, A., Higuera-Ciapara, I., Velázquez-Fernández, J. B., & Vallejo-Cardona, A. A., (2019). Nanomedicine review: Clinical developments in liposomal applications. *Cancer Nanotechnology, 10*(1), 1–40.
6. Bertrand, N., Wu, J., Xu, X., Kamaly, N., & Farokhzad, O. C., (2014). Cancer nanotechnology: The impact of passive and active targeting in the era of modern cancer biology. *Advanced Drug Delivery Reviews, 66*, 2–25. https://doi.org/10.1016/j.addr.2013.11.009.
7. Bwatanglang, I. B., Mohammad, F., Yusof, N. A., Abdullah, J., Alitheen, N. B., Hussein, M. Z., Abu, N., et al., (2016). *In vivo* tumor targeting and anti-tumor effects of 5-fluororacil loaded, folic acid targeted quantum dot system. *Journal of Colloid and Interface Science, 480*, 146–158. https://doi.org/10.1016/j.jcis.2016.07.011.
8. Chen, F., Hong, H., Goel, S., Graves, S. A., Orbay, H., Ehlerding, E. B., Shi, S., et al., (2015). *In vivo* tumor vasculature targeting of Cus@MSN based theranostic nanomedicine. *ACS Nano, 9*(4), 3926–3934. https://doi.org/10.1021/nn507241v.
9. Chen, W. H., Luo, G. F., Lei, Q., Cao, F. Y., Fan, J. X., Qiu, W. X., Jia, H. Z., et al., (2016). Rational design of multifunctional magnetic mesoporous silica nanoparticle for tumor-targeted magnetic resonance imaging and precise therapy. *Biomaterials, 76*, 87–101. https://doi.org/10.1016/j.biomaterials.2015.10.053.
10. Dai, Y., Su, J., Wu, K., Ma, W., Wang, B., Li, M., Sun, P., et al., (2019). Multifunctional thermosensitive liposomes based on natural phase-change material: Near-infrared light-triggered drug release and multimodal imaging-guided cancer combination therapy. *ACS Applied Materials and Interfaces, 11*(11), 10540–10553. https://doi.org/10.1021/acsami.8b22748.

11. Danhier, F., Feron, O., & Préat, V., (2010). To exploit the tumor microenvironment: Passive and active tumor targeting of nanocarriers for anticancer drug delivery. *Journal of Controlled Release, 148*(2), 135–146. https://doi.org/10.1016/j.jconrel.2010.08.027.
12. Das, M., Datir, S. R., Singh, R. P., & Jain, S., (2013). Augmented anticancer activity of a targeted, intracellularly activatable, theranostic nanomedicine based on fluorescent and radiolabeled, methotrexate-folic acid-multiwalled carbon nanotube conjugate. *Molecular Pharmaceutics, 10*(7), 2543–2557. https://doi.org/10.1021/mp300701e.
13. Das, M., Duan, W., & Sahoo, S. K., (2015). Multifunctional nanoparticle-EpCAM aptamer bioconjugates: A paradigm for targeted drug delivery and imaging in cancer therapy. *Nanomedicine: Nanotechnology, Biology, and Medicine, 11*(2), 379–389. https://doi.org/10.1016/j.nano.2014.09.002.
14. Ding, L., Ren, F., Liu, Z., Jiang, Z., Yun, B., Sun, Q., & Li, Z., (2020). Size-dependent photothermal conversion and photoluminescence of theranostic NaNdF4 nanoparticles under excitation of different-wavelength lasers. *Bioconjugate Chemistry, 31*(2), 340–351. https://doi.org/10.1021/acs.bioconjchem.9b00700.
15. Duman, F. D., Akkoc, Y., Demirci, G., Bavili, N., Kiraz, A., Gozuacik, D., & Acar, H. Y., (2019). Bypassing pro-survival and resistance mechanisms of autophagy in EGFR-positive lung cancer cells by targeted delivery of 5FU using theranostic Ag_2S quantum dots. *Journal of Materials Chemistry. B, 7*(46), 7363–7376. https://doi.org/10.1039/c9tb01602c.
16. Gao, N., Yang, W., Nie, H., Gong, Y., Jing, J., Gao, L., & Zhang, X., (2017). Turn-on theranostic fluorescent nanoprobe by electrostatic self-assembly of carbon dots with doxorubicin for targeted cancer cell imaging, *in vivo* hyaluronidase analysis, and targeted drug delivery. *Biosensors and Bioelectronics, 96*, 300–307. https://doi.org/10.1016/j.bios.2017.05.019.
17. Gholibegloo, E., Mortezazadeh, T., Salehian, F., Forootanfar, H., Firoozpour, L., Foroumadi, A., Ramazani, A., & Khoobi, M., (2019). Folic acid decorated magnetic nanosponge: An efficient nanosystem for targeted curcumin delivery and magnetic resonance imaging. *Journal of Colloid and Interface Science, 556*, 128–139. https://doi.org/10.1016/j.jcis.2019.08.046.
18. Groult, H., Poupard, N., Herranz, F., Conforto, E., Bridiau, N., Sannier, F., Bordenave, S., et al., (2017). Family of bioactive heparin-coated iron oxide nanoparticles with positive contrast in magnetic resonance imaging for specific biomedical applications. *Biomacromolecules, 18*(10), 3156–3167. https://doi.org/10.1021/acs.biomac.7b00797.
19. Guo, Y., Wang, X. Y., Chen, Y. L., Liu, F. Q., Tan, M. X., Ao, M., Yu, J. H., Ran, H. T., & Wang, Z. X., (2018). A light-controllable specific drug delivery nanoplatform for targeted bimodal imaging-guided photothermal/chemo synergistic cancer therapy. *Acta Biomaterialia, 80*, 308–326. https://doi.org/10.1016/j.actbio.2018.09.024.
20. Hao, Y., Zhang, B., Zheng, C., Ji, R., Ren, X., Guo, F., Sun, S., et al., (2015). The tumor-targeting core-shell structured DTX-loaded PLGA@Au nanoparticles for chemo-photothermal therapy and X-ray imaging. *Journal of Controlled Release, 220*(A), 545–555. https://doi.org/10.1016/j.jconrel.2015.11.016.
21. Jenkins, S. V., Nima, Z. A., Vang, K. B., Kannarpady, G., Nedosekin, D. A., Zharov, V. P., Griffin, R. J., et al., (2017). Triple-negative breast cancer targeting and killing by EpCAM-directed, plasmonically active nanodrug systems. *NPJ Precision Oncology, 1*(1), 27. https://doi.org/10.1038/s41698-017-0030-1.

22. Kang, M. K., Mao, W., Lee, J. B., & Yoo, H. S., (2017). Epidermal growth factor (EGF) fragment-guided anticancer theranostic particles for pH-responsive release of doxorubicin. *International Journal of Pharmaceutics, 519*(1, 2), 104–112. https://doi.org/10.1016/j.ijpharm.2017.01.017.
23. Kono, K., Nakashima, S., Kokuryo, D., Aoki, I., Shimomoto, H., Aoshima, S., Maruyama, K., et al., (2011). Multi-functional liposomes having temperature-triggered release and magnetic resonance imaging for tumor-specific chemotherapy. *Biomaterials, 32*(5), 1387–1395. https://doi.org/10.1016/j.biomaterials.2010.10.050.
24. Lee, G. Y., Qian, W. P., Wang, L., Wang, Y. A., Staley, C. A., Satpathy, M., Nie, S., et al., (2013). Theranostic nanoparticles with controlled release of gemcitabine for targeted therapy and MRI of pancreatic cancer. *ACS Nano, 7*(3), 2078–2089. https://doi.org/10.1021/nn3043463.
25. Li, L., Lu, Y., Jiang, C., Zhu, Y., Yang, X., Hu, X., Lin, Z., et al., (2018). Actively targeted deep tissue imaging and photothermal-chemo therapy of breast cancer by antibody-functionalized drug-loaded x-ray-responsive bismuth Sulfide@Mesoporous silica core-shell nanoparticles. *Advanced Functional Materials, 28*(5), 1–24. https://doi.org/10.1002/adfm.201704623.
26. Lim, E. K., Kim, T., Paik, S., Haam, S., Huh, Y. M., & Lee, K., (2015). Nanomaterials for theranostics: Recent advances and future challenges. *Chemical Reviews, 115*(1), 327–394. https://doi.org/10.1021/cr300213b.
27. Liu, T., Zhang, P., Huang, X., Chi, X., Li, Z., Zhang, Z., Guo, D. S., & Yang, X., (2019). Magnetic core-shell S-nitrosothiols nanoparticles as tumor dual-targeting theranostic platform. *Colloids and Surfaces. B, Biointerfaces, 181*, 400–407. https://doi.org/10.1016/j.colsurfb.2019.05.075.
28. Luong, D., Sau, S., Kesharwani, P., & Iyer, A. K., (2017). Polyvalent folate-dendrimer-coated iron oxide theranostic nanoparticles for simultaneous magnetic resonance imaging and precise cancer cell targeting. *Biomacromolecules, 18*(4), 1197–1209. https://doi.org/10.1021/acs.biomac.6b01885.
29. Nan, X., Zhang, X., Liu, Y., Zhou, M., Chen, X., & Zhang, X., (2017). Dual-targeted multifunctional nanoparticles for magnetic resonance imaging-guided cancer diagnosis and therapy. *ACS Applied Materials and Interfaces, 9*(11), 9986–9995. https://doi.org/10.1021/acsami.6b16486.
30. Pasha, S. S., Fageria, L., Climent, C., Rath, N. P., Alemany, P., Chowdhury, R., Roy, A., & Laskar, I. R., (2018). Evaluation of novel platinum(II) based AIE compound-encapsulated mesoporous silica nanoparticles for cancer theranostic application. *Dalton Transactions, 47*(13), 4613–4624. https://doi.org/10.1039/c7dt04232a.
31. Prabhakar, U., Maeda, H., Jain, R. K., Sevick-Muraca, E. M., Zamboni, W., Farokhzad, O. C., Barry, S. T., et al., (2013). Challenges and key considerations of the enhanced permeability and retention effect for nanomedicine drug delivery in oncology. *Cancer Research, 73*(8), 2412–2417. https://doi.org/10.1158/0008-5472.CAN-12-4561.
32. Rosenblum, D., Joshi, N., Tao, W., Karp, J. M., & Peer, D., (2018). Progress and challenges towards targeted delivery of cancer therapeutics. *Nature Communications, 9*(1), 1410. https://doi.org/10.1038/s41467-018-03705-y.
33. Shi, J., Yu, X., Wang, L., Liu, Y., Gao, J., Zhang, J., Ma, R., Liu, R., & Zhang, Z., (2013). Pegylated fullerene/iron oxide nanocomposites for photodynamic therapy, targeted drug delivery and MR imaging. *Biomaterials, 34*(37), 9666–9677. https://doi.org/10.1016/j.biomaterials.2013.08.049.

34. Thomas, E., Menon, J. U., Owen, J., Skaripa-Koukelli, I., Wallington, S., Gray, M., Mannaris, C., et al., (2019). Ultrasound-mediated cavitation enhances the delivery of an EGFR-targeting liposomal formulation designed for chemo-radionuclide therapy. *Theranostics, 9*(19), 5595–5609. https://doi.org/10.7150/thno.34669.
35. Tian, J., Luo, Y., Huang, L., Feng, Y., Ju, H., & Yu, B. Y., (2016). Pegylated folate and peptide-decorated graphene oxide nanovehicle for *in vivo* targeted delivery of anticancer drugs and therapeutic self-monitoring. *Biosensors and Bioelectronics, 80*, 519–524. https://doi.org/10.1016/j.bios.2016.02.018.
36. Vallet-Regí, M., Colilla, M., Izquierdo-Barba, I., & Manzano, M., (2017). Mesoporous silica nanoparticles for drug delivery: Current insights. *Molecules, 23*(1), 1–19. https://doi.org/10.3390/molecules23010047.
37. Wong, X. Y., Sena-Torralba, A., Álvarez-Diduk, R., Muthoosamy, K., & Merkoçi, A., (2020). Nanomaterials for nanotheranostics: Tuning their properties according to disease needs. *ACS Nano, 14*(3), 2585–2627. https://doi.org/10.1021/acsnano.9b08133.
38. Wu, M., Zhang, H., Tie, C., Yan, C., Deng, Z., Wan, Q., Liu, X., et al., (2018). MR imaging tracking of inflammation-activatable engineered neutrophils for targeted therapy of surgically treated glioma. *Nature Communications, 9*(1), 4777. https://doi.org/10.1038/s41467-018-07250-6.
39. Xie, J., Mei, L., Huang, K., Sun, Y., Iris, A., Ma, B., Qiu, Y., et al., (2019). A photo-inducible protein-inorganic nanoparticle assembly for active targeted tumor theranostics. *Nanoscale, 11*(13), 6136–6144. https://doi.org/10.1039/c9nr01120j.
40. Yao, H., Su, L., Zeng, M., Cao, L., Zhao, W., Chen, C., Du, B., & Zhou, J., (2016). Construction of magnetic-carbon-quantum-dots-probe-labeled apoferritin nanocages for bioimaging and targeted therapy. *International Journal of Nanomedicine, 11*, 4423–4438. https://doi.org/10.2147/IJN.S108039.
41. Yue, L., Wang, J., Dai, Z., Hu, Z., Chen, X., Qi, Y., Zheng, X., & Yu, D., (2017). pH-responsive, self-sacrificial nanotheranostic agent for potential *in vivo* and *in vitro* dual-modal MRI/CT imaging, real-time, and *in situ* monitoring of cancer therapy. *Bioconjugate Chemistry, 28*(2), 400–409. https://doi.org/10.1021/acs.bioconjchem.6b00562.
42. Zhao, M. X., & Zhu, B. J., (2016). The research and applications of quantum dots as nano-carriers for targeted drug delivery and cancer therapy. *Nanoscale Research Letters, 11*(1), 1–9.

CHAPTER 12

APPLICATION OF BIONANOMATERIALS FOR CANCER THERAPY

SHAGUFTA RIAZ, ADEEL RIAZ, AYESHA YOUNUS, MUHAMMAD MUBIN, and MUNIR ASHRAF

ABSTRACT

Cancer is a generic term that contains a large group of diseases affecting anybody organ so rapidly. It can generate abnormal cells that grow beyond their boundaries and spread to other body organs, invading the adjoining parts. According to WHO: International Agency for Research on Cancer, the mortality rate is very high due to cancer responsible for about 10 million deaths in 2020. Among many others, breast cancer (2.26 million) was the leading cause of death in females. According to the American Cancer Society, in 2021, the projected new cases were 1.9 million diagnosed with this lethal disease, and cancer deaths were nearly 608,570 in the USA. Numerous nanomedicine platforms like micelles, polymeric nanoparticles, liposomes, and dendrimers have been applied for targeted drug delivery because the controlled release is beneficial due to less drug toxicity to healthy cells. Though nanomaterials are advantageous due to their large surface area and high surface energy, they have some disadvantages. Therefore, green synthesis of nanomaterials (nanoparticles and nanofibers) based on plants, microbes, and natural biopolymers is an emerging field with potential medical applications like bio-imaging and magnetic responsive drug delivery anti-cancer therapy and photothermal therapy. Bionanomaterials are desired for sustainable anti-cancer therapy without affecting the other body organs and the environment due to low cost, easy synthesis, autologous pharmaceutical, biocompatibility, biosafety, biodegradability, and ecofriendly nature.

Sustainable Nanomaterials for Biomedical Engineering: Impacts, Challenges, and Future Prospects.
Junaid Ahmad Malik, Megh R. Goyal & Mohamed Jaffer M. Sadiq (Eds.)

In this chapter, different therapies and their harmful side effects have been discussed.

12.1 INTRODUCTION

Cancer has been the most destructive disease in current times; unfortunately, the spread of this harmful disease increases day by day. According to the World Health Organization (WHO) survey, by the end of 2030, the people diagnosed with this grave taken disease are anticipated to skyrocket to 21.6 million – a stark 53% increase from the most recent stated numbers. For human development, such disastrous diseases are a significant obstacle that should tackle with great responsibility. Without exceptions, these facts and figures regarding cancer inflation are alarming and signal urgent actions to alleviate this disaster confronted by human beings worldwide. Until 2019, the projected cancer cases were 1762.450, and deaths due to cancer were 606.880 cancer. According to the American Cancer Society, there was a decline of about 2% in male patients, but this lethal disease was stable in female patients. At the same time, the death rate due to cancer was declined from 2007 to 2016 yearly by 1.4% and 1.8%, respectively, due to the development of new diagnosis and treatment techniques in medical science [37]. Considering this condition, researchers have been striving hard to find novel treatment methods and carriers to deliver the anti-cancer drug at the targeted site [14, 36] The traditional techniques like surgery, laser or radiation therapy, chemotherapy, biological, and hormone therapies have been implemented successfully to increase the life span of an infected person with many other disadvantages [13]. Therefore, medical researchers aim to develop the carriers with selective toxicity and delivery of antitumor or anticancer drug to cancerous or tumor cells with no or least hazardous effect on normal cells [77].

Advanced research accelerated by these issues has resulted in new findings [90]. Various polymers in singular or composite and nanocomposites in diverse forms like liposomes, hydrogels, micelles, polymersomes, lipid-solid hybrid particles, and dendrimers have been implemented as a potential candidate to diagnose and treat cancer. Nanocomposites, nanomaterials, nanofibers have grabbed the medical personnel due to extraordinary properties like eco-friendly nature, ease of fabrication, unique design capacity, high drug retaining capacity, selective toxicity, sustainable, and controlled drug release, and cost-effectiveness. In polymeric composites, the collaboration of polymeric matrix and the nano reinforcement is critically essential for attaining hybrid formation's particular and tunable properties [54]. Multi-stimuli responsive polymers mainly offer a unique drug release mechanism for sustainable and controlled drug delivery in a balanced and biologically effective way [82].

In this chapter, traditional therapies and their shortcomings have been reviewed. At the same time, the bionanomaterials are considered best for advanced cancer therapy, and their mechanism of action against cancer cells has been discussed in detail.

12.1.1 CANCER AND ITS TYPES

Cancer is a diverse group of more than 100 diseases that can grow in every part of the body. In cancer, the cells undergo specific changes to attain the potential of uncontrolled division and growth with spread to any other part of the body. The cell is the basic unit of life, and all living organisms are made up of cells. New cells generate from existing cells through the cell cycle, and various checkpoints control this cell cycle. Cancer development occurs when these checkpoints are altered to invade abnormal cells continuously, thus leading to cancer formation. Cancers are classified in two ways, based on histology, and based on site of involvement. Based on histology, cancers are classified as carcinoma, sarcoma, lymphoma, myeloma, leukemia, and mixed types [107].

12.1.2 CAUSES OF CANCER

Anything that can alter the genetic makeup, potentiating it to proliferate abnormally, can cause cancer development. An individual's genetic makeup of cells may have an inherited predisposition for cancer. Some of the causes of cancer are described below, and the list is not all-inclusive, but only significant causes have been mentioned [12]:

- Ionizing radiation like alpha, beta, gamma, and X-rays from different sources like uranium, radon, etc., and UV light from the sun.
- Viruses like hepatitis B, Epstein-Barr virus, C viruses, and human papillomavirus, etc., and bacteria like helicobacter pylori.
- Behavioral activities like tobacco consumption, alcohol, obesity, unsafe sexual practices, etc.

- Exposure to chemicals like benzene, asbestos, cadmium, nickel, aflatoxins, pesticides, fertilizers, etc.
- Immunosuppressed states like chronic steroid use or HIV infection.
- Inherited genetic mutations like BRCA1, BRCA2, TP53, PTEN, etc.

12.1.3 CANCER THERAPIES

The goal of cancer treatment is to cure the disease, and if a cure is not possible, then control the disease [29]. There are several different treatment options for cancer cure and control. Depending upon the stage and grade of the disease, one or a combination of more than one of these treatment options can be used for each patient. Cancer treatment can be deployed as primary, adjuvant, neoadjuvant, and palliative intent.

12.1.3.1 SURGERY

Surgery is the oldest therapeutic strategy of cancer. The tumor and nearby tissue are resected either fully or partially to cure or control the disease [116].

12.1.3.2 CHEMOTHERAPY

Chemotherapy uses different drugs that essentially work on the same principle of destroying the abnormally growing cancer cells or slowing their growth. Hence chemotherapy reduces the tumor size making other modalities like surgery and radiation work better [21].

12.1.3.3 RADIATION THERAPY

Radiation therapy is the second most effective treatment option after surgery, and results vary from person to person. In radiation therapy, different types of radiation are used to target the cancer tissue to kill the cancer cells [103].

12.1.3.4 BIOLOGICAL THERAPY

Biological therapy uses the body's defense system to recognize and target the cancer cells, achieving biological control. Cancer cells develop an ability to hide from the immune system and go unnoticed. With biological therapy, either cancer cells are made easily recognizable for the immune system, or the immune system is induced to detect cells [38].

12.1.3.5 HORMONE THERAPY

The growth of some cancer types is dependent on the hormonal drive, for example, breast, and prostate cancer. Hormone therapy is the use of hormones and other substances that block and change these hormones, thus causing slowing or stopping of cancer cells growth [2].

12.1.3.6 PHOTODYNAMIC THERAPY

Photodynamic therapy is a type of cancer treatment that uses light to activate a specific photosensitizer drug at a specific site, thus achieving localized cancer control. Specific photosensitizers (PSs) are taken up by the target tissue, which upon exposure to specific light wavelengths produces reactive oxygen species killing the cancer cells [24].

12.1.3.7 LASER THERAPY

Laser therapy is an easy, precise, relatively safer, and less time-consuming way of treating some cancers like skin, breast, head, and neck cancers. Different lasers used are carbon dioxide laser, YAG laser, Argon laser [112].

12.1.3.8 HYPERTHERMIA

In contrast to normal cells, cancer cells are more likely to get damaged with temperatures in the range of 41–45°C. This preferential response is the result of differential blood flow to the cancer cells as well as other local factors like oxygenation, pH, etc. [91].

12.1.3.9 TRANSPLANTATION AND DONATION OR BLOOD TRANSFUSION AND BONE MARROW

Transplantation is the removal of an affected organ or tissue and its replacement with a healthy and viable organ or part of an organ from another individual. It is a definitive treatment option for some cancers like leukemias and lymphomas [32].

12.1.3.10 SIDE EFFECTS OF TRADITIONAL CANCER TREATMENTS

Though traditional therapeutics are effective against cancer and can destroy the tumor cells, specific issues are related to these therapies. All these therapies are complex procedures, due to which they are expensive. The drugs used in chemotherapy cannot distinguish cancer cells from healthy cells. Therefore, the normal cells can also be destroyed along with the malignant ones. The drug has poor solubility, and they are incapable of targeting the particular site, abrupt release, short biological half-life, and increased multi-drug resistance [72]. Stem cell transplant causes the issues like bleeding, and risk of infection. Graft-versus-host disease can occur, responsible for the devastating effect on the intestine, skin, liver, and other tissues and organs. Surgery is a painful procedure along with a severe risk of infection. There is a risk of anesthesia reaction. Due to radiation, only cancer cell proliferation is inhibited, but nearby normal cells can also be affected, causing problems like hair loss, nausea, and mouth sores. Hormone therapy can block the human body's ability to generate hormones, or hormones' normal functioning can be interfered with. Such interference can be accountable for nausea, weak bones, fatigue, diarrhea; female vaginal dryness ad enlarged tender breasts. Due to immunotherapy, the immune system can be affected severely. The immune system, which has to jazz up to show its action against tumor or cancer cells, can also act against normal tissues and cells [45].

12.2 MATERIALS AND METHODS FOR CANCER TREATMENT OTHER THAN TRADITIONAL THERAPIES

Due to remarkable technological advancements towards a better understanding of cancer causes and treatments, new methods have been developed and practiced as cancer therapies. But still, it is difficult to treat this aggressive disease due to various reasons. The leading cause is inter- or intra- cancer heterogeneity and mutation of genes causing cancer. Secondary, this lethal disease could affect anybody organ so rapidly that the abnormal cell growth can propagate beyond its boundaries within no time and invade the adjacent parts. This is not a static disease because it evolves and progresses over time, accruing to gene mutations [64, 121].

Neither the cancer war has been won nor lost, but still, in the 21st century, the curing, clinical management, nor is survival from this disease a big challenge. With continuous research, only cancer has been transformed into a chronic disease due to which the life duration of patients has been increased, and they are living a better-quality life. As mentioned above, several traditional therapies have played an essential role in treating this devastating disease. On the other hand, various novel techniques and emerging strategies have been adopted using biopolymers and bionanomaterials (Figure 12.1) because these have shown tremendous potential for cancer therapy, reducing death rate and other side effects caused by traditional therapeutics [17, 45].

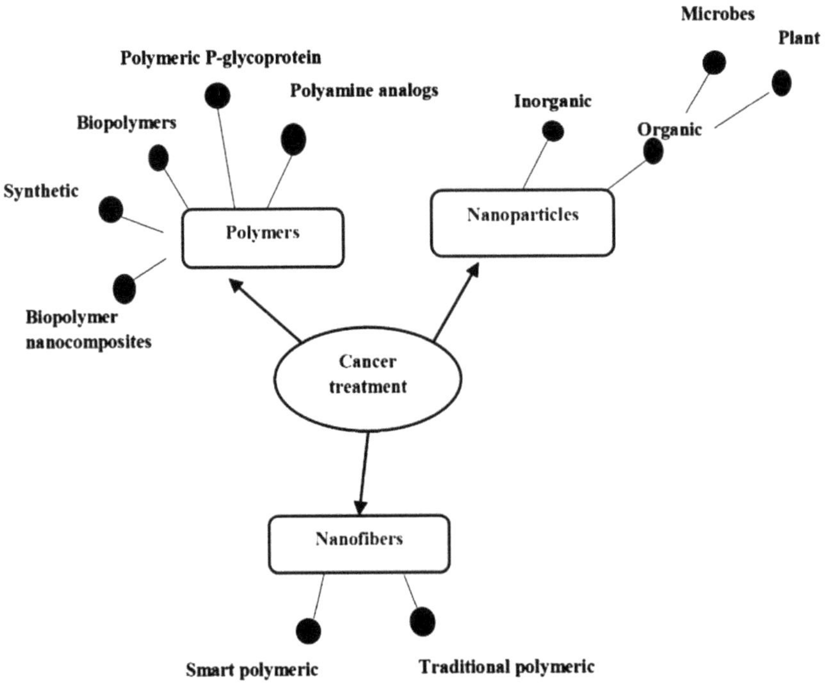

FIGURE 12.1 Various treatment materials and structures used for cancer diagnosis and treatment.

12.2.1 POLYMERIC THERAPEUTICS FOR CANCER

Nanotechnology and advancements in polymers have recently gained attention that is increasing daily for operative diagnosis and treatment of cancer lumps. Biopolymers can exhibit direct or indirect impacts to destroy the cancerous cells. Due to which these polymers have gained a lot of research interest in various arenas like biomedicines, pharmaceuticals by showing excellent cytotoxicity for different cancerous and tumor cells [57]. They are appropriate for drug delivery because of the following reasons: biodegradability, biocompatibility, sustainable drug release, subcellular size, depending upon the molecular weight of polymer the slower clearance and more excellent circulation time and less immunogenicity of drug, easy attachment of drug, increased solubility of the hydrophobic drug and longer release rate through diffusion process. The linear polymers are easy to fabricate and easily combined with various ligands targeting the cancerous site. Until recently, polymer therapeutics has been focusing mainly on two classes of bioactive polymers, i.e., polyamine analogs and polymeric p-glycoprotein inhibitors [111].

12.2.1.1 POLYAMINE ANALOGUES

Polyamine analogs can impede cellular growth and kill cancerous cells. They are now the focus in cancer therapy due to their similarity with naturally occurring polyamines. They can interfere with normal polyamine metabolic paths; interrupt the polyamine pool, and cause cancer cell destruction or death because the job of natural polyamine is to outspread from phospholipid structure of cell membrane and transmission of molecular signals DNA structure and conformational changes [59]. Like natural polyamine, polyamine analogs can easily permeate to the sites of natural polyamine function. Polyamines could be easily modified using different alkylating agents; therefore, it can do a range of structural changes to generate various polyamine analogs that can exhibit structure-dependent cytotoxicity to cancerous or tumor cells by targeting the function of polyamines [71]. These polyamine analogs can be accumulated inside the cell by a similar transport system as of natural polyamine.

12.2.1.1.1 Mechanism of Action

Polyamine analogs have expressed different cytotoxic mechanisms of action. Reported cytotoxic actions to cancer cells are like; DNA polymer analogs interaction, cell cycle regulation, tuned cell death, and cell cycle regulation. The most significant mechanism of action is given as in subsections.

12.2.1.1.2 Deoxyribonucleic Acid – Polymer Analogs Interaction

Deoxyribonucleic acid (DNA) has a vital role in the survival and proliferation of living cells by transmitting the genetic information for normal cell metabolism. Polyamine and DNA interaction is significant to stabilize the DNA through controlling the conformational transition. The polyamine analog interferes with polyamine-DNA interaction upon accumulation within the cell. Working on the same pattern as natural polyamine, polyamine analog alters or delays the normal conformational function and structure of DNA [66]. Despite standard DNA structure, the interaction of DNA with polyamine analogs provides an accumulated structure encouraging the inhibition or death of cancer cells rather than average cell growth. On breast cancer cells, T. J. Thomas and T. Thomas studied several polyamine analogs, i.e., bis(ethyl)spermine analogs. They found that these polyamine analogs effectively inhibited the growth of breast cancer cells in culture by inhibiting the escalation of estrogen receptor-positive and estrogen receptor-negative [105].

12.2.1.2 POLYMERIC P-GLYCOPROTEIN (PGP) INHIBITORS

These polymers can inhibit the normal functioning of P-glycoprotein and can eradicate multidrug resistance (MDR). Indirectly, they are responsible for improving anti-cancer drug efficacy against cancer cells due to increased intracellular accumulations of drug and reversal of MDR. Due to the resistance of drugs against various diseases, therapeutic efficacy was challenging to achieve. To inhibit the P-gp function and reduce the MDR effect and improve drug delivery and therapeutic effects, various polymeric P-glycoprotein inhibitors have been extensively investigated. Based on evolving periods, the P-glycoprotein inhibitors can be classified into three generations. The 3rd generation P-glycoprotein inhibitors are the novel molecule developed based on the structure-activity relationship. The first member of 3rd generation inhibitor was the acridone-carboxamide derivative. Another member of this generation is diaryl imidazole which was recognized as an effective P-GP inhibitor that increased the oral bioavailability of the drug paclitaxel by lowering the efflux function of P-glycoprotein in a gastrointestinal way [104].

12.2.1.2.1 Mechanism of Inhibition of P-Glycoproteins (PGPs)

The inhibition mechanism of PEG and Pluronics® was examined extensively. The mechanism of action shown was due to lowering membrane viscosity (membrane fluidization) because of the residence of Pgps within the cell membrane; secondly, Pluronic® stimulates the ATP exhaustion inside the cell, as shown in Figure 12.2. This depletion can result in an adverse effect on MDR function [84].

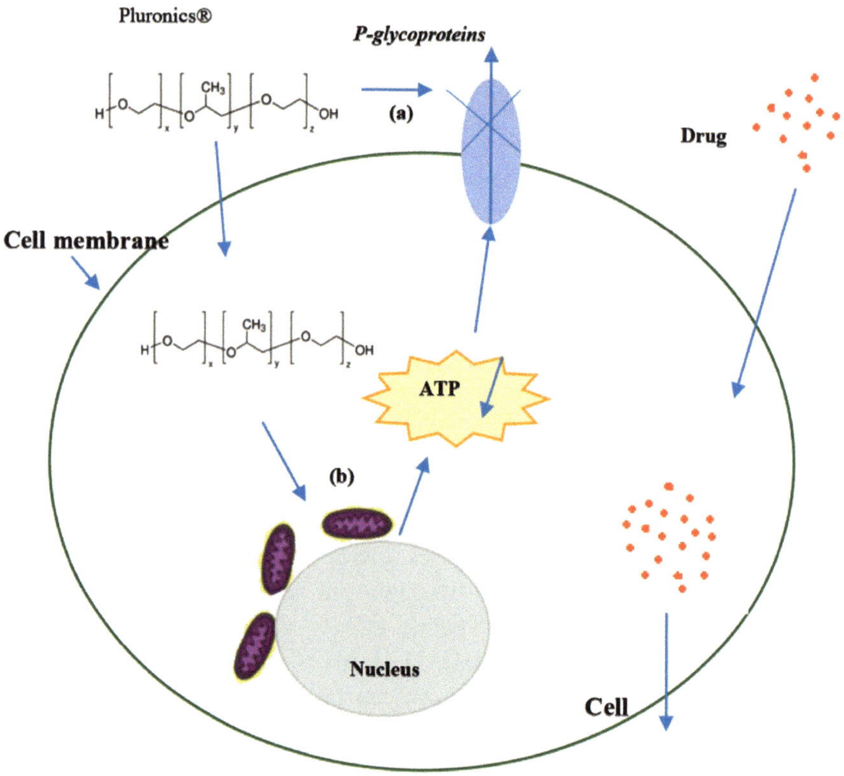

FIGURE 12.2 Schematic representation showing inhibition mechanism of P-glycoproteins on cell membrane and inside the cancer cell.

12.2.1.3 DENDRIMERS

Dendrimers are polymers with various branched monomers created from a central core radially. They are different from linear polymers due to their branched structure and have unique high-density surface functional groups. The dendrons arising from the central core define the generation of dendrimers. Higher the generation, more significant will be the dendrimers with a more significant number of branches and a higher number of surface functional groups. These branched polymers are considered promising candidates for DDS due to their perfectly branched system, good compatibility, and flexible chemistry. Dendrimers have found their applications in the medical field as passive anti-cancer nanocarriers [49]. The cytotoxicity of positive surface charge (cationic) could be a problem but, this could resolve by attaching ligands or drug molecules solely or both. The dendrons are fabricated mainly by various polymers such as poly (glycerol-succinic acid), triazine, poly-L-lysine, polyamide amine, poly(glycerol), melamine, polypropylene-mine, poly (ethylene glycol), and poly[2,2-bis(hydroxymethyl)-propionic acid]. Polyamidoamine and poly(propylene imine) are considered a potential applicants for the drug being capable of increased solubility and bioavailability of hydrophobic drugs [118].

12.2.1.4 BIOPOLYMERS AS DRUG CARRIERS

Polymers are grouped into three main categories natural or biopolymers, synthetic polymers, and polymer nanocomposites. Each type has its unique mechanism of action against cancer cells. Therefore, these polymers have been used for various cancer treatments. Synthetic polymers are fluorinated polymers, polyolefin, silicones, and polyesters that have been used widely in pharmaceutics and medicine applications. At the same time, the biopolymers are protein-based, i.e., fibroin and silk, and polysaccharides are cellulose, alginate, starches, chitosan, and hyaluronic acid. Polymer nanocomposites – the hybrid formation of macromolecular matrix and organic nanofillers have earned much attention due to their extraordinary characteristics. They can be degraded efficiently by the action of enzymes or bacteria and fungi [76].

Due to biodegradability and biocompatibility, they have been used as a drug carrier to diagnose diseases, wound healing, tissue engineering, and cancer treatment. They are best for cancer drug release because instead of immediate release, they provide controlled drug release needed for the cancer treatment. Thus, polymer nanocomposites have replaced the direct drug release system due to sustainable release, fewer side effects, the most negligible negative impact on healthy cells, and biosafety. For their projected actions, the clinically accepted nanocarriers deliver the preferential drug discharge at the targeted site [23]. Responsive or innovative polymers are a promising class of polymer nanoparticles that require some stimuli or trigger mechanism to deliver the drug at the required site. Response to some stimulus is one of the most fundamental processes that exist in living beings. Polymer-based innovative nanomaterials (internally responsive) have been used for oncology [110].

12.2.2 NANOMATERIALS FOR CANCER THERAPY

Cancer is caused by uncontrolled cells that can spread and invade other normal cells beyond limits through the lymphatic system with devastating health effects. Researchers have been trying various materials and techniques as cancer therapy to target the malignant site with the most negligible negative impact on normal cells because the traditional cancer therapies are incapable of delivering the drug to the target site and affecting the healthy and cancerous cells. The other issue is no or less solubility due to the hydrophobic nature that made these medicines biologically objectionable. Nanomaterial, due to their tiny size, has a large surface-to-volume ratio and high surface energy, due to which they acquire excellent attributes making them feasible for various applications. Therefore, nanomedicines have been developed and tried to deliver at the targeted site with improved efficacy against cancer cells. Nanomaterials have grabbed great interest in several areas of medicine, including cancer. In recent times, these nanomaterials have displayed a more significant potential for oncology treatment.

Cancer nanotechnology has been used for cancer identification, treatment, and post-curing management, focusing on modifying biological designs [3]. With selective toxicity, these nanomaterials drug therapy is advantageous because of delivering anti-cancer medicine to cancerous cells. Though it has been used to treat several diseases, nanomedicine is still an emerging field with many needed improvements. It would change the efficacy of medicine/drugs due to the uniqueness of tumors or various cancer cells. Blood flow can affect the rate of nanoparticles' travel. Therefore, using different materials with different shapes, structures, and sizes, self-therapeutic nanoparticles are synthesized through various techniques that exhibit different chemical and physical characteristics used for cancer therapeutics (Figure 12.3) [3]. Diagnosis of cancer generally constitutes two parts – one is the identification or diagnosis of cancer; the second is to identify the extent or stage of cancer. The diagnosis is realized through various testing techniques like urine or blood test, CT scans, MRI, Ultrasound, imaging via X-rays, and biopsies.

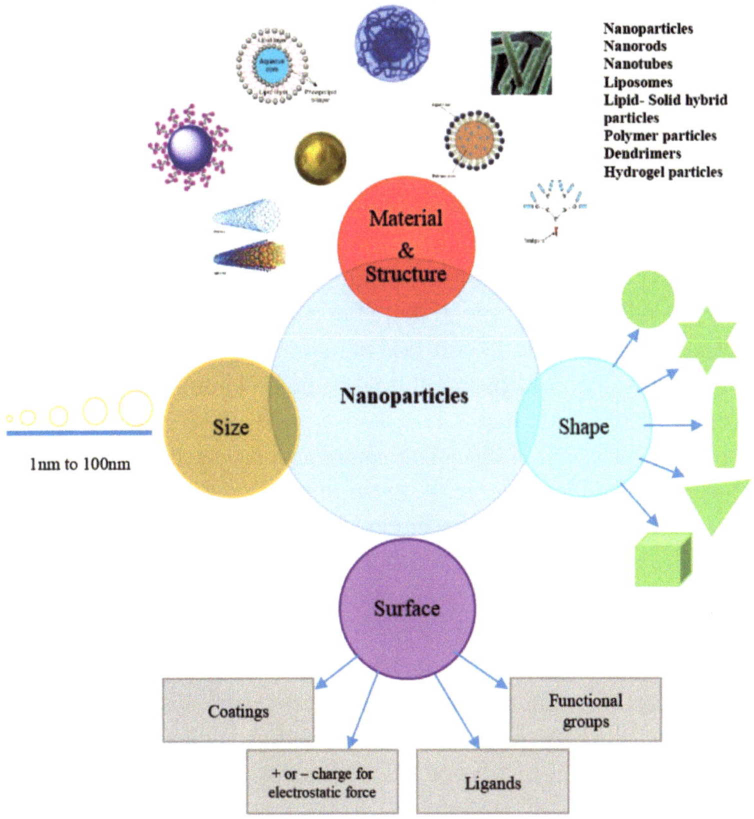

FIGURE 12.3 Different nanoparticles as a potential candidate for cancer therapy.

A final image is obtained through diagnostic imaging to conclude whether cancerous mass exists in the body or not. Such conclusions are sometimes wrong. To address such issues, biopsies define the positive or negative results of cancer. The second stage is done to see whether cancer has spread or is at the initial stage. The doctor defines the stage in roman numerals. Lack of sophisticated equipment is a serious issue due to which accurate detection/ diagnosis is not possible. Though X-rays and CT scans made it possible for doctors to define the cancer stage, the true reflection of depth or exact location of the tumor (proper mapping) is still a problem to be solved.

Nanotechnology made it possible to diagnose and determine the stage of cancer. Correct diagnostic imaging is a tool now doctors must conclude the progression of cancer and best suitable treatment according to the extent of cancer. If the diagnosis is ambiguous, then it is difficult to suggest a proper treatment. Due to nanotechnology, DDS was developed with a reduced negative impact on healthy cells, increased tumor accumulation, and an augmented delivery system. Surface functionalization of nanoparticles made it possible to target cancerous cells directly, allowed imaging and anti-cancer drugs to be delivered directly to affected cells [15]. The nanoparticle complexes are fabricated for cancer imaging and therapy that involves encapsulation, covalent, and non-covalent cargo binding to diagnose, allow imaging, drug delivery at the target site, and kill cancerous cells [75].

The Food and Drug Administration America (FDA) approved a few nanoparticles for DDS. Some of them have been used at present with tremendous clinical results. The 1st certified nanomaterial was Doxil, functionalized by Polyethylene glycol liposomal doxorubicin. About 46 nanoparticles have been under various medical trials, and around 26 nanoparticles have been certified by EMA and FDA. The nanoparticles used for DDS should be biocompatible, biodegradable, with minimum burst but sustainable release, maximum accumulation at the target site by active or passive targeting by escaping the immune system. Nanoparticles mostly used are inorganic such as quantum dots (QDs) [96], silver [119], gold [101], carbon-based nanoparticles [19], magnetic nanoparticles [102], silica, and zirconium oxide [31] while organic nanoparticles are polymer particles [114], hydrogel [26], lipid solid particles [10], dendrimers [92], micelles [50], liposomes, and exosomes [73] as shown in Figure 12.3. Inorganic nanoparticles have grabbed attention for the treatment of several DDSs because of the unique physical and chemical attributes they acquire due to their tiny size, high surface energy, and high surface area to volume ratio.

12.2.2.1 GREEN NANOMATERIALS: BIOCOMPATIBLE FOR CANCER TREATMENT

The synthesis of nanoparticles used for various applications by chemical routes may involve various hazardous chemicals and toxic substrates, producing harmful wastes affecting the ecosystem while using a large amount of energy, giving a low yield [113]. Also, these methods are not sustainable;

therefore, some other sustainable routes have always been considered to get high productivity with a minimum negative impact on the environment. Green synthesis of nanomaterials is one of them. Because natural resources and materials are eventually more cost-effective, sustainable, and produce more efficient products. In this regard, various plant parts and their extracts, microorganisms [60] like bacteria [93], actinomycetes [63], algae [48], yeast, and fungi [65], have been stated for the fabrication of various nanomaterials that are used proficiently for different applications as shown in Figure 12.4.

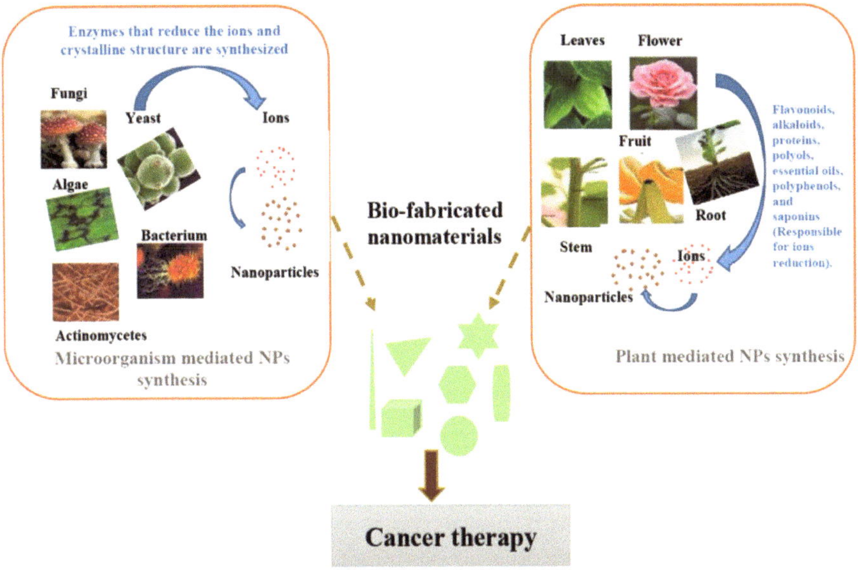

FIGURE 12.4 Plants and microorganisms' mediated synthesis of nanomaterials used for cancer treatment.

12.2.2.1.1 Mechanism of Action of Nanoparticles for Cancer Treatment

Nanoparticles' cancer therapy increased the use of naturally occurring materials due to improved bioavailability at the targeted site. This therapy increased the stability of the drug. It mitigated the exposure of the drug to healthy cells and tissues, due to which this therapy is more effective as compared to other cancer therapies [25] The inorganic nanoparticle like gold and silver [70], TiO_2 [68], ZnO [115], iron oxide and FePt [6] can destroy the tumor cells without any conjugated drug or ligand. These nanoparticles could make their way into tumor cells by endocytosis and determine their location inside the cell. These nanoparticles can enter the mitochondria where they react with H_2O_2 produced by mitochondrial respiration and generate reactive oxygen species responsible for lipid oxidation of cell membrane, causing cell death. In short, the cell toxicity of nanoparticles can lead to RNA and DNA damages, mitochondrial destruction, stimulation to apoptosis, and oxidative stress of cancerous cells. The schematic as Figure 12.5 exhibits the mechanism of anti-cancer action of nanoparticles on malignant cells.

Also, these nanoparticles are responsible for affecting the functionality of vascular endothelial growth factor or vascular permeability factor causing angiogenesis within the tumor cells. Thus, nanoparticles are potential anti-cancer therapy that is more effective as compared to the traditional therapies [48].

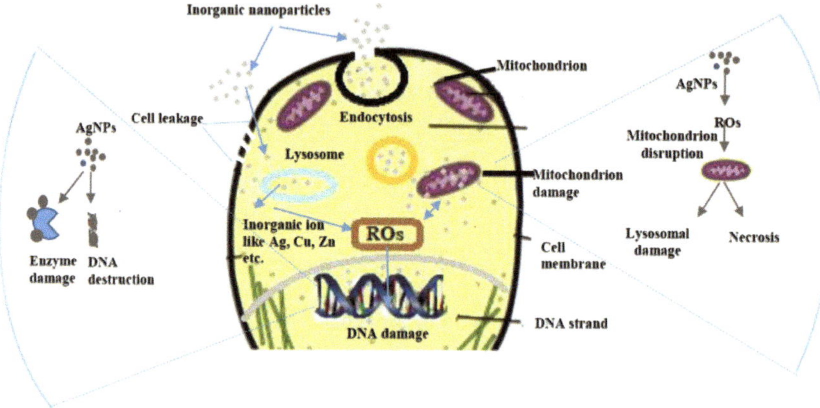

FIGURE 12.5 Schematic representation showing the mechanism of action of nanoparticles by generation of ROs causing cancer cell death.

12.2.2.1.2 Plant-Based Nanoparticles

Several researches have been conducted to isolate the chloroplast to catalyze nanomaterials' biofabrication [33, 78, 123]. The exact mechanics of biosynthesis of nanomaterials is not well evident yet. Still, it is believed that some enzymes are involved in the bio-reduction of metal ions into crystalline nanostructures [56]. For plant-mediated fabrication and stabilization of nanoparticles, different phytochemicals in plant extracts like flavonoids, alkaloids, proteins, polyols, essential oils, polyphenols, and saponins are implicated in the bio-reduction of metal ions to synthesize nanoparticles [42]. Plant-mediated biosynthesis of AgNPs has been discussed in detail by Muhammad et.al. [62].

12.2.2.1.3 Microbes-Based Nanoparticles for Anticancer Activity

The first interaction occurs between positively charged metal ions and negatively charged bacterial cells for microbial-mediated metal nanoparticles. The enzymes inside or at the surface of the cell are responsible for the bio-reduction of metal ions. In contrast, some bacterial cell wall components, alcoholic compounds, and different proteins may also be involved in the fabrication and stabilization of nanoparticles. Different types of nanoparticles used for cancer treatment, bio-sourced for their synthesis, and size are given in Table 12.1.

TABLE 12.1 Biobased Nanoparticles Synthesized by Using Different Plants and Microbes

Biosynthesis of NPs Using Different Plant Parts Used to Treat the Cancer Cells				
Nanoparticles	**Source**	**Plant Part for Extraction**	**Size of NPs nm)**	**References**
Plant-based Nanoparticles				
Ag NPs	*Aloe vera*	Leaf	70–192 nm	[22]
Ag NPs	*Alternathera tenella*	Leaf	48 nm	[89]
Cu NPs	*Brocolli*	Whole plant	4.8 nm	[81]
Au NPs	*Cajanus cajan*	Seed coat	9–41 nm	–
ZnO	*Delonix regia*	Petal	65–184 nm	[1]
Pd	*Oringa oleifera*	Flower	2–18 nm	[7]
Fe_3O_4	*Sargassum muticum*	Seaweed	20 nm	[69]
TiO_2	*Vigna radiata (green gram)*	Seed	–	–
ZnO/Ag	*Zingiber zerumbet*	Rhizome	23 nm	[9]
Microbes-based Nanoparticles				
TiO_2	*Bacillus cereus*	Bacteria	69–140 nm	[98]
Ag NPs	*A. flavus*	Fungus	33.5 nm	[97]
Cu NPs	*Bacillus cereus*	Bacteria	11–33 nm	[106]
Au NPs	*Humicola* spp.	Fungus	18–24 nm	[4]
Cr (III)	*Bacillus subtilis*	Bacteria	20 nm	[47]
Ag NPs	*Cryptococcus laurentii (BNM 0525) Y*	Yeast	35 nm	[74]
Tb_2O_3 NPs	*Fusarium oxysporum*	Fungus	10 nm	[41]
SeO_2 NPs	*Halococcus salifodinae BK18*	Bacteria	28 nm	[95]
SeO_2 NPs	*Bacillus* sp. MSh-1	Bacterium	80–220 nm	[28]
Au NPs	*Aspergillus foetidus*	Fungus	30–50 nm	[85]
Au NPs	*Bacillus flexus*	Bacteria	20 nm	[67]
Ag NPs	*Humicola* sp.	Fungus	5–25 nm	[100]

12.2.2.2 NANOFIBERS: A BIOCOMPATIBLE THERAPY FOR CANCER TREATMENT

To deal with diseases like cancer, the persistent and controlled release drug to cancerous cells is very much needed to avoid the ruinous effect to the normal cells because antitumor or anti-cancer drugs used for traditional treatments showed severe side effects. Due to the blessings of nanotechnology, it was made possible to target only malignant cells due to controlled and sustainable drug release through nanoparticles, micelles, and nanofibers [11, 99]. The nanofibers are very small-sized fibers in diameter along one dimension; they are less than 1,000 nm. Due to their small size along one dimension, their surface is per unit mass is very large compared to conventional fibrous structures, which are responsible for their carry unique characteristics, making them applicable in various fields of science [18].

There are different ways to fabricate these nanofibers like blend spinning, co-axial electrospinning, emulsion, gas jet electrospinning, phase separation, drawing, flash spinning, melt-blown, force spinning, and bicomponent spinning using two or more composites, i.e., polymer/polymer, polymer/nanoparticles, polymer/organic reagents and polymer/inorganic salts. Due to the simple spinning method and tunable characteristics by controlling the solution and process parameters, they are grabbing much attention by researchers in various fields of medicine and surgery [5]. These nanofibers are potential medical textiles for cancer treatment due to their large surface area, high surface energy, tunable pore structure, and measured/controlled drug release. The sustainable and tunable release showed sound effects due to the entrapment of anti-cancer drugs in the nanofibers. A controlled amount could be released as required by some triggering mechanism.

Most polymers are water-soluble, and researchers have been trying to find ways for the green synthesis of nanofibers to reduce the toxicity of the fibers. They could be made human and environmentally friendly with a minor effect on human health [18, 88]. Schematic showing the fabrication of nanofibers, the controlled drug release from these nanofibers is shown in Figure 12.6. The nanofibers as a nanocarrier for anti-cancer drug release are of greater importance with benefits higher than other nanomaterials. The other nanocarriers give burst release of drug that is not a choice for cancer treatment.

Nanofibers encapsulate the drug and release a sustainable amount for a more extended period than the other nanomaterials [53]. Also, due to this low encapsulation, high release profile, and ease of fabrication with tunable characteristics by controlling parameters, the nanofibers are advantageous to be marketed commercially [34]. Researchers have concluded these results by conducting many research studies. Both nanoparticles and electrospun fibers were studied for drug release and explored that both approaches with increased drug loading concentration can deliver hydrophilic/hydrophobic drugs for a more extended period due to controlled and sustainable drug release [80].

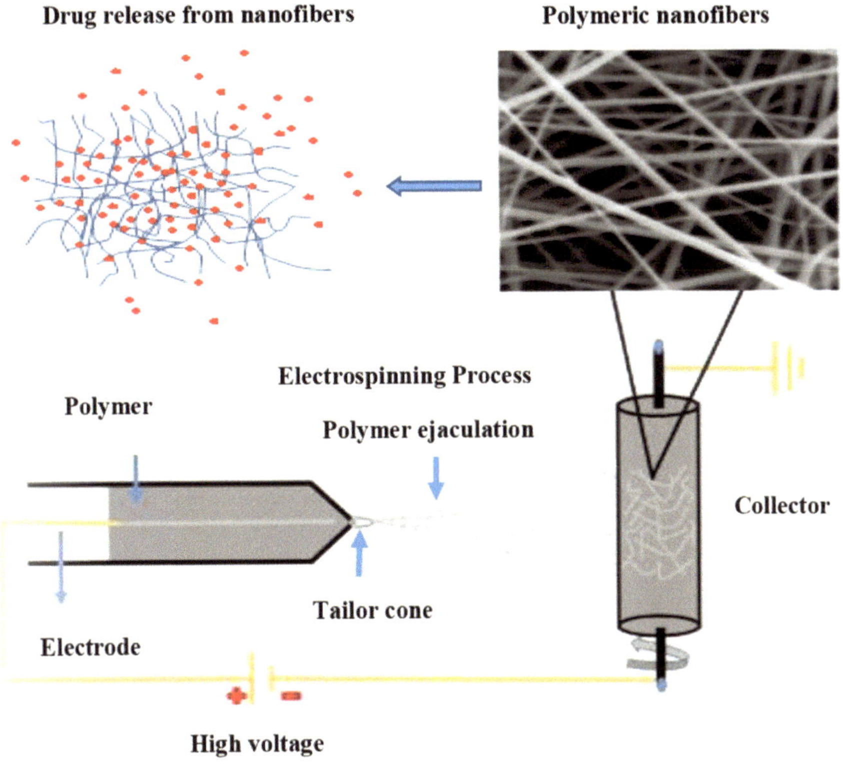

FIGURE 12.6 Schematic showing electrospinning process and release of drug from nanofibers.

A comparative study was conducted for drug release from core-sheath nanoparticles and nanofiber structures [89]. The Bovine serum albumin was entrapped as core in a sheath of methoxy poly(ethylene glycol)-Polylactic acid and poly (L-lactic acid), concluded that both methods were safe to release drug to cells. Still, coaxial electrospinning-produced nanofibers are a better way to retard the abrupt release of the drug. It could achieve different characteristics of fibers by changing different parameters for different applications [79].

12.2.2.2.1 Therapeutic Drug Delivery System (TDDS)

The traditional drugs do not have preferred active targeting capacity. They can release faster from the body without providing the required therapeutic effectiveness or toxic to other normal cells due to nonspecific targeting. Also, most of them were hydrophobic, due to which they could not dissolve properly and have poor bio-distribution [41]. Therefore, nanofibers are gaining much attention due to target specificity and sustainable drug release. These nanofibers in the form of web or scaffolds can be preferred as therapeutic nanocarriers with attributes like; biosafety, biodegradability, controlled therapeutic delivery, and high drug loading capacity [12]. Different ways of administration have been employed using nanofiber carriers like buccal, rectal, ocular, vaginal, parental oral, nasal, transdermal, or inhalation has shown in Figure 12.7 using different therapeutics.

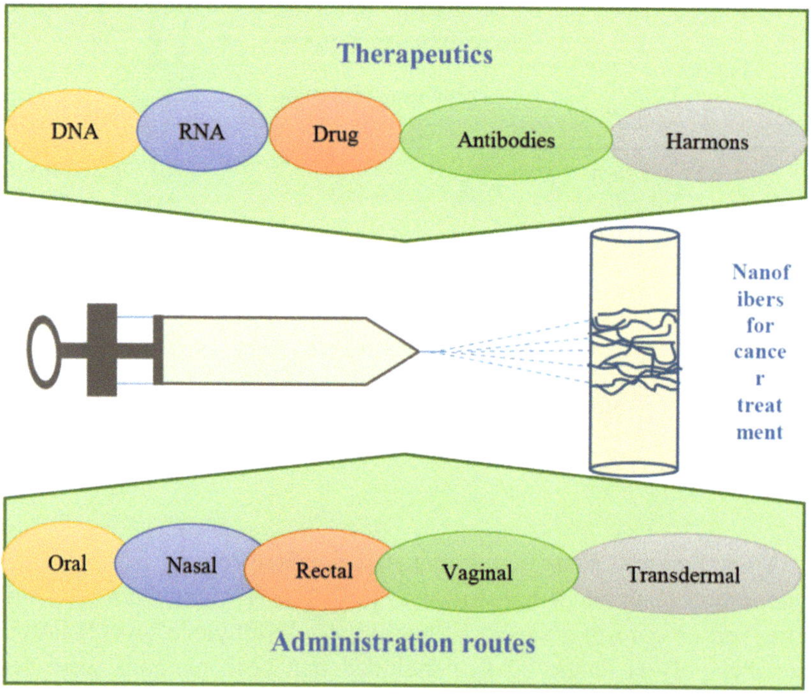

FIGURE 12.7 Types of electrospinning, different therapeutics-loaded nanofibers, and their route of administration.

12.2.2.2.2 Release Mechanism of Drug from Nanofibers

Several therapeutic drugs can efficiently be delivered from nanomedicines to macromolecules comprising diverse proteins and nucleic acids. The drug from carriers can be released by diffusion, swelling, affinity-based mechanism, or degradation [108]. The drug release from nanofibers can be tuned by controlling different factors like morphology, nature of polymers constituents, blend, pour, structure, compatibility of polymers and drug (entrapment and sustainable release in cancer treatment), the concentration of the loaded drug, nature, and type of drug to be released from fibers [39, 43, 46, 122, 124].

12.2.2.2.3 Applications of Nanofibers in Cancer Treatment

Due to various shortcomings such as lack of selective toxicity, poor drug solubility, less accumulation in cancerous cells, less functionality, low drug loading, and abrupt drug release, researchers found new treatment strategies. Nanofibers-based sustainable drug release for cancer therapy could be a promising anti-cancer therapeutic strategy. Versatile drug nanocarriers, superior drug entrapment, transfer, and mitigate side effects due to drug accretion at the targeted site with improved patient compliance, sustainable, and tunable drug release at the site without affecting the normal cells made the nanofibers'-based treatment a potential candidate for cancer treatment [27]. Several polymers have been reported to synthesize nanofibers under different solution and process parameters for encapsulation and controlled drug release at the cancer site. Mainly drug release occurs through diffusion or degradation of the polymer and can have a cytotoxic effect on the cancer cells, as shown in Figure 12.8.

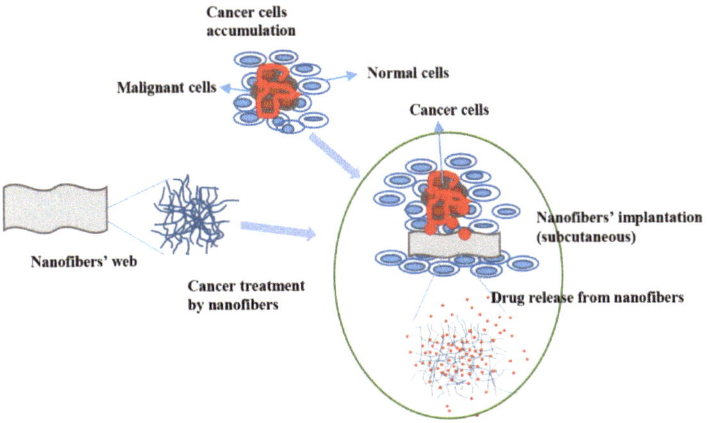

FIGURE 12.8 Schematic showing the subcutaneous implantation of nanofiber, release of drug inhibiting the growth of cancer cells.

Polymer, its surface morphology, a drug to be entrapped, fabrication technique, and post-fiber treatment for proper treatment. Polymers like Polyvinylpyrrolidone, chitosan [83], poly(caprolactone) [35] poly(lactic acid) [16, 27, 120], cellulose acetate [49], Poly(hydroxy alkanoates) [44], poly(vinyl alcohol) [40], polystyrene, and peptides [109] have been implied for nanofibers' fabrication. These polymers were used solely and as composites [86] to develop single, core-sheath, and bi-component fibers. Recently, to upgrade the anti-cancer drug solubility, electrospun nanofibers have been developed and assessed. The developed nanofibers can restrict and crystallize the agent inside the nanofibers. The Paclitaxel anti-cancer drug was loaded in folic acid-modified mesoporous hollow SnO_2 nanofibers. The *in vitro* drug release profile showed the improved dissolution rate of an anti-cancer drug (Paclitaxel), i.e., 8.34 times greater than pure Paclitaxel in 5 minutes. *In-vitro* experiments showed that these fibers could inhibit the growth of cancer cells in the liver [61]. The fast-dissolving release profile of the drug was shown by the core-sheath structure of nanofibers developed by co-axial electrospinning. Polyvinylpyrrolidone K90 or Polycaprolactone was used as a core, while SDS surfactant and Polyvinylpyrrolidone K10 were used as hydrophilic moieties to form the shell along with insoluble drugs like Quercetin/Tamoxifen citrate [52]. The *in vitro* experimentation showed the faster release profile of the drug, possibly due to uniform and homogenous distribution of the anti-cancer drug in the outermost layer of extremely thinned nanofibers with greater exposed surface area and short diffusion distance nanofibers [58]. Therefore, ultra-thin core-shell nanofibers showed a dramatic increase in the release profile of water-insoluble drugs due to modification in nanofibers' structure regardless of the nature of the drug. Nanofibers 'scaffolds with a honeycomb structure and larger space have greater drug loading capacity. Therefore, different nanofiber fabrication techniques such as emulsion, co-axial, surface modification, or blending can tune the loading capacity.

By changing the solution and process parameters such as poly (lactic acid) concentration, applied potential, distance of collector and needle, flow rate, and nano-sheet concentration, the scaffolds of biodegradable polymers were fabricated to treat cancer without having intense side effects. For sustainable drug release, the scaffolds of nanofibers with imparted drugs after post-thermal treatment showed improved stability and controlled release due to changes in the microstructure of the scaffold. After *in vivo* experimentation, the biocompatible scaffold was approved for its bio-applicability to treat cancer and showed more significant cancer cell cytotoxicity than the control, pure drug, and film with entrapped drug [55], as shown in Figure 12.9.

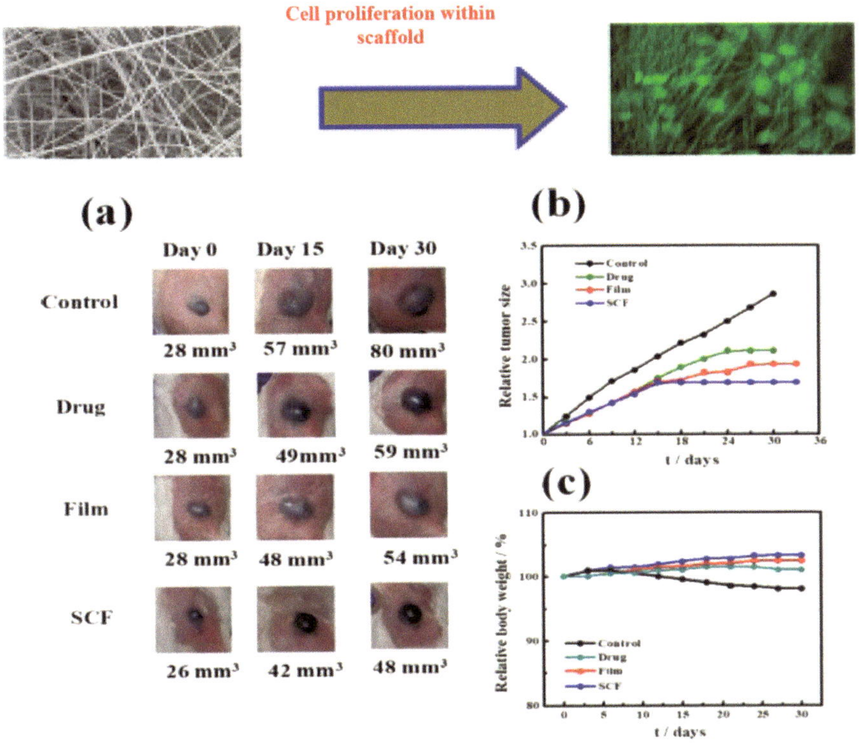

FIGURE 12.9 (a) Tumor size with respect to time for different treatment patches; (b) size of mice tumor treated with control, pure drug, film with drug and scaffold with entrapped drug; (c) relative body weight of mice with respect to time for different treatment patches.

Source: Reprinted with permission from Kumar, S., Singh, A. P., Senapati, S., & Maiti, P. © 2019 American Chemical Society.

The nanofibers are also promising candidates to address another issue of drug loading capacity. Other nanocarriers might load less amount of the drug. Still, nanofibers scaffolds can load a higher amount of drug that could be tuned to be released for a more extended period even after cancer cells cessation. A conjugate of Paclitaxel and succinic acid were synthesized that self-assembled within the nanofibers and became carrier-free with Paclitaxel as the drug carrier [117, 118]. The loading efficacy of this nano assembly was more than 89%, reported highest loading capacity of Paclitaxel. In both *in vitro* and *in vivo* assessment, it was examined that the sustainable release of the drug inhibited the growth of lung adenocarcinoma cells. It was concluded that as-synthesized nanofibers' scaffold had higher cancer cell cytotoxicity. Chitosan/PLA/GO/TiO_2/Dox fibers were developed by electrospinning proved biocompatible during *in-vivo* examination with minor toxicity to the healthy cells. The developed nanofibers' scaffolds under acidic pH released higher drugs at the specific tumorous tissues. Furthermore, the intelligent nanofibers' drug delivery system is usually mediated by several biological triggers like temperature, UV or visible light, voltage, and pH, etc., as shown in Figure 12.10.

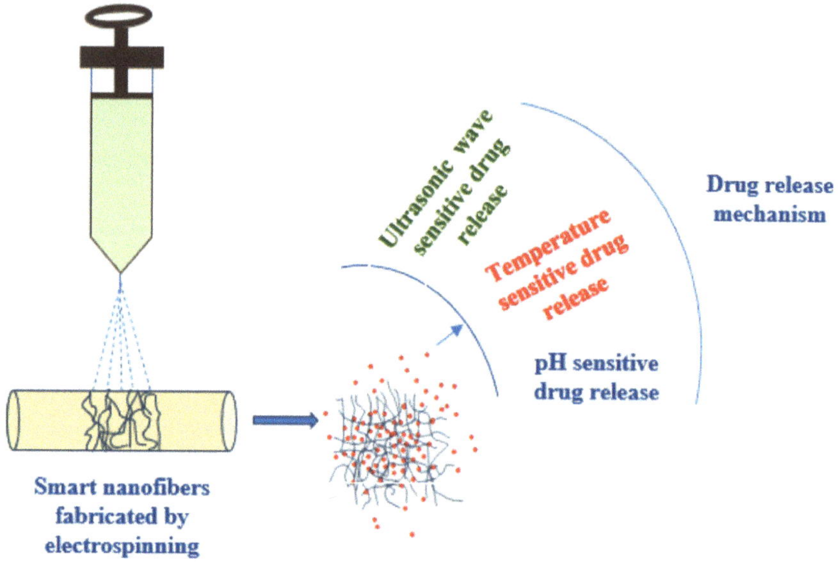

FIGURE 12.10 Different release mechanisms for drug release from smart nanofibers.

Such fibers with stimuli responsiveness are called smart nanofibers and have been extensively used for drug release in different therapies, including cancer [30]. The thermally responsive copolymer of N-hydroxymethyl acrylamide and N-isopropyl acrylamide were used for nanofibers' synthesis and loaded with anti-cancer drug DOX magnetic nanoparticles of γ-Fe_2O_3 and Fe_3O_4 for the cancer apoptosis. Nearly 70% of cancerous cells were destroyed under the fast action of heat and drugs [51]. In another study, chitosan/PLA/GO/TiO_2/Dox fibers were developed by electrospinning proved biocompatible during *in-vivo* examination with minor toxicity to the healthy cells [86, 94]. The developed nanofibers' scaffolds under acidic pH released higher drugs at the specific tumorous tissues. But to synthesize stimuli-responsive nanofibers is challenging because a slight alteration in solution and process can affect the polymer's response.

12.3 ADVANTAGES TO USE BIO-BASED NANOMATERIALS FOR CANCER THERAPY

Bio-based nanomaterials are advantageous for nanoparticle synthesis due to easy fabrication methods, cost-effectiveness, handling safety, biocompatibility, cost-effectiveness, environmental or eco-friendliness. Different metabolites catalyzing the nanoparticles' synthesis also act as capping agents responsible for stable nanostructures and provide a flexible control on the shape and size of NPs. Single-step processing is used to fabricate metal and metal oxide NPs like cobalt, zinc, silver, iron, titanium, selenium, copper, and cobalt used to treat a disease like cancer.

12.4 LIMITATIONS OF BIONANOMATERIALS' CANCER THERAPY

There are certain limitations of the use of bionanomaterials. The heterogeneity and intricacy of various tumors or cancer cells are different to target. Primarily, the *in-vivo* testing is performed on mice. The targets of nanoparticles are mainly reliant upon the blood circulation and protein-nanoparticle interaction, which can be different for humans and mice due to differences in size. The results obtained by mice cannot be ignored, but clinical trials must be done to improve nanoparticles' reliability. Due to the human immune system, when a foreign body, i.e., nanomaterial, without any antibody enters the human body can be easily recognized, resulting in various undesirable responses. Though nanoparticles have given 90% results for cancer patients, no nanocarrier or nanoparticle still has given 100% results [20].

12.5 CONCLUSIONS AND FUTURE PERSPECTIVES

Bio-based or green nanomaterials have been under extensive research for cancer diagnosis and treatment due to their efficacy and biocompatibility.

The targeted approach is the future of oncology for diagnosis and cancer treatment. The NPs are compared to the "Ehrlich's magic bullets" only to hit the cancerous cells, not the healthy ones. After a series of experimentation and testing, and assessing the positive and negative effects of nanomedicines, it is concluded that this approach is attainable. The bionanomaterials with minor negative impacts will be involved in cancer diagnosis more than the treatment. Irrespective of the time frame and the usage, the research on nanomedicines is rising day by day. Hopefully, patients will get more effective diagnoses and treatments due to endless future research possibilities in nanomedicines letting patients live a better quality of life throughout the world. In the near future, green nanomaterials will be among the best therapeutics for cancer diagnosis and treatment.

12.6 SUMMARY

Cancer is a group of more than 100 diseases with the possibility of growing in every part of the body. The cancer cells grow beyond their boundaries and affect the normal and healthy cells. Different therapies have been used significantly to cure this lethal disease, like chemotherapy, radiation, laser, hormone therapies, surgery, and many more. Though they are effective against cancer, these therapies are costly, have many side effects, and cannot distinguish normal cells from cancerous or tumor cells. Therefore, many new techniques have been tried for the diagnosis and curing of tumors or cancer. Nanotechnology is a potential candidate in the field of medicine. FDA has approved several nanomaterials to diagnose and treat cancer, and many others are in clinical trials. In this chapter, different biopolymers (in the form of nanoparticles, micelles, liposomes, and hydrogels), biobased nanofibers, and nanoparticles have been discussed used for bioimaging, diagnosis, and sustainable drug release for cancer and tumor treatment. The mod of action and release mechanisms have been described briefly. Hopefully, using biobased nanomaterials, the patient could live a better life in the future with minor hazardous health effects. In nearby future, the researchers will find more green nanomaterials to be best for cancer therapy.

KEYWORDS

- **bionanomaterials**
- **cancer**
- **enhanced permeability and retention**
- **microbes**
- **nanofibers**
- **polymers**
- **therapy**

REFERENCES

1. Abbasi, T., Anuradha, J., Ganaie, S. U., & Abbasi, S. A., (2015). Biomimetic synthesis of nanoparticles using aqueous extracts of plants (botanical species). *Journal of Nano Research, 31*(1), 138–202.
2. Abraham, J., & Staffurth, J., (2011). Hormonal therapy for cancer. *Medicine, 39*(12), 723–727.
3. Adeel, M., Duzagac, F., Canzonieri, V., & Rizzolio, F., (2020). Self-therapeutic nanomaterials for cancer therapy: A review. *ACS Applied Nano Materials,* (6), 4962–4971.
4. Ahmad, A., Senapati, S., Khan, M. I., Kumar, R., & Sastry, M., (2003). Extracellular biosynthesis of monodisperse gold nanoparticles by a novel extremophilic actinomycete, *Thermomonospora* sp. *Langmuir, 19*(8), 3550–3553.
5. Al-Enizi, A. M., Zagho, M. M., & Elzatahry, A. A., (2018). Polymer-based electrospun nanofibers for biomedical applications. *Nanomaterials, 8*(4) 259–280.
6. Alphandéry, E., (2020). Bio-synthesized iron oxide nanoparticles for cancer treatment. *International Journal of Pharmaceutics, 586* (30) 119472–119514.
7. Anand, K., Tiloke, C., Phulukdaree, A., Ranjan, B., Chuturgoon, A., Singh, S., & Gengan, R. M., (2016). Biosynthesis of palladium nanoparticles by using *Moringa oleifera* flower extract and their catalytic and biological properties. *Journal of Photochemistry and Photobiology B: Biology, 165*(12), 87–95.
8. Avramović, N., Mandić, B., Savić-Radojević, A., & Simić, T., (2020). Polymeric nanocarriers of drug delivery systems in cancer therapy. *Pharmaceutics, 12*(4) 298–315.
9. Azizi, S., Mohamad, R., Rahim, R. A., Moghaddam, A. B., Moniri, M., Ariff, A., & Namvab, F., (2016). ZnO-Ag core-shell nanocomposite formed by green method using essential oil of wild ginger and their bactericidal and cytotoxic effects. *Applied Surface Science, 384*, 517–524.
10. Bayón-Cordero, L., Alkorta, I., & Arana, L., (2019). Application of solid lipid nanoparticles to improve the efficiency of anti-cancer drugs. *Nanomaterials, 9*(3), 474–493.

11. Bharali, D. J., & Mousa, S. A., (2010). Emerging nanomedicines for early cancer detection and improved treatment: Current perspective and future promise. *Pharmacology and Therapeutics, 128*(2), 324–335.
12. Blackadar, C. B., (2016). Historical review of the causes of cancer. *World Journal of Clinical Oncology, 7*(1), 54–86.
13. Cagnacci, A., & Venier, M., (2019). The controversial history of hormone replacement therapy. *Medicina (Lithuania), 55*(9), 602–613.
14. Chaudhary, Z., Ahmed, N., Rehman, A., & Khan, G. M., (2017). *Lipid Polymer Hybrid Carrier Systems for Cancer Targeting: A Review, 67*(2), 86–100.
15. Chen, H. H., Lu, I. L., Liu, T. I., Tsai, Y. C., Chiang, W. H., Lin, S. C., & Chiu, H. C., (2019). Indocyanine green/doxorubicin-encapsulated functionalized nanoparticles for effective combination therapy against human MDR breast cancer. *Colloids and Surfaces B: Biointerfaces, 177*(2019), 294–305.
16. Chen, Z., Chen, Z., Zhang, A., Hu, J., Wang, X., & Yang, Z., (2016). Electrospun nanofibers for cancer diagnosis and therapy. *Biomaterials Science, 4*(6), 922–932.
17. Chui, P., (2019). Cancer- and chemotherapy-related symptoms and the use of complementary and alternative medicine. *Asia-Pacific Journal of Oncology Nursing, 6*(1), 4–6.
18. Contreras-Cáceres, R., Cabeza, L., Perazzoli, G., Díaz, A., López-Romero, J. M., Melguizo, C., & Prados, J., (2019). Electrospun nanofibers: Recent applications in drug delivery and cancer therapy. *Nanomaterials, 9*(4), 656–679.
19. Darrigues, E., Dantuluri, V., Nima, Z. A., Vang-Dings, K. B., Griffin, R. J., Biris, A. R., Gosh, A., & Biris, A. S., (2017). Raman spectroscopy using plasmonic and carbon-based nanoparticles for cancer detection, diagnosis, and treatment guidance. Part 2: Treatment., *Drug Metabolism Reviews, 49*(2), 253–283.
20. De Jong, W. H., & Borm, P. J., (2008). Drug delivery and nanoparticles: Applications and hazards. *International Journal of Nanomedicine, 3*(2), 133–149.
21. DeVita, V. T., & Chu, E., (2008). A history of cancer chemotherapy. *Cancer Research, 68* (21), 8643–8653.
22. Dilshad, E., Sojla, T. S., Bashir, Z., Hamza, I. A., Saddique, K. A., Zahra, N., & Shabbir, M., (2019). Silver nanoparticles of *Aloe vera* gel as strong therapeutic candidates. *International Journal of Bioscience, 15*(1), 360–373.
23. Din Ud, F., Aman, W., Ullah, I., Qureshi, O. S., Mustapha, O., Shafique, S., & Zeb, A., (2017). Effective use of nanocarriers as drug delivery systems for the treatment of selected tumors. *International Journal of Nanomedicine, 12*(1), 7291–7309.
24. Dolmans, D. E. J. G. J., Fukumura, D., & Jain, R. K., (2003). Photodynamic therapy for cancer. *Nature Reviews Cancer, 3*(5), 380–387.
25. Ediriwickrema, A., & Saltzman, W. M., (2015). Nanotherapy for cancer: Targeting and multifunctionality in the future of cancer therapies. *ACS Biomaterials Science & Engineering, 1*(2), 64–78.
26. Fan, D. Y., Tian, Y., & Liu, Z. J., (2019). Injectable hydrogels for localized cancer therapy. *Frontiers in Chemistry, 7*(11), 675–685.
27. Fan, Y., & Moon, J. J., (2015). Nanoparticle drug delivery systems designed to improve cancer vaccines and immunotherapy. *Vaccines, 3*(3), 662–685.
28. Forootanfar, H., Adeli-Sardou, M., Nikkhoo, M., Mehrabani, M., Amir-Heidari, B., Shahverdi, A. R., & Shakibaie, M., (2014). Antioxidant and cytotoxic effect of

biologically synthesized selenium nanoparticles in comparison to selenium dioxide. *Journal of Trace Elements in Medicine and Biology, 28*(1), 75–79.
29. Fridman, W. H., Teillaud, J. L., Sautès-Fridman, C., Pagès, F., Galon, J., Zucman-Rossi, J., Tartour, E., et al., (2011). The ultimate goal of curative anti-cancer therapies: Inducing an adaptive anti-tumor immune response. *Frontiers in Immunology, 2* (12), 66–69.
30. Galaev, I. Y., & Mattiasson, B., (1999). "Smart" polymers and what they could do in biotechnology and medicine. *Trends in Biotechnology, 17*(8), 335–340.
31. Generalov, R., Kuan, W. B., Chen, W., Kristensen, S., & Juzenas, P., (2015). Radio sensitizing effect of zinc oxide and silica nanocomposites on cancer cells. *Colloids and Surfaces B: Biointerfaces, 129*(1), 79–86.
32. Goubran, H. A., Elemary, M., Radosevich, M., Seghatchian, J., El-Ekiaby, M., & Burnouf, T., (2016). Impact of transfusion on cancer growth and outcome. *Cancer Growth and Metastasis, 9*(2016), 1–8.
33. Gour, A., & Jain, N. K., (2019). Advances in green synthesis of nanoparticles. *Artificial Cells, Nanomedicines and Biotechnology, 47*(1), 844–851.
34. Goyal, R., Macri, L. K., Kaplan, H. M., & Kohn, J., (2016). Nanoparticles and nanofibers for topical drug delivery. *Journal of Controlled Release, 240*(1), 77–92.
35. Hadjianfar, M., Semnani, D., & Varshosaz, J., (2018). Polycaprolactone/chitosan blend nanofibers loaded by 5-fluorouracil: An approach to anti-cancer drug delivery system. *Polymers for Advanced Technologies, 29*(12), 2972–2981.
36. Hallan, S. S., Kaur, P., Kaur, V., Mishra, N., & Vaidya, B., (2016). Lipid polymer hybrid as emerging tool in nanocarriers for oral drug delivery. *Artificial Cells, Nanomedicine and Biotechnology, 44*(1), 334–349.
37. Henley, S. J., Ward, E. M., Scott, S., Ma, J., Anderson, R. N., Firth, A. U., Thomas, C. C., et al., (2020). Annual report to the nation on the status of cancer, part I: National cancer statistics. *Cancer, 126*(10), 2225–2249.
38. Hersh, E. M., & Stopeck, A. T., (1997). Advances in the biological therapy and gene therapy of malignant disease. *Clinical Cancer Research, 3*(12 II), 2623–2629.
39. Hrib, J., Sirc, J., Hobzova, R., Hampejsova, Z., Bosakova, Z., Munzarova, M., & Michalek, J., (2015). Nanofibers for drug delivery – incorporation and release of model molecules, influence of molecular weight and polymer structure. *Beilstein Journal of Nanotechnology, 6*(1), 1939–1945.
40. Iqbal, H., Khan, B. A., Khan, Z. U., Razzaq, A., Khan, N. U., Menaa, B., & Menaa, F., (2020). Fabrication, physical characterizations and *in vitro* antibacterial activity of cefadroxil-loaded chitosan/poly(vinyl alcohol) nanofibers against *Staphylococcus aureus* clinical isolates. *International Journal of Biological Macromolecules, 144*(1), 921–931.
41. Iram, S., Khan, S., Ansary, A. A., Arshad, M., Siddiqui, S., Ahmad, E., Khan, R. H., & Khan, M. S., (2016). Biogenic terbium oxide nanoparticles as the vanguard against osteosarcoma. *Spectrochimica Acta Part A: Molecular and Biomolecular Spectroscopy, 168*(5), 123–131.
42. Ishak, N. A. I. M., Kamarudin, S. K., & Timmiati, S. N., (2019). Green synthesis of metal and metal oxide nanoparticles via plant extracts: An overview. *Materials Research Express, 6*(11), 112004–112037.
43. Jing, Z., Lixin, Y., Qizhi, L., Xuefei, Z., Huili, G., Xiuling, X., Xuesi, C., & Xiabin, J., (2005). Influence of the drug compatibility with polymer solution on the release kinetics

of electrospun fiber formulation. *Journal of Controlled Release : Official Journal of the Controlled Release Society, 105*(1, 2), 43–51.
44. Jain, R., & Tiwari, A., (2015). Role of polyhydroxyalkanoates in cancer and other drug delivery systems. *Journal of Cancer Research and Therapeutics, 11*(2), 494–495.
45. Jiang, G. M., Tan, Y., Wang, H., Peng, L., Chen, H. T., Meng, X. J., Li, L. L., et al., (2019). The relationship between autophagy and the immune system and its applications for tumor immunotherapy. *Molecular Cancer, 18*(1), 1–22.
46. Kamaly, N., Yameen, B., Wu, J., & Farokhzad, O. C., (2016). Degradable controlled-release polymers and polymeric nanoparticles: Mechanisms of controlling drug release. *Chemical Reviews, 116*(4), 2602–2663.
47. Kanakalakshmi, A., Janaki, V., Shanthi, K., & Kamala-Kannan, S., (2017). Biosynthesis of Cr(III) nanoparticles from electroplating wastewater using chromium-resistant *Bacillus subtilis* and its cytotoxicity and antibacterial activity. *Artificial Cells, Nanomedicine and Biotechnology, 45*(7), 1304–1309.
48. Karmous, I., Pandey, A., Haj, K. B., & Chaoui, A., (2019). Efficiency of the green synthesized nanoparticles as new tools in cancer therapy: Insights on plant-based bioengineered nanoparticles, biophysical properties, and anti-cancer roles. *Biological Trace Element Research, 196*(1), 330–342.
49. Kesharwani, P., & Iyer, A. K., (2015). Recent advances in dendrimer-based nanovectors for tumor-targeted drug and gene delivery. *Drug Discovery Today, 20*(5), 536–547.
50. Keskin, D., & Tezcaner, A., (2017). Micelles as a delivery system for cancer treatment. *Current Pharmaceutical Design, 23*(35) 5230–5241.
51. Kim, Y. J., Ebara, M., & Aoyagi, T., (2013). A smart hyperthermia nanofiber with switchable drug release for inducing cancer apoptosis. *Advanced Functional Materials, 23*(46), 5753–5761.
52. Kim, Y. J., Park, M. R., Kim, M. S., & Kwon, O. H., (2012). Polyphenol-loaded polycaprolactone nanofibers for effective growth inhibition of human cancer cells. *Materials Chemistry and Physics, 133*(2, 3), 674–680.
53. Kuang, G., Zhang, Z., Liu, S., Zhou, D., Lu, X., Jing, X., & Huang, Y., (2018). Biphasic drug release from electrospun polyblend nanofibers for optimized local cancer treatment. *Biomaterials Science, 6*(2), 324–331.
54. Kumar, S., Sarita, N. M., Dilbaghi, N., Tankeshwar, K., & Kim, K. H., (2018). Recent advances and remaining challenges for polymeric nanocomposites in healthcare applications. *Progress in Polymer Science, 80*(1), 1–38.
55. Kumar, S., Singh, A. P., Senapati, S., & Maiti, P., (2019). Controlling drug delivery using nanosheet-embedded electrospun fibers for efficient tumor treatment. *ACS Applied Bio Materials, 2*(2), 884–894.
56. Kuppusamy, P., Yusoff, M. M., Maniam, G. P., & Govindan, N., (2016). Biosynthesis of metallic nanoparticles using plant derivatives and their new avenues in pharmacological applications – An updated report. *Saudi Pharmaceutical Journal, 24*(4), 473–484.
57. Lee, H. J., Koo, A. N., Kwon, I. K., & Lee, S. C., (2013). Polymer therapeutics for treating cancer. In: *Biomaterials for Cancer Therapeutics: Diagnosis, Prevention and Therapy* (pp. 113–133). New York: Woodhead Publishing.
58. Li, J. J., Yang, Y. Y., Yu, D. G., Du, Q., & Yang, X. L., (2018). Fast dissolving drug delivery membrane based on the ultra-thin shell of electrospun core-shell nanofibers. *European Journal of Pharmaceutical Sciences, 122*(15), 195–204.

59. Li, J., Meng, Y., Wu, X., & Sun, Y., (2020). Polyamines and related signaling pathways in cancer. *Cancer Cell International, 20*(539), 1–16.
60. Li, X., Xu, H., Chen, Z. S., & Chen, G., (2011). Biosynthesis of nanoparticles by microorganisms and their applications. *Journal of Nanomaterials, 2011*(8), 1–16.
61. Lv, H., Wu, C., Liu, X., Bai, A., Cao, Y., Shang, W., Hu, L., & Liu, Y., (2018). Folate-functionalized mesoporous hollow SnO_2 nanofibers as a targeting drug carrier to improve the antitumor effect of Paclitaxel for liver cancer therapy. *BioMed Research International, 2018*, 1–11.
62. Muhammad, O., Ali T. K., Abida R., Muhammad, A. K., Irshad, A., Nazar, U. I., Muthupandian, S., Muhammad, F. U., Muhammad, A., & Zabta, K. S., (2016). Green synthesis of silver nanoparticles via plant extracts: beginning a new era in cancer theranostics. *Nanomedicine (London, England), 11* (23), 3157–3177.
63. Manimaran, M., & Kannabiran, K., (2017). Actinomycetes-mediated biogenic synthesis of metal and metal oxide nanoparticles: Progress and challenges. *Letters in Applied Microbiology, 64*(6), 401–408.
64. MJ, D., (2013). The war on cancer: Are we winning? *Tumor Biology : The Journal of the International Society for Oncodevelopmental Biology and Medicine, 34*(3), 1275–1284.
65. Moghaddam, A. B., Namvar, F., Moniri, M., Tahir, P. M., Azizi, S., & Mohamad, R., (2015). Nanoparticles biosynthesized by fungi and yeast: A Review of their preparation, properties, and medical applications. *Molecules, 20*(9), 16540–16565.
66. Moschou, P. N., & Roubelakis-Angelakis, K. A., (2014). Polyamines and programmed cell death. *Journal of Experimental Botany, 65*(5) 1285–1296.
67. Murugan, M., Anthony, K. J. P., Jeyaraj, M., Rathinam, N. K., & Gurunathan, S., (2014). Biofabrication of gold nanoparticles and its biocompatibility in human breast adenocarcinoma cells (MCF-7). *Journal of Industrial and Engineering Chemistry, 20*(4), 1713–1719.
68. Nadeem, M., Tungmunnithum, D., Hano, C., Abbasi, B. H., Hashmi, S. S., Ahmad, W., & Zahir, A., (2018). The current trends in the green syntheses of titanium oxide nanoparticles and their applications. *Green Chemistry Letters and Reviews, 11*(4), 492–502.
69. Namvar, F., Rahman, H. S., Mohamad, R., Baharara, J., Mahdavi, M., Amini, E., Chartrand, M. S., & Yeap, S. K., (2014). Cytotoxic effect of magnetic iron oxide nanoparticles synthesized via seaweed aqueous extract. *International Journal of Nanomedicine, 9*(1), 2479–2488.
70. Nethi, S. K., Mukherjee, A., & Mukherjee, S., (2020). Biosynthesized gold and silver nanoparticles in cancer theranostics. In: *Handbook of Nanomaterials and Nanocomposites for Energy and Environmental Applications* (pp. 1–15). England: Springer.
71. Novita, S. I., Setiawan, T., Seock, K. K., Toni, W. Y., Won, C. K., & Young, K. H., (2021). Metabolism and function of polyamines in cancer progression. *Cancer Letters, 519*(10), 91–104.
72. Nurgali, K., Jagoe, R. T., & Abalo, R., (2018). Editorial: Adverse effects of cancer chemotherapy: Anything new to improve tolerance and reduce sequelae? *Frontiers in Pharmacology, 9*(3), 245–248.
73. Olusanya, T. O. B., Ahmad, R. R. H., Ibegbu, D. M., Smith, J. R., & Elkordy, A. A., (2018). Liposomal drug delivery systems and anti-cancer drugs. *Molecules, 23*(4), 907–924.

74. Ortega, F. G., Fernández-Baldo, M. A., Fernández, J. G., Serrano, M. J., Sanz, M. I., Diaz-Mochón, J. J., Lorente, J. A., & Raba, J., (2015). Study of antitumor activity in breast cell lines using silver nanoparticles produced by yeast. *International Journal of Nanomedicine, 10*(2015), 2021–2031.
75. Park, K., Lee, S., Kang, E., Kim, K., Choi, K., & Kwon, I. C., (2009). New generation of multifunctional nanoparticles for cancer imaging and therapy. *Advanced Functional Materials, 19*(10), 1553–1566.
76. Pathak, V. M., & Navneet, (2017). Review on the current status of polymer degradation: A microbial approach. *Bioresources and Bioprocessing, 4*(1), 1–31.
77. Patra, A., Satpathy, S., Shenoy, A. K., Bush, J. A., Kazi, M., & Hussain, M. D., (2018). Formulation and evaluation of mixed polymeric micelles of quercetin for treatment of breast, ovarian, and multidrug-resistant cancers. *International Journal of Nanomedicine, 13*(5), 2869–2881.
78. Peralta-Videa, J. R., Huang, Y., Parsons, J. G., Zhao, L., Lopez-Moreno, L., Hernandez-Viezcas, J. A., & Gardea-Torresdey, J. L., (2016). Plant-based green synthesis of metallic nanoparticles: Scientific curiosity or a realistic alternative to chemical synthesis? *Nanotechnology for Environmental Engineering, 1*(1), 1–29.
79. Pillay, V., Dott, C., Choonara, Y. E., Tyagi, C., Tomar, L., Kumar, P., Du Toit, L. C., & Ndesendo, V. M. K., (2013). A review of the effect of processing variables on the fabrication of electrospun nanofibers for drug delivery applications. *Journal of Nanomaterials, 2013,* 1–22.
80. Prabhu, P., (2019). Nanofibers for medical diagnosis and therapy. In: *Handbook of Nanofibers* (pp. 831–867). Springer, Cham.
81. Prasad, P. R., Kanchi, S., & Naidoo, E. B., (2016). *In-vitro* evaluation of copper nanoparticles cytotoxicity on prostate cancer cell lines and their antioxidant, sensing and catalytic activity: One-pot green approach. *Journal of Photochemistry and Photobiology B: Biology, 161*(8), 375–382.
82. Priya, J. H., John, R., Alex, A., & Anoop, K. R., (2014). Smart polymers for the controlled delivery of drugs–a concise overview. *Acta Pharmaceutica Sinica B, 4*(2), 120–127.
83. Rahimi, M., Jafari, A., Tabaei, S. J. S., Rahimi, M., Taranejoo, S., & Ghanimatdan, M., (2020). Herbal extract incorporated chitosan based nanofibers as a new strategy for smart anti-cancer drug delivery system: An *in vitro* model. *Worl. Cancer Research Journal, 7*, 1–11.
84. Robinson, K., & Tiriveedhi, V., (2020). Perplexing role of p-glycoprotein in tumor microenvironment. *Frontiers in Oncology, 10*, 265–274.
85. Roy, S., Das, T. K., Maiti, G. P., & Basu, U., (2016). Microbial biosynthesis of nontoxic gold nanoparticles. *Materials Science and Engineering B: Solid-State Materials for Advanced Technology, 203*, 41–51.
86. Samadi, S., Moradkhani, M., Beheshti, H., Irani, M., & Aliabadi, M., (2018). Fabrication of chitosan/poly(lactic acid)/graphene oxide/TiO_2 composite nanofibrous scaffolds for sustained delivery of doxorubicin and treatment of lung cancer. *International Journal of Biological Macromolecules, 110*(2018), 416–424.
87. Sathishkumar, P., Vennila, K., Jayakumar, R., Yusoff, A. R. M., Hadibarata, T., & Palvannan, T., (2016). Phyto-synthesis of silver nanoparticles using *Alternanthera tenella* leaf extract: An effective inhibitor for the migration of human breast adenocarcinoma (MCF-7) cells. *Bioprocess and Biosystems Engineering, 39*(4), 651–659.

88. Sedghi, R., Shaabani, A., Mohammadi, Z., Samadi, F. Y., & Isaei, E., (2017). Biocompatible electrospinning chitosan nanofibers: A novel delivery system with superior local cancer therapy. *Carbohydrate Polymers, 159,* 1–10.
89. Shan, X., Liu, C., Li, F., Ouyang, C., Gao, Q., & Zheng, K., (2015). Nanoparticles vs. nanofibers: A comparison of two drug delivery systems on assessing drug release performance *in vitro*. *Designed Monomers and Polymers, 18*(7), 678–689.
90. Siegel, R. L., Miller, K. D., & Jemal, A., (2019). Cancer statistics. *A Cancer Journal for Clinicians, 69*(1), 7–34.
91. Siemann, D. W., & Horsman, M. R., (2015). Modulation of the tumor vasculature and oxygenation to improve therapy. *Pharmacology and Therapeutics, 153*(2015), 107–124.
92. Singh, J., Jain, K., Mehra, N. K., & Jain, N. K., (2016). Dendrimers in anti-cancer drug delivery: Mechanism of interaction of drug and dendrimers. *Artificial Cells, Nanomedicine, and Biotechnology, 44*(7), 1626–1634.
93. Sivakumar, B., Aswathy, R. G., Sreejith, R., Nagaoka, Y., Iwai, S., Suzuki, M., Fukuda, T., et al., (2014). Bacterial exopolysaccharide based magnetic nanoparticles: A versatile nanotool for cancer cell imaging, targeted drug delivery and synergistic effect of drug and hyperthermia mediated cancer therapy. *Journal of Biomedical Nanotechnology, 10*(6), 885–899.
94. Son, Y. J., Kim, Y., Kim, W. J., Jeong, S. Y., & Yoo, H. S., (2015). Antibacterial nanofibrous mats composed of Eudragit for pH-dependent dissolution. *Journal of Pharmaceutical Sciences, 104*(8), 2611–2618.
95. Srivastava, P., Braganca, J. M., & Kowshik, M., (2014). *In vivo* synthesis of selenium nanoparticles by *Halococcus salifodinae* BK18 and their anti-proliferative properties against HeLa cell line. *Biotechnology Progress, 30*(6), 1480–1487.
96. Sui, X., Luo, C., Wang, C., Zhang, F., Zhang, J., & Guo, S., (2016). Graphene quantum dots enhance the anti-cancer activity of cisplatin via increasing its cellular and nuclear uptake. *Nanomedicine: Nanotechnology, Biology, and Medicine, 12*(7), 1997–2006.
97. Sulaiman, G. M., Hussien, H. T., & Saleem, M. M. N. M., (2015). Biosynthesis of silver nanoparticles synthesized by *Aspergillus flavus* and their antioxidant, antimicrobial and cytotoxicity properties. *Bulletin of Materials Science, 38*(3), 639–644.
98. Sunkar, S., Nachiyar, C. V., Lerensha, R., & Renugadevi, K., (2014). Biogenesis of TiO_2 nanoparticles using endophytic *Bacillus cereus*. *Journal of Nanoparticle Research, 16*(11), 1–11.
99. Sutradhar, K. B., & Amin, M. L., (2014). Nanotechnology in cancer drug delivery and selective targeting. *ISRN Nanotechnology, 2014* (ID 939378), 1–12.
100. Syed, A., Saraswati, S., Kundu, G. C., & Ahmad, A., (2013). Biological synthesis of silver nanoparticles using the fungus *Humicola* sp. and evaluation of their cytoxicity using normal and cancer cell lines. *Spectrochimica Acta-Part A: Molecular and Biomolecular Spectroscopy, 114*(10), 144–147.
101. Sztandera, K., Gorzkiewicz, M., & Klajnert-Maculewicz, B., (2018). Gold nanoparticles in cancer treatment. *Molecular Pharmaceutics, 16*(1), 1–23.
102. Tarantash, M., Nosrati, H., Kheiri, M. H., & Baradar, K. A., (2018). Preparation, characterization and *in vitro* anti-cancer activity of Paclitaxel conjugated magnetic nanoparticles. *Drug Development and Industrial Pharmacy, 44*(11), 1895–1903.
103. Thariat, J., Hannoun-Levi, J. M., Sun, M. A., Vuong, T., & Gérard, J. P., (2013). Past, present, and future of radiotherapy for the benefit of patients. *Nature Reviews Clinical Oncology, 10*(1), 52–60.

104. Thomas, H., & Coley, H. M., (2020). Overcoming multidrug resistance in cancer: An update on the clinical strategy of inhibiting p-glycoprotein. *Cancer Control, 10*(2), 159–165.
105. Thomas, T., (2018). Cellular and animal model studies on the growth inhibitory effects of polyamine analogues on breast cancer. *Medical Sciences, 6*(1), 24–37.
106. Tiwari, M., Jain, P., Chandrashekhar, H. R., Narayanan, K., Bhat, K. U., Udupa, N., & Rao, J. V., (2016). Biosynthesis of copper nanoparticles using copper-resistant *Bacillus cereus*, a soil isolate. *Process Biochemistry, 51*(10), 1348–1356.
107. Torpy, J. M., Lynm, C., & Glass, R. M., (2010). Cancer: The basics. *Journal of the American Medical Association, 304*(14), 1626–1634.
108. Torres-Martinez, E. J., Cornejo, B. J. M., Serrano, M. A., Pérez, G. G. L., & Villarreal, G. L. J., (2018). A summary of electrospun nanofibers as drug delivery system: Drugs loaded and biopolymers used as matrices. *Current Drug Delivery, 15*(10), 1360–1374.
109. Wagh, A., Singh, J., Qian, S., & Law, B., (2013). A short circulating peptide nanofiber as a carrier for tumoral delivery. *Nanomedicine: Nanotechnology, Biology, and Medicine, 9*(4), 449–457.
110. Wagner, A. M., Spencer, D. S., & Peppas, N. A., (2018). Advanced architectures in the design of responsive polymers for cancer nanomedicine. *Journal of Applied Polymer Science, 135*(0), 46154–46173.
111. Wallace, H. M., & Niiranen, K., (2007). Polyamine analogues: An update. *Amino Acids, 33*(2), 261–265.
112. Wieman, T. J., (1986). Lasers and the surgeon. *The American Journal of Surgery, 151*(4), 493–500.
113. Wilson, N., (2018). Nanoparticles: Environmental problems or problem solvers? *BioScience, 68*(4), 241–246.
114. Wong, K. H., Lu, A., Chen, X., & Yang, Z., (2020). Natural ingredient-based polymeric nanoparticles for cancer treatment. *Molecules, 25*(16), 3620–3644.
115. Wu, H., & Zhang, J., (2018). Chitosan-based zinc oxide nanoparticle for enhanced anti-cancer effect in cervical cancer: A physicochemical and biological perspective. *Saudi Pharmaceutical Journal, 26*(2), 205–210.
116. Wyld, L., Audisio, R. A., & Poston, G. J., (2015). The evolution of cancer surgery and future perspectives. *Nature Reviews Clinical Oncology, 12*(2), 115–124.
117. Xu, H., Lu, X., Li, J., Ding, D., Wang, H., Li, X., & Xie, W., (2017). Superior antitumor effect of extremely high drug loading self-assembled paclitaxel nanofibers. *International Journal of Pharmaceutics, 526*(1, 2), 217–224.
118. Yang, H., (2016). Targeted nanosystems: Advances in targeted dendrimers for cancer therapy. *Nanomedicine: Nanotechnology, Biology, and Medicine, 12*(2), 309–316.
119. Yuan, Y. G., Peng, Q. L., & Ggurunathan, S., (2017). Silver nanoparticles enhance the apoptotic potential of gemcitabine in human ovarian cancer cells: Combination therapy for effective cancer treatment. *International Journal of Nanomedicine, 12*(2017), 6487–6502.
120. Zhiyun, Z., Shi, L., Yanxin, Q., Dongfang, Z., Zhigang, X., & Xiabin, J., (2016). Time-programmed DCA and oxaliplatin release by multilayered nanofiber mats in prevention of local cancer recurrence following surgery. *Journal of Controlled Release : Official Journal of the Controlled Release Society, 235*(10), 125–133.
121. Zehir, A., Benayed, R., Shah, R. H., Syed, A., Middha, S., Kim, H. R., & Berger, M. F., (2017). Mutational landscape of metastatic cancer revealed from prospective clinical sequencing of 10,000 patients. *Nature Medicine, 23*(6), 703–713.

122. Zhang, X., Tang, K., & Zheng, X., (2016). Electrospinning and crosslinking of Col/PVA nanofiber-microsphere containing salicylic acid for drug delivery. *Journal of Bionic Engineering, 13*(1), 143–149.
123. Zhang, Y. X., Zheng, J., Gao, G., Kong, Y. F., Zhi, X., Wang, K., Zhang, X. Q., & Cui, D. X., (2011). Biosynthesis of gold nanoparticles using chloroplasts. *International Journal of Nanomedicine, 6*(2011), 2899–2906.
124. Zupančič, Š., Sinha-Ray, S., Sinha-Ray, S., Kristl, J., & Yarin, A. L., (2015). Long-term sustained ciprofloxacin release from PMMA and hydrophilic polymer blended nanofibers. *Molecular Pharmaceutics, 13*(1), 295–305.

PART III

NANOMATERIALS FOR ANTI-MICROBIAL AND ANTI-BACTERIAL APPLICATIONS IN MEDICINE

CHAPTER 13

ROLE OF NANOMATERIALS IN MICROBIAL STUDIES

SHUBHJEET MANDAL, PIYUSH KUMAR TIWARI, ANKITA KUMARI, NITIN BAYAL, ANCHAL ANCHAL, and KUSUM UPADHYAY

ABSTRACT

Microbiology and nanomaterial sciences have contributed to science and technology by developing innovative solutions for the healthcare, environmental, and agricultural sectors. Moreover, because repeated utilization of antibiotics has resulted in MDR (multiple drug resistance) in microorganisms and the delivery of metal nanoparticles affects the food chain, it is now strongly recommended to develop interdisciplinary approaches combining microbiology and nanomaterials to combat secondary human health, environmental, and ecological damage. Nanotechnology impacts various areas of microbiology, including medical/health, food, and environmental. It enables the investigation and observation of a process at the molecular assembly level. The logical use of various areas in a synergistic manner yields new and long-term solutions. This chapter focuses on the link between nanomaterials and microbiological disciplines, emphasizing the enormous potential that may be realized through interdisciplinary study.

13.1 INTRODUCTION

Antibiotic resistance, the capability of microorganisms to surmount the effect of any drug used to stop or kill them, has become a very evident public health threat. In 2019, US disease control and prevention centers identified

Sustainable Nanomaterials for Biomedical Engineering: Impacts, Challenges, and Future Prospects.
Junaid Ahmad Malik, Megh R. Goyal & Mohamed Jaffer M. Sadiq (Eds.)

11 pathogens in the severe threat group, including *Pseudomonas aeruginosa, Streptococcus pneumoniae, Campylobacter sp., Candida sp., Salmonella sp., Shigella sp., Enterococcus vancomycin-resistant, Staphylococcus aureus,* methicillin-resistant, and extended-spectrum beta-lactamases (ESBL) producing enterobacteriaceae.

Antibiotics, such as penicillin, have proven effective in treating various ailments; nevertheless, the rise of antibiotic-resistant infectious diseases is posing a significant worldwide concern [25]. To combat this severe threat, new approaches are being developed, among which nanotechnology notably has received the primary positive response.

Nanotechnology is a branch of scientific study concerned with producing numerous materials, including particulate materials with at minimum one scale smaller than 100 nanometers. It has been considered that this field of science can regulate and modify matter at both atomic and molecular levels. Particles with this dimension have unique and vital properties [88].

Nanomaterials are of different kinds, such as nanoparticles, nano-dots, nano-cubes, nanorods, and two-dimensional materials. These have shown significant anti-microbial activity in response to various microorganisms. Of these, nanoparticles have shown unusual properties in electronics, magnets, optics, and chemistry that have effectively expanded researchers' interest in their synthesis [95]. Nanoparticles are made up of a range of essential substances and have different properties based on chemical composition, size, and shape. Various chemical and physical processes that cannot prevent toxic chemicals can synthesize nanoparticles [83]. Therefore, being developed biologically has accumulated more importance than the synthetic ones due to their eco-friendly nature with cost-effective and stable application in agriculture, electronics, and medicine. They have also arisen as a suitable solution to antibiotics, including organic toxins, to treat biological pathogens and chemical contaminants.

Throughout the years, metal-based nanoparticles have been demonstrated to be an efficient anti-bacterial agent against common pathogenic microbes. Therefore, as anti-microbials, nanoparticles like platinum, titanium dioxide, and zinc oxide attract significant publicity.

Silver nanoparticles (AgNPs) display substantial anti-microbial activity against many organisms and are thus integrated into separate matrices. Titanium oxide nanoparticles are used as an anti-microbial agent in cosmetics, in filters with stable germicidal properties, and even in combination with platinum. Zinc oxide and copper oxide nanomaterials also exhibit anti-microbial behavior for which several medical and skin coatings are used. For dermatological uses, such as in creams, lotions, and ointments, ZnO powder is used.

Individually, the fields of microbiology and nanotechnology have had a strong effect on human well-being and have also successfully achieved an environmental and ecological equilibrium. On the other hand, the excess use of antibiotics has ultimately led to the drug-resistant ability of microorganisms and nanotechnology being used as an alternative to this. Still, it fails due to the harmful effects of metals on health. Therefore, these two disciplines are being projected as a single unit for providing sustainable solutions in a profound manner [12]. This chapter explores the different aspects of nanomaterials in microbial studies.

13.2 EARLY STUDIES ON NANOPARTICLES

Nanotechnology has a long history, dating back to the 4th and 5th century BC, when Indian and Chinese practitioners successfully created gold colloids, known as Swarna Bhasma in Ayurveda and utilized for therapeutic reasons, during the manufacture of traditional medications. Paracelsus of Europe used gold colloids to treat psychiatric illnesses and syphilis in the middle-ages [26]. Nevertheless, Faraday published the very first scientific publication of gold colloids in 1857. Richard Feynman reiterated their genuine interest in this area when giving his famous speech in 1959, "There is plenty of room at the bottom." Eric Drexler popularized this idea when he discussed it in his book Engines of Development, "The Coming Era of Nanotechnology," in the 1980s.

Jatzkewitz produced a combination of polyvinylpyrrolidone and mescaline, a polymer-based drug with a tiny peptide spacer between the drug and the polymer, in 1950, which is thought to be the first synthesis of nanoparticles [38]. In the mid-1960s, after this nanoparticle-based drug discovery, Bangham discovered liposomes [13]. These two significant occurrences characterize the birth of nanocarriers, and this class is just the main nanoparticle that is promoted.

In the 1970s, Ringsdorf gave the idea of selective drug conjugates that underlined the new priorities in this area of nanoparticles. In 1972, an albumin-based nanoparticle was discovered [75]. In 2005, the US Food and Drug Administration (FDA) authorized the first class of nanoparticles to treat metastatic breast cancer [36]. Polymer-based nanoparticles, deemed to be the second class, were first published in 1976 [48].

In the 1980s, a significant discovery while researching poly (styrene-co-maleic acid) attached to the cytotoxic agent neocarzinostatin (SMS). An apparent aggregation of SMANCS conjugates at the tumor site was noticed, which gradually became a condition called Improved Permeability and Retention (EPR). In 1983, Novartis authorized Sandimmune, a combination of cyclosporine and Cremophor EL, and Cremophor, which is used in the production of paclitaxel, a cytotoxic anticancer medicine [62].

The FDA authorized an implanted version of goserelin acetate, Zoladex, by AstraZeneca in 1989 to treat certain types of prostate and breast malignancies. It was the first polymer composition controlled-release. Nanotechnology has spread to several industries, including medical, biomedical, manufacturing, food, agriculture, and environmental sectors, with diverse applications [12].

In recent years, engineered nanomaterials have attained a significant position in this field due to their high effectiveness and potential risk to the microbial [64]. It may be due to their tiny size and high surface-to-volume ratio, allowing them to attach to cellular microbial structures [64].

Alvarez and Cervantes [4] discovered that engineered nanomaterials (ENMs) such as Al_2O_3 blocked methanogenesis and decreased toxicity when coated with humic acid. Yang et al. [93] noted that methanogenesis in the landfill bioreactor was blocked by Ag nanoparticles. However, one study conducted by Nyberg et al. [63] showed little effect on methanogenesis when C60 was exposed to biosolids from anaerobic sludge for a few months.

Few studies performed by Li et al. [49]; Masrahi et al. [56]; and Zheng et al. [95] have shown that ENMs such as TiO_2 and Ag nanoparticles can also inhibit the mechanism of nitrogen removal. Graphene oxide has been found to have a detrimental effect on the disposal process for wastewater. In another study, Ag nanoparticles were proven to inhibit the nitrification process in the soil microbial community [63]. Zheng and his co-worker [95], in their study, showed that titanium dioxide reduces nitrogen and phosphorous activity.

It has been pointed out that ENMs are organism-specific which means different microorganisms show variations in the susceptibility to the ENMs. According to Sun and his partner's study, some bacteria are more tolerant of Ag NPs than others. Gram-positive bacteria were shown to change their membrane lipid content to become immune to SWCNTs in a similar study. This hypothesis of the essential existence of ENMs in animals remains uncertain and needs further research.

According to reports and surveys, worldwide nanomaterials markets are anticipated to expand at a compound annual growth rate of 20% from 2016 to 2022.

13.3 CORRELATION BETWEEN NANOMATERIAL AND MICROBIOLOGY

At different stages, microbiology applies to nanotechnology. Several microbial entities, including molecular motors like flagella and pili, are nano-machines of nature. By the method of self-assembly, bacteria often form biofilms. A fully regulated assembly of building blocks is the growth of aerial hyphae in bacteria and fungi. According to the findings, the formation of viral capsids is revealed to be a regulated method of molecular recognition and nanoscale self-assembly. The nanoscale processing and modification of organic and inorganic matter require nanotechnology. Materials with biological, physical, and chemical properties are primarily regulated by well-organized molecular structures and dynamics, contributing to nanomaterial nature. Existing agricultural genetic modification techniques are now considered nanotechnology in molecular biology. In general, nanotechnology promises much more effective and efficient technologies and other means of treating food polymers and polymer assemblies for potential development to improve the consistency and protection of foodstuffs. Nanotechnology appears to offer the production of unique and specified material characteristics and the ability for these materials to self-assemble, self-heal, and maintain themselves. It is well studied that nanoscience has an impact on many fields of microbiology. Exploration of the molecular assembly stages of a method allows for visualization and accurate quantification in microbiological studies. It assists in the identification and analysis of molecular recognition and self-assembly motifs. More specifically, there are three areas in which nanotechnologists use microbiologists' techniques:

- AFM (near/far-field microscope) visualization of spatial structure in live microorganisms;
- Poking and dragging objects on a nanoscale (laser traps, optical tweezer); and
- Single molecules imaging.

13.4 CLASSIFICATION OF NANOMATERIAL

Nanotechnology includes the processing of nanomaterial (NM) types, including nano-objects and nanoparticles (NP). Nanomaterials are defined as particles with a diameter of fewer than 100 nanometers in one dimension, nano-objects are defined as particles with a diameter of fewer than 100 nanometers in two dimensions (e.g., carbon nanotubes), and nanoparticles are defined as particles with a diameter of fewer than 100 nanometers in three dimensions [79]. Given the exponential growth of nanotechnology and the growing spectrum of production and development of nanomaterials, it is mandatory to address the possible effects on environmental and human health. Because of the smaller scale, NP has far more advanced relative surface areas than equivalent traditional types; however, the smaller scale also results in greater reactivity and modified surface properties, which can add to a variety of consumer goods such as drugs, cosmetics, food, paints, and sun cream, and other applications that specially release Nanoparticles into the environment [5]. There are many kinds of other nanoparticles you can check in Figure 13.1.

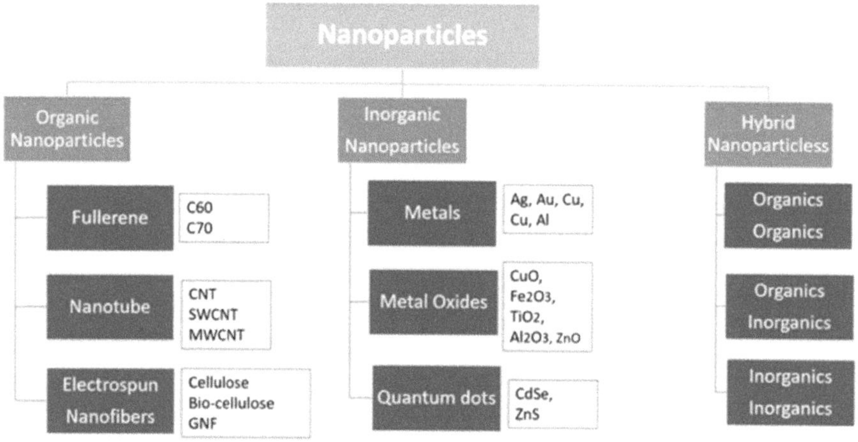

FIGURE 13.1 Classification of nanoparticles.

13.5 PRINCIPAL OF NANOMATERIAL PRODUCTION

Nanoparticles have been created utilizing various physical and chemical methods; however, few chemical approaches can completely exclude hazardous elements throughout the synthesis process. There is also a strong demand for a green nanoparticle synthesis method to be created. The biological technique of nanoparticle production utilizing microorganisms or plant extracts has shown to be a dependable and environmentally acceptable alternative to chemical and physical procedures. Microorganisms have long been used to repair hazardous metals by lowering metal ions, but there has recently been renewed interest in utilizing microbes to synthesize nanoparticles. Nanoparticles are biosynthesized using microorganisms that absorb target ions from their solutions and then accumulate the reduced metal using enzymes produced in its component form by microbial cell activity. Two different processes, intracellular, and extracellular production, may be narrowly defined [15]. The internal process invented the transferring ions into the microbial cell to create nanoparticles in the presence of cellular enzymes. Extracellular nanoparticle production necessitates metal ion trapping on the microbial cell surface and ion reduction in the presence of extracellular enzymes. Several microbes, including magnetotactic bacteria [16], S-layer bacteria, fungi, actinomycetes, and yeast, have developed nanostructured metal crystals and metallic nanoparticles and regulated the scale form, composition, and monodispersity of these particles has been studied. On the other side, the influence of nanoparticles on the microbial environment has also drawn considerable interest from nanotechnologists. Nanoparticles can sustain microbial activities. There have been a lot of random studies on the impact of nanoparticles on microbiological reaction rates [52].

13.5.1 PRODUCTION OF NANOMATERIAL BY MICROBIAL SYSTEM

According to the findings, many species, including fungi, bacteria, yeast, and actinomycetes, can synthesize metallic nanoparticles, mineral crystals, and nanoparticles, either intracellularly or extracellularly. Microorganisms have been used to synthesize nanoparticles in recent years to understand nanoparticle production mechanics better. Nanoparticle synthesis with bacteria and fungus has gotten more attention than synthesis with actinomycetes and yeast because bacteria and fungi synthesis has more sophisticated technology than yeast and actinomycetes synthesis. Table 13.1 summarizes current studies on utilizing microorganisms to synthesize nanoparticles.

TABLE 13.1 Intracellular Synthesis of Nanoparticles by Bacteria

Organism	Size (nm)	Metal/ Non-Metal	Shapes	Location of Synthesis
Idiomarina spp. PR58-8	26	Ag	–	Intracellular
Pseudomonas spp. *Bacillus*	156–265	Ag	–	Intracellular
Bacillus subtilis 168	5–25	Au	Octahedral	Intracellular
Shewanella algae	10–20	Au	–	Intracellular
Plectonema boryanum UTEX485	10	Au	Cubic	Intracellular
Corynebacterium spp. *SH09*	10–15	Ag	–	Intracellular
Lactobacillus spp. *P.*	20–50	Au, Ag, Au-Ag	Hexagonal	Intracellular
P. aeruginosa SNT1	–	Se	Spherical	Intracellular
Desulfovibrio desulfuricans	50	Pd	–	Intracellular
S. oneidensis MR-1	–	Pd	–	Intracellular
Aquaspirillum magnetotacticum	40–50	Fe_3O_4	Octahedral	Intracellular
Magnetotactic bacterium MV-1	40X40X60	Fe_3O_4	Parallelepiped	Intracellular
M. gryphiswaldense	35–120	Magnetite	Cubo-octahedral hexagonal Extracellular	Intracellular

13.5.2 PRODUCTION OF METALLIC NANOMATERIALS

During a study on the production of AgNPs, most of the particles were visibly linked to the cytoplasmic membrane surface [36]. The scientists hypothesized that silver ions were trapped on the cell surface by electrostatic activity, and then reduced to create silver nuclei in the cell wall due to extracellular enzymes. Then nanoparticles were formed on the mycelia when the nuclei accumulated. Ahmad et al. [1] demonstrated that the fungus strain *Fusarium oxysporum* produced reducing agents to decrease the ions. The experiment was conducted by mixing a pure culture of *Fusarium oxysporum*in H_2O for a few hours in the dark, then filtering the biomass solution and combining it with silver nitrate ($AgNO_3$) solution. The solution developed a yellowish-brown color, indicating the presence of elemental Ag.

13.5.3 PRODUCTION OF SYNTHETIC NANOMATERIALS

Nanoparticle production falls into majorly two categories, namely:

- Top-down production, where raw materials are vigorously transformed into smaller parts; and
- Bottom-up production where nanoparticles are synthesized from atoms and molecules.

Top-down production is widely used on industrial-scale synthesis because it is cheaper to create nanomaterials in case starting materials are cheap. However, it is a challenging way, and if the aim is to formulate uniform particles with tailored properties, then the bottom-up approach is the best method. In Table 13.2, there is a list of some Intracellular synthesis of nanoparticles by bacteria.

TABLE 13.2 Intracellular Synthesis of Nanoparticles by Bacteria

Organism	Metal/Non-Metal	Location of Synthesis	Method
Thermomonospora sp. Escherichia	Au	Extracellular	Reduction
Escherichia coli	Pd, Pt	Extracellular	Reduction
Rhodopseudomonas capsulata	Au	Extracellular	Reduction
Pseudomonas aeruginosa	Au	Extracellular	Reduction
Delftia acidovorans	Au	Extracellular	Reduction
Shewanella sp. Desulfovibrio	As	Extracellular	Reduction
Desulfovibrio desulfuricans	Pd	Extracellular	Reduction
Bacillus sphaericus JG-A12	U, Cu, Pb, Al, Cd	Extracellular	Reduction and biosorption
Klebsiella pneumonia	Ag	Extracellular	Reduction
Escherichia coli	Ag	Extracellular	Reduction
Enterobacter cloacae	Ag	Extracellular	Reduction
Lactobacillus sp.	Ag	Extracellular	Reduction and biosorption
Enterococcus faecium	Ag	Extracellular	Reduction and biosorption
Lactococcus garvieae	Ag	Extracellular	Reduction and biosorption

13.6 IMPORTANCE OF NANOMATERIAL IN MICROBIAL SYSTEM

The production of nanoparticles involves a wide range of biological, chemical, and physical processes, some of which are novel and others pretty familiar. Nature has already developed several mechanisms for the natural synthesis of nano- and micro-length scaled inorganic materials. These processes have also contributed to the growth of a relatively young and primarily untapped scientific subject centered on nanomaterial production. Developing "green chemistry" techniques that are safe, non-toxic, and environmentally friendly for the synthesis and assembly of nanoparticles would be beneficial, perhaps using organisms such as fungus, bacteria, and plants. As a result, all uni and multicellular organisms produce inorganic compounds, either intracellularly or extracellularly. Nanoparticle formulation has excellent potential in living beings. In recent years, microorganisms have been studied as a possible biofactory for producing metallic nanoparticles, including silver, cadmium sulfide, and gold [78]. Nanotechnology Investigators prefer biosynthesis as it is easier than other methods to regulate the distribution of particles obtained from this process. In addition, no environmental contamination is involved in this process, which is typically followed by other chemical methods. In the year 2000, the first bacterial production of AgNPs was published. *Pseudomonas stutzeri* bacteria (AG259) was utilized by Joerger et al. [42] to make AgNPs with a size of less than 200 nm. In 2008, the biosynthesis of silver nanocrystals by *Bacillus licheniformis* was investigated. Aqueous silver ions are converted to AgNPs when they come into contact with *Bacillus licheniformis* biomass. The color shift from whitish-yellow to brown confirmed the presence of AgNPs. The enzyme nitrate reductase most likely mediates the production of AgNPs [78].

13.6.1 IMPACT OF NANOPARTICLES STUDIES ON MICROBES

Microbiology has a vast global influence. It evolved into a field of research concerned with the form, nature, and characterization of these species and ways of regulating and using their behaviors after its origins in the 19th century. It has continued expanding into the main microbiology branches, including medical, manufacturing, forestry, food, and milk products. In the treatment of pandemics and diseases, microbiology is used. It also influences the consumption of foodstuffs and the production of primary agricultural goods. Microbiology methods are as diverse as the number of applications [20].

In recent years, nanotechnology (NT) research has made significant breakthroughs in the realm of biomedicine, with researchers focusing on the study of molecular and cellular processes. NT methods are advancing our knowledge of how bacteria are working and offering new ways to explore the functional and physical dimensions in isolation and living environments of chemicals, chemical assemblies, and intact microbial cells. However, in the emerged fields of system biology, NT instruments used in microbiology probably still have significant consequences.

In conclusion, the investigators seek to explain the interplay between coordinated modules in the regulation of cell functions. This complex and complicated challenge will teach one how to reconstitute *ex vivo* functional modules [20, 54].

NT concerns the analysis by artifacts typically smaller than 100 nm of processes in the nano-scale. Most of the NT research is concerned with the auto-assembly phenomena in which building blocks of nano dimensions are connected to complex structures.

At many stages, microbiology applies to NT. Many bacterial species, for example, are normal nanomachines such as molecular motors such as pili. By the self-assembly process, bacteria also form biofilms. Virus capsids are made using a standard molecular identification and nano-scaling self-assembly method [39, 69].

Engineered nanomaterials (ENMs) are artificial materials with at least one dimension of fewer than 100 nanometers. NP can arise spontaneously or inadvertently (e.g., large biomolecules, colloids, and ash); however, questions about the possible detrimental effects of the NPs are presented to Engineered nanomaterials.

Engineered nanoparticles are classified into four categories:

- Composites (i.e., mixtures of NPs);
- Dendrimers (e.g., nano-sized polymers);
- Metal-based materials (e.g., TiO_2 NPs); and
- Carbon-based materials (e.g., fullerenes).

Nanoparticles generally have various physical and chemical properties, such as different catalytic activity, material resistance, thermal behavior, conductivity, solubility, and optical properties relative to their respective bulk materials. The improvement in the volume-to-surface ratio is perhaps the most critical shift in the properties of NPs. Because the percentage of atoms at the particle surface rises inversely with particle size, the surface characteristics of NPs can influence the bulk material's properties. They will also transmit energy to oxygen molecules in the surrounding. Toxicity from oxygen radicals can cause cell death and damage. Because the NPs are similar in size to bio-macromolecules, including DNA, phospholipids, and proteins, they can cause molecular and cellular problems. The toxicity of NPs is influenced by several other factors, including dissolution ratio, particle shape, and surface reactivity.

13.6.2 THE IMPACT OF NMS ON THE INDIGENOUS MICROBIOME

The babies leave their mother's wombs and develop complex microbial communities on all body surfaces [24], systematically shaping their host intestines' physiology. Microbiology is increasingly recognizing the relevance of the dietary effect on the host's metabolic state [41]. Inflammatory or metabolic disease states such as type-2 diabetes [60], obesity [33], and inflammatory bowel [33] are synonymous with dysbiosis and decreased variability of commensal gut microbiota. This complex microbial community has co-evolved with the human host. Recent advances in human lifestyle and diet are primary evolutionary selection pressure on the commensal gut microbiota [40]. It is also likely that the structure and diversity of the commensal microbiome can be changed by also contact with different Nanomaterials, also for a comparatively short period.

Changes in the gastrointestinal microbiota caused by nanomaterials. In diverse settings and animal models, an increasing number of studies are attempting to determine the impact of NP exposure on microbial gut-resident populations. The daily ingestion of silver in humans was estimated to be 70–90 mg [22]. It appears to be more efficient to kill gram-negative bacteria with a thinner cell wall, which is also the case for the gut microbial ecology [19, 81]. Nanosilver has an influence that destroys a whole variety of gram-positive and gram-negative bacteria. AgNPs have been found to influence the dose-dependent Alpha and Beta Diversity of gut microbial communities, as well as the gram-positive to gram-negative ratio, in a mouse model [81]. Although there was no significant effect on nanosilver exposure in increasing doses on the wealth, the intestinal microbiota's equality (relative abundance of operating taxonomy units) decreased significantly. Weighted UniFrac distance analysis revealed the effects of the silver nanoparticle dose on the microbiological structure [81]. The dosage-dependent nanosilver-induced change in the stage of the species resulted from a reduction in abundance in the family S24-7 and *Lachnospiraceae* of the firmicutes phylum. Toxicity and morphology of the ileum and colon mucous membranes of mice were studied in doses of 4,600 ppb Nanosilver, including large doses of the nanoparticle. Particularly when exposed to (nano) silver, the ratio shift between the Bacteroides and Firmicutes depended upon the NP age and the sulfidation that happens in the environment, which reduces releases Ag^{+} ions. Although most studies have not included sufficient ion regulation, the release of Ag^{+} ions is most likely to induce dysbiosis. [81]. Since *Bacillus sp*, selective NP pressure on gut microbiota has been further confirmed. In specific, Nanosilver (Firmicutes phylum) penetration was adapted [91]. Exposure to Nanosilver will contribute to the supremacy of (nano)silver tolerant *Bacillus sp* in the commensal gut microbiota. In a rat model, silver NPs have been shown to size-dependently and dose-dependently alter the ileal mucous microbiota and increase the percentage of gram-negative bacteria [51]. However, a recent mouse study revealed that oral administration of silver NPs at a dose of 10 mg 1 per day did not affect body weight, composition, or variance of the cecal gut microbiota, independent of size or coating. In a rat model, the results of a two weeks oral administration to various formed silver NPs have been recently studied in intestinal microbiota composition of cubic and sphere-shaped NMs contributing to a decrease in selected bacteria in feces [2]. *Bacteroidesuniformis, Coprococcus eutectic, Christensenellaceae*, and *Clostridium sp*. were less abundant in cubic formed silver NPs with a diameter of 50.6 nm. In contrast, sphere-shaped silver NPs with a diameter of 45 nm and a dose of 3.6 mg/kg reduced the abundance of *Oscillospira, Dehalobacteriumsp, Peptococcae Clostridium sp*. dosage and silver NPs with a diameter of 3.6 nm. It's worth noting that NM's influence on gut microbial ecology isn't restricted to mammals; alterations in commensal gut microbiota composition have been seen in various metazoan animals exposed to nutritional NP. The composition of the intestinal microbial population has been documented to alter in zebrafish, copper, and silver NPs that possess anti-microbial propensity [19], which may be used as a means for countering antibiotic infections [74]. The diversity of midgut flora in larvae has been limited in *Drosophila melanogaster*, as *Lactobacillus brevis* has increased and *Acetobacter* has decreased.

NM exposure appears to influence the gut microbiome, and almost no studies have to date been undertaken that differentiate between the mentioned "direct" or "indirect effects. NPs may cause changes in the commensal microbiota in consumer products. Because NPs are currently employed in various foods and consumer goods and the formulation of (targeted) medications, the effects of these NPs on gut microbiota is essential to consider. The use for the oral provision of phenolic compounds has been demonstrated to modulate intestinal microbiota directly or indirectly and to increase concentrations of short-chain acids derived by microbiota in the host [56].

Although NPs have recently been shown to have a significant effect on microbial populations, most studies have concentrated on using nano-sized antifungals or better antibiotic distribution using nanocarriers [57]. Aside from the medical application of NP in antibiotic treatments, the influence of typical NP-microbe interactions on (patho)physiological and environmental processes should be considered. To begin with, the air we breathe contains a variety of fungus spores and NPs that are produced either naturally or purposefully for utilization [31]. Apart from the spores, a recent study indicates that particles contain numerous bacteria that affect the airborne bacterial and thus even human health [11], which are transported during Asian dust events.

NP-microbial complexes occur in natural habitats, but more study is required on how these hybrid bio-material systems lead to ecological or human health consequences. Until the present, most researchers have focused on sequential exposure scenarios, in which *in vitro* and *in vivo* models of lung illness were exposed to diseases first, then to NPs [1]. Mice's pre-exposure to CNTs had little effect on the early immune reaction of *Toxoplasma gondii. At the same time,* sequential CNT and Listeria monocytogenes exposure resulted in a decline in pulmonary clearance and cytokine elevation [53]. After an *in vitro* exposure to *Conidia fumigatus* and Au NPs, spores, and NPs have been found coexisting in phagosomes [1]. The microbiota-NP complex development has not been studied here. A few experiments have now been carried out, such as food intake or inhalation, and have studied the (patho) biological effects of NP-microbe contact [10]. Besides, the first proof that gastrointestinal sensitivity to Ag- or Cu NPs is likely to influence the intestinal microbiota has already been given by a recent study [21].

Furthermore, there should be no lack of the environmental effect of NPs on biological processes. For example, microbial communities have been documented to affect NPs in wastewater or soil [59]. However, when it comes to query databases, the research of NP-microbe interaction is clearly in its infancy. Surprisingly, little is known about the physical-chemical interactions between NPs and microbes and the impact of the physiological system on NP-pathogen interactions. The biomolecular corona may have a significant influence on the possible connection and the characteristics of different bacteria and NPs. Instead of sterile NPs, microorganisms most likely face corona-covered biomolecules [44]. Nanotechnology is being used to develop newer anti-microbial therapies and rationally improve NP-Microbe Assembly for various biotechnological, environmental, and biomedical applications such as biomaterials and advanced vaccine strategies, which necessitates a thorough understanding of this dynamic relationship.

One of Worldwide's leading health issues is the inability to produce new antibiotics and the growing presence of multidrug-resistant bacteria [45]. There is, therefore, an immediate need, along with enhanced targeting strategies, for the wise development of new forms of antibiotics. NP-based methods tend to be a good strengthening of common antibiotics in this regard. The use of NMs has firm hopes in combating severe bacterial infections due to their unique nano-scale physicochemical properties, such as the wide volume surface area and many surface available options. NPs are also intensively studied as intelligent diagnostic instruments, medication distribution, or intrinsic antibiotic action. As the future uses of NMs as new anti-bacterial and the proposed underlying pathways are already explored in many ways, we have only highlighted some of the significant results [57]. However, it has not been studied how NMs are implemented as possible antifungals, for example, by explicitly influencing the vitality of pathogenic fungal spores.

It has now been acknowledged that when approaching complex physiological settings, biomolecules will fastly adsorb all known NPS. Thus in physiological or natural settings, the biophysical properties and fates of NPs often vary dramatically from those of the formulated unpolluted particles during processing. Thus, in general, in diverse physiological and ecological settings, unclean NPs tend to occur only for a brief time. Usually, the macromolecule composition of the ecosystems analyzed varies considerably between the corona profiles of biomolecules [80].

Different biomolecules can be either rich or have a poor NP surface affinity. The particle content, surface characteristics, and size, exposure duration, and the relative ratios between physiological fluid and nanoparticle dispersion all played a role in determining the shape of the biomolecule [17, 28]. In addition to biomolecules, ion presence and pH can impact the (colloidal) stability of NPs, mainly metal or metal oxide NPs, and their corona signatures via modifying electrostatic interactions [7, 28]. Biochemical environments such as plasma, intestinal fluid, and lung surfactants vary in structure.

13.6.3 COMMERCIAL IMPACTS

The impact on human health of engineered nanoparticles may be few, but they are the subject of an immense conversation. Nanotechnology has enormous promise, but to guarantee that their products are safe, industry, and the general public must be more aware of the potential impact of ENPs on human health and the environment. Nanotechnology expands exponentially, and there is a need for studies to consider the potential threats.

13.6.3.1 THE FIRST TIME – THE "NEW ASBESTOS" ARE CARBON NANOTUBES (CNTS)?

Demand for CNTs is growing because of peculiar features, such as particular electricity. Its needle-like structure, however, resembles asbestos, which poses concerns regarding its protection. New research [23] of mice reveals that a particular form of CNT has effects similar to asbestos. CNTs, usually of a few nanometers in diameters, are clinic carbon molecules. They are solid, efficient to conduct heat, and possess specific electrical properties that can be used in many fields, including optics and electronics. It is anticipated that demand for CNT will increase. However, owing to their superficial similarity to asbestos, there are questions over their alleged health risks. Asbestos contamination causes a particular form of cancer called mesothelioma, and CNT is also vulnerable to cancer.

Analysis subjected mice to numerous forms of asbestos, carbon nanoparticles (CNPs), and carbohydrates in several walls (MWNTs). MWNTs are made up of several nanotubes inside. In the lining of their abdominal cavity, the mouse was exposed to the chemicals that resemble asbestos in the linings of the human chest cavity and that typically produce mesothelioma. Inflammation and the formation of scar-like formations or lesions, which are typical asbestos side effects, have been controlled.

The findings suggest that asbestos-like activity is exhibited only by long MWNTs. However, the authors warn that their fiber-specific tests might expose people to nano-carbons in the form of particles that aren't mentioned in this chapter. It illustrates the importance of using the best toxicity assessment process.

Although the findings show that there would be enough exposure to asbestos in the atmosphere or at the workplace to induce it is unclear if the long CNT causes asbestosis. It shows that more study on the levels of exposure of long CNTs is needed before their use is more disseminated.

13.6.3.2 THE EFFECTS OF SUNSCREEN NPS ON SKIN DNA

Sunscreen, particularly those containing ZnO, is one of the most frequently utilized nanotechnology-based products. ZnO stays white on the skin in its conventional form, but many sunscreens use a nano shape. The sunscreen is transparent, and ZnO NPs disperse less glare. Although its effectiveness, nothing is known about the possible relationship between DNA and ZnO NPs in the skin. Human skin cells can be damaged by ZnO NPs, according to a new study [86]. In many regions of the globe, these nanoparticles are utilized as UV filters in sunscreens. Human cells were exposed to various ZnO NPs from the epidermis in the study (the top layer of the skin). A variety of responses and cell shifts were examined at varying time intervals (between 3 and 48 hours) to determine the amount of ZnO NP toxicity that could affect cells in the skin. The data revealed that DNA in ZnO NPs was significantly damaged after 6 hours of exposure at two of the higher values (0.8 microgram per ml and 5 micrograms per ml). The results show that NP causes oxidative stress in cells even at low doses (0.008 to 0.8 metric ml). Free radicals are produced by oxidative stress, which is linked to skin cancer. These findings are significant since the studied concentrations are much smaller than sunscreens (the doses differ but can be around 160 mg/ml). The authors propose that ZnO NPs in sunscreens and during handling should be taken into account.

13.6.3.3 TESTING THE TOXICITY OF NMS

DNA destroying materials can lead to a mutation in human cells, which can ultimately lead to cancer. The recent results indicate that precautionary steps are required to eliminate these threats from manufacturers. Two commercially already available NMs were studied, CNTs, and nanofibers (NFs). CNTs, like asbestos, have been proven in the past to cause mesothelioma in mice (a form of cancer). They used NTs and NFs to process cultured human lung cells and found DNA alterations. The NMs induced DNA damage in cultured cells, and the CNT dosage was closely proportional to the damage levels.

NPs are generally securely bonded in a matrix so that inhalation is not a problem for processed goods. Though, experiments like this show that if NPs are accessible in the environment, they can be toxic. It must also implement regulations to safeguard human wellbeing against the danger of inhaling NPs.

13.6.3.4 PREDICTING THE INFLAMMATORY POTENTIAL OF NPS

With all the advantages of using NMs, the impact of these particles on human health has been increased. The propensity for inflammation of such particles in the lungs when inhaled is especially significant. The immune system's reaction to irritants is inflammation.

To present, the majority of research has focused on animal trials to see if NPs may cause inflammation. The possibility for a variety of easy *in vitro* experiments to replace animals as the toxicity test for NPs was studied in recent research [76]. The researchers compared the results of *in vitro* studies to those of rats to see if they were equivalent.

In this study, rats were exposed to a panel of metal oxides that are widely employed in industry to evaluate the lungs' inflammatory response to each NP. Four different analyzes of tissue cultures that were subjected to the multiple NPs were also carried out. Only two NPs, nickel oxide, and alumina caused significant inflammation in the lungs of rats. Alumina was expected the inflammatory potentials of nickel oxide. There were three types of NPs from alumina, from various sources, and in various sizes tested. Because only alumina two could contribute to lung inflammation, the researchers indicated it couldn't indicate all variants in one NP strain to be tested.

While individual tests were not sufficient, the study found that tests may be utilized to predict the inflammatory ability of metal oxides. This suggests that *in vitro* toxicity testing for metal oxide NPs should be developed for use in screening so that particles that require additional testing may be identified.

13.6.3.5 INHALED NPS CAN ENTER THE BLOODSTREAM

It's crucial to think about how NPs communicate with the body, when they infiltrate the circulation or different cells in the body, and whether or not NPs occur in the lungs. This insight will aid us in determining what health risks are linked to NPs. Recent research is considered all manufactured and combustion-derived forms of all NPs (defined as having less than 100 nanometers) [22].

According to the study, when absorbed into the lungs' depths, NPs may come into contact with the 140 m^2 of folded surface present in the lungs. The NPs may then pass through or translocate via the cells lining the lung's tiny blood veins. They will circulate all over the body from here. An earlier study has shown that a tiny percentage of NPs can move from the lungs to the bloodstream and transport into other sections. For example, this has been shown in rat effect studies of TiO_2 NPs. While the principle of translocation is rational, it is not entirely known the degree of translocation and its significance for human health. The researchers provided the following explanation for how NPs communicate with the body and create health issues. The following research areas should be examined:

- Is it essential for NPs to translocate into the bloodstream to cause cardiovascular problems?
- What if the body's NPs add up?
- How are the NPs translocating in the body happening?
- What are the numerous methods by which NPs enter the body's various cells, and what factors influence NP uptake by cells?

Research exploring these concerns will help to clarify the effect on human health of exposure to NPs.

13.6.3.6 THE POSSIBLE EFFECTS OF NPS ON THE ENVIRONMENT

This section focuses on the potential environmental impacts of ENPs.

13.6.3.6.1 Discovering How NPs Affect the Environment

While nanotechnology evolves early on, bacteria, algae, and fungi in natural ecosystems already communicate with ENPs. A recent study [30] finds differences in our comprehension of this relationship. The analysis reveals five main characteristics of non-existing ENPs:

- At which stages in marine, marine, and atmospheric ecosystems do ENPs become problematic? The specific NP quantity and the amounts at which ENPs are potentially toxic to species are uncertain.

- What are the chemical and physical properties of ENPs? ENPs are more aggressive and likely to bind to other molecules or ENPs because of the high surface/volume ratio, but their behavior can vary depending upon the environment. For specific purposes, ENPs are handled to keep them from being clustered into sediments with other particles and limit their availability to organisms. However, some ENPs are handled intentionally to preserve their separate status in applications like environmental conservation of water or land.
- How do I join cells with ENPs? Any small molecules can cross fungal, algae, and bacterial cell walls. Airborne ENPs can collect on the leaves under which cells can penetrate. Experiments have shown that fungi can use their roots to combine soil ENPs.
- What are the adverse consequences of ENPs? Increased ENP reactivity can affect photosynthesis and breathing. Studies have shown associations with decreased plant growth or improved permeability of bacterial cells between high concentrations of such ENPs. Increased cell weight and decreased seaweed fertilization comprise the indirect toxic effects of ENP aggregation. Photosynthesis can also be avoided by reducing nutrient absorption. It may also affect the toxicity of other contaminants.
- Are ENPs accumulating in the food chain? ENPs have been discovered to persist for long periods within bacterial cells and accumulate in more prominent species. Various conditions can induce various toxic behaviors at various food chain levels.

13.6.3.7 ASSESSING THE ECO-TOXICOLOGICAL RISKS OF NPS

According to a new study, more research on the environmental impact of NPs is needed [46]. This thesis explores ENPs such as metal oxides and NTs. While the existing legislation protects specific commercial applications of ENPs, no clear laws for these products exist worldwide. The cause for this is a lack of knowledge of ENP behavior.

The researchers suggest numerous methods to classify and detect the ENPs in environmental systems to solve this issue. Because the eco-toxicity of the NPs under research is influenced by several factors like height, structure, and shape, in evaluating the risk connected with each form of ENP, variations in one of these properties will significantly influence the environmental behavior of NPs. The clear continuum and variety of NPs allow for identifying and quantifying certain particles by no single process. Microscope-based methods, for example, may be sufficient to detect.

Although ENPs on the water is to be used for sediments, soils, including sludge, and processes, the combination of isolation approaches with advanced analytical techniques. Researchers advocate for consensus on environmental criteria that make it possible for data from various research to be better comparable and understood. This science is a child and needs scholars from several fields to collaborate and business and academia to work together.

13.6.3.8 HOW NTS COULD BE RELEASED INTO THE ENVIRONMENT

CNTs are a type of nanoparticle with distinct physical and technical properties. They are the materials of hope for many potential innovations, such as sporting equipment, textiles, and batteries. Questions regarding their welfare have been raised, though. Therefore, to enforce precautionary steps, it is necessary to consider how they can be released into the environment involuntary. A new study explored potential means of releasing CNT from goods before human exposure [9]. From a lifecycle perspective, two types of mass goods containing potential CNT were measured. There were rechargeable batteries (both used in cell phones) and synthetic fabrics (used inexpensive sportswear).

While CNT is unlikely to be emitted during normal usage, it will be emitted by researchers throughout the processing, recycling, and disposal phases of their life cycle. When CNT is mishandled, it can be released into the air as dust. Incineration of leftovers from battery and metal recycling, in particular, may lead to occupational CNT exposure. The CNT batteries would certainly not be decreased by batteries mixed with household waste. There may also be chemical exposure if batteries are disposed of in waste disposal areas. During use, NTs cannot be exempt from use. CNT may be released by wear and tear, contributing to human exposure while wearing the fabrics near the body. CNT may be released into the environment through the recycling and disposal of old textiles. The hazardous waste control shall not extend to the textiles used.

They are either sold as second-hand clothing or discarded as household garbage. Although the destiny of old textiles in industrialized nations is unknown, the researchers believe that they can be disposed of in those places by open burning. It will result in CNT pollution, as CNT only is eliminated by incineration above 850°C. The required temperatures could only be attained to degrade CNT if current waste incinerators were working correctly.

These examples show how CNT may be produced uncontrollably at various stages along the product life cycle. Human exposure constitutes a concern since the negative health consequences of CNT may be mitigated. Researchers follow the precautionary theory and assume it should be avoided that CNT is released from goods. Responsible product production is critical for ensuring the safety of employees and consumers, and it should be implemented early in the nanotechnology innovation process.

13.6.3.9 MANAGEMENT OF NP EXPOSURE AT WORK

The development and use of NMs are growing. While this produces more employment, the earliest and most publicity would possibly arise in those sectors. There is little information about the effects of NM exposure at work, but Risk Management Techniques are essential to mitigate possible damage.

A recent study discusses an established job risk assessment system and outlines alternative strategies for managing NM exposure in working environments. The research finds a well-known health and safety methodological structure for future use in handling NMs [70].

The critical risks for contamination are established in the manufacture and use of NMs, such as inhalation and touching. It also includes future jobs and duties that are more likely to emerge due to the manufacturing of NM goods, such as machining, sanding, or boiling NP-containing materials. Recommendations have been provided to monitor the sensitivity to NMs within this system.

However, regulations for NPs may need to be more rigid as they have a greater possible toxic impact for one weight than for larger particles. The first proposed control mechanism is to delete or replace NMs. Since NMs are designed with unique characteristics, however, covering the particle with a less dangerous substance or modifying its shape may be a more feasible solution. Separating the NPs or trapping them before they are bonded in a material is another approach using airborne particle collecting ventilation systems. Management monitoring, e.g., limiting the worker's exposure to NPs, may also be carried out. Personal protection aids may also be used, such as breathing machines, masks, and protective garments. Also, environmental management and employee monitoring ensure that safeguards avoid hazardous exposure are successful.

13.7 CHALLENGES FOR NANOMATERIAL APPLICATION ON MICROBIAL STUDIES

We explain why the dynamic interaction between NMs and microorganisms is far from outdated and why it is still a "hot topic" that is developing yet not fully understood. 'where small reaches bigger' is thus increasingly relevant not only to fundamental or applied nanoscience but also to consider the effect on the human and human health of (bio)particles and biological aerosols. It's intriguing to investigate how natural NMs may (co-) regulate ecologically sustainable systems and how manufactured NMs affect (pathobiological) microbial spheres and, possibly, human health. Though still ignored, this information would explicitly encourage the notion demanded by several regulatory networks of 'healthy by nature' for applied nanotechnology. Therefore, we support the combination, employing tiered experimental and transdisciplinary collaboration in academia and industry, of the toolbox for advanced techniques and models for the study of nanotechnology and Microbiology.

Our survey indicates that while basic concepts have already been established, the link between the NM's physicochemical characteristics and microorganisms has only begun to be studied. NMs tend to be commonly able to communicate with different forms of microbes at varying degrees that are not currently accurate. It is essential to thoroughly overcome the introduction of primary chemical 'cognition components' as blocks to drive the dynamic development of NM-microbes. We have heard that stealth shift techniques allow us to minimize NM-bacteria contact, although not related to fungal spores. The sub-supermolecular mechanism underlying this discovery is determined concerning its ecological and pathobiological importance to other microorganisms, including bacteria, algae, or allergens. Also, NMs are growing in our diet, which may help form the human microbiome and deter or intensify diseases.

We have a biomolecular corona, an 'old buddy' as we work with applications in physiologically important biological micro-environments. In addition, different forms of Bio-Coronas influence the NP-microbe crosstalk and its effect on NP-target cell recognition. As a result, it's worth revisiting if the nanoformulation described in various research is still active in real-world pathophysiological circumstances, particularly for the anti-bacterial NMs that have been observed. For example, nanoparticles' bactericidal activity may rely heavily on their spontaneous connection to pathogens. As a result, pathogen-binding NMs and their intended therapeutic impact might inhibit pathophysiologic settings such as wounds, the stomach, or the bloodstream. Establishing a physicochemical strategy to solve certain limits is essential, thereby enhancing performance and reducing side effects. It's not done yet, and there are still a lot of challenges. Should we coin the word "nanoparticle corona," which covers microbes? We envision comprehensive research into the impact of NM-microorganisms on the environment and human health, including fungal and gastrointestinal diseases and allergies.

Is this even true of all; Are microbes close to an insight that pristine NMs tend only for a brief period to live in diverse environments? This knowledge is essentially essential not just for the understanding and minimization of possible "nanotoxicity" but also for the creation and advancement of NMs for potential applications in biotechnology, ecology, and biomedicine for rationally forming microorganisms.

The fact that there are not just researchers but also major publications shaking the area with a solid trans-disciplinary understanding appears to be impeding some progress in the subject. In the still young and expanding field of nano bioscience, we regularly encounter great material chemistry but a poor understanding of microbiology, and vice versa. It is not easy to recognize new and paradigm-changing results for editors and critics in these fields.

We, therefore, call for all parties to prevent comments like "I believe it is not appropriate to publish this chapter." Work completed isn't a revelation, or really? We are not persuaded that the proposed results have the potential relevance for publication we need.' We conduct the 'Consider Again, as this isn't right and have strong scientific evidence for your decisions, contrary to our Nobel Prize-winner Bob Dylan.

13.8 FUTURE PERSPECTIVE OF NANOMATERIALS IN MICROBIAL STUDIES

While a complete knowledge of the link between nanomaterial qualities and microbial toxicity is still in its early stages, broad interpretations of the material features of nanoparticles that impact microbial toxicity have been produced. For instance, their ability to release dissolved ions, their toxicity, and their ability to produce ROS may suggest possible microbial toxicity.

Furthermore, reactivity and alignment changes in surface atoms caused by nanoparticle size reductions and curve modifications frequently interact with enhanced toxicity. The surface coating, which changes those processes or prevents or enables interaction with a microbial surface, can reduce or improve toxic determinants. The surface coating of nanoparticles and its role in avoiding or increasing toxicity is a critical variable that may be utilized to direct biological interactions with nanomaterials. A tailored surface coating, for example, may be used to focus nanomaterials on a specific microorganism [34] or, as with AgNPs, to increase toxicity [61] theoretically. The chemical stability of nanoparticles is currently mistaken as surface coatings for trade, reactivity, or fouling. Such transitions are supposed to occur as the climate changes in biomedical applications or environmental exposures [66]. The compatibility of surface coatings for all planned practical applications (e.g., catalyst, conversion of solar energy, electronic devices) and biological systems are not understood. It would be necessary to understand better the apartments' surface and the stability of the engineered coatings.

Currently, general conclusions cannot be taken on species-based responses to engineered nanomaterials. Model organism tests effectively evaluate the toxicity potential of nanomaterials, even though they do not reflect the diversity of species and potential changes in natural systems. These studies involve well-known components, many species, and redundant viability and growth steps.

Experiments with model organisms can allow researchers to determine a species' molecular response to nanoparticle application. Such reactions will explicitly rely on organisms and are more prosperous than live/dead evaluations. More excellent knowledge of dose-dependent, molecular physiological modifications will contribute to detecting biomarkers to the vulnerability of nanomaterials and the predicted genetic responses. To anticipate possible effects on environmental environments, understanding species-dependent reactions to nanomaterial exposure would be critical. In natural settings, the numerous microbial communities may contribute to accidental transportation and materials processing. Chemical exposures may adversely impact unique population members or disrupted materials or energies in the atmosphere. The evidence on nanoparticles' impacts on natural environments is limited [3, 68, 85, 93]. A multi-scale knowledge at the molecular, cellular, and population-level would be necessary to anticipate the environmental consequences of nanomaterials. Such awareness has to be related to the changes in the microbial population and the ecosystem that can be environmentally and biologically influenced. The role of microorganisms in geochemical processes has been better understood because of linked multi-scale knowledge of ecological processes and contemporary technical approaches for environmental bioremediation of contaminants like uranium and mercury [84]. This focused research will benefit and direct the safe use and removal of these components for cellular and community-level responses to aromatherapies on a molecular level.

NMs have a potentially harmful effect on the nature of the soil, the ecology of agriculture, and the entire operation of the ecosystem, including human health. Understanding the relationship between engineered nanoparticles and microorganisms and the future modification of these materials' fate, transportation, and transformation is essential to the environment-friendly understanding of the diverse existing and prospective applications of nana materials. To lead viable paths for the disposal and creation of "ecologically favorable" nanomaterials in designed nanoparticles, a thorough knowledge of the relationship between the physic favorable characteristics of nanomaterials and the resulting microbial response is required.

13.9 SUMMARY

Nanotechnology's biosynthesis of nanoparticles has the tremendous interest of nanotechnologists because of its low cost of upstream and downstream processing. Biosynthesized nanoparticles are mostly uniform in size and more stable than chemically or mechanically produced. If we see the economy involved in such cases, microbially synthesized nanoparticles are ideal to go. In the industrial sector, specific nanoparticles are used as a catalyst to increase microbial reaction rates.

Many researchers are using nanoparticles to visualize the cellular dynamics of microbial systems, including yeast, Bacteria, Fungi, actinobacteria, and unicellular organisms. Using the help of UV-Viz and transmission electron microscopy, researchers visualize the microbial systematics and metabolic pathway study. Researchers are producing modified nanoparticles, specially organo-metallic nanoparticles, using a microbial system, especially in molecular medicine. Microbially produced vitamins and antibiotics are the best examples of organo-metallic nanoparticles biosynthesis. If we consider the role of microbes in wastewater treatment, then nanoparticles add value in such cases by their reducing nature. Nanoparticles show high anti-microbial activity, and it makes the best choice for food preservatives, quality enhancers, and processing aids.

KEYWORDS

- **future microbial studies**
- **microbial studies**
- **microbiology**
- **nanobiotechnology**
- **nanomaterials**
- **nanotechnology**
- **next generation microbial studies**

REFERENCES

1. Ahmad, A., Mukherjee, P., Senapati, S., Mandal, D., Khan, M. I., Kumar, R., & Sastry, M., (2003). Extracellular biosynthesis of silver nanoparticles using the fungus *Fusarium oxysporum*. *Colloids and Surfaces B: Biointerfaces, 28*(4), 313–318.
2. Aitken, R. J., Chaudhry, M. Q., Boxall, A. B. A., & Hull, M., (2006). Manufacture and use of nanomaterials: Current status in the UK and global trends. *Occupational Medicine, 56*(5), 300–306.
3. Akbarzadeh, A., Zare, D., Farhangi, A., Mehrabi, M. R., Norouzian, D., Tangestaninejad, S., & Bararpour, N., (2009). Synthesis and characterization of gold nanoparticles by tryptophane. *American Journal of Applied Sciences, 6*(4), 69–71.
4. Alvarez, L. H., & Cervantes, F. J., (2012). Assessing the impact of alumina nanoparticles in an anaerobic consortium: Methanogenic and humus reducing activity. *Applied Microbiology and Biotechnology, 95*(5), 1323–1331.
5. Andreas, R. K., Claudia, S., Aasgeir, H., & Fadri, G., (2008). Studying the potential release of carbon nanotubes throughout the application life cycle. *Journal of Cleaner Production, 16*(8, 9), 927–937.
6. Anna, A. S., James, P. F., Elena, R. K., Ashley, R. M., Jenny, R. R., Yulia, Y. T., James, M. A., et al., (2008). Sequential exposure to carbon nanotubes and bacteria enhances pulmonary inflammation and infectivity. *American Journal of Respiratory Cell and Molecular Biology, 38(5)*, 579–590.
7. Backhed, H. D., Wang, T., Hooper, L. V., Koh, G. Y., Nagy, A., Semenkovich, C. F., & Gordon, J. I., (2004). The gut microbiota as an environmental factor that regulates fat storage. *Proceedings of the National Academy of Sciences of the United States of America, 101*, 15718–15723.
8. Ball, A. S., Sayali, P., & Sarvesh, S., (2019). Introduction into nanotechnology and microbiology. *Methods in Microbiology, 46*, 1–18. Academic Press.
9. Bangham, A. D., & Horne, R. W., (1964). Negative staining of phospholipids and their structural modification by surface-active agents as observed in the electron microscope. *Journal of Molecular Biology, 8*(5), 660-IN10.
10. Batley, G. E., Kirby, J. K., & McLaughlin, M. J., (2013). Fate and risks of nanomaterials in aquatic and terrestrial environments. *Accounts of Chemical Research, 46*(3), 854–862.

11. Brem, H., Kader, A., Epstein, J. I., Tamargo, R. J., Domb, A., Langer, R., & Leong, K. W., (1989). Biocompatibility of a biodegradable, controlled-release polymer in the rabbit brain. *Selective Cancer Therapeutics, 5*(2), 55–65.
12. Burkett, S. L., & Mann, S., (1996). Spatial organization and patterning of gold nanoparticles on self-assembled biolipid tubular templates. *Chemical Communications, 3*, 321, 322.
13. Buzea, C., & Ivan, P., (2017). Nanomaterials and their classification. EMR/ESR/EPR spectroscopy for characterization of nanomaterials (pp. 3–45). Springer - New Delhi.
14. Chen, Y. C., & Yu, K. P., (2017). Enhanced anti-microbial efficacy of thermal-reduced silver nanoparticles supported by titanium dioxide. *Colloids and Surfaces B: Biointerfaces, 154*, 195–202.
15. Chen, Y., Wang, D., Zhu, X., Zheng, X., & Feng, L., (2012). Long-term effects of copper nanoparticles on wastewater biological nutrient removal and N_2O generation in the activated sludge process. *Environmental Science & Technology, 46* (22), 12452–12458.
16. Courtney, C. M., Goodman, S. M., McDaniel, J. A., Madinger, N. E., Chatterjee, A., & Nagpal, P., (2016). Photoexcited quantum dots for killing multidrug-resistant bacteria. *Nature Materials, 15*, 529–534.
17. Craig, A. P., Rodger, D., Ian, K., Andrew, M., William, A. H. W., Anthony, S., Vicki, S., Simon, B., William, M., &Ken, D., (2008). Carbon nanotubes introduced into the abdominal cavity of mice show asbestos-like pathogenicity in a pilot study. *Nature Nanotechnology, 3*(7), 423–428.
18. Dai, X., Guo, Q., Zhao, Y., Zhang, P., Zhang, T., Zhang, X., & Li, C., (2016). Functional silver nanoparticle as a benign anti-microbial agent that eradicates antibiotic-resistant bacteria and promotes wound healing. *ACS Applied Materials Interfaces, 8*, 25798–25807.
19. Das, P., Xenopoulos, M. A., Williams, C. J., Hoque, M. E., & Metcalfe, C. D., (2012). Effects of silver nanoparticles on bacterial activity in natural waters. *Environmental Toxicology and Chemistry, 31*, 122–130.
20. Dodds, D. R., (2017). Antibiotic resistance: A current epilogue. *Biochemical Pharmacology, 134*, 139–146.
21. Dykman, L., & Khlebtsov, N., (2012). Gold nanoparticles in biomedical applications: Recent advances and perspectives. *Chemical Society Reviews, 41*(6), 2256–2282.
22. Eeliu, N. D., Docter, M. H., Del, P. P., Ashraf, S., Kolosnjaj-Tabi, J., Macchiarini, P., Nielsen, P., et al., (2016). *Chemical Society Review, 45*, 2440–2457.
23. Elbourne, A., Truong, V. K., Cheeseman, S., Rajapaksha, P., Gangadoo, S., Chapman, J., & Crawford, R. J., (2019). The use of nanomaterials for the mitigation of pathogenic biofilm formation. In: *Methods in Microbiology* (Vol. 46, pp. 61–92).
24. Fakhrullin, R. F., Zamaleeva, A. I., Minullina, R. T., Konnova, S. A., & Paunov, V. N., (2012). Cyborg cells: Functionalization of living cells with polymers and nanomaterials. *Chemical Society Reviews, 41*(11), 4189–4206.
25. Gao, J., Wang, Y., Hovsepyan, A., & Bonzongo, J. C. J., (2011). Effects of engineered nanomaterials on microbial catalyzed biogeochemical processes in sediments. *Journal of Hazardous Materials, 186*(1), 940–945.
26. García, A., Delgado, L., Torà, J. A., Casals, E., González, E., Puntes, V., & Sánchez, A., (2012). Effect of cerium dioxide, titanium dioxide, silver, and gold nanoparticles on the activity of microbial communities intended in wastewater treatment. *Journal of Hazardous Materials, 199*, 64–72.

27. Garrett, W. S., Gallini, C. A., Yatsunenko, T., Michaud, M., DuBois, A., Delaney, M. L., Punit, S., et al., (2010). Enterobacteriaceae act in concert with the gut microbiota to induce spontaneous and maternally transmitted colitis. *Cell Host Microbe, 8*, 292–300.
28. Gebauer, J. S., Malissek, M., Simon, S., Knauer, S. K., Maskos, M., Stauber, R. H., Peukert, W., & Treuel, L., (2012). Impact of the nanoparticle–protein corona on colloidal stability and protein structure. *Langmuir, 28*, 9673–9679.
29. Gihring, T. M., Zhang, G., Brandt, C. C., Brooks, S. C., Campbell, J. H., Carroll, S., Criddle, C. S., et al., (2011). A limited microbial consortium is responsible for extended bioreduction of uranium in a contaminated aquifer. *Applied Environmental Microbiology, 77*, 5955–5965.
30. Goodrich, J. K., Davenport, E. R., Beaumont, M., Jackson, M. A., Knight, R., Ober, C., Spector, T. D., et al., (2016). Genetic determinants of the gut microbiome in UK twins. *Cell Host Microbe, 19*, 731–743.
31. Gradishar, W. J., Tjulandin, S., Davidson, N., Shaw, H., Desai, N., Bhar, P., Hawkins, M., & O'Shaughnessy, J., (2005). Phase III trial of nanoparticle albumin-bound paclitaxel compared with polyethylated castor oil-based paclitaxel in women with breast cancer. *Journal of Clinical Oncology, 23*(31), 7794–7803.
32. Grasso, G., Zane, D., & Dragone, R., (2020). Microbial Nanotechnology: Challenges and prospects for green biocatalytic synthesis of nanoscale materials for sensoristic and biomedical applications. *Nanomaterials, 10*(1), 2079–4991.
33. Gunawan, C., Teoh, W. Y., Marquis, C. P., & Amal, R., (2013). Induced adaptation of *Bacillus* sp. to anti-microbial nanosilver. *Small, 9*(21), 3554–3560.
34. Hadjidemetriou, M., & Kostarelos, K., (2017). Evolution of the nanoparticle corona. *Nature Nanotechnology, 12*, 288–290.
35. Hajipour, M. J., Fromm, K. M., Ashkarran, A. A., Jimenez De, A. D., De Larramendi, I. R., Rojo, T., Serpooshan, V., Parak, W. J., & Mahmoudi, M., (2012). Anti-bacterial properties of nanoparticles. *Trends Biotechnology, 30*, 499–511.
36. Han, X., Geller, B., Moniz, K., Das, P., Chippindale, A. K., &Walker, V. K., (2014). Monitoring the developmental impact of copper and silver nanoparticle exposure in *Drosophila* and their microbiomes. *Science Total Environment, 487*, 822–829.
37. Harwood, C. R., Pohl, S., Smith, W., & Wipat, A., (2013). Bacillus subtilis: Model gram-positive synthetic biology chassis. In: *Methods in Microbiology* (Vol. 40, pp. 87–117). Academic Press.
38. Sonnenburg, J. L., & Bäckhed, F., (2016). Diet–microbiota interactions as moderators of human metabolism. *Nature, 535*(7610), 56–64.
39. James, C., Regan, F., & Sullivan, T., (2012). *Nanoparticles in Anti-Microbial Materials: Use and Characterization* (Vol. 23, p. 254). Royal Society of Chemistry.
40. Jatzkewitz, H., (1954). Incorporation of physiologically-active substances into a colloidal blood plasma substitute. I. Incorporation of mescaline peptide into polyvinylpyrrolidone. *Hoppe-Seyler's Zeitschrift fur Physiologische Chemie, 297*, 149–156.
41. Javurek, B., Suresh, D., Spollen, W. G., Hart, M. L., Hansen, S. A., Ellersieck, M. R., & Rosenfeld, C. S., (2017). Gut dysbiosis and neurobehavioral alterations in rats exposed to silver nanoparticles. *Scientific Reports, 7*(1), 1–15.
42. Joerger, R., Klaus, T., & Granqvist, C. G., (2000). Biologically produced silver–carbon composite materials for optically functional thin-film coatings. *Advanced Materials, 12*(6), 407–409.

43. Kalimuthu, K., Suresh, B. R., Venkataraman, D., Bilal, M., & Gurunathan, S., (2008). Biosynthesis of silver nanocrystals by *Bacillus licheniformis*. *Colloids and Surfaces B: Biointerfaces, 65*(1), 150–153.
44. Khan, M. T., Nieuwdorp, M., & Bäckhed, F., (2014). Microbial modulation of insulin sensitivity. *Cell Metabolism, 20*(5), 753–760.
45. Kolahalam, L. A., Viswanath, I. K., Diwakar, B. S., Govindh, B., Reddy, V., & Murthy, Y. L. N., (2019). Review on nanomaterials: Synthesis and applications. *Materials Today: Proceedings, 18*, 2182–2190.
46. Kopf, H., Joshi, R. K., Soliva, M., & Speiser, P., (1976). Study on micelle polymerization in the presence of low molecular-weight drugs. 1. Production and isolation of nanoparticles, residual monomer determination, physical-chemical data. *Pharmazeutische Industrie, 38*, 281–284.
47. Kumar, N., Shah, V., & Walker, V. K., (2011). Perturbation of an arctic soil microbial community by metal nanoparticles. *Journal of Hazardous Materials, 190*(1–3), 816–822.
48. Li, D., Cui, F., Zhao, Z., Liu, D., Xu, Y., Li, H., & Yang, X., (2014). The impact of titanium dioxide nanoparticles on biological nitrogen removal from wastewater and bacterial community shifts in activated sludge. *Biodegradation, 25*(2), 167–177.
49. Li, X., Xu, H., Chen, Z. S., & Chen, G., (2011). Biosynthesis of nanoparticles by microorganisms and their applications. *Journal of Nanomaterials, 2011*, 1–16.
50. Liang, Z., Das, A., & Hu, Z., (2010). Bacterial response to a shock load of nanosilver in an activated sludge treatment system. *Water Research, 44*(18), 5432–5438.
51. Natan, M., & Banin, E., (2017). From nano to micro: Using nanotechnology to combat microorganisms and their multidrug resistance. *FEMS Microbiology Reviews, 41*, 302–322.
52. Madureira, R., Campos, D., Gullon, B., Marques, C., Rodriguez-Alcala, L. M., Calhau, C., Alonso, J. L., et al., (2016). Fermentation of bioactive solid lipid nanoparticles by human gut microflora. *Food & Function, 7*, 516–529.
53. Mahon, H. D. R., & Dawson, K. A., (2012). Stabilizing fluorescent silica nanoparticles against dissolution effects for biological studies. *Chemical Communications, 48*(64), 7970–7972.
54. Manchester, M., & Steinmetz, N. F., (2009). *Viruses and Nanotechnology* (1st edn., p. 147). Berlin: Springer.
55. Mandal, S., Piyush, K. T., Anees, M., & Anchal, D., (2020). Cancer gene therapy and its techniques in cancer biology. *International Journal of Cell Biology and Cellular Processes, 6*(1), 38–54.
56. Masrahi, A., VandeVoort, A. R., & Arai, Y., (2014). Effects of silver nanoparticle on soil-nitrification processes. *Archives of Environmental Contamination and Toxicology, 66*(4), 504–513.
57. Miller, K. P., Wang, L., Benicewicz, B. C., & Decho, A. W., (2015). Inorganic nanoparticles engineered to attack bacteria. *Chemical Society Review, 44*, 7787–7807.
58. Mishra, V. K., & Kumar, A., (2009). Impact of metal nanoparticles on the plant growth promoting rhizobacteria. *Digest Journal of Nanomaterials and Biostructures, 4*, 587–592.
59. Mohanty, A., Wu, Y., & Cao, B., (2014). Impacts of engineered nanomaterials on microbial community structure and function in natural and engineered ecosystems. *Applied Microbiology and Biotechnology, 98*(20), 8457–8468.

60. Monopoli, M. P., Aberg, C., Salvati, A., & Dawson, K. A., (2012). Biomolecular coronas provide the biological identity of nanosized materials. *Nature Nanotechnology, 7*, 779–786.
61. Mühlfeld, P. G., & Rutishauser, R., (2008). Translocation and cellular entering mechanisms of nanoparticles in the respiratory tract, *Swiss Medical Weekly, 138*(27, 28), 387–391.
62. Navarro, B. A., Behra, R., Hartmann, N. B., Filser, J., Miao, A. J., Quigg, A., Santschi, P. H., & Sigg, L., (2008). Environmental behavior and ecotoxicity of engineered nanoparticles to algae, plants and fungi, *Ecotoxicology, 7*(5), 372–386.
63. Nyberg, L., Turco, R. F., & Nies, L., (2008). Assessing the impact of nanomaterials on anaerobic microbial communities. *Environmental Science & Technology, 42*(6), 1938–1943.
64. Petros, R. A., & DeSimone, J. M., (2010). Strategies in the design of nanoparticles for therapeutic applications. *Nature Reviews Drug Discovery, 9*(8), 615–627.
65. Pietroiusti, A. M., & Campagnolo, L., (2016). New frontiers in nanotoxicology: Gut microbiota/microbiome-mediated effects of engineered nanomaterials. *Toxicology and Applied Pharmacology, 299*, 90–95.
66. Pratik, S., Maria, J., Gonzalez, & Marianne, M., (2006). Viruses and their uses in nanotechnology. *Drug Development Research, 67*(1), 23–41.
67. Ravishankar, R. V., (2011). Nanoparticles and their potential application as anti-microbials. *Science Against Microbial Pathogens: Communicating Current Research and Technological Advances, 2011*, 197–209.
68. Reinhardt, C., Bergentall, M., Greiner, T. U., Schaffner, F., Östergren-Lundén, G., Petersen, L. C., Ruf, W., & Bäckhed, F., (2012). Tissue factor and PAR1 promote microbiota-induced intestinal vascular remodeling. *Nature, 483*, 627–631.
69. Saleh, T. A., (2020). Nanomaterials: Classification, properties, and environmental toxicities. *Environmental Technology & Innovation, 20*, 1–16.
70. Scheffel, U., Rhodes, B. A., Natarajan, T. K., & Wagner, H. N., (1972). Albumin microspheres for study of the reticuloendothelial system. *Journal of Nuclear Medicine, 13*(7), 498–503.
71. Schulte, P., Geraci, C., Zumwalde, R., Hoover, M., & Kuempel, E., (2008). Occupational risk management of engineered nanoparticles. *Journal of Occupational and Environmental Hygiene, 5*(4), 239–249.
72. Seho, C., Dongwook, L., Jun, H. J., Sora, L., Dahye, Y., & Taegun, S., (2016). Alterations in the airborne bacterial community during Asian dust events occurring between February and March 2015 in South Korea. *Scientific Reports, 6*, 1–9.
73. Senlin, L., Rodger, D., Craig, P., Paul, D., Fiona, M., Ellen, D., William, M., Vicki, S., & Ken, D., (2009). Efficacy of simple short-term *in vitro* assays for predicting the potential of metal oxide nanoparticles to cause pulmonary inflammation. *Environmental Health Perspectives, 117*(2), 241–247.
74. Sharma, V. K., Yngard, R. A., & Lin, Y., (2009). Silver nanoparticles: Green synthesis and their anti-microbial activities. *Advances in Colloid and Interface Science, 145*(1, 2), 83–96.
75. Shehzad, A., Raheem, S., Hanan, A., & Ebtesam, A. A., (2020). Application of nanomaterials in treatment of microbial and viral infections. In: Khan, F. A., (eds.), *Applications of Nanomaterials in Human Health* (1st edn., pp. 173–190). Springer: Singapore.

76. Simonin, M., Colman, B. P., Tang, W., Judy, J. D., Anderson, S. M., Bergemann, C. M., Rocca, J. D., et al., (2018). Plant and microbial responses to repeated $Cu(OH)_2$ nanopesticide exposures under different fertilization levels in an agro-ecosystem. *Frontiers in Microbiology, 9*(1769), 1–17.
77. Stone, V., Nowack, B., Baun, A., Van, D. B. N., Von, D. K. F., Dusinska, M., Handy, R., et al., (2010). Nanomaterials for environmental studies: Classification, reference material issues, and strategies for Physico-chemical characterization. *Science of the Total Environment, 408*(7), 1745–1754.
78. Suresh, K., Pelletier, D. A., Wang, W., Moon, J. W., Gu, B., Mortensen, N. P., Allison, D. P., et al., (2010). Silver nanocrystallites: Biofabrication using *Shewanella oneidensis*, and an evaluation of their comparative toxicity on gram-negative and gram-positive bacteria. *Environmental Science & Technology, 44*, 5210–5215.
79. Swedin, L., Arrighi, R., Andersson-Willman, B., Murray, A., Chen, Y., Karlsson, M. C., Georen, S. K., et al., (2012). Pulmonary exposure to single-walled carbon nanotubes does not affect the early immune response against *Toxoplasma gondii. Particle and Fiber Toxicology, 9*, 16–23.
80. Tapan, K. S., & Andrey, L. R., (2009). Nonspherical noble metal nanoparticles: Colloidchemical synthesis and morphology control. *Advanced Materials, 21*, 1–24.
81. Tenzer, S., Docter, D., Kuharev, J., Musyanovych, A., Fetz, V., Hecht, R., Schlenk, F., et al., (2013). Biomolecule-corona formation confers resistance of bacteria to nanoparticle-induced killing: Implications for the design of improved nanoantibiotics. *Nature Nanotechnology, 8*, 772–781.
82. Tiede, K., Hassellöv, M., Breitbarth, E., Chaudhry, Q., & Boxall, A. B., (2009). Considerations for environmental fate and ecotoxicity testing to support environmental risk assessments for engineered nanoparticles. *Journal of Chromatography A, 1216*(3), 503–509.
83. Valsami-Jones, E., & Lynch, I., (2015). How safe are nanomaterials?. *Science, 350*(6259), 388–389.
84. Van, D. B. S., Ambroise, J., Lecloux, H., Levard, C., Soulas, R., De Temmerman, P. J., Palmai-Pallag, M., et al., (2016). Dietary silver nanoparticles can disturb the gut microbiota in mice. *Particle and Fiber Toxicology, 13*, 1–16.
85. Vilanova, O., Mittag, J. J., Kelly, P. M., Milani, S., Dawson, K. A., Radler, J. O., & Franzese, G., (2016). Understanding the kinetics of protein–nanoparticle corona formation. *ACS Nano, 10*, 10842–10850.
86. Vyom, S., Ritesh, K. S., Neha, S., Devendra, P., Mukul, D., & Alok, D., (2009). DNA damaging potential of zinc oxide nanoparticles in human epidermal cells. *Toxicology Letters, 185*(3), 211–218.
87. Walkey, D., Olsen, J. B., Song, F., Liu, R., Guo, H., Olsen, D. W., Cohen, Y., et al., (2014). Protein corona fingerprinting predicts the cellular interaction of gold and silver nanoparticles. *ACS Nano, 8*, 2439–2455.
88. Wang, Y., Yang, Y., Shi, Y., Song, H., & Yu, C., (2020). Antibiotic-free anti-bacterial strategies enabled by nanomaterials: Progress and perspectives. *Advanced Materials, 32*(18), 1–58.
89. Weiss, J., Takhistov, P., & McClements, D. J., (2006). Functional materials in food nanotechnology. *Journal of Food Science, 71*(9), R107–R116.

90. Wijnhoven, S. W. P., Peijnenburg, W. J. G. M., Herberts, C. A., Hagen, W. I., Oomen, A. G., Heugens, E. H. W., Roszek, B., et al., (2009). The safety and efficacy of dressings with silver–addressing clinical concerns. *Nanotoxicology, 3*, 109–138.
91. Wilding, L. A., Bassis, C. M., Walacavage, K., Hashway, S., Leroueil, P. R., Morishita, M., Maynard, A. D., et al., (2016). Repeated dose (28-day) administration of silver nanoparticles of varied size and coating does not significantly alter the indigenous murine gut microbiome. *Nanotoxicology, 10*, 513–520.
92. Yadav, K. K., Singh, J. K., Gupta, N., & Kumar, V. J. J. M. E. S., (2017). A review of nanobioremediation technologies for environmental cleanup: A novel biological approach. *Journal of Materials and Environmental Science, 8*(2), 740–757.
93. Yang, Y., Xu, M., Wall, J. D., & Hu, Z., (2012). Nanosilver impact on methanogenesis and biogas production from municipal solid waste. *Waste Management, 46*, 1176–1184.
94. Zhang, X., Yan, S., Tyagi, R. D., & Surampalli, R. Y., (2011). Synthesis of nanoparticles by microorganisms and their application in enhancing microbiological reaction rates. *Chemosphere, 82*(4), 489–494.
95. Zheng, C. Y., & Wu, R., (2011). Long-term effects of titanium dioxide nanoparticles on nitrogen and phosphorus removal from wastewater and bacterial community shift in activated sludge. *Environmental Science & Technology, 45*(17), 7284–7290.
96. Ziąbka, M., Mertas, A., Krol, W., Bobrowski, A., & Chłopek, J., (2012). High density polyethylene containing anti-bacterial silver nanoparticles for medical applications. *Macromolecular Symposia, 315*, 218–225.

CHAPTER 14

IMPACT OF NANOMATERIALS ON MICROBIAL COMMUNITIES: APPLICATIONS AND FUTURE PERSPECTIVES

PRASEETHA P. NAIR and SAJEENA BEEVI

ABSTRACT

Nanomaterials have exotic physicochemical properties, which make them ideal candidates for numerous prospective applications in diverse fields. They have a positive or negative impact on microorganisms through several mechanisms. The antimicrobial properties of nanomaterials can efficiently be exploited for commercial applications in water purification systems, pathogen control in agricultural industry, etc. Among the antimicrobial nanomaterials, carbon-based nanomaterials got recent attention due to their superior structures, distinct atomic arrangements, and multifarious properties. Metallic nanomaterials like silver, gold, etc., also exhibit excellent antimicrobial properties, compared to other metals, which facilitates their use in biomedical fields and environmental protection. This chapter highlights the importance and scope of antimicrobial nanomaterials, the effect of the microbial activity of nanomaterials like fullerenes and graphene, their derivatives, carbon nanotubes, nano-silver, currently relevant antimicrobial nanomaterials like Au, TiO_2, ZnO, etc. It also elucidates the mechanisms of the influence of nanomaterials on microbes, and includes the advantages, disadvantages, and potential applications of antimicrobial nanomaterials.

Sustainable Nanomaterials for Biomedical Engineering: Impacts, Challenges, and Future Prospects.
Junaid Ahmad Malik, Megh R. Goyal & Mohamed Jaffer M. Sadiq (Eds.)

14.1 INTRODUCTION

Microorganisms are an essential part of our ecosystem, and they actively participate in many processes in the environment. They are very versatile and contribute a leading role in the biogeochemical cycling, which transforms matter from one form to another form [63]. However, some microorganisms are beneficial to us, while some others are really harmful. Microorganisms are highly prominent in wastewater treatment, production of biogas, food processing, etc. Engineered nanomaterials have a great impact on microorganisms. With the advent of nanotechnology, many researches investigate on the impact of engineered nanomaterials on microbes. Several nanomaterials show excellent antimicrobial properties which can better be utilized in various environmental applications. Nanomaterials are submicron sized particles having at least one dimension in the nanoscale range. They show exotic change in properties from the parent elements from which it is formed. The unusual change in their properties are due to their small size, high specific surface area, surface structure and morphology, crystal structure, surface coating, chemical composition, surface charge, surface energy, size distribution, shape, aspect ratio, solubility, porosity, agglomeration, hydrophobicity, etc., [58]. The tiny size and high specific surface area of nanomaterials make them very reactive compared to their parent elements. The availability of surface for activity is high for nanomaterials which enhances their application as catalyst materials. Richard P Feynman, the famous American Physicist, speculated the possibility and application of nanomaterials in his renowned lecture titled "There is a plenty of room at the bottom." In fact, the potential application of materials can be exploited when the size is being converted to nano. Their broad spectrum of application are largely because of their size ranging between macroscopic bulk materials and atoms. They can enter into tissues and cells, which makes them suitable for diagnostic and therapeutic applications. They can also be used as carrier agents, which makes them suitable for medical fields as drug carriers as well as fertilizer releasing agents in the agriculture industry while, nano encapsulated material can be applied in biosensors [105]. Other than agricultural sector, nanomaterials have a broad spectrum of industrial applications as in horticulture industry, cosmetic industry, paints, and coatings, textile industry, semiconductor technology, solar cell, food, and pharmaceutical industry, water, and wastewater treatment, air purification, toxic waste remediation, pollution control and degradation, semiconductor industry, etc. Thus the peculiar features of nanomaterials revolutionize their band of efficacy and utility.

This chapter highlights an in-depth analysis of the impact of nanomaterials like carbon-based materials and other miscellaneous currently relevant nanomaterials like silver, TiO_2, ZnO, CuO, Au, Chitosan, peptides, etc., on microbial communities. A critical evaluation of various possible mechanisms of antimicrobial action, comparison of activity, qualitative, and quantitative techniques of assessment of antimicrobial action and prospective area of applications are discussed.

14.1.1 PROPERTIES OF NANOMATERIALS

The characteristic properties of nanomaterials make them superior to conventional materials. The generally required mechanical properties are tensile strength, tensile modulus, impact strength, flexural strength, flexural modulus, elastic modulus, elasticity, plasticity, toughness, hardness, yield strength, rigidity, ductility, etc. Small size, large surface-to-volume ratio, and quantum confinement of nanomaterials are the reasons for the outstanding mechanical properties of nanomaterials. Composite structures can be obtained by the addition of nanoparticles to traditional materials. They act as reinforcing fillers to the materials by modifying the grain boundary, thereby forming an intra or inter granular structure and enhances their mechanical properties [99]. Thus it widens the spectrum of engineering application of nanomaterials than macroscopic materials. The thermodynamic and thermo elastic properties of nano-systems largely dependent on size [13]. The unique catalytic, mechanical, optical, and electrical properties are due to their nano-size. The optical properties of metal nanoparticles are due to Surface Plasmon resonance, the excitation of electrons due to the electron oscillation in conduction band. This boosts the radiative properties like absorption and scattering.

Nanomaterials are extensively used in the biomedical field, especially due to its size similarity with biomolecules. The enormous range of applications in the biomedical field are targeted drug delivery, biosensing, bioimaging, antimicrobial agents, etc. Having high specific surface area and porous structure, Carbon Nano Tubes (CNTs) possess excellent adsorption properties and have high efficiency in removal of contaminants. It can adsorb many pollutants like heavy metals, antibiotics, etc. Graphene has excellent mechanical properties like high elasticity, stiffness, and strength. It has adjustable band gap with very high electron mobility and thermal conductivity. It can replace many conventional materials and bring about revolutionary changes in various technologies and current processes. High specific surface area and transparency makes it suitable to produce photocatalysts which find applications in water disinfection, disintegration of pollutants and removal of heavy metals [74]. The composites formed by the mixing of TiO_2 and graphene oxide are very effective in the removal of chromium from water. Graphene poly pyrrole composites have good adsorption capacity towards oil and the composite can degrade *Escherichia coli.* Since graphene based materials possess very high specific surface area, carrier capacity and sensitivity, it can adsorb greenhouse gases from the environment.

Fullerenes have a unique structure and it is the reason for its excellent physical and chemical properties. They are highly soluble in aromatic solvents, nonpolar, have high conductivity, possess aromatic properties, great adsorption capacity and good electron acceptor. They are able to adsorb volatile organic matter from atmosphere. The properties possessed by nanomaterials are very unusual and entirely different from their parent element from which it is formed. Carbon Nano Materials (CNMs) like fullerenes, graphene, CNTs, diamond, quantum dots (QDs) are used for biomedical applications in drug targeting and delivery, preparation of nano-bio sensors [17]. The properties of nanomaterials can be improved by combining various materials to form nanocomposites.

14.1.2 ANTIMICROBIAL PROPERTIES

The antimicrobial properties of nanoparticles can be best exploited for its environmental and medicinal applications. The conventional approaches used in water treatment and disinfection can be replaced with the help of nanotechnology. The most commonly and extensively used nanoparticles or oxides of nanoparticles for antimicrobial studies are Ag, Au, Ti, Zn, Cu, and Fe. Silver and gold nanoparticles (AuNPs) were used as antimicrobial agents from time immemorial. Silver contained ointments and bandages were used for healing wounds and also used in medicines for treating conjunctivitis in infants [87]. Silver ions are less toxic towards fungus than silver nanoparticles (AgNPs). Silica nanoparticle show toxic effect on bacteria and algae. The antimicrobial properties of silver, magnesium, zinc oxide, silicon, titanium dioxide, etc., are used in the agricultural industry for pathogen control [87]. TiO_2 is toxic towards both gram positive and gram negative bacteria. ZnO nanoparticles are effective in pathogen control and its usage adds to the fertility of soil than AgNPs. The oxides of Ti and Zn are effective against

Escherichia coli, *Bacillus subtilis*, *Staphylococcus aureus*, *Salmonella typhimurium*, etc. As TiO_2, ZnO is also toxic towards both gram positive and gram negative bacteria, and the effect increases as size decreases and as the content of nanomaterial increases [105]. CNMs like fullerenes and their derivatives, CNTs, and functionalized CNTs control contaminants in soil and promotes plant growth. Graphene and its derivatives like graphite oxide, graphene oxide and reduced graphene oxide are proved to be excellent for its antimicrobial activity. Chen et al. [15] studied the effect of nanomaterials like Ag, TiO_2 and ZnO on Medicago truncatula and connected microorganisms and found that low concentrations of nanomaterials have less toxic effect to plants. Wu et al. [98] found out that fullerenes and multiwalled carbon nanotubes affect the nutrient cycles when investigated for soil microbial community. They triggered mutilation to various natural fixation pathways and degradation of lipids, phospholipids, carbohydrates, secondary plant metabolites, etc.

Most nanomaterials are toxic to microbial community. Toxicity depends on the nature of nanomaterials and microorganisms, the size, shape, and concentration of nanomaterials, the duration of interaction of nanomaterials with microbes, etc. The most common factor which affects the microbial toxicity is the concentration of nanomaterials. As explained by Bernhardt et al. [12], different postulates are there in literature which gives a relation between both parameters. One postulate is that toxicity has a linear relation with concentration (I), i.e., as the content of nanomaterials increases, toxicity also increases [12]. Second postulate is that at very low concentration of nanoparticles there is a drastic increase in toxicity with amount of nanomaterials, then it starts decreasing with concentration and after that it shows a slow and steady increase with concentration (II). Another theory is that, at very low concentration, there is no toxicity at all. Toxicity begins only when there is a particular nanomaterial content and then it starts increasing with its content (III).

14.2 EFFECT OF VARIOUS NANOMATERIALS ON MICROBIAL COMMUNITIES

Nanotechnology has developed and created new materials to enhance its effectiveness for versatile and diverse applications. Many materials express different effect on various microbial communities on interaction. The proficient features of nanomaterials make it capable of replacing conventional and traditional materials as well as the technologies. Surface modification of nanoparticles enhance its antimicrobial activity. Various parameters which affect the antimicrobial properties are size, shape, surface to volume ratio, specific surface area, concentration, chemistry, zeta potential, surface charge, type of Nano Particle (NP) and microorganism, stability of NP, way of contact, retention time, etc. [26].

Among the various nanomaterials, CNMs are very unique in their features and uses. Due to their proficient electronic and electrical properties, conventional expensive materials in electronic equipments can be replaced with CNMs. Carbon Materials exist as graphite, diamond, CNTs, fullerenes or Buckyballs and grapheme (Figure 14.1) – the properties range from soft graphite to hard diamond depending on the arrangement of carbon atoms. The effect of various nanomaterials like CNMs, organic, and inorganic nanoparticles on microorganisms is discussed here.

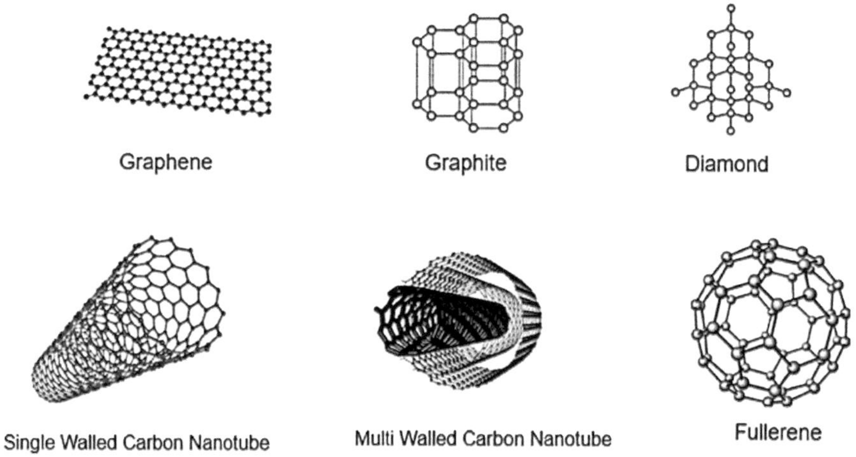

FIGURE 14.1 Allotropic forms of carbon.
Source: Modified images from Wikipedia.

14.2.1 FULLERENES

Fullerenes, discovered in 1985, are allotropes of carbon having peculiar properties with atoms connected by single and double bonds. Fullerenes exist in different categories with dissimilar properties. They are toxic to microorganisms and the effect of toxicity depends on its concentration.

They are not toxic at very low amounts, toxicity initiates only at a certain concentration depending on the microbial species [33]. At high concentration, the bio-carbon content of soil increases, and it reduces the number of microorganisms by many folds. Plants uptake fullerenes that are introduced into soil and will be accumulated in them without disintegration [8, 63]. Some studies show that anaerobic microorganisms are unaffected by the presence of fullerenes [66]. They can stimulate the production of crops, thereby increasing the consumption of nutrients from soil for its growth and thus depletes the microorganism content in soil [63]. Moreover, it is found that the presence of clay and other organic matter in soil reduces the toxic effects of fullerenes towards microbes. The toxic effect of fullerenes is less evident in aquatic microorganisms [53] and its effect varies with the category of microbes. Some microbes may have a protective layer against fullerenes, for example, it is less toxic to *Bacillus subtilis* than *Escherichia coli* [58]. Fullerenes can promote the growth of gram negative bacteria in soil [93]. Fullerenes are highly insoluble in water, whereas fullerene derivatives are soluble in water [55]. The antimicrobial effect of fullerenes are also highly specific for its derivatives. Fullerenes have toxic effect on microorganisms both in aquatic and land. The degree of toxicity depends on its concentration and the type of microbe.

14.2.2 GRAPHENE

Graphene, the newly added member to carbon family, represents a single layer of graphite and is regarded as the "hot material" of current science and technology [85]. It is a building block to all graphite forms and nanotubes. The prospective applications of graphene are due to its high aspect ratio, large surface area and outstanding properties.

Among various (Carbon Materials) CMs, graphene shows a highest rate of bio-toxicity. In soil, the contact time between graphene and microbes strongly influence the antimicrobial activity of graphene [80]. Low content of graphene in soil can be toxic to microbes and hence pollutants in soils can be removed. Studies show that some microbes will regain activities over time. High graphene content in soil can reduce microbes to the fullest extent and hence may affect the environment and ecosystem. Certain research studies prove that a small amount of graphene can endorse the growth of useful microbes in soil, for example, nitrogen fixing bacteria [24]. The organic matter and inorganic salts present in the soil may entrap graphene and reduce its mobility. In addition to graphene, its oxide also shows toxic effect on microorganisms. Similar effect of graphene is observed in aquatic environment as well. Graphene is less toxic to microorganisms with membrane structure [5]. It is less dispersible in liquid and hence its action in aquatic field is weaker than graphene composites like polymer graphene oxide. Graphene composites are highly stable and have very high microbial activity. They attack microbial cells leading to cell rupture and death. Soil incorporated with graphene can boost the activity of microbes in soil for a short period, thereby enhancing the rate of removal of pollutants from soil. But prolonged exposure can bring alteration of microbial structure leading to the tightening of soil [80]. Thus the antimicrobial activity of graphene in soil is higher than aquatic system as it is less dispersible in liquid. Moreover, graphene composites have very high toxic effect on microbial community than natural graphene.

14.2.3 CARBON NANOTUBES

Single graphene sheets rolled into hollow tubular structure and closed with caps at ends or left as open ended to form Single Walled Carbon Nano Tubes (SWCNTs). They have an approximate diameter of 1 nm to 5 nm. Multi-Walled Carbon Nano Tubes (MWCNTs) are formed by the rolling up of multiple layers of graphene sheets into concentric cylinders whose diameters can range up to 100 nm. They have high surface area and aspect ratio and fast electron transfer rates. Thermal conductivity of CNTs are higher than that of diamond. CNTs can be manufactured with different methods like arc discharge method, laser ablation, chemical vapor deposition, high pressure conversion of carbon monoxide, etc., and the mode of manufacture affects their end-use. SWCNTs can show metallic or semiconducting behavior while MWCNTs are good conductor of electricity. CNTs are not soluble in water. Solubility can be enhanced by using polymers or surfactants like sodium dodecyl benzene sulfate. As the length CNT increases, its toxicity also increases [74].

14.2.3.1 SINGLE-WALLED CARBON NANOTUBES (SWCNTS)

SWCNTs can significantly affect the presence of microorganisms in soil. The effect depends on its content in soil. The presence of microbes is unaffected by the presence of low concentration of SWCNTs in soil, while high concentration predominantly reduce the enzyme as well as microbial activity. Gram positive bacteria respond to the interaction with SWCNTs by altering the composition of intracellular membrane lipids [41]. Functionalized SWCNTs are highly toxic compared to pure ones [83]. Some microbial community may restore their original state even after prolonged exposure to (Nano Materials) NMs, while fungal community are not able to regain its original state. They diminish the soil respiration and microbial biomass. However, the organic matter content like carbon and phosphorous in soil protects the microbes from a high degree of biomass reduction [17]. The antimicrobial activity of SWCNTs are proved in aquatic environment also. In water, the toxicity of SWCNTs in dispersed state is much higher than the effect of MWCNTs since SWCNTs behave as a number of mobile "nano-darts" which can cause continuous damage to microbial cells leading to cell disruption and eventually to cell death [56]. The antimicrobial activity of SWCNTs depends on the type of microbes that the presence of SWCNTs can safeguard the phenol degrading microorganisms and considerably enhance the phenol removal rate from wastewater. Long and functionalized SWCNTs show higher toxic effect than short and natural tubes on microbes involved in sewage treatment [72]. On the other hand, as SWCNTs bring down the volume of bacteria and viruses in water, they are helpful in decontamination of water. While considering wastewater treatment, it will affect the efficiency of water treatment [57]. SWCNTs can be toxic to both gram positive and gram negative bacteria. Thus the antimicrobial activity of SWCNTs depends on their concentration, dimension functionalization, type of microbes, organic matter content in soil, etc.

14.2.3.2 MULTI-WALLED CARBON NANOTUBES (MWCNTS)

Multi-walled Carbon Nano Tubes hinder the existence of bacteria in soil, but not fungi. This property is useful in controlling the bacteria content in soil. Ge et al. [28] found out that MWCNTs at a low concentration in soil intensifies the bacterial content. Higher concentrations of MWCNTs in soil metamorphoses primary microbial communities to lenient microbial communities which can tolerate MWCNTs in soil [17]. Kerfahi et al. [47] studied the effect of acid treated MWCNTs in soil on microbes and found a tremendous increase in the production of tolerant bacteria which can withstand soil with acid treated MWCNTs. Long-time exposure to MWCNTs alleviate the presence of bacteria in the aquatic environment. Their presence [100] as well as sharp structure [31] can cause breakage of microbial cells, thereby causing change in the content of cells, even cytoplasmic loss may happen. However, low content can aid in the development of atrazine-degrading microorganisms [48]. A positive effect of MWCNTs is exhibited towards microbes which are involved in the bioremediation of petroleum [1]. The less toxic nature can be due to the presence of organic matter. Organic matter may entrap CNTs and reduce their agility thereby diminishes its antimicrobial effect. Studies show that MWCNTs enhances the growth of some species of bacteria like *Pseudomonas, Cellulomonas, Rhodococcus, and Nocardioides*, which are highly efficient for the degradation of recalcitrant aromatic organic compounds [48, 86]. MWCNTs can reduce the microbial biomass like Escherichia *coli, Staphylococcus epidermidis*, and *Pseudomonas aeruginosa* at a slow rate during the initial period, but prolonged exposure reduces their number in a drastic manner [45, 91]. MWCNTs can affect the growth of microorganisms positively or negatively. The toxic effect depends on the concentration, functionalization, time of exposure, the structure of MWCNTs, the category of microorganism and the nature of the environment. The hollow structure and high specific surface area of MWCNTs make them suitable for absorbing contaminants and microbes from aqueous environment. Large amount of MWCNTs can impact the activity of biomass in soil and alter its structure.

14.2.4 NANO-SILVER

AgNPs have antimicrobial properties which have many medicinal applications. It can be used in ointments for scar less healing of wounds. It is the most demanding and versatile nanomaterial with diverse application in the textile industry, plastic manufacture, coatings, sewage treatment, soil remediation, etc. Depending on the type of microbes the antimicrobial action may vary. For example, heterotrophic microbes (bacteria, yeast, molds) are tolerant to AgNPs than nitrifying bacteria [40], Ag NPs coated with Poly Vinyl Pyrrolidone (PVP) are toxic towards nitrifying bacteria in soil and in wastewater [60], Ag NPs shows toxicity towards methanogenic bacteria, biogas production is unaffected with low concentration of Ag NPs while high concentration reduces biogas production [102], Ag NPs inhibits nitrogen removal processes [64]. They interact with the surface of bacteria and yeast and release silver ions into solution. Ag NPs attach to cell surfaces, alter the permeability of cell membranes and other activities of cells. A mixed combination of Ag NPs and TiO_2 were effective in preventing colony formation of

Escherichia coli and *Candida albicans*. The use of TiO_2 along with Ag NPs are effective against fungi than bacteria [69]. Ag NPs were found to be effective against *Cladosporium* species [20]. Ahmed et al. [3] conducted studies on the antifungal effect of Au and Ag NPs against *Candida albicans*. They were also tested for their bioactivity against gram-positive *Staphylococcus aureus* and gram-negative *Escherichia coli*.

14.2.5 MISCELLANEOUS NANOMATERIALS

TiO_2 is a common photo-catalyst whose photosensitivity can be used for environmental applications to eliminate pollutants from air and water. It is used for the photocatalytic deactivation of microorganisms. It can be used as a toxic agent to many bacteria like gram-positive and gram-negative as well as viruses like hepatitis B, polio, etc. Alumina nanoparticles coated with humic acid shows toxicity towards methanogenic bacteria [6]. TiO_2 NPs inhibits the efficiency of nitrogen and phosphorous removal processes while Cu NPs increases the efficiency of nitrogen removal processes [64]. TiO_2 NPs are very effective against both gram-positive and gram-negative bacteria [61]. Engineered NPs of peptides, chitin, and chitin derivatives have excellent antimicrobial properties which makes them suitable for low cost water disinfection applications. The antimicrobial activity of chitosan depends on its preparation method and is affected by the presence of organic matter. Chitosan is less soluble in water and hence its disinfection action is efficient at pH less than 7. Solubility can be improved by derivatization of chitosan [55].

ZnO nanomaterials are excellent for their antibacterial activity. Various factors like UV absorption and transparency to visible like affect their activity. Conflicting results were obtained for the impact of the size of ZnO on antibacterial activity. Research conducted by Jones et al. [42] observed high toxicity for small ZnO NPs than bigger particles whereas studies conducted by Franklin et al. showed that size of ZnO NPs did not have any influence on its antibacterial activity [27]. A specific combination of ZnO-TiO_2 NPs were found to inhibit the multiplication and coverage of *Aspergillus niger* [69]. ZnO and TiO_2 NPs when doped with palladium and gold, the resultant samples Pd/ZnO and Au/TiO_2 were observed to exhibit excellent bioactivity and biofilm resistance against *Escherichia coli* [73]. Another work on a formulation of TiO_2–platinum was found to reduce the recolonization of *Escherichia coli*, *Lactobacillus acidophilus*, *Saccharomyces cerevisiae* [52]. Copper NPs are toxic to *Escherichia coli*, *Pseudomonas aeruginosa*, *Staphylococcus aureus* and *Bacillus subtilis* [10, 14, 22, 32, 38, 103]. As the antibacterial activity of copper oxide NPs is low, the concentration of CuO NP should be high to get an effective toxicity [79]. They exhibit antimicrobial properties against *Escherichia coli*, *Pseudomonas aeruginosa*, *Salmonella typhimurium*, *Proteus vulgaris*, *Klebsiella pneumoniae*, *Enterococcus faecalis*, *Shigella flexneri* and *Staphylococcus aureus*. Among them, *Escherichia coli* and *Enterococcus faecalis* are most affected while *Klebsiella pneumonia* is least affected by CuO NPs [2]. Antimicrobial activity of chitosan-based nanocomposites using nano silver, silver-zeolite ion, modified montmorillonite nanoclay and unmodified montmorillonite nanoclay were evaluated [82]. Nanocomposite with modified montmorillonite nanoclay exhibited higher antimicrobial activity than with unmodified. All composites exhibited toxic effect towards microbes to different degrees. Various nanocomposites were found to be effective against microbes. CuO_2 and SiO_2 are effective against *Saccharomyces cerevisiae* [104]. Cu NPs with polymers like acrylates and silicon act as antimicrobial agents against *Aspergillus* species, *Penicillium chrysogenum* and *Candida albicans* [25]. Modified and functionalized AuNPs are effective antimicrobial agents. Oxide of Mg possesses high antimicrobial properties. It has the same activity against gram-negative and gram-positive bacteria [36]. MgO NPs have competent antimicrobial activity against both Gram-positive and Gram-negative bacteria, spores, and viruses. Synthesis of MgO NPs is easy when compared to other metal counterparts. Manganese NPs are very effective against *Escherichia coli* and *Staphylococcus aureus* [43]. NPs of aluminum are found to prevent the growth of roots in plants. CNMs show bioactivity against gram-positive and gram-negative bacteria [95].

14.3 MECHANISM OF ANTIMICROBIAL ACTIVITY OF NANOMATERIALS

The mechanism of the antimicrobial activity of NPs is a currently researched topic of interest. Several hypotheses were developed for elucidating the exact reason for the toxic property exhibited by nanomaterials. This is a partially explored area where some postulates can be generalized but others may be organism-specific. NPs can enter into microbial cells by diffusion. Endocytosis, the process of transport of cells, part of cells or any other molecules into the cell, can also cause the entry of NPs to microbial cells.

Most commonly suggested mechanism for toxicity of nanomaterials are oxidative stress response or Reactive Oxygen Species (ROS), direct contact and interaction between nanomaterials and microbes, increase in permeability of cell membrane, penetration of nanomaterials into cell membranes, cell wall or cell membrane rupture, damage to membrane, DNA, proteins, and other cellular constituents like nucleic acid, non-oxidative mechanisms, the damage of cells due to the uptake of ions from nanomaterial solution, photocatalytic activity and visible light sensitivity. Multiple mechanisms are possible for the antimicrobial effect (Figure 14.2). The penetration of NMs into cell membrane can mutilate it, disrupt various functions, lose its integrity and can further interrupt energy transduction processes leading to cell death. The production of ROS can cause impairment to all cellular components and this leads to further production of harmful radicals. Cells have mechanisms to prevent ROS. However, when ROS level is high it causes excessive stress. Thus ROS attack lipids, disrupt membrane and cell functions. Due to the presence of NPs, various oxidation reactions happen in the microbial cell, leading to a condition of stress for cells, the oxidative stress response. This can produce free radicals like hydrogen peroxide, carbonates, superoxide, peroxy nitrate, peroxyl, hydroperoxyl, hydroxyl groups, etc. [87]. These species are very much responsible for the toxic effect of nanomaterials on microorganisms [87]. Another mechanism is during the interaction between nanomaterials and microbial cells toxic elements like ions or heavy metals can be released. These can reside inside the cells for a long time, accumulate, and impart toxicity.

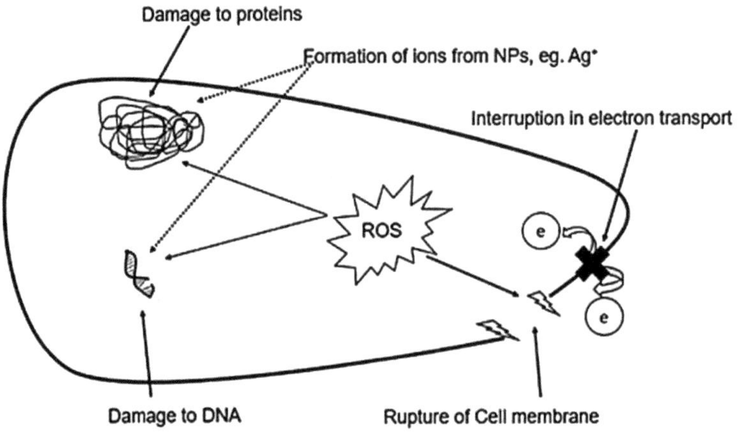

FIGURE 14.2 Antimicrobial mechanisms of nanoparticles.
Source: Modified image from Ref. [55].

14.3.1 FULLERENES

Many studies have shown that the concentration of fullerene in soil affects the microbial community, but the exact mechanism of the toxic effect of fullerenes is not fully understood [17, 87]. Many investigations suggest a mechanism that because of high electron affinity, fullerenes can hinder the electron transfer between electron donor cell and electron acceptor cell (Figure 14.2) leading to the expiry of cell [58]. Cell membranes act as the direct point of contact of NMs like fullerene. Cell to cell communications are done through cell membranes [64]. The closed cage structure of fullerenes impart it with antibacterial property by absorbing light and by generating ROS. High ROS content can destroy lipids, proteins, nucleic acid, etc., cause fatal to microorganisms. Derivatives of fullerenes are capable of binding to the microbial cell and rupturing the cell membrane. They can even cause damage to its constituents like DNA strands (Figure 14.2). Among various microbes, *Escherichia coli* [23] is mostly affected by fullerenes because of the disturbance in its electrostatic balance. Another mechanism of toxicity can be the reduction in enzyme production in microbial cells due to its presence which may lead to the inconsistency in metabolic activities [18]. Studies show that toxicity of fullerenes to eukaryotic system is due to the formation of ROS whereas for prokaryotic system it is due to direct oxidation of the cell [55]. Damage to cell membranes can disturb the energy transduction processes. This can cause the derivatives of fullerenes to inhibit the growth of *Escherichia coli* [50]. The closed cage structure of fullerene enable it for absorption of light and generation of ROS [51]. Interaction of fullerene with microbial surfaces cause DNA damage and cell membrane disruption. During interaction, electrostatic forces between fullerene and microbes play a major role in toxic mechanism. Fullerenes are highly dispersive in water and are able to cause damage to DNA [74] since it can accept electrons and exchange electrons with nucleotides. They are more toxic towards gram-positive bacteria than gram-negative species. The charges on fullerene derivatives determine its toxicity towards microbes. Positively charged ones are acutely toxic, neutral are mild toxic whereas negatively charged fullerenes are not toxic towards microbial communities.

14.3.2 GRAPHENE

Studies have shown that graphene has high oxidative stress and produce ROS in microbes. Excessive production of ROS can in turn create an oxidative stress, make imbalance in microbial cells, and reduce the microbial biomass, cause damage and destruction to cells, cell rupture and death. The toxic effect of graphene is induced once it enter into the cells. Since its size is very small, it will enter into the microbial cells easily. Moreover, its interaction with microbial cells depend on its concentration, cell growth, etc. The structure of graphene derivatives was found to affect the microbes. Studies conducted by Akhavan and Ghaderi [4] found that when *Staphylococcus aureus* and *Escherichia coli* when contacted with graphene oxide, the bacterial cells got damaged by the sharp edges of graphene oxide. Gurunathan et al. [30] reported the toxic effect of graphene oxide and reduced graphene oxide on *Pseudomonas aeruginosa* due to the formation of superoxides. Graphene can interact with lipid molecules due to dispersion and penetration into cell membranes. Li et al. [54] showed that mono layers graphene films on copper and germanium could inhibit bacterial growth while there was no effect with SiO_2. Graphene oxide adheres to the surface of *Escherichia coli* and cause damage to its cells due to high mechanical strength.

14.3.3 CARBON NANOTUBES

14.3.3.1 SINGLE WALLED CARBON NANOTUBES

Compared to MWCNTs, SWCNTs exhibit greater microbial toxicity. The major reason that causes bio-toxicity by SWCNTs is the indirect damage due to the oxidative stress response which causes destruction of cell constituents [3, 19]. Nanomaterials produce ROS in aqueous solutions which can enhance the fluidity and permeability of cell membranes. High permeability permits the penetration of NMs into cell membranes and cause intracellular damage. Length, diameter, structure, functionality, and surface chemistry of CNTs affect their antimicrobial activity. There are contradictory statements regarding the effect of length of CNTs on microbial activity. Some research works [44] say that shorter tubes are toxic than longer ones due to the high chances of interaction of short tubes with microbes. Yang et al. [101] observed that longer tubes are highly toxic compared to shorter ones because the interaction of shorter SWCNTs with microbial cell surface is lesser than longer SWCNTs. SWCNTs can enter into cells or nuclei of microbes and alter their structure, causing cell damage. But in aquatic systems, longer nanotubes perform its antimicrobial activity higher than shorter ones. This is because in liquid, shorter tubes have a tendency to undergo self-aggregation and this prevents them from interacting with large microbial communities while the self-aggregation of longer nanotubes doesn't affect its interaction with large areas of microbial cells [101]. Smaller diameter tubes interact with cell membrane easily and thereby cause damage to cells [81, 88]. The interaction of CNTs to microbes are affected by the shape of microbes also. CNTs can interact easily with spherical bacteria than rod shaped [16]. SWCNTs can be metallic or semiconducting depending on their electronic structure and it determines its antimicrobial activity. It can swing to either state, metallic or semiconductor depending on the change in tube diameter, angle, and orientation of tubes [97].

CNTs can cause denaturing of enzymes. Direct interaction can also cause toxicity. As the dispersity of CNTs in water is less, direct contact with microbes is difficult. It can be improved by functionalization. Another way of applying CNTs is by coating it on reactor surfaces or on filtration membranes. Studies show that hollow fibers incorporated with CNTs deactivated *Escherichia coli* and polio viruses [55]. The changes in intrinsic cell constituents are the reason for the toxicity of SWCNTs on *Escherichia coli.* All these mechanism of action depends very much on the type of the microorganism. Soft cell walls are more prone to the entry of NMs than hard.

14.3.3.2 MULTI WALLED CARBON NANOTUBES

Researchers have concluded with different mechanisms for the microbial activity of MWCNTs. As explained in the previous section, oxidative stress and production of ROS in the cells of microbes by MWCNTs can cause microbial damage. Their presence in microbial cells stimulate the production of more ROS. ROS are highly reactive and unstable due to the presence of unpaired electron. Their presence in microbial cells can damage the constituents of cells, eventually causing cell death can happen. Direct contact of MWCNTs with microbial cells can cause cell mutilation. The structure of MWCNTs can also impart its toxic effect on microbes [17]. Analogous mechanism of toxicity is observed for MWCNTs and SWCNTs, but the degree of toxicity is higher for SWCNTs. CNTs can be used to prevent the biofouling caused by the attachment of microorganisms on the surface of underwater pipelines.

14.3.4 NANO-SILVER

The mechanism of toxicity of silver compounds are not fully understood. Researchers have developed certain postulates for the antimicrobial property of nano forms of silver. The properties of the microbial cell membranes will be reformed by the linkage of NP to the membrane surface. They gather inside the cell membrane, makes it more permeable and infiltrates into the cell causing mutilation to DNA [89]. Another hypothesis is that silver ions which have toxicity against microorganisms will be released on the dissolution of nanosilver [65]. Silver ions inactivate important enzymes during its interaction with proteins. This can enhance the generation of ROS. Also it can disturb the porousness and assembly of cell membrane and stop DNA replication, outflow of cellular constituents and eventually death of cell. Some research shows that silver less than 10 nm are toxic to *Escherichia coli*, *Pseudomonas aeruginosa* and to certain viruses [55]. Shape and structure also influence the antimicrobial nature of nanosilver because it was found that nanosilver in the form of triangular platelets are very toxic than its rod and sphere forms [71]. The penetration of nanosilver into the cell disables the performance of enzymes which causes death of cells [67]. Research should be conducted to find out the dependence of size, shape, structure, and composition of AgNPs on toxicity. Ag NPs penetrate through the cell membrane of *Escherichia coli* and destruct it as evident from the holes on its surface [94]. The presence of sulfur containing proteins on cell wall have high affinity towards Ag which causes hole to the cell wall and enhances the permeability of cell membrane. Constituents of the cell like cytoplasmic materials get leaked through the membrane and causes the death of the organism. Ag NPs generates ROS under UV irradiation and active against *Escherichia coli* through the production of hydroxyl and superoxide radicals. The mechanisms of antimicrobial activity of nanoparticles are summarized (Table 14.1).

TABLE 14.1 Antimicrobial Mechanism of Various Nanoparticles

Nanomaterials	Proposed Mechanisms for Antimicrobial Activity
Fullerenes	Increase in membrane permeability, production of ROS, electrostatic attraction, direct interaction with microbes
Graphene and its derivatives	Oxidative stress, charge transfer, membrane stress and damage, reactive oxygen species production, extraction of phospholipids from cell membrane
SWCNTs	Rupture of cell wall and cell membrane, leakage of cell constituents, deactivation, and loss of integrity of cells.

TABLE 14.1 *(Continued)*

Nanomaterials	Proposed Mechanisms for Antimicrobial Activity
MWCNTs	Rupture of cell wall and cell membrane, leakage of cell constituents, deactivation, and loss of integrity of cells.
Silver	Release of silver ions, production of reactive oxygen species, rupture of cell membrane by penetration
Gold	Rupture of cell membrane, prevention of active transport and protein synthesis
TiO_2	Production of ROS through photocatalytic activity, rupture of cell membrane
ZnO	Generation of ROS, generation of peroxides, rupture of cell membrane by penetration
Chitosan	Rupture of cell membrane, prevention of active transport
CuO	Generation of ROS, rupture of cell wall, rupture of cell membrane
Fe_2O_3	Generation of ROS, rupture of cell wall, rupture of cell membrane, prevention of DNA replication

14.3.5 MISCELLANEOUS NANOMATERIALS

Studies show a possibility of enhancing the permeability of negatively charged microbial cell membranes when they interact with positively charged chitosan. This leads to the uptake of chitosan by cell membrane leading to rupture and thus loss of constituents of cells [75]. Chitosan loses its efficiency when the pH of is above six since amino groups are absent [75]. Antimicrobial activities of derivatives of chitosan with amino groups are stronger than natural chitosan. Chitosan forms chelating compounds with trace metals and obstructs the activity of enzymes. They penetrate into fungal cells, fix with DNA and hinder RNA synthesis [76]. The toxicity effect of nano form of titanium dioxide is due to oxidative stress response where when irradiated by UV, hydroxyl free radicals are formed by oxidative pathways and peroxide radicals by reductive pathways [49]. Anatase and rutile forms of TiO_2 are active in antimicrobial property, and among them, anatase is the best because of its photocatalytic activity. But in darkness both mineral forms possess the same anti-bioactivity. The various factors on which the antimicrobial activity of TiO_2 depends are the nature of microbe, amount of TiO_2, intensity of light, pH, temperature, contact time between microbe and TiO_2. Antimicrobial activity increases with concentration of TiO_2, contact time and light intensity [36]. An interesting work by Gelover et al. [29] showed the complete deactivation of fecal coliforms in 15 minutes when treated with water taken in TiO_2 coated containers, while the same deactivation in uncoated containers took 60 mins. The photocatalytic activity of TiO_2 is achieved by sunlight and metal doping. Production of ROS through photocatalytic activity is suggested as a mechanism of toxicity of TiO_2 [69]. ROS generation interrupt the cell activities, damage proteins, prevent DNA replication. Excitation of TiO_2 can be achieved by doping with noble metals, especially silver, and it can enhance the photocatalytic deactivation of microbes [70, 78]. Studies show that when silver doping is done the mechanism of enhancement in the photocatalytic activity of TiO_2 is due to the electron-hole separation which provides large surface area for adsorption [92]. Also the combination of Ag and TiO_2 can induce electron transfer to TiO_2 due to surface plasmon resonance by silver resulting in photocatalytic activation by charge separation [84]. Like TiO_2, ZnO NPs exhibit high efficiency for absorption of UV radiation and transparency to visible light. The principal mechanism of antibacterial activity of ZnO is the generation of hydrogen peroxide by photocatalytic excitation. Another mechanism is that the structure of the cell membrane of the bacteria will be disordered due to the infiltration of NPs into the cell wall and affects the growth of bacteria. Also, it is suggested an extension in lag growth phase of the microbes happen due to the adhesion of zinc ions to the cell membranes [7]. ZnO NPs penetrate through the cell membrane and destruct it [94]. Au NPs generates ROS under UV irradiation and active against *Escherichia coli* through the production of singlet oxygen species. The use of metals like silver, strontium, etc., along with TiO_2 enhances its photocatalytic activity. The high solubility of aluminum NPs results in phytotoxic effects [50]. Al_2O_3, CuO, SiO_2 and Fe_2O_3 produces ROS under UV radiation. High quantity of ROS are evident in the destruction of *Escherichia coli* cells. Si and Ni NPs generates ROS under UV irradiation and active against *Escherichia coli* through the production of singlet oxygen species. MgO also exhibit excellent antimicrobial activity, whose mechanisms are interaction with microbes, retention time, specific surface area of particles, presence of active oxygen on the surface, ROS production, adsorption, and penetration into the membrane [36].

All engineered nanomaterials have dissimilar toxic effects on different microbial communities. CNMs have the less toxic effect to microbes compared to other NMs. Among the various carbon nanomaterials, fullerenes have less effect on aquatic microorganisms whereas graphene has great effect. Fullerenes can alter the structure of microbial community, thus decreasing the rate of removal of pollutants from soil and also it harmfully affect the disintegration of contaminants in the environment. The degree of microbial toxicity is higher for SWCNTs than MWCNTs even though the toxicity mechanism are more or less similar for both. This can be due to the dissimilar size and specific surface area of both. During water treatment, *Pseudomonas aeruginosa, Escherichia coli* and *Staphylococcus epidermidis* are affected by MWCNTs and prolonged exposure increases the degradation rate. At low concentration, both SWCNTs and MWCNTs enhances the growth of microbes capable of decomposing petroleum. Nanocomposites are very much effective in bioremediation than other nanomaterials [64].

14.4 BENEFICIAL AND ADVERSE EFFECTS OF ANTIMICROBIAL NANOMATERIALS

Antimicrobial properties of nanomaterials are better exploited in water and wastewater treatment systems. Conventionally available chemical disinfectants produce disinfection byproducts. The application of nanomaterials will not produce byproducts of disinfection. TiO_2 is also used for water treatment since it is not expensive and also consumption doesn't cause problem. Another feature of TiO_2 is its photocatalytic disinfection effectiveness and visible light activity which is best suited for water treatment in areas where there is scarcity in electricity [55]. However, disinfection with TiO_2 is a slow process and it can be improved by doping with nitrogen or with metals like silver. The use of CNTs in water disinfection may take longer time compared to conventional disinfectants. But it can prevent biofilm formation and biofouling on membranes used for water filtration [55]. The biofilm formation by microbes are the source of breeding of microbial communities. NPs are very effective to destroy biofilm due to its small size, shape, and large specific surface area [96].

While we discuss about the possibilities and capacities of nanomaterials in diverse arenas, we should admit the fact that every technology has limitations. The use of NMs for microbial control has serious limitations. The major drawback of the application of NMs in water treatment is the formation of aggregates. Nanomaterials of TiO_2, fullerenes, etc., can severely form aggregates in water which limits its application in water treatment. Similarly Zn ions and ZnO nanoparticles are soluble in water. Thus even if they are effective antimicrobial agents, aquatic microbes will be affected by its presence. High solubility in water limits its application in water treatment and disinfection [55]. Another problem with the usage of nanomaterials in drinking water purification system is its low retention in the process because of the high dispersity of nanoparticles in water. Costly downstream processes are required to retain and recycle nanomaterials. Carryover of nanomaterials with treated water may be highly toxic to human beings as well as the ecosystem as most of the nanomaterials are known to be carcinogenic although exceptions are there like silver, TiO_2, etc., while fullerenes and ZnO nanoparticles exhibit toxic effect to mammals at certain concentration. Continuous contact to high dose silver can cause blackening of skin. Another thought-provoking studies conducted by researchers is really contradictory to the prospective applications in the area of antimicrobial effect of NMs that we have discussed so far. Among the NMs, CNMs have low degradation capability to microbes. Their massive use for various applications can produce residuals of CNMs to the environment. Massive emission of CNMs are certainly harmful to the environment. Kang et al. [46] investigated on the ability of microbes to degrade CNMs and observed a low degradation rate. They insisted the importance of further studies to see this reverse application. Studies show that nanomaterials can inhibit the production of useful metabolites like minerals by certain microorganisms which are beneficial not only to themselves but also to other microbes [64]. The presence of graphene in soil while degrading microorganisms, it can affect the structure of microbes and thereby the nutrient cycles in soil and gradually causes tightening of soil. This affects the texture and fertility of soil.

14.5 ASSESSMENT OF ANTIMICROBIAL ACTIVITIES OF NANOMATERIALS

The sensitivity of microorganisms against nanomaterials can be tested using various susceptibility tests that are used to assess the sensitivity of microbes against antibiotics. In antibiotic susceptibility tests, microbes will be exposed to antibiotics, and the area of microbial growth is found out. The areas without microbes called zones of inhibition can be identified. From the size of the zone of inhibition, the minimum inhibitory concentration (MIC), the minimum antibiotic required to hinder the microbial expansion can be estimated. In disc diffusion method or Kirby-Bauer method, the growth of microbes is observed in the area of antibiotic impregnated disc. If the microbe is sensitive to the particular antibiotic, zone of inhibition will be formed. Depending on the diameter of the zone of inhibition, the microorganism will be categorized as sensitive or resistant to antibiotic. The data of zones of inhibition will be correlated with MIC. The modified forms of these tests are used by researchers to find out the activity of NPs towards microorganisms. Ahamed et al. [2] studied the toxicity of CuO NPs against various bacterial strains. They estimated the zone of inhibition using well diffusion assay, where different strains of bacteria were exposed to CuO NPs impregnated discs. The MIC was found out by broth microdilution method, a similar technique for determining antibiotic or nanoparticle susceptibility to microbes. Statistical analysis was conducted to compare the zone of inhibition of different strains like *Escherichia coli*, *Pseudomonas aeruginosa*, *Salmonella typhimurium*, *Proteus vulgaris*, *Klebsiella pneumonia*, *Enterococcus faecalis*, *Shigella flexneri*, and *Staphylococcus aureus*. Zones of inhibition are observed to be more for *Escherichia coli* and *Enterococcus faecalis* and least for *Klebsiella pneumonia* [2]. Thus CuO NPs exhibited excellent bioactivity against different bacterial strains. Jahangirian et al. [39] worked on the evaluation of Cu derivatives against *Escherichia coli* and *Staphylococcus aureus* using diffusion method. Results showed higher antibacterial activity by copper derivatives on *Escherichia coli* than *Staphylococcus aureus*. Ren and coworkers [79] compared the antimicrobial activity of Cu, CuO, Ag, and ZnO nanoparticles and MIC was evaluated using soya broth. CuO was found to be very effective against microbes with minimum MIC and the combination of CuO with Ag enhanced the activity. Narayanan et al. also evaluated the antimicrobial activity of ZnO NPs using disk diffusion method. Meruvu et al. [62] studied the antimicrobial activity of ZnO NPs against *Bacillus subtilis* and *Escherichia coli* by disc diffusion technique and observed that zones of inhibition increases with the dosage of ZnO and decreases with particle size [62]. Ahmed et al. [3, 9] found out the antifungal effect of Au and Ag NPs against *Candida albicans* and also against *Staphylococcus aureus* and *Escherichia coli*. Antimicrobial assay was done with modified disk diffusion method and found out larger rings which implies excellent antimicrobial effect for Ag and Au. Microdilution assay was carried out to evaluate the MIC and found out lower MIC for Ag compared to Au while the minimum dosage required for gram positive bacteria was more for both Ag and Au. Thus gram-negative bacteria are more toxic with Ag and Au NPs. Fourier transform infrared spectroscopy (FTIR) and Thermogravimetric Analysis (TGA) showed interaction between NPs and proteins.

Disk diffusion technique was opted by Varghese et al. to evaluate CNM's bioactivity against gram positive and gram negative bacteria like *Proteus refrigere*, *Pseudomonas aeruginosa*, *Staphylococcus aureus*, and *Streptococcus hemolyticus* and inhibition zones were found out [95]. The diameter of the inhibition zone was in the decreasing order of magnitude for *Proteus refrigere, Staphylococcus aureus*, and *Streptococcus hemolyticus.*

14.6 POTENTIAL APPLICATIONS OF ANTIMICROBIAL NANOMATERIALS

Nanomaterials like CNTs, graphene, gold, etc., can be combined with microbial cells for the construction of microbial-cell biosensors which are used for the detection of components. They are used for the analysis of constituents in environmental, food, and agriculture, medical field, etc. They are less expensive, highly stable and sensible, very specific, prompt detection and analysis compared to conventional analysis techniques like liquid or gas chromatography. Microorganisms can act as bio-sensing element. The production of microbial biosensor with *Pseudomonas fluorescens* and *Pseudomonas putida* was reported by Odaci and colleagues [68]. CNTs along with chitosan and bacterial cells were used to formulate a biocompatible membrane as a part of biosensor [21]. Osmium compounds combined with SWCNTs were used for the production of biosensor to monitor toxicity [77]. Hnaein et al. [35] developed a sensor for the determination of carcinogen content in soil and water by using *Pseudomonas putida* on gold microelectrodes with CNT composites. Similarly, nanoparticles of gold and silver along with microbial cells can be used for the preparation of biosensors.

Another prospective area where antimicrobial property of nanomaterials are used is in water disinfection and microbial control. Water is very essential for maintaining life and it is essential to preserve water resources. Due to population density and industrialization, availability of pure water is a big challenge faced in many parts of the world. Only solace to these problems is appropriate disinfection of water and development of technologies to control the spread of contaminants in water. The problem with conventional chemical disinfectants are the formation of disinfection by-products. Due to the large specific surface area and reactivity, antimicrobial nanomaterials are exceptionally strong oxidants. Since they are inert in water environment, there is less chance to produce by-products after disinfection and can be used to improve conventional disinfection methods [55]. Nano form of chitosan is used as flocculant in wastewater treatment systems, as disinfectant in water treatment and as an antimicrobial agent in membranes [55]. The photocatalytic disinfection effectiveness and visible light activity of TiO_2 makes it suitable for its application in water purification systems and also it is not expensive and stable in water. Application of UV radiation is the conventional technique for water disinfection. But for some microorganisms, it demands high dosage of UV radiation for the treatment to be effective. In such cases, a hybrid technology can be developed by combining the application of UV radiation with photocatalytic nanomaterials like TiO_2 for improving the conventional technology [55]. Thus this hybrid technology can be made effective in large scale water treatment systems. The same hybrid technology can be applied for the prevention of biofouling in conventional drinking water treatment systems. The major challenge in conventional water treatment systems is the clogging and damage to the filtration membranes caused by the formation of biofilms. Also, as in the previous case, it may require a high dosage of UV as well as multiple stages of treatment in conventional technique for the removal of contaminants to make the water effective for drinking purposes. Thus in the traditional technology with UV, if the membranes are incorporated with antimicrobial and photosensitive nanomaterials like TiO_2 the efficiency of pathogen removal will be improved and multiple stage treatments can be avoided. Some other antimicrobial filtration membranes developed by researchers are silver impregnated hollow fibers, composite membranes with chitosan, poly acrylic acid/polyethylene glycol layers or any other functionalization of nanoparticles [55]. CNTs act as water purification membrane for pollution removal. The shape of SWCNT make suitable for its application as a filter for removal of microbes from water.

ZnO nanoparticles can be used in cosmetics, lotions, and ointments. It is also used in mouthwashes, paints, and coatings. Its antimicrobial property is effectively used in mouthwashes, cosmetics, paints, lotions, and ointments, whereas the ability to prevent biofouling is used in surface coatings.

The broad application of fullerenes are in the area of nanoelectronics, drug delivery, nanocomposite technology, etc. Nanosilver is the most cited nanomaterial which has plenty of applications in consumer product exploiting its antimicrobial property. It is used in food industry for nutrients supplements, paints, and coating industry [55]. AgNPs incorporated socks can prevent the fouling odor caused by microorganisms. Paints and coatings incorporated with AgNPs can be used to prevent moss formation on buildings. Drinking water purifier system using silver impregnated membrane technology claims almost 99% removal of pathogens. TiO_2 is used in industrial scale water purification system due to its ability to disintegrate organic pollutants. Its antibacterial property is effectively utilized for its use in home air purifiers. Antimicrobial nanomaterials can be used to regulate fouling of membranes used in water filtration and on underwater pipelines used for water treatment and distribution [55].

In construction fields, there are potential scopes of microorganism based product to improve the performance of concrete. During aging cracks appear on concrete structures. Studies conducted by Bang et al. [11] observed that concrete treated with microbial calcite are excellent in their performance with increased strength, stiffness, and crack remediation. This serves as a surface protection from further degradation by microbes and crack formation.

Having high specific surface area and porous structure, CNTs can be used to adsorb antibiotics which form pollution in the environment. The adsorption capacity of CNT can be enhanced when it forms composites or functionalized with other materials [53]. CuO NPs exhibit excellent bioactivity against different bacterial strains which can be applied in various medical devices to prevent the formation of bacterial growth, expansion, and biofilm formation [2].

Antimicrobial nanoparticles have potential application in dentistry. ZnO NPs and Ag NPs can be incorporated into resins to form composites and can be used to reduce the biofilms in the oral cavity [90]. In root canal treatment, the use of antibacterial fillings like Ag NPs, nanocomposite forms of chitosan and calcium hydroxide, etc., can prevent infection. During dental implants Ag NPs, Cu, ZnO, Ti, etc., can be used. Other areas of application in dentistry are endodontics, orthodontics, dental prostheses, restorative dentistry, etc.

The antimicrobial activity of Ag NPs, the biocidal activity of ZnO NPs and the photocatalytic activity of TiO_2 NPs find it effective for food packaging applications [37]. Other NPs like Au, CuO, MgO, SiO_2, Alumina, Se NPs, CaO have high efficiency even at lower concentrations for food packaging sectors. Ag NPs are very prospective for applying in cancer treatment due to its anticancer activity [34]. The ROS and the Ag ions produced by Ag NPs are capable of selective killing cancer cells. Applications of antimicrobial nanomaterials are summarized and given in Table 14.2.

TABLE 14.2 Potential Applications of Various Antimicrobial Nanoparticles

Nanomaterials	Applications
Fullerenes	Nano-electronics, drug delivery
Graphene	Microbial-cell biosensors
SWCNTs	Water purification membrane, microbial-cell biosensors
MWCNTs	Water purification membrane, microbial-cell biosensors

TABLE 14.2 *(Continued)*

Nanomaterials	Applications
Silver	Antimicrobial food packaging, pathogen control, therapeutic use, pharmaceutical industry, disinfecting medical devices, water disinfection, dentistry, textile industry, food industry, paints, and coating industry
TiO_2	Food packaging applications, dentistry, industrial scale water purification system, home air purifiers
ZnO	Cosmetics, lotions, and ointments, mouthwashes, paints, and coatings, dentistry, Food packaging applications
Chitosan	Dentistry, parts of microbial-cell biosensors
CuO	Medical field, food packaging applications

14.7 SUMMARY

The exceptional and exclusive properties of NMs like size, surface area and quantum confinement makes them promising for various applications. The antimicrobial properties of engineered NMs are really fascinating and there are possibilities of opportunities in this area that are unexplored. Here we discussed the unique characteristics of NMs which make them predominant among the materials. Further explanations are given for the positive and negative impact of the antimicrobial NMs, their possible mechanisms, advantages, and disadvantages, evaluation of antimicrobial effect and potential applications. The existing bioactivity of NPs can be enhanced with functionalization and derivatization. Microbial biosensors and water disinfection and treatment are the imperative areas where the antimicrobial impact of NMs are utilized. Microbial biosensors can be effectively utilized to observe the presence of microbes in environmental samples. Innovative antimicrobial NMs can be employed for water disinfection and prevention of biofilm formation in underwater pipelines. However, there are serious limitations in the use of existing technologies. Researches have to be conducted to preclude the dispersion and aggregation of NMs. The loss of material can be prevented by retaining NPs. Novel engineered NMs are to be developed by functionalization, impregnation of NM into filter media, etc. Mechanisms of action of NMs on microbes are a partially revealed area. Suggested mechanisms concentrate on the disintegration of cells and cellular components. More explorations should be conducted on the possibilities of mutation and genetic changes, production of secondary metabolites during inter and intra cellular interactions. Interaction of NPs like alumina with microbes can transfer genetic materials called Horizontal Gene Transfer, which is a significant area where researches are to be conducted. Examinations should be steered for elucidating the inhibitory mechanisms of NMs on preventing the production of useful metabolites called public goods by microorganisms. Research can be extended to analyze the possibility of using CNTs for carrying electricity in microbial fuel cells. Investigations and pilot plant studies should be conducted to scale up the laboratory processes to industrial needs. Further research should be conducted for a safe and economic use of these novel technologies for the benefit of mankind. Many more scopes and opportunities are to be explored in this area to effectively eradicate the current limitations of existing systems.

ACKNOWLEDGMENT

The authors would like to sincerely acknowledge the support from colleagues of Government Engineering College, Thrissur for the successful completion of the work.

KEYWORDS

- **antimicrobial mechanism**
- **antimicrobial nanomaterials**
- **carbon nanomaterials**
- **minimum inhibitory concentration**
- **multi-walled carbon nanotubes**
- **reactive oxygen species**

REFERENCES

1. Abbasian, F., Lockington, R., Palanisami, T., Megharaj, M., & Naidu, R., (2016). Multiwall carbon nanotubes increase the microbial community in crude oil contaminated freshwater sediments. *Science of the Total Environment, 539*(1), 370–380. https://doi.org/10.1016/j.scitotenv.2015.09.031.
2. Ahamed, M., Alhadlaq, H. A., Khan, M. A. M., Karuppiah, P., & Al-Dhabi, N. A., (2014). Synthesis, characterization, and antimicrobial activity of copper oxide nanoparticles. *Journal of Nanomaterials, 2014*, 1–4. https://doi.org/10.1155/2014/637858, PubMed: 637858.

3. Ahmad, T., Wani, I. A., Manzoor, N., Ahmed, J., & Asiri, A. M., (2013). Biosynthesis, Structural characterization and antimicrobial activity of gold and silver nanoparticles. *Colloids and Surfaces. B, Biointerfaces, 107*, 227–234. https://doi.org/10.1016/j.colsurfb.2013.02.004.
4. Akhavan, O., & Ghaderi, E., (2010). Toxicity of graphene and graphene oxide nanowalls against bacteria. *ACS Nano, 4*(10), 5731–5736. https://doi.org/10.1021/nn101390x.
5. Akhavan, O., Ghaderi, E., Emamy, H., & Akhavan, F., (2013). Genotoxicity of graphene nanoribbons in human mesenchymal stem cells. *Carbon, 54*(2), 419–431. https://doi.org/10.1016/j.carbon.2012.11.058.
6. Alvarez, L. H., & Cervantes, F. J., (2012). Assessing the impact of alumina nanoparticles in an anaerobic consortium: Methanogenic and humus reducing activity. *Applied Microbiology and Biotechnology, 95*(5), 1323–1331. https://doi.org/10.1007/s00253-011-3759-4.
7. Atmaca, S., Kadri, G. Ü. L., & Cicek, R., (1998). The effect of zinc on microbial growth. *Turkish Journal of Medical Sciences, 28*(6), 595–598.
8. Avanasi, R., Jackson, W. A., Sherwin, B., Mudge, J. F., & Anderson, T. A., (2014). C60 fullerene soil sorption, biodegradation, and plant uptake. *Environmental Science and Technology, 48*(5), 2792–2797. https://doi.org/10.1021/es405306w.
9. Awwad, A. M., Salem, N. M., Aqarbeh, M. M., & Abdulaziz, F. M., (2020). Green synthesis, characterization of silver sulfide nanoparticles and antibacterial activity evaluation. *Chemistry International, 6*(1), 42–48.
10. Azam, A., Ahmed, A. S., Oves, M., Khan, M. S., & Memic, A., (2012). Size-dependent antimicrobial properties of CuO nanoparticles against gram-positive and -negative bacterial strains. *International Journal of Nanomedicine, 7*, 3527–3535. https://doi.org/10.2147/IJN.S29020.
11. Bang, S. S., Lippert, J. J., Yerra, U., Mulukutla, S., & Ramakrishnan, V., (2010). Microbial calcite, a bio-based smart nanomaterial in concrete remediation. *International Journal of Smart and Nano Materials, 1*(1), 28–39. https://doi.org/10.1080/19475411003593451.
12. Bernhardt, E. S., Colman, B. P., Hochella, M. F., Cardinale, B. J., Nisbet, R. M., Richardson, C. J., & Yin, L., (2010). An ecological perspective on nanomaterial impacts in the environment. *Journal of Environmental Quality, 39*(6), 1954–1965. https://doi.org/10.2134/jeq2009.0479.
13. Bhatt, J. C., & Kholiya, K., (2014). Effect of size on the elastic and thermodynamic properties of nanomaterials. *Indian Journal of Pure and Applied Physics, 52*(09), 604–608.
14. Chatterjee, A. K., Chakraborty, R., & Basu, T., (2014). Mechanism of antibacterial activity of copper nanoparticles. *Nanotechnology, 25*(13), 135101. https://doi.org/10.1088/0957-4484/25/13/135101.
15. Chen, C., Tsyusko, O. V., McNear, Jr., D. H., Judy, J., Lewis, R. W., & Unrine, J. M., (2017). Effects of biosolids from a wastewater treatment plant receiving manufactured nanomaterials on *Medicago truncatula and* associated soil microbial communities at low nanomaterial concentrations. *Science of the Total Environment, 609*, 799–806. https://doi.org/10.1016/j.scitotenv.2017.07.188.
16. Chen, H., Wang, B., Gao, D., Guan, M., Zheng, L., Ouyang, H., Chai, Z., et al., (2013). Broad-spectrum antibacterial activity of carbon nanotubes to human gut bacteria. *Small, 9*(16), 2735–2746. https://doi.org/10.1002/smll.201202792.

17. Chen, M., Sun, Y., Liang, J., Zeng, G., Li, Z., Tang, L., Zhu, Y., et al., (2019). Understanding the influence of carbon nanomaterials on microbial communities. *Environment International, 126*, 690–698. https://doi.org/10.1016/j.envint.2019.02.005.
18. Chen, M., Zeng, G., Xu, P., Lai, C., & Tang, L., (2017). How do enzymes "meet" nanoparticles and nanomaterials? *Trends in Biochemical Sciences, 42*(11), 914–930. https://doi.org/10.1016/j.tibs.2017.08.008.
19. Chen, M., Zhou, S., Zhu, Y., Sun, Y., Zeng, G., Yang, C., Xu, P., et al., (2018). Toxicity of carbon nanomaterials to plants, animals and microbes: Recent progress from 2015-present. *Chemosphere, 206*(1), 255–264. https://doi.org/10.1016/j.chemosphere.2018.05.020.
20. Chen, P., (2012). Detection of DNA damage response caused by different forms of titanium dioxide nanoparticles using sensor cells. *Journal of Biosensors and Bioelectronics, 3*(5). https://doi.org/10.4172/2155-6210.1000129.
21. Demirkol, D. O., & Timur, S., (2011). Chitosan matrices modified with carbon nanotubes for use in mediated microbial biosensing. *Microchimica Acta, 173*(3, 4), 537–542. https://doi.org/10.1007/s00604-011-0596-1.
22. Deryabin, D. G., Aleshina, E. S., Vasilchenko, A. S., Deryabina, T. D., Efremova, L. V., Karimov, I. F., & Korolevskaya, L. B., (2013). Investigation of copper nanoparticles antibacterial mechanisms tested by luminescent *Escherichia coli* strains. *Nanotechnologies in Russia, 8*(5, 6), 402–408. https://doi.org/10.1134/S1995078013030063.
23. Deryabin, D. G., Efremova, L. V., Vasilchenko, A. S., Saidakova, E. V., Sizova, E. A., Troshin, P. A., Zhilenkov, A. V., & Khakina, E. A., (2015). A zeta potential value determines the aggregate's size of penta-substituted [60]fullerene derivatives in aqueous suspension, whereas a positive charge is required for toxicity against bacterial cells. *Journal of Nanobiotechnology, 13*, 50. https://doi.org/10.1186/s12951-015-0112-6.
24. Du, J., Hu, X., & Zhou, Q., (2015). Graphene oxide regulates the bacterial community and exhibits property changes in soil. *RSC Advances, 5*(34), 27009–27017. https://doi.org/10.1039/C5RA01045D.
25. Essa, A. M., & Khallaf, M. K., (2016). Antimicrobial potential of consolidation polymers loaded with biological copper nanoparticles. *BMC Microbiology, 16*(1), 144. https://doi.org/10.1186/s12866-016-0766-8.
26. Fernando, S. S. N., Gunasekara, T. D. C. P., & Holton, J., (2018). Antimicrobial nanoparticles: Applications and mechanisms of action. *Sri Lankan Journal of Infectious Diseases, 8*(1), 2–11. https://doi.org/10.4038/sljid.v8i1.8167.
27. Franklin, N. M., Rogers, N. J., Apte, S. C., Batley, G. E., Gadd, G. E., & Casey, P. S., (2007). Comparative toxicity of nanoparticulate ZnO, bulk ZnO, and ZnCl2 to a freshwater microalga (*Pseudokirchneriella subcapitata*): The importance of particle solubility. *Environmental Science and Technology, 41*(24), 8484–8490. https://doi.org/10.1021/es071445r.
28. Ge, Y., Shen, C., Wang, Y., Sun, Y. Q., Schimel, J. P., Gardea-Torresdey, J. L., & Holden, P. A., (2018). Carbonaceous nanomaterials have higher effects on soybean rhizosphere prokaryotic communities during the reproductive growth phase than during vegetative growth. *Environmental Science and Technology, 52*(11), 6636–6646. https://doi.org/10.1021/acs.est.8b00937.

29. Gelover, S., Gómez, L. A., Reyes, K., & Teresa, L. M. T., (2006). A practical demonstration of water disinfection using TiO2 films and sunlight. *Water Research, 40*(17), 3274–3280. https://doi.org/10.1016/j.watres.2006.07.006.
30. Gurunathan, S., Han, J. W., Dayem, A. A., Eppakayala, V., & Kim, J. H., (2012). Oxidative stress-mediated antibacterial activity of graphene oxide and reduced graphene oxide in *Pseudomonas aeruginosa. International Journal of Nanomedicine, 7*, 5901–5914. https://doi.org/10.2147/IJN.S37397.
31. Hai, R., Wang, Y., Wang, X., Du, Z., & Li, Y., (2014). Impacts of multiwalled carbon nanotubes on nutrient removal from wastewater and bacterial community structure in activated sludge. *PLOS One, 9*(9), e107345. https://doi.org/10.1371/journal.pone.0107345.
32. Harikumar, P. S. (2016). Antibacterial activity of copper nanoparticles and copper nanocomposites against *Escherichia Coli* bacteria. *International Journal of Sciences, 2*(2), 83–90. https://doi.org/10.18483/ijSci.957.
33. Hao, Y., Ma, C., Zhang, Z., Song, Y., Cao, W., Guo, J., Zhou, G., Rui, Y., Liu, L., & Xing, B., (2018). Carbon nanomaterials alter plant physiology and soil bacterial community composition in a rice-soil-bacterial ecosystem. *Environmental Pollution, 232*, 123–136. https://doi.org/10.1016/j.envpol.2017.09.024.
34. Hembram, K. C., Kumar, R., Kandha, L., Parhi, P. K., Kundu, C. N., & Bindhani, B. K., (2018). Therapeutic perspective of plant-induced silver nanoparticles: Application as antimicrobial and anticancer agent. *Artificial Cells, Nanomedicine, and Biotechnology, 46*(sup3), S38–S51. https://doi.org/10.1080/21691401.2018.1489262.
35. Hnaien, M., Lagarde, F., Bausells, J., Errachid, A., & Jaffrezic-Renault, N., (2011). A new bacterial biosensor for trichloroethylene detection based on a three-dimensional carbon nanotubes bioarchitecture. *Analytical and Bioanalytical Chemistry, 400*(4), 1083–1092. https://doi.org/10.1007/s00216-010-4336-x.
36. Hossain, F., Perales-Perez, O. J., Hwang, S., & Román, F., (2014). Antimicrobial nanomaterials as water disinfectant: Applications, limitations and future perspectives. *Science of the Total Environment, 466, 467*(1), 1047–1059. https://doi.org/10.1016/j.scitotenv.2013.08.009.
37. Hoseinnejad, M., Jafari, S. M., & Katouzian, I., (2018). Inorganic and metal nanoparticles and their antimicrobial activity in food packaging applications. *Critical Reviews in Microbiology, 44*(2), 161–181. https://doi.org/10.1080/1040841X.2017.1332001.
38. Jamshidi, A., & Jahangiri-, M., (2014). Synthesis of copper nanoparticles and its antibacterial activity against *Escherichia coli. Asian Journal of Biological Sciences, 7*(4), 183–186. https://doi.org/10.3923/ajbs.2014.183.186.
39. Jahangirian, H., Haron, M. J., Ismail, M. H. S., Moghaddam, R. R., Hejri, L. A., Abdollahi, Y., Rezayi, M., & Vafaei, N. (2013). Well diffusion method for evaluation of antibacterial activity of copper phenyl fatty hydroxamate synthesized from canola and palm kernel oils. *Digest Journal of Nanomaterials and Biostructures, 8*(3), 1263–1270.
40. Jeong, E., Im, W. T., Kim, D. H., Kim, M. S., Kang, S., Shin, H. S., & Chae, S. R., (2014). Different susceptibilities of bacterial community to silver nanoparticles in wastewater treatment systems. *Journal of Environmental Science and Health. Part A, Toxic/Hazardous Substances and Environmental Engineering, 49*(6), 685–693. https://doi.org/10.1080/10934529.2014.865454.
41. Jin, L., Son, Y., DeForest, J. L., Kang, Y. J., Kim, W., & Chung, H., (2014). Single-walled carbon nanotubes alter soil microbial community composition. *Science of the Total Environment, 466, 467*(1), 533–538. https://doi.org/10.1016/j.scitotenv.2013.07.035.

42. Jones, N., Ray, B., Ranjit, K. T., & Manna, A. C., (2008). Antibacterial activity of ZnO nanoparticle suspensions on a broad spectrum of microorganisms. *FEMS Microbiology Letters, 279*(1), 71–76. https://doi.org/10.1111/j.1574-6968.2007.01012.x.
43. Kamran, U., Bhatti, H. N., Iqbal, M., Jamil, S., & Zahid, M., (2019). Biogenic synthesis, characterization and investigation of photocatalytic and antimicrobial activity of manganese nanoparticles synthesized from *Cinnamomum verum* Bark extract. *Journal of Molecular Structure, 1179*(5), 532–539. https://doi.org/10.1016/j.molstruc.2018.11.006.
44. Kang, S., Herzberg, M., Rodrigues, D. F., & Elimelech, M., (2008). Antibacterial effects of carbon nanotubes: Size does matter! *Langmuir, 24*(13), 6409–6413. https://doi.org/10.1021/la800951v.
45. Kang, S., Mauter, M. S., & Elimelech, M., (2009). Microbial cytotoxicity of carbon-based nanomaterials: Implications for river water and wastewater effluent. *Environmental Science and Technology, 43*(7), 2648–2653. https://doi.org/10.1021/es8031506.
46. Kang, S., Zhu, Y., Chen, M., Zeng, G., Li, Z., Zhang, C., & Xu, P., (2019). Can microbes feed on environmental carbon nanomaterials? *Nano Today, 25*, 10–12. https://doi.org/10.1016/j.nantod.2019.01.001.
47. Kerfahi, D., Tripathi, B. M., Singh, D., Kim, H., Lee, S., Lee, J., & Adams, J. M., (2015). Effects of functionalized and raw multi-walled carbon nanotubes on soil bacterial community composition. *PLOS One, 10*(3), e0123042. https://doi.org/10.1371/journal.pone.0123042.
48. Khodakovskaya, M. V., Kim, B. S., Kim, J. N., Alimohammadi, M., Dervishi, E., Mustafa, T., & Cernigla, C. E., (2013). Carbon nanotubes as plant growth regulators: Effects on tomato growth, reproductive system, and soil microbial community. *Small, 9*(1), 115–123. https://doi.org/10.1002/smll.201201225.
49. Kikuchi, Y., Sunada, K., Iyoda, T., Hashimoto, K., & Fujishima, A., (1997). Photocatalytic bactericidal effect of TiO2 thin films: Dynamic view of the active oxygen species responsible for the effect. *Journal of Photochemistry and Photobiology A, 106*(1–3), 51–56. https://doi.org/10.1016/S1010-6030(97)00038-5.
50. Klaine, S. J., Alvarez, P. J., Batley, G. E., Fernandes, T. F., Handy, R. D., Lyon, D. Y., Mahendra, S., et al., (2008). Nanomaterials in the environment: Behavior, fate, bioavailability, and effects. *Environmental Toxicology and Chemistry, 27*(9), 1825–1851. https://doi.org/10.1897/08-090.1.
51. Kleandrova, V., Luan, F., Speck-Planche, A., & Cordeiro, M. N. D. S., (2015). Review of structures containing fullerene-C60 for delivery of antibacterial agents. Multitasking model for computational assessment of safety profiles. *Current Bioinformatics, 10*(5), 565–578. https://doi.org/10.2174/1574893610666151008011628.
52. La Russa, M. F., Ruffolo, S. A., Rovella, N., Belfiore, C. M., Palermo, A. M., Guzzi, M. T., & Crisci, G. M., (2012). Multifunctional TiO2 coatings for cultural heritage. *Progress in Organic Coatings, 74*(1), 186–191. https://doi.org/10.1016/j.porgcoat.2011.12.008.
53. Lawrence, J. R., Swerhone, G. D. W., Dynes, J. J., Hitchcock, A. P., & Korber, D. R., (2016). Complex organic corona formation on carbon nanotubes reduces microbial toxicity by suppressing reactive oxygen species production. *Environmental Science: Nano, 3*(1), 181–189. https://doi.org/10.1039/C5EN00229J.
54. Li, J., Wang, G., Zhu, H., Zhang, M., Zheng, X., Di, Z., Liu, X., & Wang, X., (2014). Antibacterial activity of large-area monolayer graphene film manipulated by charge transfer. *Scientific Reports, 4*(1), 4359. https://doi.org/10.1038/srep04359.

55. Li, Q., Mahendra, S., Lyon, D. Y., Brunet, L., Liga, M. V., Li, D., & Alvarez, P. J., (2008). Antimicrobial nanomaterials for water disinfection and microbial control: Potential applications and implications. *Water Research, 42*(18), 4591–4602. https://doi.org/10.1016/j.watres.2008.08.015.
56. Liu, S., Wei, L., Hao, L., Fang, N., Chang, M. W., Xu, R., Yang, Y., & Chen, Y., (2009). Sharper and faster "nano darts" kill more bacteria: A study of antibacterial activity of individually dispersed pristine single-walled carbon nanotube. *ACS Nano, 3*(12), 3891–3902. https://doi.org/10.1021/nn901252r.
57. Luongo, L. A., & Zhang, X. J., (2010). Toxicity of carbon nanotubes to the activated sludge process. *Journal of Hazardous Materials, 178*(1–3), 356–362. https://doi.org/10.1016/j.jhazmat.2010.01.087.
58. Lyon, D. Y., Brunet, L., Hinkal, G. W., Wiesner, M. R., & Alvarez, P. J., (2008). Antibacterial activity of fullerene water suspensions (nC60) is not due to ROS-mediated damage. *Nano Letters, 8*(5), 1539–1543. https://doi.org/10.1021/nl0726398.
59. Manjunatha, S. B., Biradar, D. P., & Aladakatti, Y. R., (2016). Nanotechnology and its applications in agriculture: A review. *Journal of Farm Sciences, 29*(1), 1–13.
60. Masrahi, A., VandeVoort, A. R., & Arai, Y., (2014). Effects of silver nanoparticle on soil-nitrification processes. *Archives of Environmental Contamination and Toxicology, 66*(4), 504–513. https://doi.org/10.1007/s00244-013-9994-1.
61. Matsunaga, T., Tomoda, R., Nakajima, T., & Wake, H., (1985). Photoelectrochemical sterilization of microbial cells by semiconductor powders. *FEMS Microbiology Letters, 29*(1, 2), 211–214. https://doi.org/10.1111/j.1574-6968.1985.tb00864.x.
62. Meruvu, H., Vangalapati, M., Chippada, S. C., & Bammidi, S. R., (2011). Synthesis and characterization of zinc oxide nanoparticles and its antimicrobial activity against *Bacillus subtilis* and *Escherichia coli*. *Rasayan Journal of Chemistry, 4*(1), 217–222.
63. Monica, R. C., & Cremonini, R., (2009). Nanoparticles and higher plants. *Caryologia, 62*(2), 161–165. https://doi.org/10.1080/00087114.2004.10589681.
64. Mohanty, A., Wu, Y., & Cao, B., (2014). Impacts of engineered nanomaterials on microbial community structure and function in natural and engineered ecosystems. *Applied Microbiology and Biotechnology, 98*(20), 8457–8468. https://doi.org/10.1007/s00253-014-6000-4.
65. Morones, J. R., Elechiguerra, J. L., Camacho, A., Holt, K., Kouri, J. B., Ramírez, J. T., & Yacaman, M. J., (2005). The bactericidal effect of silver nanoparticles. *Nanotechnology, 16*(10), 2346–2353. https://doi.org/10.1088/0957-4484/16/10/059.
66. Nyberg, L., Turco, R. F., & Nies, L., (2008). Assessing the impact of nanomaterials on anaerobic microbial communities. *Environmental Science and Technology, 42*(6), 1938–1943. https://doi.org/10.1021/es072018g.
67. Ocsoy, I., Paret, M. L., Ocsoy, M. A., Kunwar, S., Chen, T., You, M., & Tan, W., (2013). Nanotechnology in plant disease management: DNA-directed silver nanoparticles on graphene oxide as an antibacterial against *Xanthomonas* perforans. *ACS Nano, 7*(10), 8972–8980. https://doi.org/10.1021/nn4034794.
68. Odaci, D., Timur, S., & Telefoncu, A., (2008). Bacterial sensors based on chitosan matrices. *Sensors and Actuators. Part B, 134*(1), 89–94. https://doi.org/10.1016/j.snb.2008.04.013.
69. Ortega-Morales, B. O., Reyes-Estebanez, M. M., Gaylarde, C. C., Camacho-Chab, J. C., Sanmartín, P., Chan-Bacab, M. J., Granados-Echegoyen, C. A., & Pereañez-Sacarias, J. E., (2018). Chapter 13; Antimicrobial properties of nanomaterials used to control

microbial colonization of stone substrata. In: Hosseini, M., & Karapanagiotis, I., (eds.), *Advanced Materials for the Conservation of Stone* (pp. 277–298). Springer International Publishing.

70. Page, K., Palgrave, R. G., Parkin, I. P., Wilson, M., Savin, S. L. P., & Chadwick, A. V., (2007). Titania and silver–titania composite films on glass—Potent antimicrobial coatings. *J. Mater. Chem., 17*(1), 95–104. https://doi.org/10.1039/B611740F.
71. Pal, S., Tak, Y. K., & Song, J. M., (2007). Does the antibacterial activity of silver nanoparticles depend on the shape of the nanoparticle? A study of the gram-negative bacterium *Escherichia coli. Applied and Environmental Microbiology, 73*(6), 1712–1720. https://doi.org/10.1128/AEM.02218-06.
72. Parise, A., Thakor, H., & Zhang, X., (2014). Activity inhibition on municipal activated sludge by single-walled carbon nanotubes. *Journal of Nanoparticle Research, 16*(1), 2159. https://doi.org/10.1007/s11051-013-2159-3.
73. Pathak, T. K., Kroon, R. E., Craciun, V., Popa, M., Chifiriuc, M. C., & Swart, H. C., (2019). Influence of Ag, Au and Pd noble metals doping on structural, optical and antimicrobial properties of zinc oxide and titanium dioxide nanomaterials. *Heliyon, 5*(3), e01333. https://doi.org/10.1016/j.heliyon.2019.e01333.
74. Peng, Z., Liu, X., Zhang, W., Zeng, Z., Liu, Z., Zhang, C., Liu, Y., et al., (2020). Advances in the application, toxicity and degradation of carbon nanomaterials in environment: A review. *Environment International, 134*, 105298. https://doi.org/10.1016/j.envint.2019.105298.
75. Qi, L., Xu, Z., Jiang, X., Hu, C., & Zou, X., (2004). Preparation and antibacterial activity of chitosan nanoparticles. *Carbohydrate Research, 339*(16), 2693–2700. https://doi.org/10.1016/j.carres.2004.09.007.
76. Rabea, E. I., Badawy, M. E. T., Stevens, C. V., Smagghe, G., & Steurbaut, W., (2003). Chitosan as antimicrobial agent: Applications and mode of action. *Biomacromolecules, 4*(6), 1457–1465. https://doi.org/10.1021/bm034130m.
77. Rawson, F. J., Garrett, D. J., Leech, D., Downard, A. J., & Baronian, K. H., (2011). Electron transfer from *Proteus vulgaris* to a covalently assembled, single-walled carbon nanotube electrode functionalized with osmium bipyridine complex: Application to a whole cell biosensor. *Biosensors and Bioelectronics, 26*(5), 2383–2389. https://doi.org/10.1016/j.bios.2010.10.016.
78. Reddy, K. M., Feris, K., Bell, J., Wingett, D. G., Hanley, C., & Punnoose, A., (2007). Selective toxicity of zinc oxide nanoparticles to prokaryotic and eukaryotic systems. *Applied Physics Letters, 90*(213902), 2139021–2139023. https://doi.org/10.1063/1.2742324, PubMed: 18160973.
79. Ren, G., Hu, D., Cheng, E. W., Vargas-Reus, M. A., Reip, P., & Allaker, R. P., (2009). Characterization of copper oxide nanoparticles for antimicrobial applications. *International Journal of Antimicrobial Agents, 33*(6), 587–590. https://doi.org/10.1016/j.ijantimicag.2008.12.004.
80. Ren, W., Ren, G., Teng, Y., Li, Z., & Li, L., (2015). Time-dependent effect of graphene on the structure, abundance, and function of the soil bacterial community. *Journal of Hazardous Materials, 297*(30), 286–294. https://doi.org/10.1016/j.jhazmat.2015.05.017.
81. Rhiem, S., Barthel, A. K., Meyer-Plath, A., Hennig, M. P., Wachtendorf, V., Sturm, H., Schäffer, A., & Maes, H. M., (2016). Release of 14C-labelled carbon nanotubes from polycarbonate composites. *Environmental Pollution, 215*, 356–365. https://doi.org/10.1016/j.envpol.2016.04.098.

82. Rhim, J. W., Hong, S. I., Park, H. M., & Ng, P. K., (2006). Preparation and characterization of chitosan-based nanocomposite films with antimicrobial activity. *Journal of Agricultural and Food Chemistry, 54*(16), 5814–5822. https://doi.org/10.1021/jf060658h.
83. Rodrigues, D. F., Jaisi, D. P., & Elimelech, M., (2013). Toxicity of functionalized single-walled carbon nanotubes on soil microbial communities: Implications for nutrient cycling in soil. *Environmental Science and Technology, 47*(1), 625–633. https://doi.org/10.1021/es304002q.
84. Seery, M. K., George, R., Floris, P., & Pillai, S. C., (2007). Silver doped titanium dioxide nanomaterials for enhanced visible light photocatalysis. *Journal of Photochemistry and Photobiology A, 189*(2, 3), 258–263. https://doi.org/10.1016/j.jphotochem.2007.02.010.
85. Šefčovičová, J., & Tkac, J., (2015). Application of nanomaterials in microbial-cell biosensor constructions. *Chemical Papers, 69*(1), 42–53. https://doi.org/10.2478/s11696-014-0602-2.
86. Shrestha, B., Acosta-Martinez, V., Cox, S. B., Green, M. J., Li, S., & Cañas-Carrell, J. E., (2013). An evaluation of the impact of multiwalled carbon nanotubes on soil microbial community structure and functioning. *Journal of Hazardous Materials, 261*, 188–197. https://doi.org/10.1016/j.jhazmat.2013.07.031.
87. Singh, J., Vishwakarma, K., Ramawat, N., Rai, P., Singh, V. K., Mishra, R. K., Kumar, V., et al., (2019). Nanomaterials and microbes' interactions: A contemporary overview. *3 Biotech*, 9(3), 68. Biotech Publishing. https://doi.org/10.1007/s13205-019-1576-0.
88. Smith, S. C., & Rodrigues, D. F., (2015). Carbon-based nanomaterials for removal of chemical and biological contaminants from water: A review of mechanisms and applications. *Carbon, 91*, 122–143. https://doi.org/10.1016/j.carbon.2015.04.043.
89. Sondi, I., & Salopek-Sondi, B., (2004). Silver nanoparticles as antimicrobial agent: A case study on *E. coli* as a model for gram-negative bacteria. *Journal of Colloid and Interface Science, 275*(1), 177–182. https://doi.org/10.1016/j.jcis.2004.02.012.
90. Song, W., & Ge, S., (2019). Application of antimicrobial nanoparticles in dentistry. *Molecules, 24*(6), 1033. https://doi.org/10.3390/molecules24061033.
91. Srivastava, A., Srivastava, O. N., Talapatra, S., Vajtai, R., & Ajayan, P. M., (2004). Carbon nanotube filters. *Nature Materials, 3*(9), 610–614. https://doi.org/10.1038/nmat1192.
92. Sung-Suh, H. M., Choi, J. R., Hah, H. J., Koo, S. M., & Bae, Y. C., (2004). Comparison of Ag deposition effects on the photocatalytic activity of nanoparticulate TiO2 under visible and UV light irradiation. *Journal of Photochemistry and Photobiology A, 163*(1, 2), 37–44. https://doi.org/10.1016/S1010-6030(03)00428-3.
93. Tong, Z., Bischoff, M., Nies, L., Applegate, B., & Turco, R. F., (2007). Impact of fullerene (C60) on a soil microbial community. *Environmental Science and Technology, 41*(8), 2985–2991. https://doi.org/10.1021/es061953l.
94. Ul-Islam, M., Shehzad, A., Khan, S., Khattak, W. A., Ullah, M. W., & Park, J. K., (2014). Antimicrobial and biocompatible properties of nanomaterials. *Journal of Nanoscience and Nanotechnology, 14*(1), 780–791. https://doi.org/10.1166/jnn.2014.8761.
95. Varghese, S., Kuriakose, S., & Jose, S., (2013). Antimicrobial activity of carbon nanoparticles isolated from natural sources against pathogenic gram-negative and gram-positive bacteria. *Journal of Nanoscience, 2013*, 1–5. https://doi.org/10.1155/2013/457865, PubMed: 457865.

96. Varier, K. M., Gudeppu, M., Chinnasamy, A., Thangarajan, S., Balasubramanian, J., Li, Y., & Gajendran, B., (2019). Chapter 12; Nanoparticles: Antimicrobial applications and its prospects. In: Naushad, M., Rajendran, S., & Gracia, F., (eds.), *Advanced Nanostructured Materials for Environmental Remediation* (pp. 321–355). Springer.
97. Vecitis, C. D., Zodrow, K. R., Kang, S., & Elimelech, M., (2010). Electronic-structure-dependent bacterial cytotoxicity of single-walled carbon nanotubes. *ACS Nano, 4*(9), 5471–5479. https://doi.org/10.1021/nn101558x.
98. Wu, F., You, Y., Werner, D., Jiao, S., Hu, J., Zhang, X., Wan, Y., Liu, J., Wang, B., & Wang, X., (2020). Carbon nanomaterials affect carbon cycle-related functions of the soil microbial community and the coupling of nutrient cycles. *Journal of Hazardous Materials, 390*, 122144. https://doi.org/10.1016/j.jhazmat.2020.122144.
99. Wu, Q., Miao, W., Zhang, Y., Gao, H., & Hui, D., (2020). Mechanical properties of nanomaterials: A review. *Nanotechnology Reviews, 9*(1), 259–273. https://doi.org/10.1515/ntrev-2020–0021.
100. Yadav, T., Mungray, A. A., & Mungray, A. K., (2016). Effect of multiwalled carbon nanotubes on UASB microbial consortium. *Environmental Science and Pollution Research International, 23*(5), 4063–4072. https://doi.org/10.1007/s11356-015-4385-y.
101. Yang, C., Mamouni, J., Tang, Y., & Yang, L., (2010). Antimicrobial activity of single-walled carbon nanotubes: Length effect. *Langmuir, 26*(20), 16013–16019. https://doi.org/10.1021/la103110g.
102. Yang, Y., Xu, M., Wall, J. D., & Hu, Z., (2012). Nanosilver impact on methanogenesis and biogas production from municipal solid waste. *Waste Management, 32*(5), 816–825. https://doi.org/10.1016/j.wasman.2012.01.009.
103. Yoon, K. Y., Hoon, B. J. H., Park, J. H., & Hwang, J., (2007). Susceptibility constants of *Escherichia coli* and *Bacillus subtilis* to silver and copper nanoparticles. *Science of the Total Environment, 373*(2, 3), 572–575. https://doi.org/10.1016/j.scitotenv.2006.11.007.
104. Zarzuela, R., Carbú, M., Gil, M. L. A., Cantoral, J. M., & Mosquera, M. J., (2017). CuO/SiO2 nanocomposites: A multifunctional coating for application on building stone. *Materials and Design, 114*(15), 364–372. https://doi.org/10.1016/j.matdes.2016.11.009.
105. Zhang, L., Jiang, Y., Ding, Y., Povey, M., & York, D., (2007). Investigation into the antibacterial behavior of suspensions of ZnO nanoparticles (ZnO nanofluids). *Journal of Nanoparticle Research, 9*(3), 479–489. https://doi.org/10.1007/s11051-006-9150-1.

CHAPTER 15

ANTIMICROBIAL POTENTIAL OF METALLIC NANO-STRUCTURES: SYNTHESIS, TYPES, APPLICATIONS, AND FUTURE PROSPECTS

SABEEN ASLAM, RAFIA REHMAN, MUHAMMAD USMAN ALVI, MARGHOOB AHMED, GHULAM MUSTAFA, MUHAMMAD SHAHID, SVETLANA IGNATOVA, and AFSAR BANO

ABSTRACT

Nanomaterials have been promising owing to their distinctive properties, including unique surface morphology and high surface area, and for their effective utilization in eco-friendly technologies. In the recent past, due to the extensive manifestation of multidrug-resistant microorganisms, there has been expanding attention to the employment of innovative nanostructured materials for antimicrobials properties. Nanomaterial-based microbial sensors have been designed for the direct determination of pathogenic bacteria. Moreover, nanomaterial–microbe interaction has a key role in the treatment of microbial diseases. Inorganic nanomaterials have been noticeably more stable in severe conditions than organic nanomaterials and turned out to be more exploited in antimicrobial studies. Hence, significant impacts have been reported for nanomaterials involving metals such as silver, copper, iron, gold, etc., against pathogens such as viruses, bacteria, and fungal species. The current chapter covers these aspects with major contents, including (i) introduction, (ii) scope of nanomaterials as antimicrobial agents, (iii) metal-based nanomaterial in microbial studies. Subheadings of this chapter include (a) transition metal-based nano-antimicrobial agents, (b)

Sustainable Nanomaterials for Biomedical Engineering: Impacts, Challenges, and Future Prospects.
Junaid Ahmad Malik, Megh R. Goyal & Mohamed Jaffer M. Sadiq (Eds.)

Alkaline earth metal-based nano-antimicrobial agents, (c) metal derivatives as nano-antimicrobial agents, (iv) nanomaterial-based microbial sensors, (v) impact of nanomaterials on microbial ecosystem, and (vi) summary.

15.1 INTRODUCTION

Nanomaterial–microbe interaction plays a major role in the treatment of microbial diseases [36]. Inorganic nanomaterials have been noticeably more stable in severe conditions than organic nanomaterials and turned to be more utilized in antimicrobial studies [42]. Additives of metal as well as metals oxide NPs (nanoparticles) are broadly used in numerous areas, including textiles, coating-based usages, fiber-reinforced plastic products and for biomedical purposes due to remarkable antimicrobial applications. Metal, and other related antimicrobial NPs also possess a variety of Nano medicinal applications. For the future of medicine and pharmaceutics, the production of fast and inexpensive inorganic antimicrobial drugs, including metal and oxide of metal NPs as alternatives to conventional antibiotics could be potent [7].

Using biomolecular interactions with biosensors and nanosensors based on nanomaterials transforms the information into observable boundaries for sensitive detection of pathogens of food and, thus, it's likely to be proposed that nanosensors are integrated to ensure food safety standards [36]. Moreover, nanomaterial-based microbial sensors have been designed for direct determination of pathogenic bacteria [13].

The current chapter describes the antimicrobial potential of nanomaterials with special focus on metals based nanomaterials. The antimicrobial potential of transition metals, alkaline earth metals, and some other metals based nanomaterials has been explored here with some details about nano-biosensors based on these metals. This chapter will be very helpful for researchers as well as students working in the field of nanotechnology, nano-biotechnology, and medical sciences.

15.2 AN INTRODUCTION TO MICROBES

The microbial world is diverse, vast, and dynamic. Bacteria, fungi, fungal spore, viruses, protozoa, microalgae, archaea, a wide variety of zoo-/ phytoplankton and many more organisms having different lifestyles and morphologies are included in microbes [3, 34]. All other life-forms, including the human body depends on the metabolic activity of microbes [3].

Microbes are often dreaded as they may cause diseases [3]. Pathogenic microorganisms can cause disease in humans or animals [26]. Micro-sized bio-particles are not only composed of socio-economical resistant pathogens, such as bacteria or viruses (drug-resistant) but also include the huge variety of zoo-/phytoplankton or pathogenic fungal spores (shown in Figure 15.1, in relation to human cells and nanoparticles) [42]. A short description of some microorganisms is given in the following points.

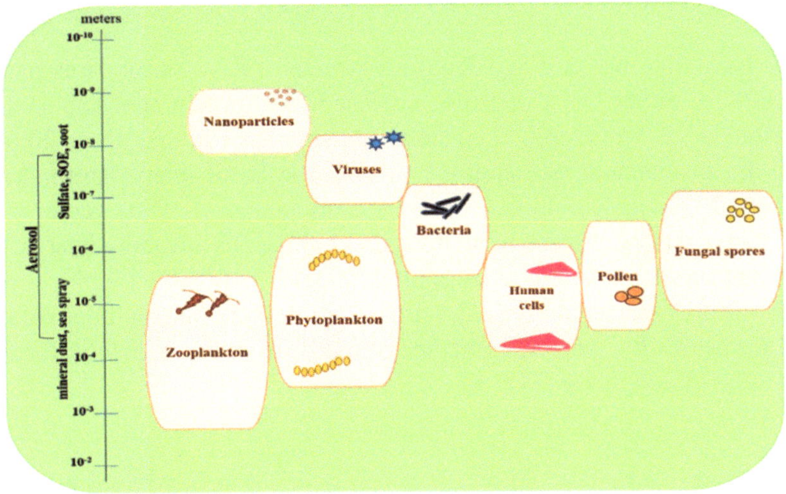

FIGURE 15.1 Size ranges of nanosized, microsized (bio)particles and (bio)aerosols.
Source: Modified after Ref. [42].

15.2.1 BACTERIA

Bacteria are smaller, unicellular, and not as much complex than eukaryotic cells, e.g., mammalian red blood cells. They exist as cocci, rods, helical forms and rarely as branching filaments [26].

15.2.2 BACTERIAL VIRUSES

A virus (description given below), inserts its genome into bacterium host, starting production of viral DNA and new viruses. Bacteriophages or bacterial viruses are viruses that pass on a disease to bacteria [3].

15.2.3 FUNGI

A large group of eukaryotic non-photosynthetic organisms, fungi that may be unicellular or multicellular include such various forms as, water molds, slime molds, mushrooms, puffballs, molds, bracket fungi and yeasts [3]. Fungi produce mycotoxins if present in stored food or on crops, e.g., nuts or grains, and some fungi invade tissues, thus causing disease in animals and humans [26].

15.2.4 VIRUSES

Viruses, unlike bacteria and fungi are not cells. A virion or virus particles consists of either DNA or RNA, enclosed in a protein coat and some are surrounded by envelopes. Typically, viruses range in size from 20 nm to 300 nm in diameter. Structurally, viruses consist of many shapes such as spherical, elongated, bullet-shaped or brick-shaped. Pathogenic viruses can cause serious disease in animals and humans by invading host cells. In humans and animals, few viruses are etiologically involved in the development of malignant tumors [26]. Viruses are obligate intracellular parasites. All forms of life comprising members of archae, eucarya, and bacteria may be diseased by viruses [3].

15.2.5 ALGAE

Algae, a physiologically and morphologically vast group of organisms are usually considered plant-like as they have chlorophyll. In water, many algae are free-living; some grows in the surface of rocks and also on various structures in the environment [26].

15.2.6 PROTOZOA

Protozoa are also a large group of microorganisms that can cause diseases, but they are usually harmless. For example, *Plasmodium falciparum* leads to malaria, Cryptosporidium, waterborne Giardia, and few species of Entamoeba be able to cause diarrhea [37.

15.3 BACKGROUND OF ANTIMICROBIAL STUDIES

Antimicrobial agents include antiseptics, disinfectants, and antibiotics [41]. Previously, diseases that caused morbidity and mortality on a large scale were brought under control and have changed the modern world [9].

Thousands of years ago, antimicrobial agents were applied in healing the wounds. During the 19th century, the detection of disinfectants and chemical preservatives and a better understanding of the nature of inflammation and infection allowed greater control on infections [5]. During the 20th century, the discovery and progress of antibiotics provided effective antimicrobial agents due to advanced medical therapy and marked the inadequacies of several previous remedies [5].

A number of antibiotics are available to control different diseases caused by human pathogenic microbes [33]. The topical antimicrobials often used in modern wound-healing are iodine and silver-containing products. The natural and chemicals compounds like acetic acid, honey, sodium hypochlorite, proflavine, chlorhexidine, hydrogen peroxide, and potassium permanganate have been used as antimicrobial agents [5].

Among these, plant-based products traditionally known to fight microbial infections are estimated to play a huge role incidentally [33]. An ancient remedy, honey has been re-discovered due to therapeutic properties such as antibacterial activity and the ability to cure wounds. Hydrogen peroxide (H_2O_2) has also been extensively used as a disinfectant and antiseptic [5].

It has been observed that the invented medicines are not much active in treatment, as the pathogenic microorganisms are becoming unaffected and adopting themselves into multidrug-resistant pathogens. To resist against these pathogenic microbes, there is a need of innovative antibiotics [33].

15.4 SCOPE OF NANOMATERIALS AS ANTIMICROBIAL AGENTS

To combat the pathogen infections, research mainly focuses on resistant pathogens which turned out to be the discovery of innovative methods. The use of nanotechnology in identifying how well the modification of catalyst for the production of NPs can be used for the diagnosis and treatment of pathogenic infections, has become one of the new fields evolving in response to this demanding threat. Research on the application of nanomaterials in the healthcare and biomedical areas is a fascinated field against resistant pathogens owing to factors such as doping effect, size effect and economic advantages along with prolonged and stable shelf-life for durable storage [41].

15.4.1 NANOMATERIALS

Nanomaterials refer to have at least one dimension ranging from 0.2–100 nm in the nanoscale and are named as nanomaterials or bionanomaterials if used in biomedicine [14, 25, 40].

15.4.2 NANOMATERIAL AND OTHER ANTIMICROBIAL AGENTS

Resistance to nanomaterials is tough to create as nano-antimicrobials are involved in several biological pathways of pathogenic microorganisms. These nanomaterials offer advantages over conventional antibiotics of being cost-effective, less toxic, a long shelf-life, prolonged effectiveness and safety for long-term storage [22, 27].

One of the innovative exploitation of nanomaterials as antimicrobial agents is against viruses, bacteria, and other pathogenic microbes [15]. The use of nanoparticles consists predominantly of metals and their oxides to kill pathogens, e.g., zinc, silver, copper, and titanium and are widely studied due to their stability, non-toxicity, effective biological properties, etc. [15, 41]. Nanomaterials have the potential to be utilized against microbes causing infections in different body parts of human (an overview is given in Figure 15.2).

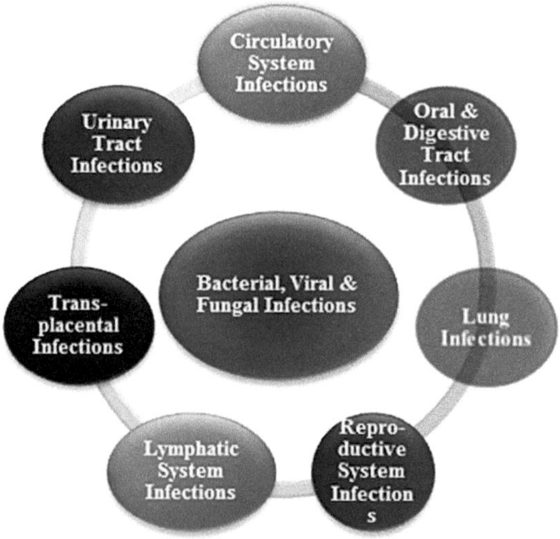

FIGURE 15.2 Potential of nano-antimicrobials in fighting infections in different parts of the human body.
Source: Adapted from: Ref. [19].

15.4.3 TYPES OF NANOMATERIALS

Nanomaterials are classified as organic and inorganic nanomaterials. Nanomaterials of both natures have shown effectiveness by disrupting microbial growth through different mechanisms. Both organic and inorganic nanomaterials have shown against microbial [40].

Different types of nanomaterials include carbon-based nanomaterials (carbon nanotubes and fullerenes), nanocomposites, dendrimers, metal-based nanoparticles, and natural nanoparticles [27]. Some of the types of nanomaterials are shown in Figure 15.3.

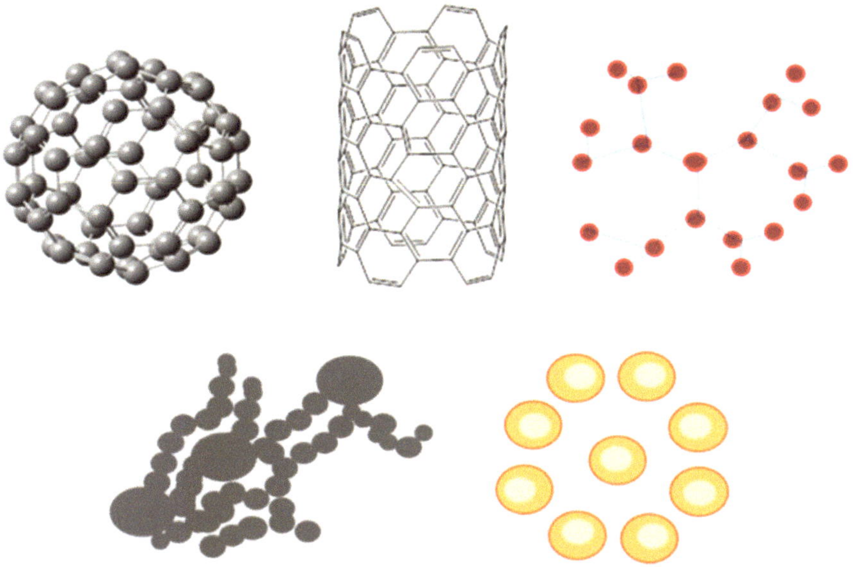

FIGURE 15.3 Schematic structures for different types of nanomaterials: (i) fullerene; (ii) SWCNT; (iii) dendrimer; (iv) nanocomposites; and (v) metal-based nanoparticles.
Source: Modified after Ref. [27].

15.4.3.1 ORGANIC NANOMATERIALS

Among organic nanomaterials, carbon-based nanoparticles have revealed potent microbicidal properties. Fullerenes, graphene oxide (GO) and single-walled carbon nanotubes (SWCNTs) nanoparticles showed high antimicrobial activity as reported earlier [8].

Organic antibacterial materials have less stable nature at high temperature, which ultimately cause barriers in designing products. Some studies have shown that carbon nanotubes (CNT) exploited the antibacterial properties against gram-negative and gram-positive bacteria; however, they are not widely applied due to their poor water dispersion. Similarly, Chitosan (Ch) NPs have demonstrated broad spectrum antifungal, antibacterial, and antiviral activities [40].

15.4.3.2 INORGANIC NANOMATERIALS

Inorganic nanomaterials like gold, zinc, silver, titanium, magnesium, alginate, copper, copper oxide and iron oxide have exhibited promising antimicrobial activities [4, 40]. As inorganic nanomaterials are more stable in harsh conditions, so they have been more frequently employed in antimicrobial materials [40].

15.4.4 SYNTHESIS OF NANOMATERIALS

The metallic nanoparticles and other nanomaterials synthesis can be accomplished by using chemical, biologically or physical pathways [36]. For the chemical synthesis of metallic nanoparticles, numerous methods have been published so for. Many of them require high temperatures and pressures. The physical methods involve arc discharge method, spray pyrolysis, mechanical grinding and many others. The use of natural extracts and microorganisms is an alternative approach for the biosynthesis of nanomaterials at neutral pH, ambient pressure and temperature [26]. In biosynthesis of nanoparticles, bottom-up technique is involved in which reducing and stabilizing agents are involved in the process of synthesis [25] while other methods involve bottom-down approach. A brief overview of all these methods is shown in Figure 15.4.

15.4.5 PREFERENCE OF INORGANIC NANOMATERIALS OVER ORGANIC NANOMATERIALS

Inorganic metal oxides have been progressively used as an antimicrobial properties. In comparison with organic oxides, inorganic oxides have much better effectiveness on drug-resistant microbial pathogens such as robustness, long shelf life and their stability. The important antimicrobial activity and selective toxicity to biological systems of inorganic metal oxide based nanomaterials such as SiO_2, MgO, TiO_2, ZnO, indicate their possible use in surgical devices, diagnostics, therapeutics, and nanotechnology-based antimicrobial drugs [28].

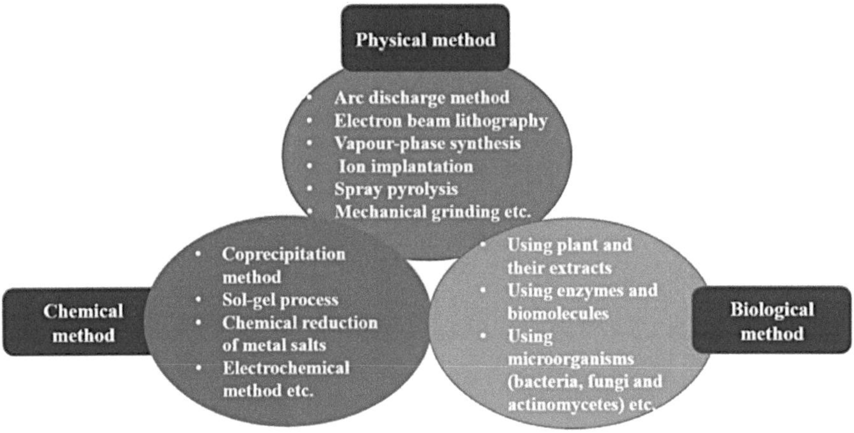

FIGURE 15.4 An overview of different approaches used for the synthesis of nanoparticles. *Source:* Adapted after Ref. [25].

Incidentally, how can combined approaches be employed on both variety of elements (fullerenes, Cu, Fe, Au, Ag, etc.), and biologics (antibiotic compounds and bacteriophage) along with shapes (nanorods, nanoplates, nanodarts, etc.), offer the best opportunity to design sustainable nano antimicrobials [6].

15.4.6 MECHANISM OF ACTION OF NANOMATERIALS AGAINST MICROBES

Nanomaterials are mainly proven to possess antifungal, antibacterial, and anti-viral properties. Nanomaterials are attacked on microbes either by direct attachment with microbial cell membrane, or they alter the membrane potential followed by damage of DNA, proteins, inhibition of binding of tRNA to ribosome along with decrease of ATP level or by generating ROS that eventually kill the cells (Figure 15.5) [30, 40].

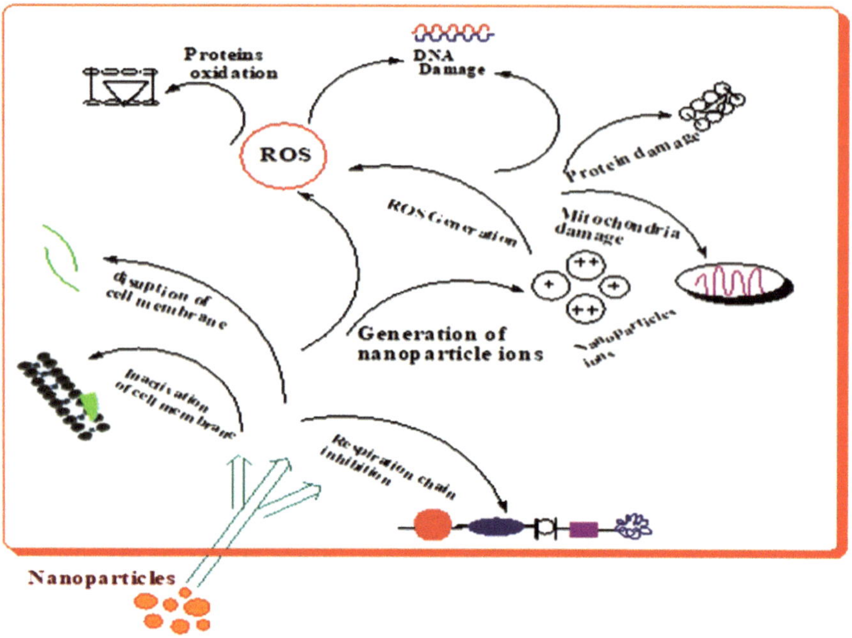

FIGURE 15.5 Diagrammatic representation of nanoparticles interactions with cells and their diverse antimicrobial mechanisms.

Both microbes and NMs have different physicochemical varieties and shapes. Moreover, microbial shapes demonstrated by rod-like bacteria versus coccal, may also impact interactions of NM–microbe [35, 40].

These diagrams point out that their attraction varies greatly from the microbial surface. For example, microbes are linked with silica nanoparticles in the subsequent manner: spores of fungal are larger than those of bacteria and microalgae [34]. It was therefore suggested, on the basis of colloidal electrostatics, positively charged nanomaterials may attach much more effectively to microbes than negatively charge nanomaterials. But, latest studies have revealed that several negatively charged NMs were able to sturdily bind to microbes such as algae, bacteria, or fungal spores, exhibiting negative surface charge [35]. Nanomaterials appeared to communicate with all sorts of bacteria [34]. Overall, nanomaterials having the antimicrobial activity generally depends on following two factors: (i) bacteria types for example gram-positive or gram-negative; and (ii) physicochemical characteristics of NMs, such as shape, size, and surface area (Figure 15.6) [22, 42].

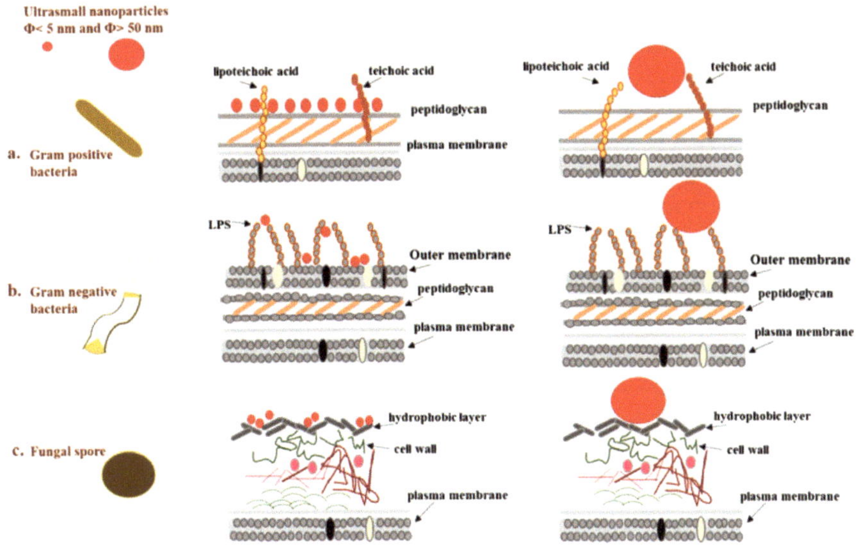

FIGURE 15.6 NM interaction depends on NP size and surface microbial structures. (a) Gram-positive bacteria; (b) gram-negative bacteria; (c) fungal spores. [Objects are not drawn to scale].
Source: Modified after Ref. [42].

15.5 METAL-BASED NANOMATERIAL IN MICROBIAL STUDIES

The antimicrobial effects of metals including gold (Au), copper (Cu), silver (Ag), titanium (Ti), and zinc (Zn), all having numerous properties, spectra of action and potencies, have been well-known for decades [7]. An innovative and modern approach of drug improvement is by using metallic NPs as novel designs of antimicrobial agents which have potency for sustainability solutions to health and infection care [28, 39].

Both intracellular and extracellular metal and metal oxide nanoparticles have been known for being influential antimicrobial agents that exploit on a wide range of target sites with excellent results, specific, and rapid action along with high biodegradability and bioavailability levels [27]. Nanoparticles consisting of metal ions are frequently detected to be non-toxic or inert. For example palladium, silver sulfide, iron, gold, and platinum have show poor solubility [36]. Recently it was observed that various metal nanomaterials, e.g., CaO, ZnO, MgO, Au, Cu, CuO, Ag, Ag_2O, Si, SiO_2, TiO_2, Fe_3O_4, Al_2O_3, and chitosan were known to exhibit antimicrobial activities [7, 10].

15.5.1 ANTIFUNGAL AND BACTERIAL MECHANISM

The inhibition mechanism of NPs versus various fungi and bacteria involves removal of metal ions that interfere with constituents of cells through several pathways which include the generation of reactive oxygen species (ROS) generation, DNA damage, creation of pores in cell the membranes, cell cycle arrest and cell wall damage and finally prevent the growth of cells [34].

15.5.2 ANTIVIRAL MECHANISM

Metal-based nanoparticles are representing remarkable antiviral effect against pathogens and for clinical purposes [7, 17]. Various viruses such as hepatitis B virus, HIV-1, influenza virus, the respiratory syncytial virus, the tacaribe virus, the monkeypox virus and herpes simplex virus type-1 (HSV-1) are killed by metal NPs [29].

In recent times, development in using metal nanoparticles as antiviral agents have been progressed rapidly due to the capability of a multi-target attack on viruses along with negligible influence on resistance development. Antiviral activity and the efficacy of metal nanoparticles influence on their metal ions, size, and form. Among all metal-based nanoparticles, Au, and Ag have established to show antiviral activity [17].

15.6 TRANSITION METAL-BASED NANO-ANTIMICROBIAL AGENTS

Recently, the transition metals NPs have been used frequently in nano-biotechnology [29]. Usually, the metals that are increasingly considered to be antimicrobial agents are inside the d-block transition metals (Au, Hg V, Co, Ti, Zn, Tb, Cr, Ni, Cu, W, Ag, Cd) [39].

Contrary to other antibacterial agents, transition metal NPs have shown a wide spectrum of antibacterial, antifungal, and antiviral activity [38]. Some of such transition metal nanoparticles are described in detail in the following points.

15.6.1 SILVER NANOPARTICLES (AGNPS)

The most common inorganic nanoparticles used as antimicrobial agents are Ag nanoparticles [7]. Among all metals and metal oxide-based nanoparticles, Ag_2O strongly kill all microbes such as viruses, fungi, and bacteria [7, 11, 22]. Silver nanoparticles (AgNPs) are also used as anti-inflammatory, anti-angiogenesis, antiplatelet, and anticancer agents due to their unique physicochemical properties and superior biological functions [11].

The Ag nanoparticles are active in killing and preventing the growth of bacteria like *Micrococcus luteus* and *E. coli* [34]. The antibacterial action of Ag nanoparticles results from the damage to the bacterial cell membrane. Ag NPs may cause bacterial membrane pits and gaps and then disrupt the cell [7]. Ag nanoparticles are shape-dependent antibacterial properties as in three different ways: truncated triangular, rod-shaped, and spherical [7]. Moreover, it is obvious that the smaller is its size, the greater the concentration and surface area of the silver NPs, higher is its antibacterial effect [27].

Antifungal activity of Ag NPs have been applied on different fungi comprising *Botrytis cinerea, Penicillium expansum, Trichophyton rubrum, Phomopsis spp., Candida albicans, Saccharomyces cerevisiae* and *Candida tropicalis* [34]. Ag nanoparticles have also been examined for their effect versus certain pathogenic yeasts. For example, pathogenic *Candida spp* shows higher antifungal activities by AgNPs [34]. AgNPs also revealed antiviral and preventive effects against H_3N_2 influenza virus infection [11].

15.6.2 GOLD NANOPARTICLES (AUNPS)

Gold NPs are clustered or colloidal particles consist of a gold core, a biocompatible and inert compound [31]. AuNPs are being modified to introduce antimicrobial properties [24]. It appears that Au nanoparticles are safer due to the ROS-independent mechanism for antimicrobial activity of mammalian cells than the other metals NPs [7].

Au nanoparticles are known to be more valuable due to their ability to functionalize as polyvalent effects, nontoxicity, photothermal activity and ease of detection [7]. The antibacterial activities of AuNPs functionalized with 5-fluorouracil against *S. aureus*, *Micrococcus luteus*, *E. coli* and *P. aeruginosa*. Au nanoparticle's antibacterial activity was due to the following factors: (i) These NPs are bound to the bacterial membrane, accompanied by membrane potential alteration and decrease in ATP level; (ii) inhibition of tRNA attachment to the ribosome [7].

AuNPs well-designed with 5-fluorouracil has shown antifungal activities against *Aspergillus fumigates* and *Aspergillus niger*. The Au nanoparticles stabilized with PEG inhibit virus fusion and have an antiviral effect against HIV-1; however, the exact mechanism is unclear. Studies have shown that AuNPs attach with gp120, inhibit viral entrance and avoid attachment to CD4 [23].

15.6.3 MANGANESE NANOPARTICLES

Manganese, in Manganese dioxide (MnO_2) and trimanganese tetroxide (Mn_3O_4) forms has also been reported to have antimicrobial properties [41].

Among the metals studied in nanotechnology, manganese dioxide (MnO_2) is one of the most desirable inorganic materials, presenting photocatalysis and antimicrobial behavior. This is primarily owing to its vast chemical and physical properties, including its ion exchange properties, imaging contrast, energy storage and medicinal applications [18].

The bio-functionalized MONPs have demonstrated superior activity against pathogenic microorganisms such as fungal strains, gram negative and gram-positive bacteria than non-functionalized manganese oxide nanoparticles and even standard drugs. Therefore, it can be concluded that functionalized manganese oxide nanoparticles have displayed higher bactericidal activity through bacterial growth resistance than with non-functionalized nanoparticles and against some funguses including *Candida albicans*, *Aspergillus, niger Curvularia lunata* and *Trichophyton simii* [12].

It has been evaluated that Mn_3O_4 NPs are more sensitive in terms of bacterial activity against *S. aureus* and *E. coli* bacteria [41].

15.6.4 IRON NANOPARTICLES

Iron oxide (III) crystallizes in hexagonal shape and is present as hematite α-Fe_2O_3 mineral nature, a stable oxide. Different forms of nanostructured iron oxide are nanotubes, nanowires, nanospheres, etc. The bactericidal activity of Fe_2O_3 NPs against *E. coli* and *S. aureus* has been reported, wherever an increase of such effect is detected as iron oxide NPs concentration is increased [41]. Fe_3O_4 NPs showed antimicrobial activity against several bacteria such as *S. epidermidis*, *E. coli, S. aureus*, *Xanthomonas*, and *P. vulgaris* have been established by many groups [29].

ROS, singlet oxygen ($^1O_2$), oxidative stress, H_2O_2, superoxide radicals ($O_2^{\cdot-}$) or OH^- can be the modes of antimicrobial action [42].

15.6.5 COPPER NANOPARTICLES

Cooper is considered as an excellent applicant for the manufacture of metal-based NPs. In addition to being extremely heat resistant, it is stable, inexpensive, robust, and simple to synthesize as well [31]. Cu NPs are of great interest to the scientists due to their excellent antimicrobial activities against microbes such as bacteria, viruses, algae, and fungi, along with improved physical, chemical, and biological activities [4, 7].

Cu nanoparticles have considerable potential as a bactericidal agent [27]. Copper, in copper oxide (CuO) form is much cheaper than Ag and other bactericidal NMs [24].

The antifungal and antibacterial activities of CuNPs assessed on many microbes, comprising *P. aeruginosa*, *Salmonella choleraesuis*, *B. subtilis*, methicillin resistant *S. aureus*, *C. albicans*, *E. faecalis*, *P. aeruginosa*, *K. pneumonia* and *E. coli*, The effects showed the excellent ability of these NPs as antimicrobial agents [7]. Among the pathogens here, *E. faecalis* and *E. coli* exhibited the maximum sensitivity to CuO NPs, whereas *K. pneumoniae* were sensitive to these NPs.

The attachment of NPs to bacterial cell walls is the mechanism of the antibacterial behavior of CuO NPs due to opposite electrical charges, which results in a reduction in the bacterial cell wall. Furthermore, generate ROS by Cu^{2+} ions, resulting in bacterial oxidative stress-induced DNA and damage membrane [7].

15.6.6 ZINC NANOPARTICLES

Zinc is involved in the catalytic action of many enzymes and is broadly distributed all over the body tissue in the organism [31]. It has been reported that Zinc, in the form of Zn nanoparticles (ZnNPs) and Zn oxide nanoparticles (ZnO NPs) exhibit antimicrobial activities against broad spectrum fungi and bacteria [24].

Antibacterial potential of ZnO nanoparticles against *Bacillus subtilis, Escherichia coli, Salmonella typhimurium,* and *Staphylococcus aureus* and found effective antibacterial activities of ZnO NPs against gram-negative and gram-positive bacteria, that increased with, increasing surface area, concentration, decreasing particle size and temperature fluctuations [40]. Comparative analysis of antimicrobial effect of ZnO nanoparticles against *E. coli* and *P. aeruginosa* (Gram-negative) and *S. aureus* and *B. subtilis* (Gram-positive) bacteria [7].

ZnO NPs mechanism of antimicrobial activity is associated to distraction of bacterial cell membrane, weakening hydrophobicity of cell surface, and the reduce regulation of resistance genes in bacteria and also enhance ROS production to promote bacterial death [29].

15.6.7 TITANIUM NANOPARTICLES

Titanium in the form of TiO_2 is a non-toxic and have important antimicrobial activity against definite microorganisms [29]. Photocatalytic TiO_2 have antimicrobial effect against *S. aureus*, fungi, *E. coli* and *Aspergillus niger* [29].

The ability of TiO_2 has been well recognized against bacteria [24]. Antimicrobial activity of TiO_2 depends on shape, size, crystal structure and thickness of pathogen's cellular membrane [7, 22, 23]. The nanocomposites involving TiO_2, DNA/RNA and poly-L-lysine (PL) to reveal antiviral properties against influenza A viruses (H_3N_2 strain) have been reported [29].

15.6.8 NICKLE NANOPARTICLES

Antibacterial activities against multidrug-resistant *E. coli* and *K. pneumonia* are well demonstrated by Nickel and $Ni(OH)_2$ nanoparticles. Ni-based NPs have a weaker antimicrobial potential than Ag and Si but are stronger than Au NPs. Innovative platinum-on-flower-like Nickel NPs have been synthesized using UV irradiation to stimulate ROS aggregation that ultimately increased antibacterial properties [19].

15.6.9 PALLADIUM NANOPARTICLES

Nanoparticles of Palladium have been synthesized and have shown size dependent antimicrobial activities against *S. aureus* and *E. coli* by chemical methods. Palladium NMs are therefore extremely poisonous to bacteria due to their greater reactivity, especially when they are considerably reduced in size [19].

15.7 ALKALINE EARTH METAL-BASED NANO-ANTIMICROBIAL AGENTS

Alkaline earth elements have antimicrobial potential and are adsorbed on the cell wall surface of microbes and interrupt the cell's respiration, blocking the protein synthesis that restrict further development of the organisms. These metal(II) ions are essential for the growth-inhibitor effect [13].

MgO and CaO significantly enhance antibacterial effect associated to active oxygen species and alkalinity [7]. These properties of alkaline earth metals are observed to enhance when synthesized on nanoscale in the form of oxides. Some of these metal oxide NPs are discussed in subsections.

15.7.1 CALCIUM OXIDE (CAO) NANOPARTICLES

CaO-nanoparticles have considerable antimicrobial activity. In an analysis, microwave irradiated CaO-NPs were used to investigate their antimicrobial efficacy against pathogenic yeast, gram positive and gram-negative bacteria. CaO-NPs showed antimicrobial effect towards tested organisms, including *Staphylococcus epidermidis, Pseudomonas aeruginosa* and *Candida tropicalis* in decreasing order [28].

The antibacterial mechanism of CaO NPs have been verified by the production of superoxide on the particle's surface. Mentioned results showed that CaO nanoparticles prove excellent antibacterial effects alone or in combination with other disinfectants [7].

15.7.2 MAGNESIUM OXIDE (MGO) NANOPARTICLES

MgO and hybrid Mg materials are most commonly used to tackle infection. Pure Mg is rarely used as an antimicrobial agent [19]. MgO is an antimicrobial nanoparticle and considered as a light metal that can be synthesized and completely resorbed in the body [21].

Antibacterial activity of MgO nanoparticles have been reviewed against endospore-forming bacteria [14–18] gram-positive, and gram-negative bacteria. The effects of MgO NPs on nine several types of pathogens in biofilms or planktonic forms, including gram-positive and gram-negative bacteria, their resistant strains and yeast were investigated [21].

MgO nanoparticles are reported to rupture the cellular membrane and then cause the intracellular material to leak, which ultimately kills bacterial cells [7]. The association between MgO and the cell membrane or cell wall may be the main mechanism for the noxious properties of MgO on phytoplankton bacteria [21].

15.7.3 STRONTIUM OXIDE (SRO) NANOPARTICLES

The SrO nanoparticles have demonstrated an important antibacterial effect against Gram-negative bacteria, for example, *P. vulgaris, K. pneumonia, P. aeruginosa*, and *M. morganii* than Gram-positive bacteria [1].

15.8 MISCELLANEOUS METALS DERIVATIVE NANOMATERIALS

A few other metals, metalloids, and metals derivative NPs including Ce, Si, I, Ba, Bi, Al, and Se are being increasingly studied as antimicrobial agents.

15.8.1 CERIUM (IV) OXIDE NANOPARTICLES

A good example of a redox-active nanoparticle is cerium (IV) oxide, with the potential to flip between 3+ and 4+ oxidation states. These nanoparticles have intense antimicrobial activity by direct and indirect interactions against Gram-negative and Gram-positive bacteria [19].

15.8.2 COPPER IODIDE NANOPARTICLES

Iodine and copper react to form copper iodide (CuI) nanoparticles. These NPs demonstrated strong antiviral effect against *F. calicivirus*, which might be attributed to the generation of Cu^+ ROS accompanied by capsid protein oxidation [29].

15.8.3 SILICON NANOPARTICLES

The non-toxicity of Si NPs is being used in the clinical field as antimicrobial agents. A number of studies indicate that the production of Si compounds along with metals, specifically their nano-composites demonstrates great capability to promote antimicrobial agents [7].

15.8.4 SILICON DIOXIDE (SIO_2) NANOPARTICLES

Silicon dioxide (SiO_2) has strong antimicrobial activity due to increased surface area, which is becoming more fundamental at the nano-scale. In comparison to traditional materials, silver zeolite has shown stronger antimicrobial properties of the SiO_2 nanocomposite against a wide variety of microbes. In a synthesized and explored the antimicrobial activity of nanostructures Au/SiO_2 and Ag/SiO_2 [7].

15.8.5 BARIUM ZIRCONATE TITANATE NANOPARTICLES

Barium zirconate titanate nano-powders are also included as inorganic materials and their antibacterial effect was observed on both types of bacteria. The data confirmed that nano-powders had adequate antibacterial properties along with mild hemolytic activity which is likely to make them a candidate in Decisive Data System (DD systems) as potential antibacterial agents [20].

15.8.6 BISMUTH NANOPARTICLES

Antimicrobial activity is demonstrated by Bismuth nanoparticles against several bacterial, fungal, and viral infections. Bismuth compounds are widely used to treat gastrointestinal infections [29].

At lower concentrations, BiNPs demonstrate antifungal (2 mM) and antibacterial (<1 mM) activity. Bi nanoparticles have the ability to treat stimulant bacteria and prevent the development of *Helicobacter pylori* by altering the Krebs cycle and the metabolism of nucleotides and amino acids. Therefore, BiNPs are promising materials to tackle many infections [29].

15.8.7 ALUMINUM NANOPARTICLES

Aluminum oxide (Al_2O_3), also known as corundum (a crystalline type of alumina) or alumina is a multi-phase white oxide and is thermally more stable over a wide temperature range [29]. Nanoparticles of aluminum oxide are broadly used in explosives, alloys, sensors, rocket fuel, personal care products, energy, ship wear-resistant coatings, and drug delivery systems [19].

The antimicrobial properties of AlNPs have been discussed in some studies. There was greater toxicity of AlNPs against *Pseudomonas fluorescens*, *B. subtilis* and *E. coli*, than the bulk products. Bacterial cell walls are interrupted by AlNPs, leading to the death of cells by ROS. Makvandi et al. [19]; and Rudramurthy et al. [29] have investigated the growth-inhibitory effect of aluminum oxide NPs over a wide-ranging concentration on *E. coli*.

15.8.8 SELENIUM NANOPARTICLES

Se NPs have anti-fungal properties; probiotics comprising Se nanoparticles have been shown to have a sharper decrease in the viability of *Candida albicans* than those lacking Se-enrichment. Such nanoparticles exhibit effective antibacterial action against *P. aeruginosa, E. coli, Streptococcus pyogenes, S. aureus,* and antifungal action against *Aspergillus clavatus*. Moreover, their antibacterial effectiveness is equivalent to that of ampicillin antibiotics [19].

15.9 NANOMATERIAL-BASED MICROBIAL SENSORS

In multiple sectors of life, the microorganism detection is of great importance due to health and safety concerns. Various nanomaterials have the properties to bind with the target biochemical released by microorganisms and then to translate these bindings into detectable signals. Fist property of nanomaterials, i.e., target binding with analyte is owing to their high surface area. These properties make them useful to be used not only as capturing probe but also as signal transducers [37].

This is due to these facts that a number of microorganism biosensors have been developed which involve nanomaterials [37]. In the food industry, these highly specific and sensitive nanoparticle-based biosensors have been finding extensive usage to detect and identify toxins due to pathogenic fungus and bacteria as well as chemicals present in the food samples as contaminants. Hence providing an effective substitute the existing laborious and time-consuming traditional methods [34]. Some examples of metal nanomaterials being used in biosensors are discussed here.

Gold nanomaterials are not only being used in common food packing material and as antimicrobial additive but have also been used as biosensors to detect main pollutants in foodstuffs [34]. The formerly used metal oxide semiconducting (MOS) gas sensors can detect microbial spoilage by incorporation of conducting polymers or by a blend of different types of sensors, including MOSFET (metal oxide semiconducting field effect transistors), infrared carbon dioxide sensors and MOS [16].

The NMs based sensor technology is advancing to search elements with distinctive physical and chemical characteristics, which could have fast response recognition as well high sensitivity [2]. These research studies would have the potential to enable researchers to design such nanomaterials which would have specific binding ability to certain microorganisms only. Also, such nanomaterials-based specific and selective sensors may also be designed to not only explore lively changes during the life cycle of microbes but also their structure and chemistry [35].

15.10 IMPACT OF NANOMATERIALS ON MICROBIAL ECOSYSTEM

It is evident from the discussion in the above sections that antimicrobial nanomaterials are very helpful in combating pathogenic microorganisms and have advantages of being able to overcome microbial resistance more effectively than other antibiotics, and are economically viable as well as able to administer antibiotic safely. However, the extensive utilization of nanomaterials in numerous commercial products as well as in the various fields of medical and biological sciences may result in the leakage and accumulation of these NMs or metal ions in the environment.

There are a number of microorganisms which are valuable for the environment, ecosystem, and hence humans. The beneficial roles of these microorganisms involve cycling of elements for plant growth, nitrogen fixation and bioremediation [22]. Hence, in recent years, numerous researchers have evaluated the toxicity of metal as well as non-metal nanomaterials to these microorganisms. The findings of these researches have indicated that most of the types of nanomaterials including metallic NMs have toxicity to the microbial cells. These findings lead to the fear that this nano-toxicity may affect the ecosystem by disturbing the normal nutrient and biological cycles through the unwanted death of microbes [32]. Therefore, researchers continue to examine the negative impacts of nanomaterials on the ecosystem, environment as well as human health [22]. There is a dire need to use these antimicrobial metallic nanomaterials in a controlled manner as well as to develop these materials with high selectivity and specificity for targeted microbe.

15.11 SUMMARY

Additives of metal and metal oxide NPs (nanoparticles) are broadly used in numerous fields such as textiles, coating-based usages, fiber reinforced plastic products and for biomedical purposes due to their remarkable antimicrobial applications. Metal and other related antimicrobial NPs also possess a variety of nano medicinal applications. For the future of medicine and pharmaceutics, the production of fast and inexpensive inorganic antimicrobial drugs, including metal and oxide of metal NPs as alternatives to conventional antibiotics could be potent.

Using bio-molecular interactions with biosensors and nano-sensors based on nanomaterials transforms the information into observable limits for sensitive detection of food pathogens, and thus it is likely to be proposed that nano-sensors are integrated to ensure food safety standards. Through this perspective, an attractive feature of food nanotechnology will be the advancement of nano-sensors to detect microbes and pollutants. In fact, we also anticipate that portable nanomaterial-based biosensors for microbial toxins will soon be realized with techniques that enable on-site application without carrying the samples to the laboratories with many functional principles of compact detection systems.

Some potential has been provided by NPs/nano-composites for the cure of many diseases which are difficult to manage/treat by recently developed techniques. However, certain issues must be resolved safely and effectively before using NPs/nano-composites. These concerns include a more thorough understanding of the mechanism of action of nano-composites, the enhancement of environmentally sustainable methods and their use for the production, environmental, and social consequences. A toxicity evaluation of nanoparticles is a necessary research field at the moment, focusing on the interaction of nanoparticles with biological systems. Indeed, it is not yet possible to investigate the real molecular targets and pathways for nanoparticles on cell membranes and cellular organelles.

The next generation of diagnostic techniques for pathogen detection of pathogens is being developed due to innovations in nanotechnology. In addition, it is strongly recommended to detect multiple agents in a single sample, sample processing, as well as to increase the selectivity and sensitivity of the therapeutics for significant application to challenging environmental microbes.

KEYWORDS

- **alkaline earth metals**
- **antimicrobial agents**
- **microbes**
- **nano-bio sensors**
- **nanomaterials**
- **transition metals**

REFERENCES

1. Apsana, G., Devanna, N., & Yuvasravana, R., (2018). Biomimetic synthesis and antibacterial properties of strontium oxide nanoparticles using *Ocimum sanctum* leaf extract. *Asian Journal of Pharmaceutical and Clinical Research, 11*(3), 384–389. https://doi.org/10.22159/ajpcr.2018.v11i3.20858.
2. Berekaa, M. M., (2016). Nanotechnology in wastewater treatment; influence of nanomaterials on microbial systems. *International Journal of Current Microbiology and Applied Sciences, 5*(1), 713–726. https://doi.org/10.20546/ijcmas.2016.501.072.
3. Bisen, P., Debnath, M., & Prasad, G., (eds.). (2012). *Microbes: Concepts and applications* (1st edn., p. 699). Wiley Online Library.
4. Cioffi, N., & Rai, M., (2012). *Nano-antimicrobials: Progress and Prospects* (1st edn., p. 556). Springer Science+Business Media.
5. Cooper, R., (2004). A review of the evidence for the use of topical antimicrobial agents in wound care. *World Wide Wounds, 1*, 1–8.
6. Díez-Pascual, A. M., (2018). *Antibacterial Activity of Nanomaterials* (1st edn., p. 328). Multidisciplinary Digital Publishing Institute.
7. Dizaj, S. M., Lotfipour, F., Barzegar-Jalali, M., Zarrintan, M. H., & Adibkia, K., (2014). Antimicrobial activity of the metals and metal oxide nanoparticles. *Materials Science and Engineering C, Materials for Biological Applications, 44*, 278–284. https://doi.org/10.1016/j.msec.2014.08.031.
8. Maleki, D. S. M., Mennati, A., Jafari, S., Khezri, K., & Adibkia, K., (2015). Antimicrobial activity of carbon-based nanoparticles. *Advanced Pharmaceutical Bulletin, 5*(1), 19–23. https://doi.org/10.5681/apb.2015.003.
9. Drusano, G. L., (2003). Prevention of resistance: A goal for dose selection for antimicrobial agents. *Clinical Infectious Diseases, 36*(Suppl 1), S42–S50. https://doi.org/10.1086/344653.
10. Graves, J. L., Thomas, M., & Ewunkem, J. A., (2017). Antimicrobial nanomaterials: Why evolution matters. *Nanomaterials, 7*(10), 283. https://doi.org/10.3390/nano7100283.
11. Gurunathan, S., Qasim, M., Choi, Y., Do, J. T., Park, C., Hong, K., Kim, J. H., & Song, H., (2020). Antiviral potential of nanoparticles-can nanoparticles fight against coronaviruses? *Nanomaterials, 10*(9), 1645. https://doi.org/10.3390/nano10091645.

12. Haneefa, M., Jayandran, M., Balasubramanian, M., (2017). Evaluation of antimicrobial activity of green-synthesized manganese oxide nanoparticles and comparative studies with curcuminaniline functionalized nanoform. *Asian Journal of Pharmaceutical and Clinical Research, 10*, 347–352.
13. Ibrahim, M. M., Al-Juaid, S., Mohamed, M. A., & Yassin, M. H., (2012). Synthesis, characterization, and antimicrobial activities of barbital-based alkaline earth metal complexes: The x-ray crystal structure of [ba2h (barb) 5](barb= 5, 5-diethyl barbiturate). *Journal of Coordination Chemistry, 65*(17), 2957–2971. https://doi.org/10.1080/00958972.2012.705833.
14. Kalashnikova, I., Das, S., & Seal, S., (2015). Nanomaterials for wound healing: Scope and advancement. *Nanomedicine, 10*(16), 2593–2612. https://doi.org/10.2217/NNM.15.82.
15. Kumar, R., Umar, A., Kumar, G., & Nalwa, H. S., (2017). Antimicrobial properties of ZnO nanomaterials: A review. *Ceramics International, 43*(5), 3940–3961. https://doi.org/10.1016/j.ceramint.2016.12.062.
16. Lakade, A. J., Sundar, K., & Shetty, P. H., (2017). Nanomaterial-based sensor for the detection of milk spoilage. *LWT, 75*, 702–709. https://doi.org/10.1016/j.lwt.2016.10.031.
17. Maduray, K., & Parboosing, R., (2021). Metal nanoparticles: A promising treatment for viral and arboviral infections. *Biological Trace Element Research, 199*(8), 3159–3176. https://doi.org/10.1007/s12011-020-02414-2.
18. Mahlangeni, N. T., Magura, J., Moodley, R., Baijnath, H., & Chenia, H., (2020). Biogenic synthesis, antioxidant and antimicrobial activity of silver and manganese dioxide nanoparticles using. *Cussonia zuluensis* strey. *Chemical Papers, 74*, 4253–4265.
19. Makvandi, P., Wang, C. Y., Zare, E. N., Borzacchiello, A., Niu, L. N., & Tay, F. R., (2020). Metal-based nanomaterials in biomedical applications: Antimicrobial activity and cytotoxicity aspects. *Advanced Functional Materials, 30*(22). https://doi.org/10.1002/adfm.201910021, PubMed: 1910021.
20. Mohseni, S., Aghayan, M., Ghorani-Azam, A., Behdani, M., & Asoodeh, A., (2014). Evaluation of antibacterial properties of barium zirconate titanate (BZT) nanoparticle. *Brazilian Journal of Microbiology, 45*(4), 1393–1399. https://doi.org/10.1590/s1517-83822014000400033.
21. Nguyen, N. T., Grelling, N., Wetteland, C. L., Rosario, R., & Liu, H., (2018). Antimicrobial activities and mechanisms of magnesium oxide nanoparticles (NMGO) against pathogenic bacteria, yeasts, and biofilms. *Scientific Reports, 8*(1), 16260. https://doi.org/10.1038/s41598-018-34567-5.
22. Omar, R. A., Afreen, S., Talreja, N., Chauhan, D., Ashfaq, M., & Srituravanich, W., (2019). Chapter 6; Impact of nanomaterials on the microbial system. In: Prasad, R., (ed.), *Microbial Nanobionics* (pp. 141–158). Springer.
23. Ortega-Morales, B., Reyes-Estebanez, M., Gaylarde, C., Camacho-Chab, J., Sanmartín, P., Chan-Bacab, M., Granados-Echegoyen, C., & Pereañez-Sacarias, J., (2018). Chapter 13; Antimicrobial properties of nanomaterials used to control microbial colonization of stone substrata. In: Hosseini, M., & Karapanagiotis, I., (eds.), *Advanced Materials for the Conservation of Stone* (pp. 277–298). Springer.
24. Pereira, L., Mehboob, F., Stams, A. J., Mota, M. M., Rijnaarts, H. H., & Alves, M. M., (2015). Metallic nanoparticles: Microbial synthesis and unique properties for biotechnological applications, bioavailability and biotransformation. *Critical Reviews in Biotechnology, 35*(1), 114–128. https://doi.org/10.3109/07388551.2013.819484.

25. Peters, R. J. B., Bouwmeester, H., Gottardo, S., Amenta, V., Arena, M., Brandhoff, P., Marvin, H. J. P., et al., (2016). Nanomaterials for products and application in agriculture, feed and food. *Trends in Food Science and Technology, 54*, 155–164. https://doi.org/10.1016/j.tifs.2016.06.008.
26. Quinn, P. J., Markey, B. K., Leonard, F. C., Hartigan, P., Fanning, S., & Fitzpatrick, E., (eds.). (2011). *Veterinary Microbiology and Microbial Disease* (2nd edn., p. 928). John Wiley & Sons.
27. Rai, M., Ingle, A. P., Gaikwad, S., Gupta, I., Gade, A., & Silvério Da, S. S., (2016). Nanotechnology-based anti-infectives to fight microbial intrusions. *Journal of Applied Microbiology, 120*(3), 527–542. https://doi.org/10.1111/jam.13010.
28. Roy, A., Gauri, S. S., Bhattacharya, M., & Bhattacharya, J., (2013). Antimicrobial activity of CaO nanoparticles. *Journal of Biomedical Nanotechnology, 9*(9), 1570–1578. https://doi.org/10.1166/jbn.2013.1681.
29. Rudramurthy, G. R., Swamy, M. K., Sinniah, U. R., & Ghasemzadeh, A., (2016). Nanoparticles: Alternatives against drug-resistant pathogenic microbes. *Molecules, 21*(7), 836. https://doi.org/10.3390/molecules21070836.
30. Saleh, M., Abdel-Baki, A. A., Dkhil, M. A., El-Matbouli, M., & Al-Quraishy, S., (2017). Antiprotozoal effects of metal nanoparticles against *Ichthyophthirius multifiliis*. *Parasitology, 144*(13), 1802–1810. https://doi.org/10.1017/S0031182017001184.
31. Sánchez-López, E., Gomes, D., Esteruelas, G., Bonilla, L., Lopez-Machado, A. L., Galindo, R., Cano, A., et al., (2020). Metal-based nanoparticles as antimicrobial agents: An overview. *Nanomaterials, 10*(2), 292. https://doi.org/10.3390/nano10020292.
32. Sardoiwala, M. N., Kaundal, B., & Choudhury, S. R., (2018). Toxic impact of nanomaterials on microbes, plants and animals. *Environmental Chemistry Letters, 16*(1), 147–160. https://doi.org/10.1007/s10311-017-0672-9.
33. Shaik, G., Sujatha, N., & Mehar, S. K., (2014). Medicinal plants as source of antibacterial agents to counter *Klebsiella pneumoniae*. *Journal of Applied Pharmaceutical Sciences, 4*(1), 135–147.
34. Singh, J., Vishwakarma, K., Ramawat, N., Rai, P., Singh, V. K., Mishra, R. K., Kumar, V., Tripathi, D. K., & Sharma, S., (2019). Nanomaterials and microbes' interactions: A contemporary overview. *3 Biotech, 9*(3), 68. https://doi.org/10.1007/s13205-019-1576-0.
35. Stauber, R. H., Siemer, S., Becker, S., Ding, G. B., Strieth, S., & Knauer, S. K., (2018). Small meets smaller: Effects of nanomaterials on microbial biology, pathology, and ecology. *ACS Nano, 12*(7), 6351–6359. https://doi.org/10.1021/acsnano.8b03241.
36. Suresh, A. K., Pelletier, D. A., & Doktycz, M. J., (2013). Relating nanomaterial properties and microbial toxicity. *Nanoscale, 5*(2), 463–474. https://doi.org/10.1039/c2nr32447d.
37. Sutarlie, L., Ow, S. Y., & Su, X., (2017). Nanomaterials-based biosensors for detection of microorganisms and microbial toxins. *Biotechnology Journal, 12*(4). https://doi.org/10.1002/biot.201500459.
38. Trchounian, A., Gabrielyan, L., & Mnatsakanyan, N., (2018). Nanoparticles of various transition metals and their applications as antimicrobial agents. In: Saylor, Y., & Irby, V., (eds.), *Metal Nanoparticles: Properties, Synthesis and Applications* (pp. 161–211). Nova Science Publishers.
39. Turner, R. J., (2017). Metal-based antimicrobial strategies. *Microbial Biotechnology, 10*(5), 1062–1065. https://doi.org/10.1111/1751-7915.12785.

40. Ul-Islam, M., Shehzad, A., Khan, S., Khattak, W. A., Ullah, M. W., & Park, J. K., (2014). Antimicrobial and biocompatible properties of nanomaterials. *Journal of Nanoscience and Nanotechnology, 14*(1), 780–791. https://doi.org/10.1166/jnn.2014.8761.
41. Vega-Jiménez, A. L., Vázquez-Olmos, A. R., Acosta-Gío, E., & Álvarez-Pérez, M. A., (2019). Chapter 2; *In vitro* antimicrobial activity evaluation of metal oxide nanoparticles. In: Koh, K., & Wong, V., (eds.), *Nanoemulsions-Properties, Fabrications and Applications* (pp. 1–18). IntechOpen.
42. Westmeier, D., Hahlbrock, A., Reinhardt, C., Fröhlich-Nowoisky, J., Wessler, S., Vallet, C., Pöschl, U., et al., (2018). Nanomaterial-microbe cross-talk: Physicochemical principles and (patho) biological consequences. *Chemical Society Reviews, 47*(14), 5312–5337. https://doi.org/10.1039/c6cs00691d.

CHAPTER 16

ANTIBACTERIAL POTENTIAL OF METALLIC NANOMATERIALS VERSUS BACTERIA

HALEEMA SADIA, RABEEA MUZAFFAR, RAFIA REHMAN, RIAZ HUSSAIN, SANA ULLAH, GHULAM MUSTAFA, PETER HEWITSON, and AFSAR BANO

ABSTRACT

In the last few years, nanomaterials have gained popularity due to their antibacterial properties. Different research studies have revealed that the effectiveness of the nanomaterials against microbes is increased if they are comprising of metals like Silver (Ag), Gold (Au), Copper (Cu), Zinc (Zn), Iron (Fe), Titanium (Ti), Magnesium (Mg), etc. Such materials can be synthesized by chemical, physical or biological methods. However, no matter what mode is adopted to prepare these nanomaterials, they possess a high biocidal impact against gram-positive and gram-negative bacteria as described by various studies. The current chapter is mainly focused on these aspects of the nanomaterials and especially elaborates the antibacterial activity of metallic nanomaterials (for metals including Silver (Ag), Gold (Au), Copper (Cu), Zinc (Zn), Iron (Fe) Titanium (Ti), Magnesium (Mg), etc.). The chapter will be helpful for researchers in building a correlation between metallic nanomaterial and their impact on MDR bacterial infections. Hence, finding multiple ways to produce antibacterial nanomaterials with desired properties and possible outcomes of being selective, cost-effective, efficient, non-toxic, environment-friendly, easier to prepare. However, more

Sustainable Nanomaterials for Biomedical Engineering: Impacts, Challenges, and Future Prospects.
Junaid Ahmad Malik, Megh R. Goyal & Mohamed Jaffer M. Sadiq (Eds.)

elaborative studies are required to explore the true potential and selective activity of nanomaterial against specific bacteria.

16.1 INTRODUCTION

Bacteria are the microscopic creatures that cause diseases in humans and other organisms. They have been the deadliest organism before the 1940s killing millions of humans [2, 36]. Novel antibiotics between the 1950s–1970s fought well against bacteria but soon these microbes developed resistance against these remedies [7]. As pathogenic bacteria are getting stronger every day by developing resistance against common antibiotics, stronger antibacterial agents are also required urgently [10]. In this regard, nanomaterials have shown great potential as antibacterial agent due to their unique properties. They are useful as an alternative to already available antibiotics and for killing the multi-drug resisting (MDR) bacteria [20]. Metals like Silver (Ag), Gold (Au), Copper (Cu), Zinc (Zn), Iron (Fe), Titanium (Ti), Magnesium (Mg), etc., have huge potential against different bacteria when used in nanoforms. These metallic nanomaterials can either be in pure form or in the form of oxides, sulfides, metal composites, and complexes [32]. They are effective against both gram-positive and gram-negative strains bacteria [41]. There are several methods to synthesize these materials which ultimately control the physical and chemical characteristics of the final product [5]. Similarly, there is a number of pathways that are adopted by nanomaterial to kill bacteria [32]. Metallic nanomaterials are also being successfully used in biosensors for bacterial cell detection [14].

This chapter describes the potential of metal-based nanomaterials (for metals including Silver (Ag), Gold (Au), Copper (Cu), Zinc (Zn), Iron (Fe), Titanium (Ti), Magnesium (Mg), etc.), against different classes of bacteria. Moreover, synthesis approaches and mechanistic pathways of metal nanomaterial are also highlighted. Information about nanomaterials, which can be/are being used as detecting agents for bacteria is also provided.

16.2 IMPORTANCE OF ANTIBACTERIAL STUDIES

16.2.1 BACTERIA: THE MICROSCOPIC MONSTERS

Bacteria are present everywhere around us and can be transmitted by air, water, food, insects, and contact with animals, humans, or anything contaminated with bacteria [36]. Bacteria can cause disease either by releasing toxins or inducing sensitivity in the body due to their presence. They can be classified based on their shape (spherical, rod-like, helical), growth response to oxygen (aerobic, anaerobic, facultative aerobic), nutrition source (heterotrophs, autotrophs), staining (gram-positive, gram-negative, mycobacteria) and have microscopic size from ~100 μm to 0.75 mm [15, 36]. A representative cell structure of a bacteria is shown in Figure 16.1.

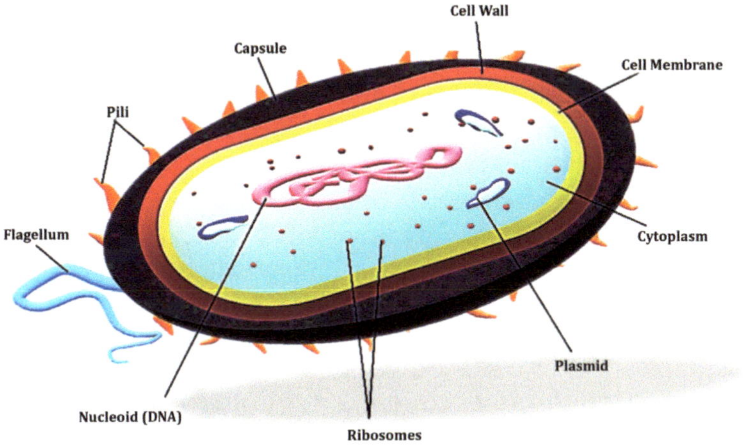

FIGURE 16.1 A representative cell structure of a bacteria.
Source: Adapted after Ref. [36].

They have been known to cause numerous diseases in humans through history and even to date [2]. After the development of vaccines and antibacterial drugs, many of these diseases have found a cure, but by the days, bacteria are becoming antibiotic-resistant, hence a lingering threat for us [36]. This fact makes it necessary to search for new possible remedies as the fight between humans and bacteria is still on.

16.3 REMEDIES AVAILABLE FOR BACTERIAL DISEASES

16.3.1 ANTIBIOTICS, AS PROMISING ANTI-BACTERIALS

Antibiotics are the molecules that fight against bacteria by either inhibiting their growth or killing them instantly. Antibiotics, which inhibit bacterial growth, e.g., chloramphenicol are said to be bacteriostatic. Antibiotics, which kill bacteria, e.g., penicillin, as a result, bacterial cells decrease in number, are said to be bactericidal. Antibiotics can show bacteriostatic or bactericidal action depending upon the circumstances, but the net effect is sufficient to stop the infection [43].

16.3.2 CHALLENGES WITH ANTIBIOTICS TODAY

With the development of antibiotics against bacteria, bacterial resistance against these antibiotics is also getting stronger. Many bacterial strains such as *Acinetobacter baumannii*, *Klebsiella pneumonia,* and *Pseudomonas aeruginosa* now, have developed multidrug resistance. There are newly emerged multi-drug-resistant (MDR) clinical isolates that are susceptible to one class of antibacterial agent [10]. Infections by MDR bacteria have high mortality rates [7]. The seriousness of this issue can be imagined by the fact that the WHO has issued a priority list of MDR bacteria that has categorized bacteria into medium, high, and critical groups based on the urgency of new antibiotics required as shown in Table 16.1.

16.4 NANOMATERIAL AS ANTIBACTERIAL AGENT

16.4.1 ANTIBIOTIC RESISTANCE IN BACTERIA

Reckless use of antibiotics has resulted in bacterial strains that are multidrug-resistant (super-bugs). Epidemic spread of bacteria among patients adds fuel to the fire, consequently, more clinical isolates are appearing with increasing resistance rapidly. The measures to control the bacterial spread and infections have not been fruitful so far. It has been observed around the globe that bacteria have become highly resistant against frequently used antibiotics which worked fine against common bacterial infections such as sepsis, diarrhea, UTIs, and STDs, urging us to opt for more effective drugs. As available chemotherapeutics are becoming less and less effective against bacteria, there is an urgent need to find more powerful alternative antibacterial agents [22].

16.4.2 WHY NANOMATERIAL?

Nanomaterials, due to their unique physio-chemical properties may serve as an antibacterial agent for multidrug-resistant (MDR) bacteria. It becomes

TABLE 16.1 The WHO List of Drug-Resistant Bacteria

Priority	Critical		High		Medium	
Bacterial Strains Resistant Against Mentioned Antibiotics	*Bacterial Strains*	*Antibiotics*	*Bacterial Strains*	*Antibiotics*	*Bacterial Strains*	*Antibiotics*
	Acinetobacter baumannii	Carbapenem-resistant	*Enterococcus faecium*	Vancomycin-resistant	*Streptococcus pneumoniae*	Penicillin-non-susceptible
	Pseudomonas aeruginosa	Carbapenem-resistant	*Staphylococcus aureus*	Methicillin-resistant, vancomycin-intermediate, and resistant	*Haemophilus influenzae*	Ampicillin-resistant

challenging for bacteria to grow resistant against these chemicals because nanomaterials can adopt multiple pathways to kill bacteria. The mechanistic pathways depend upon the size, structure, surface chemistry, shape, and core material of nanomaterial scaffolds. Moreover, due to their small size, nanomaterial is highly penetrable in the biological membrane. They can be also be loaded with antibacterial drugs which makes them an excellent carrier of the drug to the infection site. Finally, the property to modulate their interaction with bacterial cells plays a crucial role in enhancing the therapeutic effect of treatment. Antibiotics comprising of nanomaterial work against both free-floating (planktonic) and bacterial community (biofilm) microbes [20].

16.5 METAL-BASED NANOMATERIALS AS ANTIBACTERIAL AGENT

16.5.1 ANTIBACTERIAL PROPERTIES OF METAL NANOMATERIALS

Many metal nanoparticles such as Ag, Cu, Au, Zn, Fe, Ti, Mg, and their oxides have extensively been used in controlling infectious bacteria and their biological activities. These metals are active against microbes due to their nano-level structure. Metal NPs have different and unique chemical facets such as large surface area, structure, high reactivity, and cytotoxicity compared with bulky materials., thereby easily reacting with micro-organisms, hence offering a better replacement for antibiotic drugs [32].

16.5.1.1 SYNTHESIS MODES

Nanoparticles can be synthesized generally by two approaches. In "top-down" approach the size of bulk material of our interest is reduced to nano-level by physical or chemical mean. While in "Bottom-up" approach nanomaterial is built from atoms, molecules, or monomers forming clusters. It can be done via chemical or biological synthesis. The former is a catabolic approach while the latter is an anabolic approach [5] (Figure 16.2). Different synthesis processes like biological synthesis, photochemical, chemical batch deposition, chemical, electrochemical, radioactivity, radiolytic, and X-rays have been utilized to synthesize metal NPs [32].

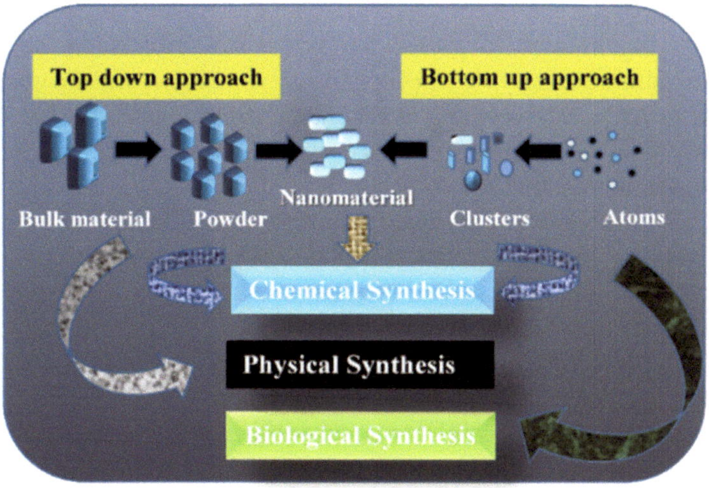

FIGURE 16.2 Synthetic of metallic nanomaterials using top-down and bottom-up approach.
Source: Adapted after Ref. [5].

Literature has revealed that biological synthesis specifically green synthesis (synthesis using plants) of nanoparticles has many advantages over other methods. Nanomaterial by green synthesis have properties like eco-friendly, non-toxic, cost-efficient, easy preparation, and enhanced biocidal effect against bacteria.

16.5.2 SILVER NANOMATERIAL AS ANTIBACTERIAL AGENT

Silver for its antibacterial properties was being studied even before penicillin or any other antibiotics discovery. Recently, bacterial resistance against multiple drugs has brought us back towards silver for the treatment of bacterial diseases. Recently, silver nanoparticles (AgNPs) are studied widely because they possess multidimensional antibacterial properties against human pathogenic bacteria. Different studies have reported the antibacterial activities of silver (AgNPs) against MDR (multi-drug resistant) bacteria, including both gram-positive and gram-negative strains. The antimicrobial activity of AgNPs has been proven to be size-dependent. The comparison between silver ions and AgNPs revealed that silver Nps have enhanced anti-pathogenic activity for bacteria which is 2.4–2.9 folds more than silver ions [16].

The AgNPs have also possessed remarkable antimicrobial activity against silk bacteria. The bacterial growth in *Bombyx mori* infected larvae gradually inhibited after feeding them mulberry leaves containing AgNPs as compared to ordinary mulberry leaves. Antimicrobial assays against food and clinical isolates of human pathogens such as *Staphylococcus aureus* and *Escherichia coli* by various standard methods have shown the growth inhibition impact of AgNPs. The minimum inhibitory concentration (MIC) value of nanoparticles depends on various factors, one of which is the synthetic route opted for the preparation of nanoparticles. MIC values of AgNPs against *Escherichia coli* by different studies have been recorded to be 0.08 and 1.5 nM while for *Bacillus cereus* it has been 0.0015 mM [41]. According to literature silver NPs have exhibited astonishing biocidal size and shape-dependent activities against clinically isolated MDR human pathogens like *Pseudomonas aeruginosa*, *Staphylococcus epidermidis,* and *Bacillus megaterium.* This makes AgNPs a potential candidate for therapeutic application s like bio-adhesives and clinical wound dressing [30].

According to Guzman et al. [21] different studies on antibacterial activity of AgNPs provide different results for different bacteria. Shape and size of nanoparticles play a pivotal role in determining for such differences to arise. For instance, AgNPs against *Staphylococcus aureus* and *Staphylococcus aureus* MRSA (both gram-positive), as well as *E. coli* and *P. aeruginosa* (both gram-negative) showed >90% growth inhibition of bacterial cells. Cell inhibition had a correlation to antibiotic activity (ciprofloxacin and oxacillin) for Gram-negative bacteria and gram-positive respectively [27]. While evaluated that AgNPs have the lowest antimicrobial activity for *Staphylococcus aureus* and *Escherichia coli.* Undoubtedly, the strongest efficacy was observed against *P. aeruginosa* [34]. Highest antibacterial activity of AgNPs against, *P. aeruginosa* (16 mm) in comparison with other pathogens against which these nanoparticles had lower activity such as *K. pneumoniae, S. aureus*, *E. coli* [28].

In another report, AgNps have displayed biological activity against multiple bacterial strains. AgNps had maximum inhibition against *Proteus* bacteria and better towards *P. mirabilis, S. aureus,* and *S. flexneri* while lesser active towards stains *S. typhi, E. coli, K. pneumonia, and P. aeruginosa* strains. MIC (minimum inhibition concentration) values ranged between 195–780 μg/ml for these micro-organisms [44]. It has also been observed that spherical and crystalline AgNPs are active against both grams strains of bacteria [42].

All of these studies indicated that the anti-pathogenic action of Ag-NPs against multiple strains of bacteria in size and shape-dependent. Moreover, the synthesis route also is a significant factor in determining the characteristics of synthesized particles. Table 16.2 enlists some of the bacterial strains against which AgNPs have been reported to be active by different studies.

TABLE 16.2 Antibacterial Activity of Silver Nanoparticles Against Gram-Positive and Gram-Negative Bacterial Strains Exhibited by Different Studies

AgNps Active Against Bacteria		**References**
Gram-Positive	*Bacillus megaterium, Staphylococcus epidermidis*	[16]
	Staphylococcus aureus MRSA	[21]
	Bacillus cereus	[41]
	Staphylococcus aureus	[21, 27, 41, 44]
Gram-Negative	*Pseudomonas aeruginosa*	[16, 21, 27, 41, 44]
	Escherichia coli	[27, 41, 44]
	Klebsiella pneumoniae	[27, 44]
	Proteus bacilli, Proteus mirabilis, Shigella flexneri, Salmonella typhi	[44]

16.5.3 COPPER NANOMATERIAL AS ANTIBACTERIAL AGENT

Characteristics like unique crystal morphologies and huge surface area make copper nanoparticles a promising candidate against MDR pathogenic bacterial [23]. CuO (Copper oxide) NM, like many other metal-based nanoparticles, have expressed antibacterial potency [13]. Given these properties copper NPs can be used for various applications including food packaging. It can be used against a common group of gram-negative bacteria *Pseudomonas* that has been seen spreading in processed food and has gained resistance against available antibiotics. Similarly, CuO NPs have performed well against infectious bacteria in the urinary tract [42]. Copper nanoparticles Cu-NPs have also shown efficacy against both positive and negative types of gram staining bacteria. Different reports have proven the antibacterial effect of copper NPs against pathogens, e.g., *B. subtilis, Bacillus megatherium, E. coli, M. luteus, and Staphylococcus aureus* [23].

Normally, CuO nanomaterials are considered to be less active than AgNps but in some circumstances the opposite is true. For instance, in a report, *S. aureus* and *E. coli* were more susceptible to Ag nanomaterial whereas *B. anthracis* and *B. subtilis* were more susceptible to Cu nanomaterial. The antibacterial activity of Cu-NPs have been proven against several common and MDR bacteria such as *M. luteus, P. aeruginosa, E. coli, S. aureus and K. pneumoniae.* Out of which, the highest susceptibility had been expressed by *E. coli* moderate activity by *S. aureus, Micrococcus luteus, and K. pneumoniae* in descending order but *P. aeruginosa* has been least susceptible to copper nanoparticles. The sensitivity of *B. subtilis* and *E. coli* to Cu-Nps have depicted a direct correlation between inhibitions of bacteria with the concentration of nanoparticle. This study had proven the efficacy of Cu-Nps over Ag-NPs [23]. Another comparative analysis of CuO-NM with metal-based nanomaterial apart from Ag NM reported the stronger antibacterial activity of copper NM over other nanomaterials [13].

CuO-NPs have also been effective against several fish and human bacterial strains (*Proteus mirabilis, Staphylococcus aureus, Vibrio anguillarum, Bacillus cereus, Aeromonas caviae, Edwardsiella tarda,* and *Aeromonas hydrophila).* Among these tested strains, *B. cereus* was most sensitive to CuO-NPs [28]. In another comparative study amongst *S. parauberis, E. coli, V. anguillarum* and *S. iniae,* it was revealed that CuO-NPs expressed antimicrobial activity against all microbes with considerable activity against *S. iniae* [9]. Similar results were depicted against *Streptococcus pneumoniae* bacterium [19].

It has been observed that the antimicrobial efficacy of CuO-NPs is stronger against Gram-positive pathogens than Gram-negative. According to reports, this difference can be attributed to the difference in the structure of the cell wall of these pathogens [19]. Gram-negative microbes such as *Proteus sp.* and *P. aeruginosa* have less quantity of negatively charged peptidoglycans which makes them less sensitive to these positively charged antimicrobial agents [23].

CuO-NPs coated to fabric have displayed 100% inhibition of *S. aureus* and *E. coli* after 48 hours of incubation. Therefore, this trait makes CuO-NPs nanoparticles a promising agent for surface coating of protective clothing which reduces the risk of infection proliferation [23]. Table 16.3 enlists some of the bacterial strains against which copper nanoparticles have been reported to be active by different studies.

TABLE 16.3 Antibacterial Efficacy of Copper NPs Against Bacterial Stains (Gram-Negative and Positive) as Exhibited in Different Studies

CuNps Active Against Bacteria		References
Gram-Positive	*Streptococcus parauberis, Streptococcus iniae*	[9]
	Streptococcus pneumoniae	[19]
	Bacillus anthracis, Bacillus megaterium, Bacillus subtilis, Staphylococcus aureus	[23]
	Bacillus cereus, Micrococcus luteus	[28]
Gram-Negative	*Vibrio anguillarum, Escherchia coli*	[9]
	Pseudomonas aeruginosa, Klebsiella pneumonia	[23]
	Aeromonas caviae, Aeromonas hydrophila, Proteus mirabilis, Edwardsiella tarda	[28]

16.5.4 GOLD NANOMATERIAL AS ANTIBACTERIAL AGENT

Apart from Ag and Cu NPs, Au NPs have been in attention due to their outstanding features such as non-toxicity, inert nature, bacterial detection, photo-thermal activity, and interaction with biomolecules. Due to its biocompatibility with cellular systems and inert nature gold NPs can be used in several clinical applications. Au-NPs have the advantage over other metal-based nanomaterials due to their outstanding properties [30].

It has been reviewed that different concentrations of Au-NPs in solution exhibit bacterial growth inhibition against common gram strains of bacteria. Antimicrobial activity against multidrug-resistant human pathogenic gram-negative stains, e.g., *P. aeruginosa* and *E. coli* in addition to MDR gram-positive strains of bacteria like *B. subtilis* has also been shown by AuNPs [38].

Antimicrobial assays of Au-Nps against bacteria such as *Salmonella typhi*, *S. aureus*, *Pseudomonas aeruginosa,* and *E. coli* have depicted their strong antibacterial activity. Fabricated Au-Nps have been reported to inhibit the growth of these microbes by 86% for *P. aeruginosa*, 88% for *E. coli,* and 94% for *K. pneumonia* [42]. Table 16.4 enlists some of the bacterial strains against which AuNPs have been reported to be active by different studies.

TABLE 16.4 Antibacterial Potential of Gold NPs Against Different Bacterial Gram Stains Demonstrated by Different Studies

AuNps Active Against Bacteria		References
Gram-Positive	*Bacillus subtilis*	[38]
	Staphylococcus aureus	[42]
Gram-Negative	*Escherichia coli*	[38]
	Pseudomonas aeruginosa	[38, 42]
	Klebsiella pneumonia, Salmonella Typhi	[42]

16.5.5 ZINC NANOMATERIAL AS ANTIBACTERIAL AGENT

Zinc is another metal that has shown excellent antibacterial potency due to its selectivity in killing the bacterial cell and low toxicity to the human body [30]. ZnO nanomaterial is fairly low cost, and effective against a variety of bacteria in a size-dependent manner. The use of zinc in treatment is approved by FDA, hence, it can be used as a food additive. ZnO NPs have been proven to have growth-inhibiting potential against superbugs like methicillin-resistant *Staphylococcus epidermidis* (MRSE) stains, (methicillin-sensitive) *Staphylococcus aureus* (MSSA), (methicillin-resistant) *Staphylococcus aureus* (MRSA), and other bacteria that have been resistant to such antibacterials. These include microbes like *E. coli, Listeria monocytogenes, Lactobacillus, Klebsiella pneumonia, Streptococcus mutans,* and *Salmonella enteritidis* [13].

ZnO NPs have displayed distinct modes of action against microbes from other metallic oxides. Particularly, it has been observed that bulk-sized ZnO Nps (0.1–1.0 mm) express a stronger antibacterial effect against pathogens (*E. coli, B. subtilis*, and *S.* aureus). Especially, particle size from 2 mm to 45–12 nm of ZnO NPs has been more effective against *E. coli* suggesting the stronger antibacterial activity of larger-sized particles over smaller-sized particles [16].

The antibacterial study of ZnO nanoparticles against five gram-negative strains *(Enterobacter aerogenes, Escherichia coli, Salmonella typhi, Klebsiella pneumoniae,* and *Proteus vulgaris)* and two gram-positive strains (*Bacillus subtilis* and *Staphylococcus aureus*) has exhibited strong biocidal activity against these microbes. FE-SEM imaging has depicted morphological changes in bacterial cells like cell shrinkage, cell membrane and cell wall damage, and cellular aggregation in bacterial culture. The zone of inhibition for gram-negative stains was reported to be (15–16 mm in diameter) whereas for gram-positive stains, it was (16–23 mm in diameter). These results also have demonstrated the stronger antibacterial activity of Zn nanoparticles against negative stains over positive ones [39].

In a study, the antibacterial efficacy of ZnO NPs towards *E. coli* ATCC 25922, *Staphylococcus aureus, S. aureus* ATCC 4163, and *Pseudomonas aeruginosa* ATCC 6749 increased by increasing concentration. This activity was comparable to the activity of gentamicin. Whereas in another study, ZnO nanoparticles synthesized differently have shown better efficacy of *P. aeruginosa over* gentamicin. At the same time, the zone inhibition of other strains was similar [18].

In a different study, MIC (minimum inhibition concentration) analysis of ZnO NPs against bacterial strain *S. aureus and B. subtilis* and *E. coli* have shown significant inhibition of S. *aureus* and *E. coli* bacterial culture whereas no activity was displayed for bacteria strain *B. subtilis* [35].

Likewise, Zn_2SnO_4 nanoparticles have also been tested for their antibacterial potential against different bacterial strains. Such nanoparticles have depicted considerable biocidal action against pathogens like *B. subtilis* and *S. aureus*, and pathogens like *Escherichia coli* and *Klebsiella pneumonia*. This study also indicated that the antibacterial action of Zn_2SnO_4 NPs increased by increasing nanoparticle concentration. Zn_2SnO_4 has expressed maximum antibacterial potential against gram-positive strains of *S. aureus* and *B. subtilis* while against strains of *E. coli* and *Klebsiella pneumonia* moderated activity has been observed against [25].

All of these studies have displayed a strong biocidal effect of Zn nanoparticles against several different bacteria covering both gram-negative and gram-positive strains, including several MDR bacteria. Therefore, it is needless to say that Zn nanoparticles are also one of the promising antibacterial agents among many other metal-based nanoparticles. Table 16.5 enlists some of the bacterial strains against which zinc nanoparticles have been reported to be active by different studies.

TABLE 16.5 Antibacterial Efficacy of Zinc NPs Against Different Bacterial Gram Stains Depicted by Several Studies

ZnNps Active Against Bacteria		References
Gram-Positive	*Staphylococcus epidermidis* MRSE, *Lactobacillus* bacteria, *Staphylococcus aureus* MRSA, *Listeria monocytogenes, Streptococcus mutans*	[13]
	Bacillus subtilis	[16, 25, 39]
	Staphylococcus aureus	[13, 16, 18, 25, 35, 39]
Gram-Negative	*Salmonella enteritidis*	[13]
	Klebsiella pneumoniae	[13, 25, 39]
	Escherichia coli	[13, 16, 18, 25, 35, 39]
	Pseudomonas aeruginosa	[18]
	Proteus vulgaris, Salmonella Typhi	[39]

16.5.6 IRON NANOMATERIAL AS ANTIBACTERIAL AGENT

Recently, Iron oxide (IO) nanoparticles have been the center of attention in biomedical and technological research due to their countless applications such as anti-pathogenic activity, magnetic storage capacity, cell sorting, biosensor, food preservation MRI and target drug delivery [12].

Different bacterial strains have demonstrated susceptibility for these anti-microbial particles. Stains like *P. aeruginosa, P. mirabilis, S. typhi,* and *Klebsiella pneumonia* have shown minimum MIC values for Fe-NPs while maximum MIC value was shown by *S. flexneri* strain [44].

Studies on Fe_3O_4 have reported that such nanoparticles are excellent anti-bacterial agents that are active against both gram strains of bacteria. However, different studies have shown varied susceptibility of bacterial strains to these nanoparticles. For instance, in one study, gram-negative bacteria have exhibited more sensitivity towards Iron oxides (IO-NPs) nanoparticles when compared to gram-positive pathogens. The aforementioned nanoparticles have demonstrated better performance against microbial strains (*E. coli* and *P. vulgaris*) than bacterial stains of (*S. aureus*) [34]. While in another study, evaluation of antimicrobial potential of iron oxide (IO-NPs) on 10 pathogens viz. *V. cholerae* (MTCC 3904), *Bacillus licheniformis* (MTCC 7425), *E. coli* (MTCC 1089), *Pseudomonas aeruginosa* (MTCC 1034), *Shigella flexneri* (Lab isolate), *Bacillus subtilis* (MTCC 7164), *Staphylococcus aureus*

(MTCC 1144), *Bacillus brevis* (MTCC 7404), *Streptococcus aureus* (Lab isolate) have shown moderate activity of NPs on eight pathogens (two gram-negative and six gram-positive) [12]. In some cases, FeNPs have resulted in a severe biocidal effect on human bacteria such as *MRSA*, *Salmonella, E. coli,* and *S. aureus*. The inhibition effect can be reported in descending order as; *S. aureus, Salmonella, MARSA,* and *E coli* [17].

Significant antibacterial activity of Fe-NPs has been viewed on *P. aeruginosa.* Direct correlation of concentration of Fe-NPs with growth inhibition in bacteria has been reported as bacterial inhibition gradually increases with the increase in nanoparticles concentration. These results were shown against micro-organisms like *P. vulgaris, Pseudomonas fluorescence, Vibrio fluvialis, Escherichia coli, Proteus mirabilis, Klebsiella pneumonia, Enterococcus faecalis, Micrococcus luteus, Staphylococcus aureus, B. subtilis,* and *Bacillus cereus*. Iron-nanoparticles(Fe-NPs) had a comparable inhibitory effect to inhibition activity of Norfloxacin antibiotic [8]. Table 16.6 enlists some of the bacterial strains against which iron nanoparticles have been reported to be active by different studies.

TABLE 16.6 Antibacterial Potency of Iron NPs Against Different Bacterial Strains Demonstrated by Various Studies

FeNps Active Against Bacteria		References
Gram-Positive	*Bacillus cereus, Enterococcus faecalis, Micrococcus luteus*	[8]
	Bacillus subtilis	[8, 12]
	Bacillus brevis, Bacillus licheniformis, Staphylococcus epidermidis, Streptococcus aureus	[12]
	Staphylococcus aureus	[12, 17, 29, 34]
	Staphylococcus aureus MRSA	[17]
Gram-Negative	*Pseudomonas fluorescence, Proteus vulgaris, Vibrio fluvialis*	[8]
	Proteus mirabilis, Klebsiella pneumoniae	[8, 44]
	Shigella flexneri	[8, 12, 17, 29, 34]
	Vibrio cholerae	[12]
	Pseudomonas aeruginosa	[12, 44]
	Escherichia coli, Salmonella typhi	[17, 44]

16.5.7 TITANIUM NANOMATERIAL AS ANTIBACTERIAL AGENT

Titanium dioxide (TiO_2) has also been widely studied due to its extensive application against microbes [13]. It works antagonistically towards (*Bacillus subtilis, Staphylococcus aureus*) and (*Escherichia coli*) bacterial strains [16]. Given its ability to kill *Lactobacillus acidophilus* TiO_2 has found its way into dentistry (toothbrushes, dental implants, and screws, etc.) [30]. Titanium efficacy towards bacteria (*S. epidermidis* and *S. aureus*) is influenced by controlling its characteristics such as nanoparticle size, surface chemistry, and crystallinity [42].

Antibacterial efficacy analysis of TiO_2 nanoparticles against microbes like *E. coli, S. aureus,* and *P. vulgaris* has depicted maximum activity against *Proteus vulgaris* in terms of zone inhibition. The tendency of maximum zone inhibition shifted towards *Escherichia coli* when titanium nanoparticles had been synthesized differently, indicating that they possess different traits with different synthetic processes [3].

While comparing the biocidal impact of Ti nanoparticles against selected strains, including *E. coli* and *p. aeruginosa,* it was shown that TiO_2 NPs performed better activity towards *E. coli.* Similarly, the comparison among positive strain has maximum efficacy on *B. substilis* [4]. Another comparison among *Clostridium tetani, Bacillus cereus, Vibrio cholera,* and *Escherichia coli* has exhibited the maximum effectiveness of Ti nanoparticles against *C. tetani* [11]. As described for other nanomaterials, the concentration and inhibitory effect depict the same relationship between Ti-NPs and bacteria. More concentration of nanoparticles has shown more growth inhibition in bacteria [30]. Fabricated titanium dioxide NPs have depicted antibacterial efficacy towards *S. aureus*, *P. aeruginosa,* and *E. coli* [1].

All the above studies have reported the antibacterial potency of titanium nanoparticles against various groups of bacteria. Therefore, such NPs may serve as a better antibacterial agent than less effective common antibiotics. Table 16.7 enlists some of the bacterial strains against which titanium nanoparticles have been reported to be active by different studies.

TABLE 16.7 Antibacterial Demonstration of Titanium NPs Multiple Bacterial Stains by Different Studies

Ti-Nps Active Against Bacteria		References
Gram-Positive	*Staphylococcus aureus*	[1, 3, 16, 42]
	Bacillus subtilis	[4, 16]
	Bacillus cereus, Clostridium tetani	[11]
	Lactobacillus acidophilus	[30]
	Staphylococcus epidermidis	[42]

TABLE 16.7 *(Continued)*

Ti-Nps Active Against Bacteria		References
Gram-Negative	*Pseudomonas aeruginosa*	[1, 4]
	Escherichia coli	[1, 3, 4, 11]
	Proteus vulgaris	[3]
	Vibrio cholera	[11]

16.5.8 MAGNESIUM NANOMATERIAL AS ANTIBACTERIAL AGENT

Another metal that has shown promising results against infectious bacterial strain is magnesium. MgO nanoparticles have displayed antibacterial potency for different bacteria, including gram-positive and gram-negative strains. MgO nanoparticles have an edge over other nanoparticles due to their economic synthesis from available precursors [13].

MgONPs have been evaluated towards *S. epidermidis*–MTCC-2639, *Staphylococcus aureus*–MTCC-9442, *Bacillus cereus*–MTCC-9017 and *Proteus vulgaris*–MTCC-7299, *Escherichia coli*–MTCC-9721, *K. pneumonia*–MTCC-9751 and exhibited significant efficacy against both strains of bacteria. Values for the zone of inhibition (ZOI) have been reported to be 18 mm, 17 mm, and 16 mm for *E. coli, S. aureus* and other three strains (*B. cereus, P. vulgaris,* and *K. pneumonia*) respectively [40].

Just like Fe, Zn, Cu, and Ag nanoparticles of MgO have also shown effectiveness against gram-negative foodborne micro-organism *Escherichia coli* [24]. Likewise, its biocidal efficacy against *Acidovorax oryzaestains* (RS-2) have been exhibited after 18 hours of incubation [31]. MgO NPs have also been tested against bacteria like *Bacillus subtilis* and *Pseudomonas aeruginosa* and the results have depicted strong antibacterial impact against these pathogens [33]. The concentration effect of Mg nanoparticles on bacterial growth has shown similar results to above mentioned metallic nanoparticles. An increase in the concentration of MgO nanoparticles has increased its antibacterial efficacy. Between *Bacillus* bacteria and *Escherichia coli,* the later have shown higher susceptibility against magnesium NPs. From various studies on Mg NPs it has been clear that gram-negative bacteria (*Proteus mirabilis, Salmonella species, Klebsiella pneumonia, Pseudomonas aeruginosa, Acinetobacter species, E. coli)* are more sensitive to the antibacterial action of magnesium NPs than gram-positive bacteria (*Streptococcus pneumonia, Micrococcus sp., Staphylococcus epiderimidis, Bacillus sp., Staphylococcus aureus,*) because of difference in their cell wall [6].

All of the studies conclude that such metal-based nanomaterial bears astounding anti-pathogenic characteristics and huge potential against multiple strains of human and other MDR pathogenic bacteria. Factors like synthetic process, particle size, shape, concentration, and bacterial test strains influence the efficacy of particles against a particular bacterium sample. Nevertheless, controlling these parameters may produce nanoparticles with distinct and desirable properties of targeting particular groups of bacteria. Table 16.8 enlists some of the bacterial strains against which magnesium nanoparticles have been reported to be active by different studies.

TABLE 16.8 Antibacterial Properties of Magnesium NPs Against Different Bacterial Strains Displayed by Different Studies

MgNps Active Against Bacteria		References
Gram-Positive	*Micrococcus* bacteria, *Streptococcus pneumonia*	[6]
	Bacillus subtilis	[6, 33]
	Bacillus cereus, Staphylococcus epidermidis, Staphylococcus aureus	[6, 40]
Gram-Negative	*Acinetobacter* bacteria, *Salmonella* bacteria, *Proteus mirabilis, Acidovorax oryzaestains*	[6]
	Klebsiella pneumoniae	[6, 40]
	Escherichia coli	[6, 24, 40]
	Pseudomonas aeruginosa	[33]
	Proteus vulgaris	[40]

Table 16.9 summarizes the antibacterial activity and potential of the metals discussed in this chapter against different bacteria reported. Highlighted box indicates the effectiveness of that metal against mentioned bacteria. Most of these bacteria are on the WHO priority list which requires new antibiotics urgently.

TABLE 16.9 The Antibacterial Efficacy and Potential of the Metals Discussed in This Chapter

Antibacterial Metal-based Nanomaterials	Ag	Cu	Au	Zn	Fe	Ti	Mg
Bacillus anthracis		✓					
Bacillus brevis					✓		
Bacillus cereus	✓	✓			✓	✓	✓
Bacillus clostridium						✓	

TABLE 16.9 *(Continued)*

	Antibacterial Metal-based Nanomaterials	Ag	Cu	Au	Zn	Fe	Ti	Mg
Gram-Positive Strains	*Bacillus licheniformis*					■		
	Bacillus megaterium	■	■					
	Bacillus subtilis		■	■	■	■	■	■
	Enterococcus faecalis					■		
	Lactobacillus acidophilus				■		■	
	Listeria monocytogenes				■			
	Micrococcus luteus		■			■		■
	Staphylococcus aureus	■	■	■	■	■	■	■
	Staphylococcus aureus MRSA	■			■	■		
	Staphylococcus epidermidi	■				■	■	■
	Staphylococcus epidermidis MRSE				■			
	Streptococcus aureus		■			■		
	Streptococcus iniae		■					
	Streptococcus mutans				■			
	Streptococcus parauberis		■					
	Streptococcus pneumoniae		■					■
Gram-Negative Strains	*Acidovorax oryzaestains*							■
	Acinetobacter species							■
	Aeromonas caviae		■					
	Aeromonas hydrophila		■					
	Edward siellatarda		■					
	Escherichia coli	■	■	■	■	■	■	■
	Klebsiella pneumoniae	■	■	■	■	■		■
	Proteus bacilli	■						
	Proteus mirabilis	■	■			■		■
	Proteus vulgaris				■	■	■	■
	Pseudomonas aeruginosa	■	■	■	■	■	■	■
	Pseudomonas fluorescence					■		
	Salmonella enteritidis				■			■
	Salmonella Typhi	■		■	■			■
	Shigella flexneri	■				■		
	Vibrio anguillarum		■					
	Vibrio cholerae					■	■	
	Vibrio fluvialis					■		

16.5.9 MECHANISM OF ACTION OF ANTIBACTERIAL METALLIC NANOMATERIALS

Some studies have indicated that metal and metallic oxides work synergistically against bacteria by multiple pathways such as generating reactive oxygen species (ROS), penetration, adhesion, signal alteration, and free radicle pathways [41].

16.5.9.1 CELLULAR MEMBRANE DISRUPTION

The antibacterial action of nanoparticles depends upon surface charges, chemical composition, structure, and size of the nanoparticles. They interact with the plasma membrane of bacteria which results in alteration of membrane properties like transportation activity and permeability. Thereafter, these nanoparticles can get penetrated into the bacterial cell and membrane disruption leads to cellular leakage followed by pathogen death. It is clear that cell wall structure plays a significant role in the interaction of nanomaterial with bacterial cells. Normally, gram-negative strains of bacteria are more sensitive towards antibacterial nanomaterial than gram-positive strains of bacteria [32].

16.5.9.2 PROTEIN DAMAGE

Similarly, nanomaterial containing Sulphur interact with cellular proteins and these Sulphur molecules bring about change in electrostatic forces and cellular permeability of lipid bilayer. This again leads to cellular disruption and micro-organism's death [32].

16.5.9.3 ROS GENERATION

Another pathway nanomaterial can adopt to kill bacteria is the generation of (ROS) reactive oxygen species like superoxide ions, hydroxyl radicles, hydrogen peroxides, and singlet oxygen. These species interact antagonistically with cellular structures (such as DNA, ribosomes, enzymes, etc.), and inhibit their activity consequently kill the bacteria. Generally, bacteriostatic or bactericidal action take place by following mechanisms: (i) generation of ROS; (ii) plasma membrane rupture; (iii) electron-transport disturbance; (iv) enzymes inhibitions; and (v) DNA disruption/synthesis interruption (Figure 16.3) [32].

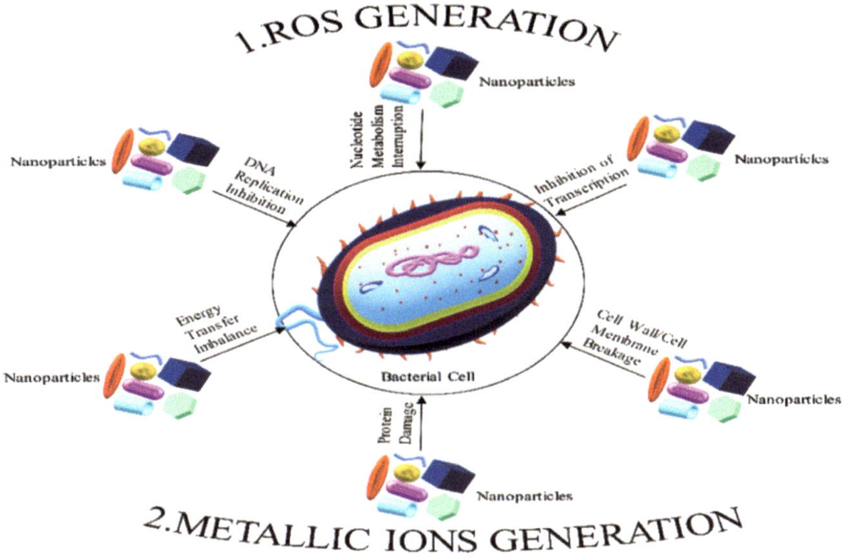

FIGURE 16.3 Possible mechanistic pathways by metallic nanoparticles against bacteria.
Source: Adapted after Ref. [32].

16.5.10 FACTOR AFFECTING THE ANTIBACTERIAL ACTIVITY OF METAL NPS

Several factors influence the antibacterial activity of metallic nanoparticles, such as the scale of the particle (bulky/nano), different mechanisms result in different antibacterial action, peptidoglycan layer in the cell wall, carrier, size, and shape and pH of NPs play a pivotal role in their pathogenic activity [32].

It has been observed that the penetration of NPs in gram-positive bacteria depends upon cell wall thickness and its interaction with nanoparticles. In the case of negative strains, this interaction depends upon the shape, core material, surface charge, and size of bacteria [29, 32].

Conversely, analysis of the antibacterial action towards both types of bacterial strains (Gram-negative and Gram-positive) is crucial as many metallic NPs exhibit stronger effect against Gram-positive strain and show lesser activity towards Gram-negative strain and vice versa [32]. It has been a common observation that gram-negative bacteria are less sensitive towards positively charged metallic nanoparticles because they possess less peptidoglycan that has negative charges on it [23]. While in the case of gram-positive bacteria they contain thick peptidoglycan; hence more metallic ions get trapped in bacterial cell wall. Prolong interaction leads to more penetration and more antibacterial activity is demonstrated by such types of bacteria [6].

16.6 NANOMATERIAL AS DETECTING AGENTS FOR BACTERIA

16.6.1 PRINCIPLES OF BACTERIAL DETECTION

Bacterial cell detection depends on the morphological characteristics or structure chemistry of the cell. The detection methods are based on the following principles [14].

16.6.1.1 PHENOTYPIC METHOD

Microscopic features such as cellular structure, gram staining, shape (rod, spherical, cocci, capsule, and spiral), growth pattern, speed, and colony formation are determined. Biochemical and physiological features are also analyzed in this method [14].

16.6.1.2 IMMUNOLOGICAL/SEROLOGICAL METHOD

Interaction of bacterial antigen with its selective antibody serves as a base for bacterial detection. Test for properties like particle agglomeration, direct/indirect immuno-fluorescence, optical immunoassays, and immuno-chromatography are performed [14].

16.6.1.3 GENOTYPIC METHOD

In this method tests for gene detection, genetic material (RNA or DNA), particular gene code, or nucleic acid sequence are performed for final definitive detection [14].

16.6.1.4 METALLIC NPS-BASED METHOD

Metallic NPs show remarkable potential in detecting multiple bacteria through different target sites with high sensitivity. Due to their extremely small size NPs provide large surface area, as a result, immobilized biological molecules on the surface increase in the number that maximize the binding events. Therefore, the enhanced signal coupled with sensitive electrical or optical transducers provides ultrafine biochemical detection. Metallic nanoparticles also help in detecting bacterial cells in complex samples [37].

Metallic NPs can also serve as a carrier to deliver recognition elements to the biological target site as they can penetrate bacterial cells more quickly due to their small size. This can be done by fabricating Metallic NPs with target analytes or detection elements. Metallic nanoparticles incorporated with other diagnostic materials opens door to synthesize simple, rapid, low-cost, and sensitive biosensors for multiple pathogenic detections. To design such a system not only for detection but for target delivery and imagining, the interaction between nanomaterial and cellular structure must be well comprehended. The following aspects should be focused on while designing these sensors i. Selective biomarker on cellular membrane ii. Physical and chemical characteristics of nanoparticles. While the essentials to design an effective biosensor include: (i) detection element that interacts with target; (ii) signal transduction component that produces a detectable signal when it binds to the analyte's surface; (iii) a signal read-out device [26].

16.7 SUMMARY

From ancient times bacterial infections have been a great challenge for public health. After the discovery of antibiotics, the uproar about the bacterial threat was hushed for some time, but later it came out with a louder bang when bacteria started fighting these antibiotics and emerged as super-bugs having resistance against multiple drugs. With the evolution of multi-drug resistant (MDR) pathogenic bacteria, a dire need of novel and stronger antibacterial agents was also aroused. As old methods of treatment and antibiotics became less and less effective, metal-based nanomaterials showed substantial potential against these bacteria. Many studies on metal-based (Fe, Ti, Ag, Cu, Au, Zn, and Mg) nanomaterial reported their antibacterial activity by multiple pathways. They were effective against most of the MDR strains with almost no or low toxicity to human cells. As nanoparticles have shown to target multiple sites in bacteria, it is difficult for these microbes to develop resistance against such antibacterial agents.

Although, there are some limitations in the utilization of existing studies about these nanomaterials. For instance, the unavailability of unified standards and several varied parameter such as NP type produced, synthesis processes, action time, different test strains, and concentration of nanoparticles makes it complex to have the comprehensive and comparative evaluation of nanoparticles with the reproducible result against different bacteria. However, generally, metal-based nanoparticles have numerous biological applications against bacteria either alone or in combination with other antibacterial agents (e.g., antibiotics). More study and research is required to understand and achieve the best possible results from metallic nanoparticles.

Finally, keeping in view all the available reports, it may be concluded that such nanomaterial has a promising future as a powerful antibacterial agent.

KEYWORDS

- **iron oxides nanoparticles**
- **methicillin-resistant staphylococcus aureus**
- **minimum inhibitory concentration**
- **multi-drug resistant**
- **reactive oxygen species**
- **silver nanoparticles**

REFERENCES

1. Abdolmajid, E., Kharazi, H., Chalaki, M., Khojasteh, M., Haghighat, S., Attar, F., Nemati, F., & Falahati, M., (2019). Titanium oxide nanoparticles fabrication, hemoglobin interaction, white blood cells cytotoxicity, and antibacterial studies. *Journal of Biomolecular Structure and Dynamics, 37*(11), 3007–3017. https://doi.org/10.1080/07391102.2018.1499555.
2. Achtman, M., (2016). How old are bacterial pathogens?. *Proceedings of the Royal Society B: Biological Sciences,* 283(1836), 20160990. https://doi.org/10.1098/rspb.2016.0990, PubMed: 20160990.
3. Ahmad, W., Jaiswal, K. K., & Soni, S., (2020). Green synthesis of titanium dioxide (TiO_2) nanoparticles by using *Mentha arvensis* leaves extract and its antimicrobial properties. *Inorganic and Nano-Metal Chemistry, 50*(10), 1032–1038. https://doi.org/10.1080/24701556.2020.1732419.

4. Ajmal, N., Saraswat, K., Bakht, M. A., Riadi, Y., Ahsan, M. J., & Noushad, M., (2019). Cost-effective and eco-friendly synthesis of titanium dioxide (TiO_2) nanoparticles using fruit's peel agro-waste extracts: Characterization, *in vitro* antibacterial, antioxidant activities. *Green Chemistry Letters and Reviews, 12*(3), 244–254. https://doi.org/10.1080/17518253.2019.1629641.
5. Al-Hakkani, M. F., (2020). Biogenic copper nanoparticles and their applications: A review. *SN Applied Sciences, 2*(3), 1–20. https://doi.org/10.1007/s42452-020-2279-1.
6. Al-Hazmi, F., Alnowaiser, F., Al-Ghamdi, A. A., Al-Ghamdi, A. A., Aly, M. M., Al-Tuwirqi, R. M., & El-Tantawy, F., (2012). A new large-scale synthesis of magnesium oxide nanowires: Structural and antibacterial properties. *Superlattices and Microstructures, 52*(2), 200–209. https://doi.org/10.1016/j.spmi.2012.04.013.
7. Aminov, R. I., (2010). A brief history of the antibiotic era: Lessons learned and challenges for the future. *Frontiers in Microbiology, 1*, 134. https://doi.org/10.3389/fmicb.2010.00134.
8. Amutha, S., & Sridhar, S., (2018). Green synthesis of magnetic iron oxide nanoparticle using leaves of *Glycosmis mauritiana* and their antibacterial activity against human pathogens. *Journal of Innovations in Pharmaceutical and Biological Sciences, 5*, 22–26.
9. Ananth, A., Dharaneedharan, S., Heo, M. S., & Mok, Y. S., (2015). Copper oxide nanomaterials: Synthesis, characterization and structure-specific antibacterial performance. *Chemical Engineering Journal, 262*, 179–188. https://doi.org/10.1016/j.cej.2014.09.083.
10. Anderson, R., Groundwater, P. W., Todd, A., & Worsley, A., (eds.). (2012). *Antibacterial Agents: Chemistry, Mode of Action, Mechanisms of Resistance and Clinical Applications* (1st edn., p. 378). John Wiley & Sons Ltd.
11. Aswini, R., Murugesan, S., & Kannan, K., (2021). Bio-engineered TiO_2 nanoparticles using *Ledebouria revoluta* extract: Larvicidal, histopathological, antibacterial and anticancer activity. *International Journal of Environmental Analytical Chemistry, 101*(15), 2926–2936. https://doi.org/10.1080/03067319.2020.1718668.
12. Behera, S. S., Patra, J. K., Pramanik, K., Panda, N., & Thatoi, H., (2012). Characterization and evaluation of antibacterial activities of chemically synthesized iron oxide nanoparticles. *World Journal of Nano Science and Engineering, 2*(4), 196–200. https://doi.org/10.4236/wjnse.2012.24026.
13. Beyth, N., Houri-Haddad, Y., Domb, A., Khan, W., & Hazan, R., (2015). Alternative antimicrobial approach: Nano-antimicrobial materials. *Evidence-Based Complementary and Alternative Medicine: ECAM, 2015*, 246012. https://doi.org/10.1155/2015/246012.
14. Bohara, R. A., & Pawar, S. H., (2015). Innovative developments in bacterial detection with magnetic nanoparticles. *Applied Biochemistry and Biotechnology, 176*(4), 1044–1058. https://doi.org/10.1007/s12010-015-1628-9.
15. Chen, D., & Qian, X., (2018). *A Brief History of Bacteria: The Everlasting Game Between Humans And Bacteria* (1st edn., p. 296). World Scientific Publishing Pte. Limited.
16. Cioffi, N., & Rai, M., (2012). *Nano-antimicrobials: Progress and Prospects* (1st edn., p. 556). Springer Science+Business Media.
17. Da'na, E., Taha, A., & Afkar, E., (2018). Green synthesis of iron nanoparticles by *Acacia nilotica* pods extract and its catalytic, adsorption, and antibacterial activities. *Applied Sciences, 8*(10), 1922. https://doi.org/10.3390/app8101922.

18. Dobrucka, R., & Długaszewska, J., (2016). Biosynthesis and antibacterial activity of ZnO nanoparticles using *Trifolium pratense* flower extract. *Saudi Journal of Biological Sciences, 23*(4), 517–523. https://doi.org/10.1016/j.sjbs.2015.05.016.
19. Ghasemi, N., Jamali-Sheini, F., & Zekavati, R., (2017). CuO and ag/CuO nanoparticles: Biosynthesis and antibacterial properties. *Materials Letters, 196*, 78–82. https://doi.org/10.1016/j.matlet.2017.02.111.
20. Gupta, A., Mumtaz, S., Li, C. H., Hussain, I., & Rotello, V. M., (2019). Combatting antibiotic-resistant bacteria using nanomaterials. *Chemical Society Reviews, 48*(2), 415–427. https://doi.org/10.1039/c7cs00748e.
21. Guzman, M., Dille, J., & Godet, S., (2012). Synthesis and antibacterial activity of silver nanoparticles against gram-positive and gram-negative bacteria. *Nanomedicine: Nanotechnology, Biology, and Medicine, 8*(1), 37–45. https://doi.org/10.1016/j.nano.2011.05.007.
22. Hemeg, H. A., (2017). Nanomaterials for alternative antibacterial therapy. *International Journal of Nanomedicine, 12*, 8211–8225. https://doi.org/10.2147/IJN.S132163.
23. Ingle, A. P., Duran, N., & Rai, M., (2014). Bioactivity, mechanism of action, and cytotoxicity of copper-based nanoparticles: A review. *Applied Microbiology and Biotechnology, 98*(3), 1001–1009. https://doi.org/10.1007/s00253-013-5422-8.
24. Jeevanandam, J., San, C. Y., & Danquah, M. K., (2019). Evaluating the antibacterial activity of MgO nanoparticles synthesized from aqueous leaf extract. *Med One, 4*, e190011. https://doi.org/10.20900/mo.20190011.
25. Jeronsia, J. E., Joseph, L. A., Jaculine, M. M., Vinosha, P. A., & Das, S. J., (2016). Hydrothermal synthesis of zinc stannate nanoparticles for antibacterial applications. *Journal of Taibah University for Science, 10*(4), 601–606. https://doi.org/10.1016/j.jtusci.2015.12.003.
26. Jiang, Z., Le, N. D. B., Gupta, A., & Rotello, V. M., (2015). Cell surface-based sensing with metallic nanoparticles. *Chemical Society Reviews, 44*(13), 4264–4274. https://doi.org/10.1039/c4cs00387j.
27. Logeswari, P., Silambarasan, S., & Abraham, J., (2013). Ecofriendly synthesis of silver nanoparticles from commercially available plant powders and their antibacterial properties. *Scientia Iranica, 20*(3), 1049–1054.
28. Nabila, M. I., & Kannabiran, K., (2018). Biosynthesis, characterization, and antibacterial activity of copper oxide nanoparticles (CuO NPs) from actinomycetes. *Biocatalysis and Agricultural Biotechnology, 15*, 56–62. https://doi.org/10.1016/j.bcab.2018.05.011.
29. Naseem, T., & Farrukh, M. A., (2015). Antibacterial activity of green synthesis of iron nanoparticles using *Lawsonia inermis and Gardenia Jasminoides Leaves extract. Journal of Chemistry, 2*, 1–7.
30. Naskar, A., & Kim, K. S., (2019). Nanomaterials as delivery vehicles and components of new strategies to combat bacterial infections: Advantages and limitations. *Microorganisms, 7*(9), 356. https://doi.org/10.3390/microorganisms7090356.
31. Ogunyemi, S. O., Zhang, F., Abdallah, Y., Zhang, M., Wang, Y., Sun, G., Qiu, W., & Li, B., (2019). Biosynthesis and characterization of magnesium oxide and manganese dioxide nanoparticles using *Matricaria Chamomilla* L. extract and its inhibitory effect on *Acidovorax oryzae* strain RS-2. *Artificial Cells, Nanomedicine, and Biotechnology, 47*(1), 2230–2239. https://doi.org/10.1080/21691401.2019.1622552.
32. Omar, R. A., Afreen, S., Talreja, N., Chauhan, D., Ashfaq, M., & Srituravanich, W., (2019). Chapter 6; Impact of nanomaterials on the microbial system. In: Prasad, R.,

(Ed.), *Microbial Nanobionics, Nanotechnology in the Life Sciences* (pp. 141–158). Springer.

33. Palanisamy, G., & Pazhanivel, T., (2017). Green synthesis of MgO nanoparticles for antibacterial activity. *International Research Journal of Engineering and Technology, 4*(9), 137–141.
34. Prabhu, Y. T., Rao, K. V., Kumari, B. S., Kumar, V. S. S., & Pavani, T., (2015). Synthesis of Fe_3O_4 nanoparticles and its antibacterial application. *International Nano Letters, 5*(2), 85–92. https://doi.org/10.1007/s40089–015–0141-z.
35. Ravichandran, V., Sumitha, S., Ning, C. Y., Xian, O. Y., Kiew Yu, U., Paliwal, N., Shah, S. A. A., & Tripathy, M., (2020). Durian waste mediated green synthesis of zinc oxide nanoparticles and evaluation of their antibacterial, antioxidant, cytotoxicity and photocatalytic activity. *Green Chemistry Letters and Reviews, 13*(2), 102–116. https://doi.org/10.1080/17518253.2020.1738562.
36. Sanchez, G. V., Fleming-Dutra, K. E., Roberts, R. M., & Hicks, L. A., (2016). Core elements of outpatient antibiotic stewardship. *MMWR. Recommendations and Reports: Morbidity and Mortality Weekly Report. Recommendations and Reports, 65*(6), 1–12. https://doi.org/10.15585/mmwr.rr6506a1.
37. Sanvicens, N., Pastells, C., Pascual, N., & Marco, M. P. (2009). Nanoparticle-based biosensors for detection of pathogenic bacteria. *TrAC Trends in Analytical Chemistry, 28*(11), 1243–1252. https://doi.org/10.1016/j.trac.2009.08.002.
38. Senthilkumar, A., Kashinath, L., Ashok, M., & Rajendran, A., (2017). Antibacterial properties and mechanism of gold nanoparticles obtained from *Pergularia daemia* Leaf extract. *Journal of Nanomedicine Research, 6*(1). https://doi.org/10.15406/jnmr.2017.06.00146.
39. Shanmugasundaram, T., & Balagurunathan, R., (2017). Bio-medically active zinc oxide nanoparticles synthesized by using extremophilic actinobacterium, *Streptomyces* sp.(Ma30) and its characterization. *Artificial Cells, Nanomedicine, and Biotechnology, 45*(8), 1521–1529. https://doi.org/10.1080/21691401.2016.1260577.
40. Sharma, G., Soni, R., & Jasuja, N. D., (2017). Phytoassisted synthesis of magnesium oxide nanoparticles with *Swertia* chirayaita. *Journal of Taibah University for Science, 11*(3), 471–477. https://doi.org/10.1016/j.jtusci.2016.09.004.
41. Some, S., Kumar, S. I. K., Mandal, A., Aslan, T., Ustun, Y., Yilmaz, E. Ş., Katı, A., Demirbas, A., Mandal, A. K., & Ocsoy, I., (2018). Biosynthesis of silver nanoparticles and their versatile antimicrobial properties. *Materials Research Express, 6*(1), 012001. https://doi.org/10.1088/2053-1591/aae23e.
42. Vimbela, G. V., Ngo, S. M., Fraze, C., Yang, L., & Stout, D. A., (2017). Antibacterial properties and toxicity from metallic nanomaterials. *International Journal of Nanomedicine, 12*, 3941–3965. https://doi.org/10.2147/IJN.S134526.
43. Walsh, C., (2003). *Antibiotics: Actions, Origins, Resistance* (1st edn., p. 355). American Society for Microbiology (Australian Society for Microbiology) Press.
44. Yadav, J. P., Kumar, S., Budhwar, L., Yadav, A., & Yadav, M., (2016). Characterization and antibacterial activity of synthesized silver and iron nanoparticles using *Aloe vera*. *J. Nanomed Nanotechnol., 7*(384), 2. https://doi.org/10.4172/2157-7439.1000384, PubMed: 1000384.

CHAPTER 17

SIZE AND SHAPE RELIANT ANTI-MICROBIAL APPLICATIONS OF SILVER NANOPARTICLES

SIVAKALA SAROJAM

ABSTRACT

Metal nanoparticles exhibit tunable properties depending on their morphology, composition, and crystallinity. Tweaking their shape enables tuning of their properties. Silver nanoparticles (AgNPs) have gained increased interest among the reported nano metals because of its ease of preparation and excellent properties. AgNPs are widely employed for disinfection purposes. They act as prospective antimicrobial agents and attacks varieties of active sites in the microbes. Nowadays, special emphasis was given to their size and shape-controlled preparation. Even though the studies on this area are in the budding phase, the great space is there in the fact that fine-tuning of size and shape can produce incredible properties for these materials and their applications in bio-medical field. The present chapter encompasses an analysis of shape and size reliant properties exclusively antimicrobial activities of AgNPs.

17.1 INTRODUCTION

Research on metal nanoparticles had received considerable status recently due to size and shape tunable properties which were substantively different from bulk counterparts [5, 53]. Greater efforts have been made by the

Sustainable Nanomaterials for Biomedical Engineering: Impacts, Challenges, and Future Prospects.
Junaid Ahmad Malik, Megh R. Goyal & Mohamed Jaffer M. Sadiq (Eds.)

scientific community towards developing stable dispersion of metal nanoparticles in the past three decades [10, 34, 36]. Nanoparticles of noble metals was reported to be have significant uses in the field of catalysis applications [30], cinema industry [18], electronic industry [22], photonic materials [50], and optoelectronics [29], biological labeling [38], imaging, and sensing [7]. Like other semiconducting materials, the inherent properties of nano metal particles can be tuned by regulating their morphology, composition, and crystallinity. Nipping their morphology of these properties of them has received substantial devotion from researchers due to the requirements to control their properties. Modulating morphology and size of the nano-entities leads to the acceptable tweaking of almost all the properties of these materials. The shape-dependent absorption studies were computationally predicted [11]. Studies on the effect of aspect ratio on the absorption spectra of metal nanorods exhibited a red shift [21]. In addition, triangular Nps of silver exhibited SERS in the spectral range from 700–800 nm, while spherical Nps of silver in the range of 530–570 nm [5, 26, 37].

The size and shape-controlled synthesis of the nanoparticles gives particles with a wide range of distribution of size and shape, which makes it difficult to investigate their properties [46]. Also, there is poor reproducibility in the results and are influenced by the presence of certain ionic species and other foreign species as contaminants in the system. Investigation has just started on the role(s) of such inorganic species in shape controlled synthesis of metal nanoparticles.

Hassle-free method of synthesis through reduction was one of the attractions of silver nanoparticles (AgNPs). Moreover, they possess additional advantages like robust surface plasmon resonance energy (SPRE), reasonable excitation efficiency, good electrical, optical, magnetic, and thermal properties. Besides, they exhibit size and shape trusting properties.

Among the metal nanoparticles, silver (Ag) has gained great research interest because of its easy reduction method of synthesis, strong surface plasmon resonance energy (SPRE) in the visible region, high excitation efficiency, unique electrical, optical, magnetic, and thermal properties. Moreover, they exhibit properties tuneable properties with their size and shape.

This present chapter describes briefly on the size and shape specific synthesis of AgNPs, the mechanism of the cytotoxicity and their shape dependent antimicrobial properties. Major emphasis is given to the features of anisotropic nanostructures of AgNPs and their antibacterial applications. Even though this kind of work was in its growing phase, greater significance is there in the current scenario of spreading of diseases caused by pathogens worldwide. AgNPs and its anisotropic crystalline forms was a prosecutable candidate for fighting against potent microorganisms in future.

17.2 GENERAL STRATEGIES FOR PREPARATION OF AGNPS

AgNPs can be generated through top down and bottom up strategies. The development of anisotropic NPs involves methods like the seed-mediated method, polyol synthesis, electro/photochemical techniques, etc. The solution phase methods possess an ability to grow anisotropic nanostructures in high yield. Synthesis of metal nanoparticles with controllable shapes using self-assembled micelles or mesophases from surfactants, were widely reported.

Metallic gold nano colloids was first prepared by Michael Faraday in 1856 by reduction of aqueous gold chloride. As prepared nano gold colloid was in ruby red color and was still available in London faradays museum. Later plenty of fabrication procedures have been in practice for generating gold and silver nanostructures. However, there remains a herculean task of solving the major issues like irregularity, poly-dispersity, and poor stability of the resulting structures. The reaction conditions have to be judiciously controlled to minimize the issues in size and shape of the products. Method of preparation and precise experimental conditions plays a vital role in their properties. The ideal techniques for imparting reproducibility in results and to get specific command on the morphology with mono dispersity are yet to be developed. Green chemistry principles are greatly followed by scientists for the synthesis of AgNPs nowadays. Chemical methods includes chemical reduction methods [4, 35], electrochemical methods (electrolysis) [2], and radiochemical methods [28, 44].

Shape and size tuned preparation of metal nanoparticles with reproducible results remains as the most difficult chore of the nanotechnology research. Introducing shape anisotropy at the nanoscale has emerged as a compelling way to impart new properties and functionality to the resulting materials. It enables the development of intricate nanomaterials for a series of applications. Anisotropy, is the fundamental aspect giving complexity in the properties of materials. It is behind the existence of complex systems in nature. The complexities in the behavior of a cell arises from its anisotropy however isotropy or equilibrium results in cell death so as it is with living things and the Universe. Nature provides us with plenty of examples where anisotropy leads to complexities in behavior and properties. Some examples were the self-cleaning property of lotus leaves and the fascinating colors on feathers of butterfly and peacock, gecko's feet, woodpeckers beak, etc. Researchers have been actively involved in exploring these complexities found in nature for many years in an effort to emulate and access ever more complex materials through bio mimicking for exciting applications in the current era of advanced technologies.

Solvo thermal method using the principles of controlled crystallization has resulted in a tremendous array of size and shape tuned nanoparticles. Here the crystallization follows the LaMer model under highly specified conditions of temperature, purity, and pH. However, there exist practical difficulty in controlling the above parameters which is the greatest disadvantage of the method. The powerful alternative method for the solvo thermal method was developed, and it is the seed mediated method.

Seed mediated growth is an excellent technique to prepare size and shape tuned metal nanoparticles. It is a colloidal chemical method in which seed particles act as nucleation centers and undergoes further growth in the presence of metal ions and other additives. Here the metal ions are first reduced at the seed then grow via heterogeneous nucleation. Size and shape controlled synthesis of the nanoparticles is possible in this method. [13, 22, 32].

Overcoming the Van der Waals attractive force existing between colloidal metal nanoparticles and to prevent the agglomeration which extends stability to metal nanoparticles was the great challenge in their synthesis. Capping agents are used to balance these attractive forces by electrostatic or steric stabilization. Here the capping agents can adsorb on to the different facets of the growing crystal thereby changing the surface energies of the facets and reins shape of particles by minimizing surface energy. Seed growth followed by homogeneous nucleation favors the formation of isotropic morphologies like spheres [3]. The crystal morphology and the final shape of the nanoparticle was according to growth kinetics, by in which the fastest growing facets having high surface energies vanishes, leaving behind the facets with low surface energies. Here capping agents play a vital role by changing the free energies of the different facets of the growing crystal by varying their growth rates.

Template assisted synthesis is another efficient method for making noble metal nanoparticles. This technique involves the use of spatially confined assemblies as reaction enclosures which acts as templates and produces metal nanoparticles with morphologies complementary to them. The templates can be both soft and hard. Self-assembled structure like micelles, reverse micelles, viruses, etc., acts as the soft templates [14, 25, 27]. Recently, Liquid crystalline phases are widely being employed in synthesizing anisotropic nano entities. Nanowires of AgNPs were successfully obtained using meso-phases formed by micelles from surfactants in long cylindrical channels of water [40, 56].

Hard template technique uses porous structures as structure directing agents. In this technique precursor solutions are being permeated inside the templates, further reduction followed by removal of template gives metallic NPs. The post processing to remove the template was the major disadvantage for the hard template route since it hampers the morphology of the materials. In electroless metal decomposition generated electrons from reductant was used to deposit a metal from solution onto a specific surface. Anisotropic structures like tubes, rods, and wires can be successfully developed by using these methods.

Self-assembled meso structures from Surfactant molecules are employed in the soft template route for the successful production of metal nano entities with controlled size and shape. Here the mesostructures interact with the precursor solution and the precursor will form adducts with them through non-covalent interactions. Then reduction of the ions results in nano-sized metal particles with the morphology of the soft mesostructures. The post-processing to remove the template is not needed here. It was reported to be a most competent technique for producing nano metal particles, especially arrays of nono wires, tubes, and rods on specific substrate surfaces.

Electrochemical method using applied potential was reported as an proficient method for the morphology controlled synthesis of AgNPs. The synthesis of nano metal particles with and without the use of hard templates through this method has been reported [54]. Hard templates consist of nanoporous materials such as imprinted polymer templates, anodic alumina, etc. [15]. Electrochemical deposition with a thin metal film fabricated on one side of the template, which act as cathode for electroplating was carried within these templates. Then the metal ions are reduced and deposited inside the pores of the template [19]. Controlling the electrode parameters, such as current, voltage, deposition time and self-assembly in deposition offers great scope in tuning of morphologies of metal nanoparticles [43, 48]. Finally, the NPs are obtained by removing the template through post synthetic Physico-chemical reactions [47].

Photochemical reduction involving irradiation of the metal precursor solution was a relatively easy route for the preparation of metal nanoparticles. This technique has been successfully applied to develop shape controlled synthesis of Ag, Au, and Pt NPs [1, 36]. This method has been reported to be successful for the development of AgNPs with optimized size and shape.

Biological templates like enzymes, proteins, plant extract, fungi, bacteria, etc., have been used for the formation of NPs through reduction of salts [12]. This method have the potential to be cost efficient, simple, and environmentally friendly for the preparation of NPs with diverse morphologies from spheres to cubes to wires to rods to cubes, etc.

17.3 CHARACTERIZATION OF AGNPS AND THEIR PROPERTIES

Figure 17.1 shows the general characterization techniques with the respective properties obtained from them. Where; TEM: transmission electron microscopy; SEM: scanning electron microscopy; AFM: atomic force microscopy; XRD: X-ray diffraction technique; UV-Vis: UV-visible spectroscopy; HRTEM: high resolution electron microscope; SQUID: superconductive quantum interference device; and VSM: vibrating sample magnetometer.

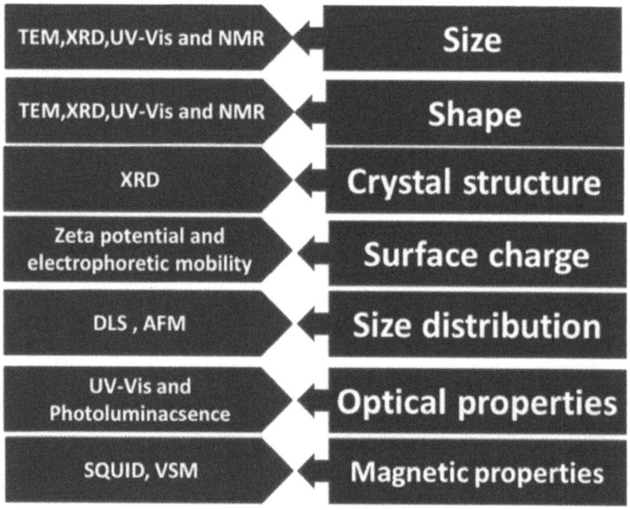

FIGURE 17.1 Characterization of AgNPs for their properties.

17.4 SURFACE PLASMON RESONANCE ENERGY (SPRE)

Silver nanostructures are container for surface plasmons: The electro-optical properties of noble metal nao materials are solely determined by their conduction electrons. Surface plasmon resonance (SPRE) in metal nanoparticles are due to the combined wavering of conduction electrons in phase with the incident radiation it is otherwise known as plasmonic property [8]. On excitation the induced charges propagate as an electro-magnetic wave over the metal surface. The confinement of oscillating charges and their enhancement on metal surface is called the SPRE band. By tuning morphologies of AgNPs, it is possible to tune their plasmonic behaviors. Similarly, scattered light, absorbing lights and enhancement in the local electric field can also be controlled. SPRE enables gold and silver to absorb or scatter radiation in visible region resulting in their color [42, 49]. There are several theoretical models proposed by scientists explaining the mechanism of electrical conduction in nanostructures. One among these is the Drude Lorentz model. In this model, metal is considered as plasma with an equal number of fixed positive ions and mobile conduction electrons. Under the irradiation of electromagnetic wave, the unbounded electrons coherently oscillates at a plasma frequency of ωp with respect to the lattice of positive ions.

For a bulk metal with infinite sizes in all three dimensions, ωp can be expressed as:

$$\omega p = (Ne^2/E_0\ Me)^{1/2}$$

where; N is the number density of electron; E_0 is the dielectric constant of vacuum; M and e are their effective mass and charge of an electron.

17.5 APPLICATIONS OF AGNPS

The major fields of applications of AgNPs are shown in Figure 17.2.

17.5.1 SIZE AND SHAPE RELIANT ANTIMICROBIAL APPLICATIONS OF AGNPS

Plenty of literatures are available discussing the bactericidal activity of AgNPs. This is taken into account for their medical uses. From ancient era onwards, silver had been in use as an antiseptic in wound dressings and disinfectant in medical appliances.

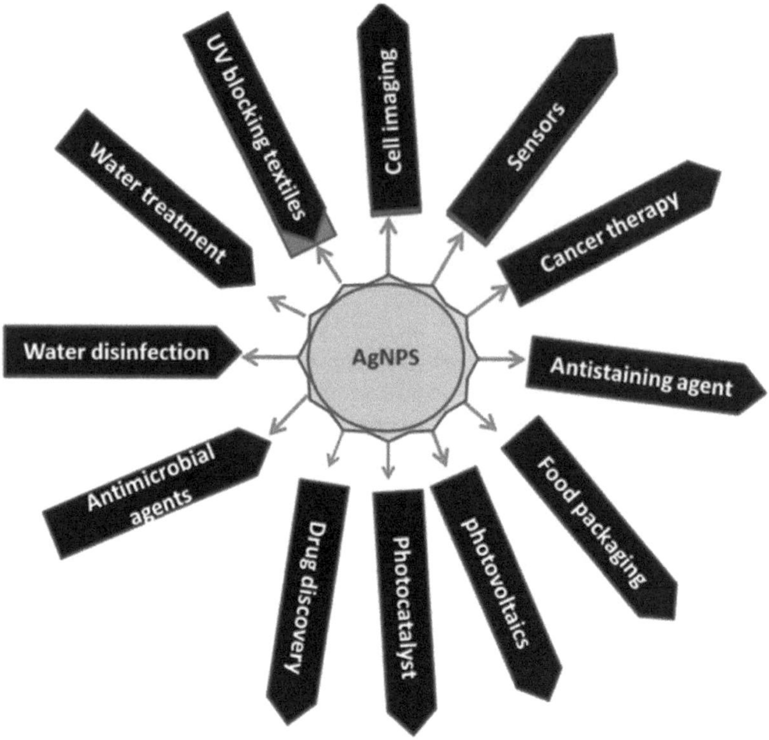

FIGURE 17.2 Major applications of AgNPs.

17.5.2 ANTIMICROBIAL APPLICATIONS

The medicinal applications of silver had been known for centuries. B.C. onwards silver was used to make water potables. After the discovery of penicillin in 1940, their use as antimicrobial agents was minimized. In 1960s Moyer introduced the antibacterial property of 0.5% silver nitrate against various bacterial species [51]. In 1968, sulfadiazine cream was developed from silver nitrate and sulfonamide to form silver sulfadiazine cream as a broad-spectrum antibacterial agent.

Silver is comparatively less toxic to humans at lower doses and were comparably cheap with respect to other existing antimicrobial medications [55]. AgNPs was reported as successful candidate for inhibiting microbial growth while protecting the growth of the host cells. The mechanism of antimicrobial action of AgNPs was determined by their mechanism of action and several other factors. Gram +ve bacteria are reported to be less susceptible to

Ag^+ than gram-ve bacteria since of peptidoglycan present in it may inactivate silver ions greatly than in gram-ve bacteria [23, 52].

The mechanism of action can be through any of the listed four pathways:

- Affecting the membrane structure of the bacterial cell, its permeability and transport activity by adhering on to it;
- Interacting with DNA or destabilizing and denaturing of the proteins through penetrating into the cell nucleus;
- Oxidises proteins, destabilizes ribosomes, mitochondria, and endoplasmic reticulum by reactive oxygen species (ROS) generation and cell toxicity;
- Genotoxicity: either through mutation or by inhibiting transcription and cell replication.

In one way, the adhesion of AgNPs on to the cell membranes leads to the damaging of the metabolic reactions of the cell which constitutes the antibacterial activity. In another way, Ag NPs produces ROS and the Ag^+ ions. They cause the disruption of cell membrane through oxidative dissolution and also react with the functional proteins. The Ag^+ ion has a greater tendency to fix with bioactive molecules, especially to groups like NH_2, phosphates, and mercapto groups, forming a virtual-covalent bond between Ag^+ ion and protein molecules in the cell [31, 39]. It finally leads to irretrievable agglomeration of the proteins and other molecules involved in the life processes of the cell. Thus, Ag^+ ions pointedly incapacitate DNA/RNA replication and life reactions, which eventually leads to death of the cell. The damage to these leads to the condition called apoptosis. Due to the complex nature of their antimicrobial mechanism the antimicrobial activity of AgNps were reported to be highly reliant on size and shape. As the size of NPs decreases, greater will be the bactericidal activity. Pal et al. [45] have studied the effect of morphology of AgNPs on the disruption of bacterial membrane using energy filtering TEM technique. In the study it was observed that triangular silver nanoplates have greater activity than spheres, cubo-octahedral, and quasi-spherical shapes [24, 41]. Liu et al. [37] have arrived at the probable mechanisms for antibacterial action of AgNPs that the released Ag^+ ions adhere on bacterial deteriorating the cell membrane. Ag^+ ions can also form complexes with the functional proteins and nucleic acids [45]. Surface modification and heterogeneity of the nanoparticles have a profound influence on the bactericidal activities apart from their size and shape.

Generally three main methods are in use for antimicrobial susceptibility studies. They are Diffusion technique (Kirby-Bauer and Stokes), dilution method (minimum inhibitory concentration), and diffusion and dilution (E-Test method). The most commonly used method in the clinical laboratory is the disc diffusion method. It was developed by Bauer and co-workers (the so-called Kirby-Bauer method). This method is well established and standard zones of inhibition have been determined for susceptible values in comparison with a standard reference.

AgNPs were also reported to have potential antiviral and antifungal activities. Studies on the antiviral activity have attracted great research interest in recent years due to the unexpected outbreak of novel viruses. World is now amidst of novel SARS covid 19, a deadly pathogen who took the lives of several million people. Even though the antiviral activity studies of AgNPs are in the budding stage, several reports are available with initial studies. The complex of mercaptoethanol sulfonate with Ag NPs was used as an antiviral agent towards *HSV-1* virus. The antiviral activity of Ag NPs using [4-(2-hydroxyethyl)-1-piperazineethanesulfonic acid] buffer on *HIV-1* was reported by Orłowski et al. [44]. The chitosan composite with Ag NPs were reported to have antiviral properties against *H1N1 influenza A virus* [9]. Elechiguerra et al. [20] studied the size-reliant antiviral activity of Ag NPs with *HIV-1*. It is reported that Ag NPs showed affinity towards the gp120 glycoprotein of the virus, which competes with the tiring of a virus to the host cells. Orłowski et al. [44] studied the effect of size of AgNPs on the antiviral activity against *HSV 2 in vitro* and *in vivo*. There is a recent suggestion from scientific reports that AgNPs can bind to the spike glycoprotein of the virus *SARS Covid 19*, thus inhibiting their binding to the cells, which provides acidic pH in respiratory epithelium which is antagonistic towards the virus.

Plenty of literature is available concerning the antifungal activities of Ag NPs. Kim et al. [35] developed anti-fungal AgNPs against *C. albicans* and *Trichophyton mentagrophytes*. Ag NPs using N,N,N′,N′-tetramethylethylenediamine as a reducing agent showed antifungal activities against *P. aeruginosa*, and *C. albicans* [33]. Ag NPs derived from ribose exhibited antifungal activity against *Candida tropicalis* and *C. albicans* [17].

In general, bactericidal activities of AgNPs were based on their distribution, uptake, and activation on the cells of bacteria which were reported to be highly influenced by properties like surface charge, area, morphology, functionalization, and size [6, 16]. The shape-dependent behavior mainly arises due to the difference in active surface areas of the anisotropic shapes.

The higher the active surface area higher will be the antimicrobial effects. Recently several reports have come focusing on the size and shape-dependent antimicrobial activities of AgNPs. The significant contributions were listed in Table 17.1. The mechanistic implications of antimicrobial activities can be summarized as it involves either of the four pathways a) destabilization of ribosomes, b) disruption of cell membrane, c) complexation with DNA bases and d) formation of reactive oxygen free radicals. The main issues in the described topic are that the existing literature on the size and shape dependent antimicrobial effect and on the mechanisms of action are inhomogeneous. It makes the integration and analysis of available data rather tedious and difficult. However, there exist harmony in the use of AgNPs against microbes, with appealing emerging applications in non-medical and environmental applications.

TABLE 17.1 Shape Dependent Antimicrobial Activity by AgNPs

Shapes Compared	Microorganism	Shape Dependency	References
Silver nanocubes, spheres, and wires.	*Escherichia coli*	Silver nanocubes > silver nanospheres	[28]
Silver nanospheres, rods truncated triangular plates	*E. coli*	Triangular nano plates > sphere > rod.	[45]
Nanospheres, nanoplatelets, nanocubes, and nanorods	*S. aureus*	Nanoplatelets > nano-spheres > nano cubes > nanorods	[27]
Silver triangular nanoplates nanospheres nanocubes and nanorods	*Escherichia coli*	Truncated plates > spheres > rods.	[45]
Truncated octahedral nano silver and nanospheres	*E. coli and Enterococcus faecium*	Truncated octahedral > spherical AgNPs (AgNS)	[1]
Silver nanocubes, platelets, and spheres	*S. aureus*	Platelets > spheres glucose synthesis > spheres > cubes	[27]
Ag nanocubes, spheres, and wires	*E. coli*	Spherical Ag NPs > cubes > wires	[28]
Nano-platelets, nanospheres	*S. aureus*	Platelets > spherical	[37]
Small nano-spheres, large nanospheres and tri angular nanoplates	*Pseudomonas aeruginosa* and *Escherichia coli*	Small spheres > platelets > large spherical particles	[46]

17.6 SUMMARY

The astonishing access to drug-resistant pathogens in the current era was a frightening spectacle. Currently, the life-threatening outcome of COVID-19 has deceased lakhs of people all over the world. Moreover, bacterial infections signify a recurrent cause of death. The research was now focussing on the development of novel and potent antibacterial agents. AgNPs have been used in the medical field for disinfection purposes. AgNPs displays size and shapes dependent properties. This chapter summarizes the antimicrobial applications of AgNPs, their mechanism of action, and their shape-dependent behaviors. The antibacterial action by AgNPS occurs through genotoxicity, cell toxicity, cell penetration, or membrane adsorption. Anisotropic nanostructures were superior in antimicrobial properties over nanospheres in the suggested four pathways. However, as the size of the nano-spheres decreases, they exhibited improved performance due to increased active surface area. Anisotropic structures possess increased electron density and sharp vertices that can penetrate easily through the bacterial membrane and are responsible for their augmented activity. AgNPs was a prospective candidate as the antimicrobial agent. But they possess the limitations of their tendency of oxidation as well as aggregation. Also, there have been a couple of contentious concerns of AgNPs involving their toxicity and environmental impacts. The environmental impacts of AgNPs remain vague due to technological constraints. However, there are few reports concerning the negative impact on the soil microbiome in the long run. Despite these difficulties, the antiviral and antibacterial properties of AgNP can be exploited for meeting the infection problems due to the unexpected entry of pathogens worldwide.

KEYWORDS

- **antibacterial application**
- **antimicrobial applications**
- **antiviral effects**
- **reactive oxygen species**
- **silver nanoparticles**
- **size dependent**
- **surface plasmon resonance energy**

REFERENCES

1. Alshareef, A., Laird, K., & Cross, R. B. M., (2017). Shape-dependent antibacterial activity of silver nanoparticles on *Escherichia coli* and *Enterococcus faecium* Bacterium. *Applied Surface Science, 424*, 310–315. https://doi.org/10.1016/j.apsusc.2017.03.176.
2. Ardalan, S., Sadeghi, B., Garmaroudi, F. S., Hashemi, M, Nezhad, H. R., & Nasrollahi, A., (2012). Comparison of the antibacterial activity on the nanosilvernshapes: nanoparticles, nanorods and nanoplates. *Advanced Powder Technology, 23*, 22–26.
3. Asharani, P. V., Low, K. M. G. L. K., Hande, M. P., & Valiyaveettil, S., (2009). Cytotoxicity and genotoxicity of silver nanoparticles in human cells. *ACS Nano, 3*(2), 279–290. https://doi.org/10.1021/nn800596w.
4. Astilean, S., Potara, M., Jakab, E., Damert, A., Popescu, O., & Canpean, V., (2011). Synergistic antibacterial activity of chitosan-silver nanocomposites on *Staphylococcus aureus*. *Nanotechnology, 22*, 135101.
5. Bailey, S., Kim, S., Liu, M., & Yan, R., (2017). Decoupling co-existing surface plasmon polariton (SPP) modes in a nanowire plasmonic waveguide for quantitative mode analysis. *Nano Research, 10*, 2395–2404.
6. Baker, C., Pradhan, A., Pakstis, L., Pochan, D. J., & Shah, S. I., (2005). Synthesis and antibacterial properties of silver nanoparticles. *Journal of Nanoscience and Nanotechnology, 5*(2), 244–249. https://doi.org/10.1166/jnn.2005.034.
7. Berti, L., Alessandrini, A., & Facci, P., (2004). DNA-templated photoinduced silver deposition. *Journal of the American Chemical Society, 14*, 2124–2147.
8. Bhattacharya, R., & Mukherjee, P., (2008). Biological properties of "naked" metal nanoparticles. *Advanced Drug Delivery Reviews, 60*(11), 1289–1306. https://doi.org/10.1016/j.addr.2008.03.013.
9. Brad, M., Angelgraeme, E., Batleychad, V., Jarolimeknicola, J., & Rogers, J., (2013). The impact of size on the fate and toxicity of nanoparticulate silver in aquatic systems *Chemosphere, 93*, 359–365.
10. Seo, D., Park, J. C., & Song, H., (2006). Polyhedral gold nanocrystals with oh symmetry: From octahedra to cubes. *Journal of the American Chemical Society, 128*(46), 14863–14870. https://doi.org/10.1021/ja062892u.
11. Chen, J., Niu, W., Li, Z. Y., Shi, L., Liu, X., Li, H., Han, S., & Xu, G., (2008). Seed-mediated growth of nearly monodisperse palladium nanocubes with controllable sizes. *Crystal Growth & Design, 8*, 4440–4447.
12. Chen, S., & Carroll, D. L., (2002). Synthesis and characterization of truncated triangular silver nanoplates. *Nano Letters, 2*(9), 1003–1007. https://doi.org/10.1021/nl025674h.
13. Chen, X., & Schluesener, H. J., (2008). Nanosilver: Product in medical application. *Applied Pharmacology Letters, 176*, 1–12.
14. Choi, M., Wu, Z., & Iglesia, E., (2010). Mercaptosilane-assisted synthesis of metal clusters within zeolites and catalytic consequences of encapsulation. *Journal of the American Chemical Society, 132*(26), 9129–9137. https://doi.org/10.1021/ja102778e.
15. Choi, O., & Hu, Z., (2008). Q. Size dependent and reactive oxygen species related nanosilver toxicity to nitrifying bacteria. *Environmental Science and Technology, 42*(12), 4583–4588. https://doi.org/10.1021/es703238h.
16. Cui, L., Chen, P., Chen, S., Yuan, Z., Yu, C., Ren, B., & Zhang, K., (2013). *In situ* study of the antibacterial activity and mechanism of action of silver nanoparticles by

surface-enhanced Raman spectroscopy. *Analytical Chemistry, 85*(11), 5436–5443. https://doi.org/10.1021/ac400245j.

17. Cunha, F. A., Cunha, M. D. C. S. O., Da Frota, S. M., Mallmann, E. J. J., Freire, T. M., Costa, L. S., Paula, A. J., et al., (2018). Biogenic synthesis of multifunctional silver nanoparticles from *Rhodotorula glutinis* and *Rhodotorula mucilaginosa*: Antifungal, catalytic and cytotoxicity activities. *World Journal of Microbiology and Biotechnology, 34*(9), 127. https://doi.org/10.1007/s11274-018-2514-8.
18. Dick, L. A., Mcfarland, A. D., Haynes, C. L., & Van, D. R. P., (2002). Metal film over nanosphere (MFON) electrodes for surface-enhanced Raman spectroscopy (SERS): Improvements in surface nanostructure stability and suppression of irreversible loss. *Journal of Physical Chemistry B, 106*(4), 853–860. https://doi.org/10.1021/jp013638l.
19. Ding, X., Rahul, N., Yanzhe, G., Huijun, T., & Yaping, Z. M., (2020). Reaction-controlled synthesis of ultrathin silver nanowires and fabrication of transparent conductive electrodes with low haze. *Industrial & Engineering Chemistry Research, 59*, 13572–13577.
20. Elechiguerra, J. L., Burt, J. L., Morones, J. R., Camacho-Bragado, A., Gao, X., Lara, H. H., & Yacaman, M. J., (2005). Interaction of silver nanoparticles with HIV-1. *Journal of Nanobiotechnology, 3*, 6. https://doi.org/10.1186/1477-3155-3-6.
21. Fugetsu, B., Takeshima, T., Tada, Y., Sakaguchi, N., & Watari, F., (2015). DNA/Agvnanoparticles as antibacterial agents against gram-negative bacteria. *Nanomaterials, 5*, 284–297.
22. Giorgio, P. M. P., & Maillard, M., (2003). Tuning the size of silver nanodisks with similar aspect ratios: Synthesis and optical properties. *The Journal of Physical Chemistry B, 107*, 2466–2472.
23. González, C. E. E., Garza-Cervantes, J. A., Rodrígue, A. V., Peralta, L. Z., González, M. T., Barriga, C. E. D., Saucedo-Salazar, E. M., et al., (2018). *In vivo* antimicrobial activity of silver nanoparticles produced via a green chemistry synthesis using *Acacia rigidula* as a reducing and capping agent. *International Journal of Nanomedicine, 13*, 2349–2363.
24. Habash, M. B., Park, A. J., Vis, E. C., Harris, R. J., & Khursigara, C. M., (2014). Synergy of silver nanoparticles and aztreonam against *Pseudomonas aeruginosa* PAO1 biofilms. *Antimicrobial Agents and Chemotherapy, 58*(10), 5818–5830. https://doi.org/10.1128/AAC.03170-14.
25. Hackenberg, S., Scherzed, A., Kessler, M., Hummel, S., Technau, A., Froelich, K., Ginzkey, C., et al., (2011). Silver nanoparticles: Evaluation of DNA damage, toxicity and functional impairment in human mesenchymal stem cells. *Toxicology Letters, 201*(1), 27–33. https://doi.org/10.1016/j.toxlet.2010.12.001.
26. Hebalkar, N., Nischala, K., & Rao, T. N., (2011). Silica-silver core-shell particles for antibacterial textile application. *Colloid Surface B, 82*, 203–208.
27. Helmlinger, J., Sengstock, C., Groß-Heitfeld, C., Mayer, C., Schildhauer, T. A., Köller, M., & Epple, M., (2016). Silver nanoparticles with different size and shape: Equal cytotoxicity, but different antibacterial effects. *RSC Advances, 6*(22), 18490–18501. https://doi.org/10.1039/C5RA27836H.
28. Hong, X., Wen, J., Xiong, X., & Hu, Y., (2016). Shape effect on the antibacterial activity of silver nanoparticles synthesized via a microwave-assisted method. *Environmental Science and Pollution Research International, 23*(5), 4489–4497. https://doi.org/10.1007/s11356-015-5668-z.

29. Ivask, A., & Kahru, A., (2013). Mapping the dawn of nano ecotoxicological research. *Accounts of Chemical Research, 46*, 823–833.
30. Jana, N. R., & Murphy, C. J., (2002). Controlling the aspect ratio of inorganic nanorods and nanowires. *Advanced Materials, 14*, 80–87.
31. Jun, J., Yuan-Yuan, D., Shao-Hai, W., Shao-Feng, Z., & Andzhongyi, W., (2007). Preparation and characterization of antibacterial silver-containing nanofibers for wound dressing applications. *Journal of US-China Medical Science, 4*, 52–54.
32. Jung, W. K., Koo, H. C., Kim, K. W., Shin, S., Kim, S. H., & Park, Y. H., (2008). Antibacterial activity and mechanism of action of the silver ion in *Staphylococcus aureus* and *Escherichia coli. Applied and Environmental Microbiology, 74*(7), 2171–2178. https://doi.org/10.1128/AEM.02001-07.
33. Kailasa, S. K., Park, T., Rohit, J. V., & Koduru, J. D., (2019). Chapter 9; Antimicrobial activity of silver nanoparticles. In: Grumezescu, A. M., (ed.), *Nanoparticles in Pharmacotherapy* (pp. 461–480). William Andrew Publishing. Elsevier, Inc.
34. Kelly, T. R., Lazarides, L., & Schatz, A., (1999). Electrodynamics of noble metal nanoparticles and nanoparticle clusters. *Journal of Cluster Science, 10*, 295–302.
35. Kim, J. Y., Sungeun, K., Kim, J., Jongchan, L., & Andyoon, J., (2005). The biocidal activity of nano-sized silver particles compared with silver ion. *Korean Society of Environmental Engineering, 27*, 771–776.
36. Li, M., Ma, Z., Zhu, Y., Xia, H., Yao, M., Chu, X., Wang, X., et al., (2016). Toward a molecular understanding of the antibacterial mechanism of copper-bearing titanium alloys against *Staphylococcus aureus. Advanced Healthcare Materials, 5*(5), 557–566. https://doi.org/10.1002/adhm.201500712.
37. Liu, W., Wu, Y., Wang, C., Li, H. C., Wang, T., Liao, C. Y., Cui, L., et al., (2010). Impact of silver nanoparticles on human cells effect of particle size. *Nanotoxicology, 4*(3), 319–330. https://doi.org/10.3109/17435390.2010.483745.
38. Lkhagvajav, N., Yasab, I. C., Elikc, E., Koizhaiganova, M., & Saria, O., (2011). Antimicrobial activity of colloidal silver nanoparticles prepared by sol-gel method. *Digest Journal of Nanomaterials and Biostructures, 6*, 149–154.
39. Luyun, J., Ibon, S., & John, F., (2020). High-yield electrochemical synthesis of silver nanoparticles by enzyme-modified boron-doped diamond electrodes. *Langmuir, 36*(22), 6089–6094.
40. Marambio-Jones, C., & Hoek, E. M. V., (2010). A review of the antibacterial effects of silver nanomaterials and potential implications for human health and the environment. *Journal of Nanoparticle Research, 12*(5), 1531–1551. https://doi.org/10.1007/s11051-010-9900-y.
41. Métraux, G. S., Xue, C., & Millstone, J. E., (2008). Mechanistic study of photomediated triangular silver nanoprism growth C had A. Mirkin. *Journal of the American Chemical Society, 130*, 8337–8342.
42. Morones, J. R., Elechiguerra, J. L., Camacho, A., Holt, K., Kouri, J. B., Ramírez, J. T., & Yacaman, M. J., (2005). The bactericidal effect of silver nanoparticles. *Nanotechnology, 16*(10), 2346–2353. https://doi.org/10.1088/0957-4484/16/10/059.
43. Murthy, P. S., Murali, M. Y., Varaprasad, K., Sreedhar, B., & Mohana, R. K., (2008). First successful design of semi-IPN hydrogel-silver nanocomposites: A facile approach for antibacterial application. *Journal of Colloid and Interface Science, 318*(2), 217–224. https://doi.org/10.1016/j.jcis.2007.10.014.

44. Orłowski, P., Kowalczyk, A., Tomaszewska, E., Ranoszek-Soliwoda, K., Węgrzyn, A., Grzesiak, J., Celichowski, G., et al., (2018). Antiviral activity of tannic acid modified silver nanoparticles: Potential to activate immune response in *Herpes genitalis*. *Viruses, 10*(10), 524. https://doi.org/10.3390/v10100524.
45. Pal, S., Tak, Y. K., & Song, J. M., (2007). Does the antibacterial activity of silver nanoparticles depend on the shape of the nanoparticle? A study of the gram-negative bacterium *Escherichia coli*. *Applied and Environmental Microbiology, 73*(6), 1712–1720. https://doi.org/10.1128/AEM.02218–06.
46. Park, M. V., Neigh, A. M., Vermeulen, J. P., De La Fonteyne, L. J., Verharen, H. W., Briedé, J. J., Van, L. H., & De Jong, W. H., (2011). The effect of particle size on the cytotoxicity, inflammation, developmental toxicity and genotoxicity of silver nanoparticles. *Biomaterials, 32*(36), 9810–9817. https://doi.org/10.1016/j.biomaterials.2011.08.085.
47. Raza, M., Kanwal, Z., Rauf, A., Sabri, A., Riaz, S., & Naseem, S., (2016). Size- and shape-dependent antibacterial studies of silver nanoparticles synthesized by wet chemical routes. *Nanomaterials, 6*(4), 74–88. https://doi.org/10.3390/nano6040074.
48. Suresh, A. K., Pelletier, D. A., & Doktycz, M. J., (2013). Relating nanomaterial properties and microbial toxicity. *Nanoscale, 5*(2), 463–474. https://doi.org/10.1039/c2nr32447d.
49. Suyue, C., Jesse, L. C., David, R. W., & Philippe, B., (2018). Elucidating the role of AgCl in the nucleation and growth of silver nanoparticles in ethylene glycol. *Crystal Growth & Design, 18*(1), 324–330.
50. Tuckman, M., Petersen, P. J., Howe, A. Y., Orlowski, M., Mullen, S., Chan, K., Bradford, P. A., & Jones, C. H., (2007). Occurrence of tetracycline resistance genes among *Escherichia coli* isolates from the phase 3 clinical trials for tigecycline. *Antimicrobial Agents and Chemotherapy, 51*(9), 3205–3211. https://doi.org/10.1128/AAC.00625-07.
51. Vivian, D. Y., & Yam, W. W., (2004). Controlled synthesis of monodisperse silver nanocubes in water. *Journal of the American Chemical Society, 26*, 41–47.
52. Xia, Y., Xiong, Y., Lim, B., & Skrabalak, S. E., (2009). Shape-controlled synthesis of metal nanocrystals: Simple chemistry meets complex physics? *Angewandte Chemie, 48*(1), 60–103. https://doi.org/10.1002/anie.200802248.
53. Xin, Q., Zihao, C., Tianyu, Y., & Kristen, A. F., (2019). Growth mechanism of five-fold twinned Ag nanowires from multiscale theory and simulations. *ACS Nano, 13*(4), 4647–4656.
54. Yang, Y., Chen, Q., Wall, J. D., & Hu, Z., (2012). Potential nanosilver impact on anaerobic digestion at moderate silver concentrations. *Water Research, 46*(4), 1176–1184. https://doi.org/10.1016/j.watres.2011.12.024.
55. You, C., Han, C., Wang, X., Zheng, Y., Li, Q., Hu, X., & Sun, H., (2012). The progress of silver nanoparticles in the antibacterial mechanism, clinical application and cytotoxicity. *Molecular Biology Reports, 39*(9), 9193–9201. https://doi.org/10.1007/s11033-012-1792-8.
56. Yu, D., & Yam, V. W., (2004). Controlled synthesis of monodisperse silver nanocubes in water. *Journal of the American Chemical Society, 126*(41), 13200, 13201. https://doi.org/10.1021/ja046037r.

INDEX

C

E

F

G

H

O

Q

R

U